W9-CRS-517

Design of Analog
CMOS Integrated Circuits

McGraw-Hill Series in Electrical and Computer Engineering

Senior Consulting Editor

Stephen W. Director, University of Michigan, Ann Arbor

Circuits and Systems

Communications and Signal Processing
Computer Engineering
Control Theory and Robotics
Electromagnetics
Electronics and VLSI Circuits
Introductory
Power
Antennas, Microwaves, and Radar

Previous Consulting Editors

Ronald N. Bracewell, Colin Cherry, James F. Gibbons, Willis W. Harman, Hubert Heffner, Edward W. Herold, John G. Linvill, Simon Ramo, Ronald A. Rohrer, Anthony E. Siegman, Charles Susskind, Frederick E. Terman, John G. Truxal, Ernst Weber, and John R. Whinnery

Design of Analog CMOS Integrated Circuits

Behzad Razavi

Professor of Electrical Engineering

University of California, Los Angeles

Tata McGraw-Hill Publishing Company Limited

NEW DELHI

McGraw-Hill Offices

New Delhi New York St Louis San Francisco Auckland Bogotá Caracas
Kuala Lumpur Lisbon London Madrid Mexico City Milan Montreal
San Juan Santiago Singapore Sydney Tokyo Toronto

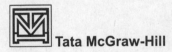
Tata McGraw-Hill

DESIGN OF ANALOG CMOS INTEGRATED CIRCUITS

Copyright © 2001 by The McGraw-Hill Companies, Inc. All rights reserved. No part of this publication may be reproduced or distributed in any form or by any means, or stored in a database or retrieval system, without the prior written consent of The McGraw-Hill Companies, Inc., including, but not limited to, in any network or other electronic storage or transmission, or broadcast for distance learning.

Some ancillaries, including electronic and print components, may not be available to customers outside the United States.

Tata McGraw-Hill Edition 2002

Ninth reprint 2005
RCLYCDLBRQXBQ

Reprinted in India by arrangement with The McGraw-Hill Companies Inc.,
New York

Sales territories: India, Pakistan, Nepal, Bangladesh, Sri Lanka and Bhutan

Library of Congress Cataloging-in-Publication Data

Razavi, Behzad.
 Design of analog CMOS integrated circuits/Behzad Razavi.
 p. cm.
 ISBN 0-07-238032-2
 1. Linear integrated circuit—Design and construction. 2. Metal oxide semiconductors,
Complimentary. I. Title.
TK 7874.654 R39 2001
621.39'732—dc21   00-044789

ISBN 0-07-052903-5

Published by Tata McGraw-Hill Publishing Company Limited,
7 West Patel Nagar, New Delhi 110 008, and printed at
Rajkamal Electric Press, Delhi 110 033

The McGraw·Hill Companies

To the memory of my parents

About the Author

Behzad Razavi received the B.Sc. degree in electrical engineering from Sharif University of Technology in 1985 and the M.Sc. and Ph.D. degrees in electrical engineering from Stanford University in 1988 and 1992, respectively. He was with AT&T Bell Laboratories and subsequently Hewlett-Packard Laboratories until 1996. Since September 1996, he has been an Associate Professor and subsequently a Professor of electrical engineering at University of California, Los Angeles. His current research includes wireless transceivers, frequency synthesizers, phase-locking and clock recovery for high-speed data communications, and data converters.

Professor Razavi served as an Adjunct Professor at Princeton University, Princeton, NJ, from 1992 to 1994, and at Stanford University in 1995. He is a member of the Technical Program Committees of the Symposium on VLSI Circuits and the International Solid-State Circuits Conference (ISSCC), in which he is the chair of the Analog Subcommittee. He has also served as Guest Editor and Associate Editor of the IEEE Journal of Solid-State Circuits, IEEE Transactions on Circuits and Systems, and International Journal of High Speed Electronics.

Professor Razavi received the Beatrice Winner Award for Editorial Excellence at the 1994 ISSCC, the best paper award at the 1994 European Solid-State Circuits Conference, the best panel award at the 1995 and 1997 ISSCC, the TRW Innovative Teaching Award in 1997, and the best paper award at the IEEE Custom Integrated Circuits Conference in 1998. He is the author of *Principles of Data Conversion System Design* (IEEE Press, 1995), and *RF Microelectronics* (Prentice Hall, 1998), and the editor of *Monolithic Phase-Locked Loops and Clock Recovery Circuits* (IEEE Press, 1996).

Preface

In the past two decades, CMOS technology has rapidly embraced the field of analog integrated circuits, providing low-cost, high-performance solutions and rising to dominate the market. While silicon bipolar and III-V devices still find niche applications, only CMOS processes have emerged as a viable choice for the integration of today's complex mixed-signal systems. With channel lengths projected to scale down to 0.03 μm, CMOS technology will continue to serve circuit design for probably another two decades.

Analog circuit design itself has evolved with the technology as well. High-voltage, high-power analog circuits containing a few tens of transistors and processing small, continuous-time signals have gradually been replaced by low-voltage, low-power systems comprising thousands of devices and processing large, mostly discrete-time signals. For example, many analog techniques used only ten years ago have been abandoned because they do not lend themselves to low-voltage operation.

This book deals with the analysis and design of analog CMOS integrated circuits, emphasizing fundamentals as well as new paradigms that students and practicing engineers need to master in today's industry. Since analog design requires both intuition and rigor, each concept is first introduced from an intuitive perspective and subsequently treated by careful analysis. The objective is to develop both a solid foundation and methods of analyzing circuits by inspection so that the reader learns what approximations can be made in which circuits and how much error to expect in each approximation. This approach also enables the reader to apply the concepts to bipolar circuits with little additional effort.

I have taught most of the material in this book both at UCLA and in industry, polishing the order, the format, and the content with every offering. As the reader will see throughout the book, I follow four "golden rules" in writing (and teaching): (1) I explain *why* the reader needs to know the concept that is to be studied; (2) I put myself in the reader's position and predict the questions that he/she may have while reading the material for the first time; (3) With Rule 2 in mind, I pretend to know only as much as the (first-time) reader and try to "grow" with him/her, thereby experiencing the same through process; (4) I begin with the "core" concept in a simple (even imprecise) language and gradually add necessary modifications to arrive at the final (precise) idea. The last rule is particularly important in teaching circuits because it allows the reader to observe the evolution of a topology and hence learn both analysis and synthesis.

The text comprises 18 chapters whose contents and order are carefully chosen to provide a natural flow for both self-study and classroom adoption in quarter or semester systems.

Unlike some other books on analog design, we cover only a *bare minimum* of MOS device physics at the beginning, leaving more advanced properties and fabrication details for later chapters. To an expert, the elementary device physics treatment may appear oversimplified, but my experience suggests that (a) first-time readers simply do not absorb the high-order device effects and fabrication technology before they study circuits because they do not see the relevance; (b) if properly presented, even the simple treatment proves adequate for a substantial coverage of basic circuits; (c) readers learn advanced device phenomena and processing steps much more readily *after* they have been exposed to a significant amount of circuit analysis and design.

Chapter 1 provides the reader with motivation for learning the material in this book.

Chapter 2 describes basic physics and operation of MOS devices.

Chapters 3 through 5 deal with single-stage and differential amplifiers and current mirrors, respectively, developing efficient analytical tools for quantifying the behavior of basic circuits by inspection.

Chapters 6 and 7 introduce two imperfections of circuits, namely, frequency response and noise. Noise is treated at an early stage so that it "sinks in" as the reader accounts for its effects in subsequent circuit developments.

Chapters 8 through 10 describe feedback, operational amplifiers, and stability in feedback systems, respectively. With the useful properties of feedback analyzed, the reader is motivated to design high-performance, stable op amps and understand the trade-offs between speed, precision, and power dissipation.

Chapters 11 through 13 deal with more advanced topics: bandgap references, elementary switched-capacitor circuits, and the effect of nonlinearity and mismatch. These three subjects are included here because they prove essential in most analog and mixed-signal systems today.

Chapters 14 and 15 concentrate on the design of oscillators and phase-locked loops, respectively. In view of the wide usage of these circuits, a detailed study of their behavior and many examples of their operation are provided.

Chapter 16 is concerned with high-order MOS device effects and models, emphasizing the circuit design implications. If preferred, this chapter can directly follow Chapter 2 as well. Chapter 17 describes CMOS fabrication technology with a brief overview of layout design rules.

Chapter 18 presents the layout and packaging of analog and mixed-signal circuits. Many practical issues that directly impact the performance of the circuit are described and various techniques are introduced.

The reader is assumed to have a basic knowledge of electronic circuits and devices, e.g., *pn* junctions, the concept of small-signal operation, equivalent circuits, and simple biasing. For a senior-level elective course, Chapters 1 through 8 can be covered in a quarter and Chapters 1 through 10 in a semester. For a first-year graduate course, Chapters 1 through 11 plus one of Chapters 12 through 15 can be taught in one quarter, and the first 16 chapters in one semester.

The problem sets at the end of each chapter are designed to extend the reader's understanding of the material and complement it with additional practical considerations. A solutions manual is available for instructors.

Behzad Razavi

July 2000

Acknowledgments

Writing a book begins with a great deal of excitement. However, after two years of relentless writing, drawing, and revising, when the book exceeds 600 pages and it is almost impossible to make the equations and subscripts and superscripts in the last chapter consistent with those in the first, the author begins to feel the streaks of insanity, realizing that the book will never finish without the support of many other people.

This book has benefited from the contributions of many individuals. A number of UCLA students read the first draft and the preview edition sentence by sentence. In particular, Alireza Zolfaghari, Ellie Cijvat, and Hamid Rafati meticulously read the book and found several hundred errors (some quite subtle). Also, Emad Hegazi, Dawei Guo, Alireza Razzaghi, Jafar Savoj, and Jing Tian made helpful suggestions regarding many chapters. I thank all.

Many experts in academia and industry read various parts of the book and provided useful feedback. Among them are Brian Brandt (National Semiconductor), Matt Corey (National Semiconductor), Terri Fiez (Oregon State University), Ian Galton (UC San Diego), Ali Hajimiri (Caltech), Stacy Ho (Analog Devices), Yin Hu (Texas Instruments), Shen-Iuan Liu (National Taiwan University), Joe Lutsky (National Semiconductor), Amit Mehrotra (University of Illinois, Urbana-Champaign), David Robertson (Analog Devices), David Su (T-Span), Tao Sun (National Semiconductor), Robert Taft (National Semiconductor), and Masoud Zargari (T-Span). Jason Woo (UCLA) patiently endured and answered my questions about device physics. I thank all.

Ramesh Harjani (University of Minnesota), John Nyenhius (Purdue University), Norman Tien (Cornell University), and Mahmoud Wagdy (California State University, Long Beach) reviewed the book proposal and made valuable suggestions. I thank all.

My wife, Angelina, has made many contributions to this book, from typing chapters to finding numerous errors and raising questions that made me reexamine my own understanding. I am very grateful to her.

The timely production of the book was made possible by the hard work of the staff at McGraw-Hill, particularly, Catherine Fields, Michelle Flomenhoft, Heather Burbridge, Denise Santor-Mitzit, and Jim Labeots. I thank all.

I learned analog design from two masters: Mehrdad Sharif-Bakhtiar (Sharif University of Technology) and Bruce Wooley (Stanford University) and it is only appropriate that I express my gratitude to them here. What I inherited from them will be inherited by many generations of students.

Behzad Razavi
July 2000

Brief Contents

A/D { 9 10 *(handwritten annotation)*

Jacob Baker — ch 22, 24, 26 . *(handwritten annotation)*

Contents

Chapter 1

Introduction to Analog Design

1.1 Why Analog?

It was in the early 1980s that many experts predicted the demise of analog circuits. Digital signal processing algorithms were becoming increasingly more powerful while advances in integrated-circuit (IC) technology provided compact, efficient implementation of these algorithms in silicon. Many functions that had traditionally been realized in analog form were now easily performed in the digital domain, suggesting that, with enough capability in IC fabrication, all processing of signals would eventually occur digitally. The future looked quite bleak to analog designers and they were seeking other jobs.

But, why are analog designers in such great demand today? After all, digital signal processing and IC technologies have advanced tremendously since the early 1980s, making it possible to realize processors containing millions of transistors and performing billions of operations per second. Why did this progress not confirm the earlier predictions?

While many types of signal processing have indeed moved to the digital domain, analog circuits have proved *fundamentally* necessary in many of today's complex, high-performance systems. Let us consider a few applications where it is very difficult or even impossible to replace analog functions with their digital counterparts regardless of advances in technology.

Processing of Natural Signals   Naturally occurring signals are analog—at least at a macroscopic level. A high-quality microphone picking up the sound of an orchestra generates a voltage whose amplitude may vary from a few microvolts to hundreds of millivolts. The photocells in a video camera produce a current that is as low as a few electrons per microsecond. A seismographic sensor has an output voltage ranging from a few microvolts for very small vibrations of the earth to hundreds of millivolts for heavy earthquakes. Since all of these signals must eventually undergo extensive processing in the digital domain, we observe that each of these systems consists of an analog-to-digital converter (ADC) and a digital signal processor (DSP) [Fig. 1.1(a)]. The design of ADCs for high speed, high precision, and low power dissipation is one of many difficult challenges in analog design.

In practice, the electrical version of natural signals may be prohibitively small for direct digitization by the ADC. The signals are also often accompanied by unwanted, out-of-band

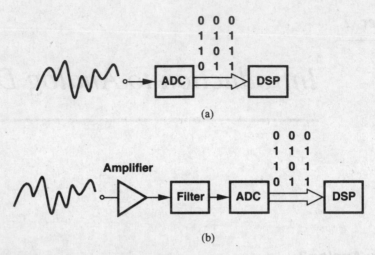

(a)

(b)

Figure 1.1   (a) Digitization of a natural signal, (b) addition of amplification and filtering for higher sensitivity.

interferers. The front end of Fig. 1.1(a) may therefore be modified as shown in Fig. 1.1(b), where an amplifier boosts the signal level and an analog filter suppresses the out-of-band components. The design of high-performance amplifiers and filters is also a topic of active research today.

Digital Communications   Binary data generated by various systems must often be transmitted over long distances. For example, computer networks in large office buildings may transmit the data over cables that are hundreds of meters long.

What happens if a high-speed stream of binary data travels through a long cable? As illustrated in Fig. 1.2, the signal experiences both attenuation and "distortion," no longer resembling a digital waveform. Thus, a receiver similar to that of Fig. 1.1(b) may be necessary here.

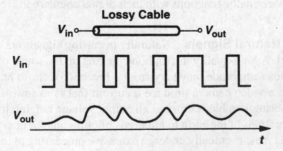

Figure 1.2   Attenuation and distortion of data through a lossy cable.

In order to improve the quality of communication, the above system may incorporate "multi-level"—rather than binary—signals. For example, if, as shown in Fig. 1.3, every two consecutive bits in the sequence are grouped and converted to one of four levels, then

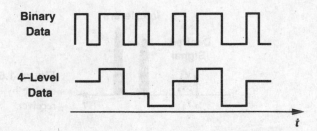

Figure 1.3 Use of multi-level signalling to reduce the required bandwidth.

each level is twice as long as a bit period, demanding only *half* the bandwidth required for transmission of the binary stream. Utilized extensively in today's communication systems, multi-level signals necessitate a digital-to-analog converter (DAC) in the transmitter to produce multiple levels from the grouped binary data and an ADC in the receiver to determine which level has been transmitted. The key point here is that increasing the number of levels relaxes the bandwidth requirements while demanding a higher precision in the DAC and the ADC.

Disk Drive Electronics The data stored magnetically on a computer hard disk is in binary form. However, when the data is read by a magnetic head and converted to an electrical signal, the result appears as shown in Fig. 1.4. The amplitude is only a few millivolts, the noise content is quite high, and the bits experience substantial distortion.

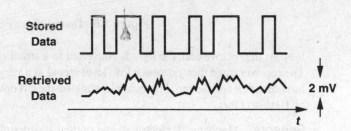

Figure 1.4 Data stored in and retrieved from a hard disk.

Thus, as illustrated in Fig. 1.1, the signal is amplified, filtered, and digitized for further processing. Depending on the overall system architecture, the analog filter in this case may in fact serve to remove a significant portion of the noise and the distortion of the signal. The design of each of these building blocks poses great challenges as the speed of computers and their storage media continues to increase every year. For example, today's disk drives require a speed of 500 Mb/s.

Wireless Receivers The signal picked up by the antenna of a radio-frequency (RF) receiver, e.g., a pager or a cellular telephone, exhibits an amplitude of only a few microvolts and a center frequency of 1 GHz or higher. Furthermore, the signal is accompanied by large

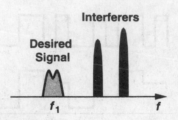

Figure 1.5 Signal and interferers received by the antenna of a wireless receiver.

interferers (Fig. 1.5). The receiver must therefore amplify the low-level signal with minimal noise, operate at a high frequency, and withstand large unwanted components. Note that these requirements are necessary even if the desired signal is not in "analog" form. The trade-offs between noise, frequency of operation, tolerance of interferers, power dissipation, and *cost* constitute the principal challenge in today's wireless industry.

Optical Receivers   For transmission of high-speed data over very long distances, cables generally prove inadequate because of their limited bandwidth and considerable attenuation. Thus, as illustrated in Fig. 1.6, the data is converted to light by means of a laser diode and transmitted over an optical fiber, which exhibits an extremely wide band and a very low

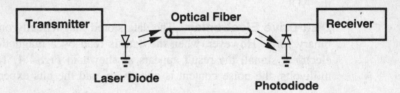

Figure 1.6   Optical fiber system.

loss. At the receive end, the light is converted to a small electrical current by a photodiode. The receiver must then process a low-level signal at a very high speed, requiring low-noise, broadband circuit design. For example, state-of-the-art optical receivers operate in the range of 10 to 40 Gb/s.

Sensors   Mechanical, electrical, and optical sensors play a critical role in our lives. For example, video cameras incorporate an array of photodiodes to convert an image to current and ultrasound systems use an acoustic sensor to generate a voltage proportional to the amplitude of the ultrasound waveform. Amplification, filtering, and A/D conversion are essential functions in these applications.

An interesting example of sensors is the accelerometers employed in automobiles to activate air bags. When the vehicle hits an obstacle, the drop in the speed is measured as acceleration and, if exceeding a certain threshold, it triggers the air bag release mechanism. Modern accelerometers are based on a variable capacitor consisting of a fixed plate and a deflectable plate [Fig. 1.7(a)]. The deflection and hence the value of the capacitor are proportional to the acceleration, requiring a circuit that accurately measures the change in capacitance. The design of such interface circuits is quite difficult because for typical

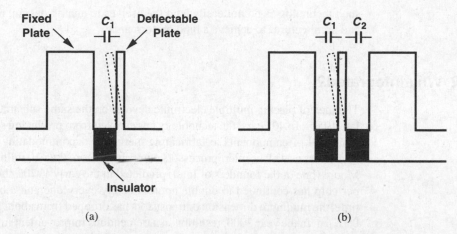

Figure 1.7 (a) Simple accelerometer, (b) differential accelerometer.

accelerations, the interplate capacitance may change by less than 1%, demanding a high precision in the measurement. In practice, the structure of Fig. 1.7(b) is used to provide two capacitors that change in opposite directions, reducing the task to the measurement of the *difference* between two capacitances rather than the absolute value of one.

Microprocessors and Memories   Today's microprocessors and memories draw upon a great deal of analog design expertise. Many issues related to the distribution and timing of data and clocks across a large chip or among chips mandate that high-speed signals be viewed as analog waveforms. Furthermore, nonidealities in signal and power interconnects on the chip as well as package parasitics require a solid understanding of analog design. In addition, semiconductor memories employ high-speed "sense amplifiers" extensively, necessitating many analog techniques. For these reasons, it is often said "high-speed digital design is in fact analog design."

The foregoing applications demonstrate the wide and inevitable spread of analog circuits in modern industry. But, why is analog design difficult? We make the following observations. (1) Whereas digital circuits entail primarily one trade-off between speed and power dissipation, analog design must deal with a multi-dimensional trade-off consisting of speed, power dissipation, gain, precision, supply voltage, etc. (2) With the speed and precision required in processing analog signals, analog circuits are much more sensitive to noise, crosstalk, and other interferers than are digital circuits. (3) Second-order effects in devices influence the performance of analog circuits much more heavily than that of digital circuits. (4) The design of high-performance analog circuits can rarely be automated, usually requiring that every device be "hand-crafted." By contrast, many digital circuits are automatically synthesized and laid out. (5) Despite tremendous progress, modeling and simulation of many effects in analog circuits continue to pose difficulties, forcing the designers to draw upon experience and intuition when analyzing the results of a simulation. (6) An important thrust in today's semiconductor industry is to design analog circuits in mainstream IC technologies used to fabricate digital products. Developed and characterized for digital applications, such technologies do not easily lend themselves to analog design, requiring novel circuits and architectures to achieve a high performance.

1.2 Why Integrated?

The idea of placing multiple electronic devices on the same substrate was conceived in the late 1950s. In 40 years, the technology has evolved from producing simple chips containing a handful of components to fabricating memories accommodating more than one billion transistors as well as microprocessors comprising more than 10 million devices. As Gordon Moore (one of the founders of Intel) predicted in the early 1970s, the number of transistors per chip has continued to double approximately every one and a half years. At the same time, the minimum dimension of transistors has dropped from about 25 μm in 1960 to about 0.18 μm in the year 2000, resulting in a tremendous improvement in the speed of integrated circuits.

Driven by primarily the memory and microprocessor market, integrated-circuit technologies have also embraced analog design extensively, affording a complexity, speed, and precision that would be impossible to achieve using discrete implementations. Analog and mixed analog/digital integrated circuits containing tens of thousands of devices now routinely appear in consumer products. We can no longer build a discrete prototype to predict the behavior and performance of modern analog circuits.

1.3 Why CMOS?

The idea of metal-oxide-silicon field-effect transistors (MOSFETs) was patented by J. E. Lilienfeld in the early 1930s—well before the invention of the bipolar transistor. Owing to fabrication limitations, however, MOS technologies became practical much later, in the early 1960s, with the first several generations producing only n-type transistors. It was in the mid-1960s that complementary MOS (CMOS) devices (i.e., both n-type and p-type transistors) were introduced, initiating a revolution in the semiconductor industry.

CMOS technologies rapidly captured the digital market: CMOS gates dissipated power only during switching and required very few devices, two attributes in sharp contrast to their bipolar or GaAs counterparts. It was also soon discovered that the dimensions of MOS devices could be scaled down more easily than those of other types of transistors. Furthermore, CMOS circuits proved to have a lower fabrication cost.

The next obvious step was to apply CMOS technology to analog design. The low cost of fabrication and the possibility of placing both analog and digital circuits on the same chip so as to improve the overall performance and/or reduce the cost of packaging made CMOS technology attractive. However, MOSFETs were quite slower and noisier than bipolar transistors, finding limited application.

How did CMOS technology come to dominate the analog market as well? The principal force was device scaling because it continued to improve the speed of MOSFETs. The intrinsic speed of MOS transistors has increased by more than three orders of magnitude in the past 30 years, becoming comparable with that of bipolar devices even though the latter have also been scaled (but not as fast). Multi-gigahertz analog CMOS circuits are now in production.

1.4 Why This Book?

The design of analog circuits itself has evolved together with the technology and the performance requirements. As the device dimensions shrink, the supply voltage of integrated circuits drops, and analog and digital circuits are fabricated on one chip, many design issues arise that were unimportant only a decade ago. Such trends demand that the analysis and design of circuits be accompanied by an in-depth understanding of their advantages and disadvantages with respect to new technology-imposed limitations.

Good analog design requires intuition, rigor, and creativity. As analog designers, we must wear our engineer's hat for a quick and intuitive understanding of a large circuit, our mathematician's hat for quantifying subtle, yet important effects in a circuit, and our artist's hat for inventing new circuit topologies.

This book describes modern analog design from both intuitive and rigorous angles. It also fosters the reader's creativity by carefully guiding him/her through the evolution of each circuit and presenting the thought process that occurs during the development of new circuit techniques.

1.5 General Concepts

1.5.1 Levels of Abstraction

Analysis and design of integrated circuits often require thinking at various levels of abstraction. Depending on the effect or quantity of interest, we may study a complex circuit at device physics level, transistor level, architecture level, or system level. In other words, we may consider the behavior of individual devices in terms of their internal electric fields and charge transport [Fig. 1.8(a)], the interaction of a group of devices according to their electrical characteristics [Fig. 1.8(b)], the function of several building blocks operating as a unit [Fig. 1.8(c)], or the performance of the system in terms of that of its constituent subsystems [Fig. 1.8(d)]. Switching between levels of abstraction becomes necessary in both understanding the details of the operation and optimizing the overall performance. In fact, in today's IC industry, the interaction between all groups, from device physicists to system designers, is essential to achieving a high performance and a low cost. In this book, we begin with device physics and develop increasingly more complex circuit topologies.

1.5.2 Robust Analog Design

Many device and circuit parameters vary with the fabrication process, supply voltage, and ambient temperature. We denote these effects by PVT and design circuits such that their performance remains in an acceptable range for a specified range of PVT variations. For example, the supply voltage may vary from 2.7 V to 3.3 V and the temperature from $0°$ to $70°$. Robust analog design in CMOS technology is a challenging task because device parameters vary significantly from wafer to wafer.

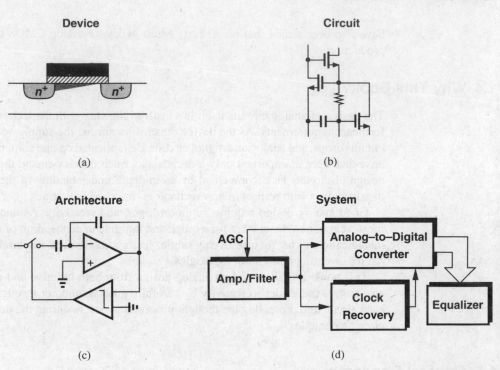

Figure 1.8 Abstraction levels in circuit design: (a) device level, (b) circuit level, (c) architecture level, (d) system level.

1.5.3 Notations

The voltages and currents in integrated circuits typically contain a bias component and a signal component. While it is desirable to employ a notation that distinguishes between these quantities, in practice other difficulties arise. For example, if the drain bias current of a transistor is denoted by I_D and the drain signal current by i_D, then the Laplace transform of i_D, $I_D(s)$, may be confused with I_D unless it is always accompanied by s. Furthermore, it is confusing to write the low-frequency gain of a circuit as $v_{out}/v_{in} = -g_m R_D$ and the high-frequency gain as $V_{out}/V_{in} = -g_m R_D/(1 + R_D C_L s)$.

In this book, we denote most voltages and currents by uppercase letters, making it clear from the context which component they represent. For example, I_D, V_{GS}, and V_X denote bias, signal, or bias+signal quantities. For input and output voltages, we use V_{in} and V_{out}, respectively.

Chapter 2

Basic MOS Device Physics

In studying the design of integrated circuits, one of two extreme approaches can be taken: (1) begin with quantum mechanics and understand solid-state physics, semiconductor device physics, device modeling, and finally the design of circuits; (2) treat each semiconductor device as a black box whose behavior is described in terms of its terminal voltages and currents and design circuits with little attention to the internal operation of the device. Experience shows that neither approach is optimum. In the first case, the reader cannot see the relevance of all of the physics to designing circuits, and in the second, he/she is constantly mystified by the contents of the black box.

In today's IC industry, a solid understanding of semiconductor devices is essential, more so in analog design than in digital design because in the former, transistors are not considered as simple switches and many of their second-order effects directly impact the performance. Furthermore, as each new generation of IC technologies scales the devices, these effects become more significant. Since the designer must often decide which effects can be neglected in a given circuit, insight into device operation proves invaluable.

In this chapter, we study the physics of MOSFETs at an elementary level, covering the bare minimum that is necessary for basic analog design. The ultimate goal is still to develop a circuit model for each device by formulating its operation, but this is accomplished with a good understanding of the underlying principles. After studying many analog circuits in Chapters 3 through 13 and gaining motivation for a deeper understanding of devices, we return to the subject in Chapter 16 and deal with other aspects of MOS operation.

We begin our study with the structure of MOS transistors and derive their I/V characteristics. Next, we describe second-order effects such as body effect, channel-length modulation, and subthreshold conduction. We then identify the parasitic capacitances of MOSFETs, derive a small-signal model, and present a simple SPICE model. We assume that the reader is familiar with such basic concepts as doping, mobility, and *pn* junctions.

2.1 General Considerations

2.1.1 MOSFET as a Switch

Before delving into the actual operation of the MOSFET, we consider a simplistic model of the device so as to gain a feeling for what the transistor is expected to be and which aspects of its behavior are important.

Shown in Fig. 2.1 is the symbol for an n-type MOSFET, revealing three terminals: gate (G), source (S), and drain (D). The latter two are interchangeable because the device is

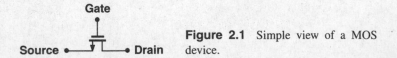

Figure 2.1 Simple view of a MOS device.

symmetric. When operating as a switch, the transistor "connects" the source and the drain together if the gate voltage, V_G, is "high" and isolates the source and the drain if V_G is "low."

Even with this simplified view, we must answer several questions. For what value of V_G does the device turn on? In other words, what is the "threshold" voltage? What is the resistance between S and D when the device is on (or off)? How does this resistance depend on the terminal voltages? Can we always model the path between S and D by a simple linear resistor? What limits the speed of the device?

While all of these questions arise at the circuit level, they can be answered only by analyzing the structure and physics of the transistor.

2.1.2 MOSFET Structure

Fig. 2.2 shows a simplified structure of an n-type MOS (NMOS) device. Fabricated on a p-type substrate (also called the "bulk" or the "body"), the device consists of two heavily-doped n regions forming the source and drain terminals, a heavily-doped (conductive) piece

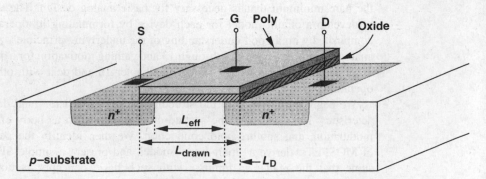

Figure 2.2 Structure of a MOS device.

of polysilicon[1] (often simply called "poly") operating as the gate, and a thin layer of silicon dioxide (SiO_2) insulating the gate from the substrate. The useful action of the device occurs in the substrate region under the gate oxide. Note that the structure is symmetric with respect to S and D.

The dimension of the gate along the source-drain path is called the length, L, and that perpendicular to the length is called the width, W. Since during fabrication the S/D junctions "side-diffuse," the actual distance between the source and the drain is slightly less than L. To avoid confusion, we write, $L_{eff} = L_{drawn} - 2L_D$, where L_{eff} is the "effective" length, L_{drawn} is the total length,[2] and L_D is the amount of side diffusion. As we will see later, L_{eff} and the gate oxide thickness, t_{ox}, play an important role in the performance of MOS circuits. Consequently, the principal thrust in MOS technology development is to reduce both of these dimensions from one generation to the next without degrading other parameters of the device. Typical values at the time of this writing are $L_{eff} \approx 0.15\ \mu$m and $t_{ox} \approx 50$ Å. In the remainder of this book, we denote the effective length by L.

If the MOS structure is symmetric, why do we call one n region the source and the other the drain? This becomes clear if the source is defined as the terminal that provides the charge carriers (electrons in the case of NMOS devices) and the drain as the terminal that collects them. Thus, as the voltages at the three terminals of the device vary, the source and the drain may exchange roles. These concepts are practiced in the problems at the end of the chapter.

We have thus far ignored the substrate on which the device is fabricated. In reality, the substrate potential greatly influences the device characteristics. That is, the MOSFET is a *four*-terminal device. Since in typical MOS operation the S/D junction diodes must be reverse-biased, we assume the substrate of NMOS transistors is connected to the most negative supply in the system. For example, if a circuit operates between zero and 3 volts, $V_{sub,NMOS} = 0$. The actual connection is usually provided through an ohmic p^+ region, as depicted in the side view of the device in Fig. 2.3.

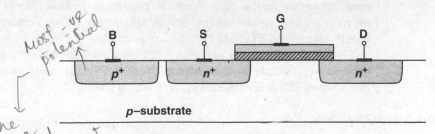

Figure 2.3 Substrate connection.

In complementary MOS (CMOS) technologies, both NMOS and PMOS transistors are available. From a simplistic view point, the PMOS device is obtained by negating all of

---

[1]Polysilicon is silicon in amorphous (non-crystal) form. As explained in Chapter 17, when the gate silicon is grown on top of the oxide, it cannot form a crystal.

[2]The subscript "drawn" is used because this is the dimension that we draw in the layout of the transistor (Section 2.4.1).

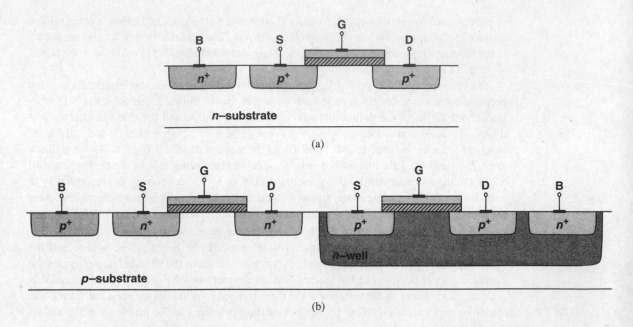

Figure 2.4   (a) Simple PMOS device, (b) PMOS inside an n-well.

the doping types (including the substrate) [Fig. 2.4(a)], but in practice, NMOS and PMOS devices must be fabricated on the same wafer, i.e., the same substrate. For this reason, one device type can be placed in a "local substrate," usually called a "well." In most of today's CMOS processes, the PMOS device is fabricated in an n-well [Fig. 2.4(b)]. Note that the n-well must be connected to a potential such that the S/D junction diodes of the PMOS transistor remain reverse-biased under all conditions. In most circuits, the n-well is tied to the most positive supply voltage. For the sake of brevity, we sometimes call NMOS and PMOS devices "NFETs" and "PFETs," respectively.

Fig. 2.4(b) indicates an interesting difference between NMOS and PMOS transistors: while all NFETs share the same substrate, each PFET can have an independent n-well. This flexibility of PFETs is exploited in some analog circuits.

2.1.3 MOS Symbols

The circuit symbols used to represent NMOS and PMOS transistors are shown in Fig. 2.5. The symbols in Fig. 2.5(a) contain all four terminals, with the substrate denoted by "B" (bulk) rather than "S" to avoid confusion with the source. The source of the PMOS device is positioned on top as a visual aid because it has a higher potential than its gate. Since in most circuits the bulk terminals of NMOS and PMOS devices are tied to ground and V_{DD}, respectively, we usually omit these connections in drawing [Fig. 2.5(b)]. In digital circuits, it is customary to use the "switch" symbols depicted in Fig. 2.5(c) for the two types, but we prefer those in Fig. 2.5(b) because the visual distinction between S and D proves helpful in understanding the operation of circuits.

$c = \varepsilon_0 \frac{A}{d}$

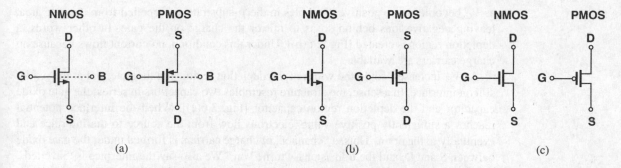

Figure 2.5   MOS symbols.

2.2 MOS I/V Characteristics

In this section, we analyze the generation and transport of charge in MOSFETs as a function of the terminal voltages. Our objective is to derive equations for the I/V characteristics such that we can elevate our abstraction from device physics level to circuit level.

2.2.1 Threshold Voltage

Consider an NFET connected to external voltages as shown in Fig. 2.6(a). What happens as the gate voltage, V_G, increases from zero? Since the gate and the substrate form a capacitor,

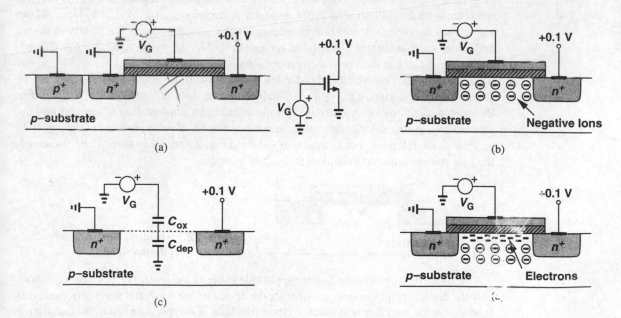

Figure 2.6   (a) A MOSFET driven by a gate voltage, (b) formation of depletion region, (c) onset of inversion, (d) formation of inversion layer.

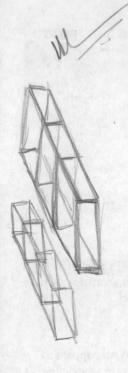

as V_G becomes more positive, the holes in the p-substrate are repelled from the gate area, leaving negative ions behind so as to mirror the charge on the gate. In other words, a depletion region is created [Fig. 2.6(b)]. Under this condition, no current flows because no charge carriers are available.

As V_G increases, so do the width of the depletion region and the potential at the oxide-silicon interface. In a sense, the structure resembles two capacitors in series: the gate oxide capacitor and the depletion region capacitor [Fig. 2.6(c)]. When the interface potential reaches a sufficiently positive value, electrons flow from the source to the interface and eventually to the drain. Thus, a "channel" of charge carriers is formed under the gate oxide between S and D, and the transistor is "turned on." We also say the interface is "inverted." The value of V_G for which this occurs is called the "threshold voltage," V_{TH}. If V_G rises further, the charge in the depletion region remains relatively constant while the channel charge density continues to increase, providing a greater current from S to D.

In reality, the turn-on phenomenon is a gradual function of the gate voltage, making it difficult to define V_{TH} unambiguously. In semiconductor physics, the V_{TH} of an NFET is usually defined as the gate voltage for which the interface is "as much n-type as the substrate is p-type." It can be proved [1] that[3]

$$V_{TH} = \Phi_{MS} + 2\Phi_F + \frac{Q_{dep}}{C_{ox}}, \qquad (2.1)$$

where Φ_{MS} is the difference between the work functions of the polysilicon gate and the silicon substrate, $\Phi_F = (kT/q)\ln(N_{sub}/n_i)$, q is electron charge, N_{sub} is the doping concentration of the substrate, Q_{dep} is the charge in the depletion region, and C_{ox} is the gate oxide capacitance per unit area. From pn junction theory, $Q_{dep} = \sqrt{4q\epsilon_{si}|\Phi_F|N_{sub}}$, where ϵ_{si} denotes the dielectric constant of silicon. Since C_{ox} appears very frequently in device and circuit calculations, it is helpful to remember that for $t_{ox} \approx 50$ Å, $C_{ox} \approx 6.9$ fF/μm^2. The value of C_{ox} can then be scaled proportionally for other oxide thicknesses.

In practice, the "native" threshold value obtained from the above equation may not be suited to circuit design, e.g., $V_{TH} = 0$ and the device does not turn off for $V_G \geq 0$. For this reason, the threshold voltage is typically adjusted by implantation of dopants into the channel area during device fabrication, in essence altering the doping level of the substrate near the oxide interface. For example, as shown in Fig. 2.7, if a thin sheet of p^+ is created, the gate voltage required to deplete this region increases.

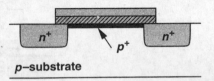

Figure 2.7 Implantation of p^+ dopants to alter the threshold.

The above definition is not directly applicable to the *measurement* of V_{TH}. In Fig. 2.6(a), only the drain current can indicate whether the device is "on" or "off," thus failing to reveal at what V_{GS} the interface is as much n-type as the bulk is p-type. As a result, the calculation

---

[3]Charge trapping in the oxide is neglected here.

[handwritten note at top: The value of gate voltage for which which the inverted region starts conducting (or) e flows from s to D is Vth.]

of V_{TH} from I/V measurements is somewhat ambiguous. We return to this point later but assume in our preliminary analysis that the device turns on *abruptly* for $V_{GS} \geq V_{TH}$.

The turn-on phenomenon in a PMOS device is similar to that of NFETs but with all of the polarities reversed. As shown in Fig. 2.8, if the gate-source voltage becomes sufficiently

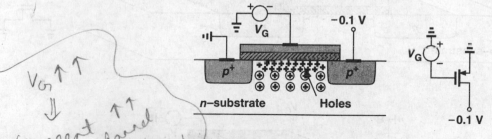

Figure 2.8 Formation of inversion layer in a PFET.

[handwritten note in margin: $V_G \uparrow \uparrow$ ⇓ Current ↑↑ as the channel charge density increases]

negative, an inversion layer consisting of holes is formed at the oxide-silicon interface, providing a conduction path between the source and the drain.

2.2.2 Derivation of I/V Characteristics

In order to obtain the relationship between the drain current of a MOSFET and its terminal voltages, we make two observations.

First, consider a semiconductor bar carrying a current I [Fig. 2.9(a)]. If the charge density along the direction of current is Q_d coulombs per meter and the velocity of the charge is v meters per second, then

$$I = Q_d \cdot v. \tag{2.2}$$

To understand why, we measure the total charge that passes through a cross section of the bar in unit time. With a velocity v, all of the charge enclosed in v meters of the bar must flow through the cross section in one second [Fig. 2.9(b)]. Since the charge density is Q_d, the total charge in v meters equals $Q_d \cdot v$. This lemma proves useful in analyzing semiconductor devices.

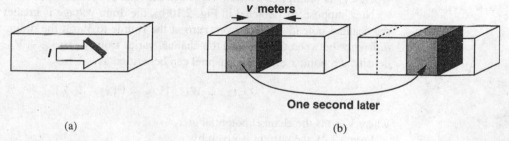

(a)   (b)

Figure 2.9 (a) A semiconductor bar carrying a current I, (b) snapshots of the carriers one second apart.

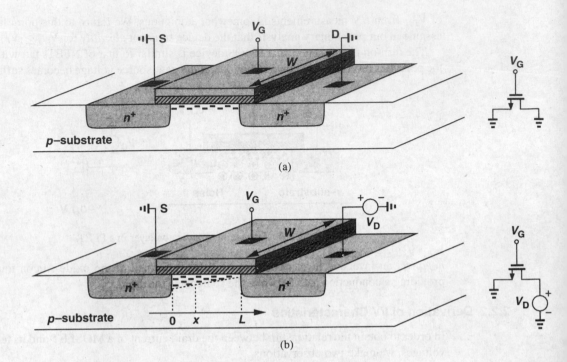

Figure 2.10 Channel charge with (a) equal source and drain voltages, (b) unequal source and drain voltages.

Second, consider an NFET whose source and drain are connected to ground [Fig. 2.10(a)]. What is the charge density in the inversion layer? Since we assume the onset of inversion occurs at $V_{GS} = V_{TH}$, the inversion charge density produced by the gate oxide capacitance is proportional to $V_{GS} - V_{TH}$. For $V_{GS} \geq V_{TH}$, any charge placed on the gate must be mirrored by the charge in the channel, yielding a uniform channel charge density (charge per unit length) equal to

$$Q_d = W C_{ox}(V_{GS} - V_{TH}), \tag{2.3}$$

where C_{ox} is multiplied by W to represent the total capacitance per unit length.

Now suppose, as depicted in Fig. 2.10(b), the drain voltage is greater than zero. Since the channel potential varies from zero at the source to V_D at the drain, the local voltage *difference* between the gate and the channel varies from V_G to $V_G - V_D$. Thus, the charge density at a point x along the channel can be written as

$$Q_d(x) = W C_{ox}[V_{GS} - V(x) - V_{TH}], \tag{2.4}$$

where $V(x)$ is the channel potential at x.

From (2.2), the current is given by

$$I_D = -W C_{ox}[V_{GS} - V(x) - V_{TH}]v, \tag{2.5}$$

where the negative sign is inserted because the charge carriers are negative and v denotes the velocity of the electrons in the channel. For semiconductors, $v = \mu E$, where μ is the mobility of charge carriers and E is the electric field. Noting that $E(x) = -dV/dx$ and representing the mobility of electrons by μ_n, we have

$$I_D = W C_{ox}[V_{GS} - V(x) - V_{TH}]\mu_n \frac{dV(x)}{dx}, \tag{2.6}$$

subject to boundary conditions $V(0) = 0$ and $V(L) = V_{DS}$. While $V(x)$ can be easily found from this equation, the quantity of interest is in fact I_D. Multiplying both sides by dV and performing integration, we obtain

$$\int_{x=0}^{L} I_D dx = \int_{V=0}^{V_{DS}} W C_{ox} \mu_n [V_{GS} - V(x) - V_{TH}]dV. \tag{2.7}$$

Since I_D is constant along the channel:

$$I_D = \mu_n C_{ox} \frac{W}{L}\left[(V_{GS} - V_{TH})V_{DS} - \frac{1}{2}V_{DS}^2\right]. \tag{2.8}$$

Note that L is the effective channel length.

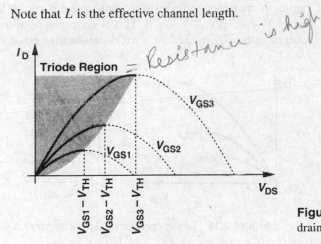

Figure 2.11 Drain current versus drain-source voltage in the triode region.

Fig. 2.11 plots the parabolas given by (2.8) for different values of V_{GS}, indicating that the "current capability" of the device increases with V_{GS}. Calculating $\partial I_D/\partial V_{DS}$, the reader can show that the peak of each parabola occurs at $V_{DS} = V_{GS} - V_{TH}$ and the peak current is

$$I_{D,max} = \frac{1}{2}\mu_n C_{ox} \frac{W}{L}(V_{GS} - V_{TH})^2. \tag{2.9}$$

We call $V_{GS} - V_{TH}$ the "overdrive voltage"[4] and W/L the "aspect ratio." If $V_{DS} \leq V_{GS} - V_{TH}$, we say the device operates in the "triode region."[5]

---

[4]Sometimes called the "effective voltage."

[5]This is also called the "linear region."

Equations (2.8) and (2.9) serve as the foundation for analog CMOS design, describing the dependence of I_D upon the constant of the technology, $\mu_n C_{ox}$, the device dimensions, W and L, and the gate and drain potentials with respect to the source. Note that the integration in (2.7) assumes μ_n and V_{TH} are independent of x and the gate and drain voltages, an approximation that we will revisit in Chapter 16.

If in (2.8), $V_{DS} \ll 2(V_{GS} - V_{TH})$, we have

$$I_D \approx \mu_n C_{ox} \frac{W}{L} (V_{GS} - V_{TH}) V_{DS}, \tag{2.10}$$

that is, the drain current is a *linear* function of V_{DS}. This is also evident from the characteristics of Fig. 2.11 for small V_{DS}: as shown in Fig. 2.12, each parabola can be approximated by a straight line. The linear relationship implies that the path from the source to the drain can be represented by a linear resistor equal to

$$R_{on} = \frac{1}{\mu_n C_{ox} \dfrac{W}{L} (V_{GS} - V_{TH})}. \tag{2.11}$$

A MOSFET can therefore operate as a resistor whose value is controlled by the overdrive voltage [so long as $V_{DS} \ll 2(V_{GS} - V_{TH})$]. This is conceptually illustrated in Fig. 2.13. Note that in contrast to bipolar transistors, a MOS device may be on even if it carries no

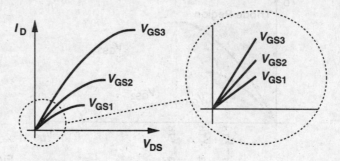

Figure 2.12   Linear operation in deep triode region.

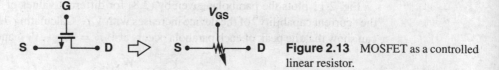

Figure 2.13   MOSFET as a controlled linear resistor.

current. With the condition $V_{DS} \ll 2(V_{GS} - V_{TH})$, we say the device operates in deep triode region.

Example 2.1

For the arrangement in Fig. 2.14(a), plot the on-resistance of M_1 as a function of V_G. Assume $\mu_n C_{ox} = 50\ \mu\text{A/V}^2$, $W/L = 10$, and $V_{TH} = 0.7$ V. Note that the drain terminal is open.

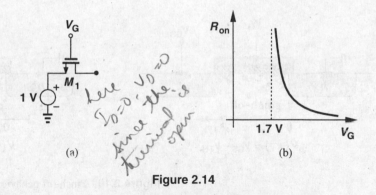

(a)   (b)

Figure 2.14

Solution

Since the drain terminal is open, $I_D = 0$ and $V_{DS} = 0$. Thus, if the device is on, it operates in the deep triode region. For $V_G < 1\,\text{V} + V_{TH}$, M_1 is off and $R_{on} = \infty$. For $V_G > 1\,\text{V} + V_{TH}$, we have

$$R_{on} = \frac{1}{50\,\mu\text{A/V}^2 \times 10(V_G - 1\,\text{V} - 0.7\,\text{V})}. \qquad (2.12)$$

The result is plotted in Fig. 2.14(b).

The utility of MOSFETs as controllable resistors and hence switches plays a crucial role in many analog circuits. This is studied in Chapter 12.

What happens if in Fig. 2.11 the drain-source voltage exceeds $V_{GS} - V_{TH}$? In reality, the drain current does *not* follow the parabolic behavior for $V_{DS} > V_{GS} - V_{TH}$. In fact, as shown in Fig. 2.15, I_D becomes relatively constant and we say the device operates in the "saturation" region.[6] To understand this phenomenon, recall from (2.4) that the local

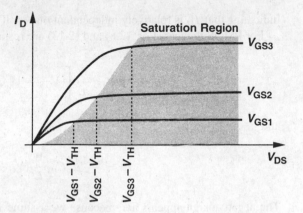

Figure 2.15 Saturation of drain current.

---

[6]Note the difference between saturation in bipolar and MOS devices.

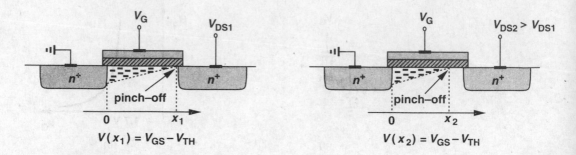

Figure 2.16 Pinch-off behavior.

density of inversion layer charge is proportional to $V_{GS} - V(x) - V_{TH}$. Thus, if $V(x)$ approaches $V_{GS} - V_{TH}$, then $Q_d(x)$ drops to zero. In other words, as depicted in Fig. 2.16, if V_{DS} is slightly greater than $V_{GS} - V_{TH}$, then the inversion layer stops at $x \leq L$, and we say the channel is "pinched off." As V_{DS} increases further, the point at which Q_d equals zero gradually moves toward the source. Thus, at some point along the channel, the local potential difference between the gate and the oxide-silicon interface is not sufficient to support an inversion layer.

With the above observations, we re-examine (2.7) for a saturated device. Since Q_d is the density of *mobile* charge, the integral on the left-hand side of (2.7) must be taken from $x = 0$ to $x = L'$, where L' is the point at which Q_d drops to zero, and that on the right from $V(x) = 0$ to $V(x) = V_{GS} - V_{TH}$. As a result:

$$I_D = \frac{1}{2}\mu_n C_{ox} \frac{W}{L'}(V_{GS} - V_{TH})^2, \tag{2.13}$$

indicating that I_D is relatively independent of V_{DS} if L' remains close to L.

For PMOS devices, Eqs. (2.8) and (2.13) are respectively written as

$$I_D = -\mu_p C_{ox} \frac{W}{L}\left[(V_{GS} - V_{TH})V_{DS} - \frac{1}{2}V_{DS}^2\right] \tag{2.14}$$

and

$$I_D = -\frac{1}{2}\mu_p C_{ox} \frac{W}{L'}(V_{GS} - V_{TH})^2. \tag{2.15}$$

The negative sign appears here because we assume I_D flows from the drain to the source, whereas holes flow in the reverse direction. Since the mobility of holes is about one-half to one-fourth of the mobility of electrons, PMOS devices suffer from lower "current drive" capability.

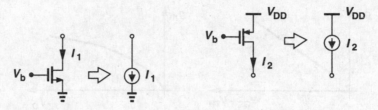

Figure 2.17 Saturated MOSFETs operating as current sources.

With the approximation $L \approx L'$, a saturated MOSFET can be used as a current source connected between the drain and the source (Fig. 2.17), an important component in analog design. Note that the current sources inject current into ground or draw current from V_{DD}. In other words, only one terminal of each current source is "floating."

Since a MOSFET operating in saturation produces a current in response to its gate-source overdrive voltage, we may define a figure of merit that indicates how well a device converts a voltage to a current. More specifically, since in processing signals we deal with the *changes* in voltages and currents, we define the figure of merit as the change in the drain current divided by the change in the gate-source voltage. Called the "transconductance" and denoted by g_m, this quantity is expressed as:

$$g_m = \frac{\partial I_D}{\partial V_{GS}}\bigg|_{VDS,\text{const.}} \tag{2.16}$$

$$= \mu_n C_{ox} \frac{W}{L}(V_{GS} - V_{TH}). \tag{2.17}$$

In a sense, g_m represents the sensitivity of the device: for a high g_m, a small change in V_{GS} results in a large change in I_D. Interestingly, g_m in the saturation region is equal to the inverse of R_{on} in deep triode region.

The reader can prove that g_m can also be expressed as

$$g_m = \sqrt{2\mu_n C_{ox} \frac{W}{L} I_D} \tag{2.18}$$

$$= \frac{2I_D}{V_{GS} - V_{TH}}. \tag{2.19}$$

Plotted in Fig. 2.18, each of the above expressions proves useful in studying the behavior of g_m as a function of one parameter while other parameters remain constant. For example, (2.17) suggests that g_m increases with the overdrive if W/L is constant whereas (2.19) implies that g_m decreases with the overdrive if I_D is constant. The concept of transconductance

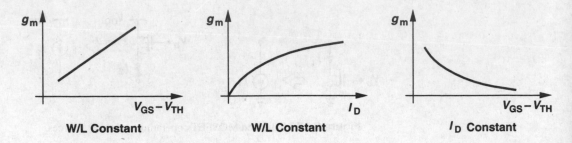

Figure 2.18 MOS transconductance as a function of overdrive and drain current.

can also be applied to a device operating in the triode region, as illustrated in the following example.

Example 2.2

For the arrangement shown in Fig. 2.19, plot the transconductance as a function of V_{DS}.

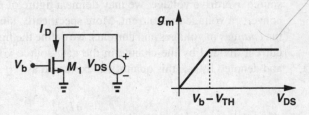

Figure 2.19

Solution

It is simpler to study g_m as V_{DS} decreases from infinity. So long as $V_{DS} \geq V_b - V_{TH}$, M_1 is in saturation, I_D is relatively constant, and, from (2.18), so is g_m. For $V_{DS} < V_b - V_{TH}$, M_1 is in the triode region and:

$$g_m = \frac{\partial}{\partial V_{GS}} \left\{ \frac{1}{2} \mu_n C_{ox} \frac{W}{L} \left[2(V_{GS} - V_{TH})V_{DS} - V_{DS}^2 \right] \right\} \tag{2.20}$$

$$= \mu_n C_{ox} \frac{W}{L} V_{DS}. \tag{2.21}$$

Thus, as plotted in Fig. 2.19, the transconductance drops if the device enters the triode region. For amplification, therefore, we usually employ MOSFETs in saturation.

The distinction between saturation and triode regions can be confusing, especially for PMOS devices. Intuitively, we note that the channel is pinched off if the difference between the gate and drain voltages is not sufficient to create an inversion layer. As depicted conceptually in Fig. 2.20, as $V_G - V_D$ of an NFET drops below V_{TH}, pinch-off occurs. Similarly,

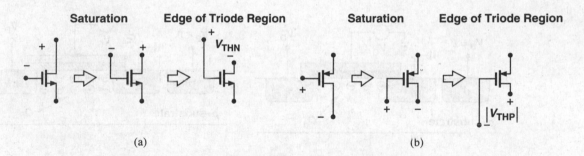

Figure 2.20 Conceptual visualization of saturation and triode regions.

if $V_D - V_G$ of a PFET is not large enough ($< |V_{THP}|$), the device is saturated. Note that this view does not require knowledge of the source voltage. This means we must know a priori which terminal operates as the drain.

2.3 Second-Order Effects

Our analysis of the MOS structure has thus far entailed various simplifying assumptions, some of which are not valid in many analog circuits. In this section, we describe three second-order effects that are essential in our subsequent circuit analyses. Other phenomena that appear in submicron devices are studied in Chapter 16.

Body Effect   In the analysis of Fig. 2.10, we tacitly assumed that the bulk and the source of the transistor were tied to ground. What happens if the bulk voltage of an NFET drops below the source voltage (Fig. 2.21)? Since the S and D junctions remain reverse-biased, we surmise that the device continues to operate properly but certain characteristics may

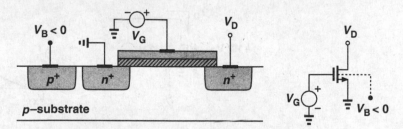

Figure 2.21 NMOS device with negative bulk voltage.

change. To understand the effect, suppose $V_S = V_D = 0$, and V_G is somewhat less than V_{TH} so that a depletion region is formed under the gate but no inversion layer exists. As V_B becomes more negative, more holes are attracted to the substrate connection, leaving a larger negative charge behind, i.e., as depicted in Fig. 2.22, the depletion region becomes wider. Now recall from Eq. (2.1) that the threshold voltage is a function of the total charge in the depletion region because the gate charge must mirror Q_d before an inversion layer is

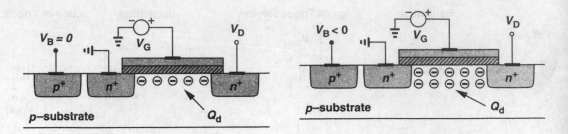

Figure 2.22   Variation of depletion region charge with bulk voltage.

formed. Thus, as V_B drops and Q_d increases, V_{TH} also increases. This is called the "body effect" or the "backgate effect."

It can be proved that with body effect:

$$V_{TH} = V_{TH0} + \gamma \left(\sqrt{|2\Phi_F + V_{SB}|} - \sqrt{|2\Phi_F|} \right), \tag{2.22}$$

where V_{TH0} is given by (2.1), $\gamma = \sqrt{2q\epsilon_{si}N_{sub}}/C_{ox}$ denotes the body effect coefficient, and V_{SB} is the source-bulk potential difference [1]. The value of γ typically lies in the range of 0.3 to 0.4 $V^{1/2}$.

Example 2.3

In Fig. 2.23(a), plot the drain current if V_X varies from $-\infty$ to 0. Assume $V_{TH0} = 0.6$ V, $\gamma = 0.4$ $V^{1/2}$, and $2\Phi_F = 0.7$ V.

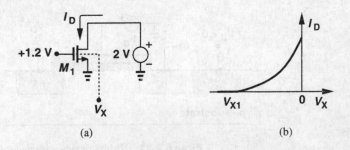

(a)   (b)

Figure 2.23

Solution

If V_X is sufficiently negative, the threshold voltage of M_1 exceeds 1.2 V and the device is off. That is,

$$1.2 \text{ V} = 0.6 + 0.4 \left(\sqrt{0.7 - V_{X1}} - \sqrt{0.7} \right). \tag{2.23}$$

and hence $V_{X1} = -4.76$ V. For $V_{X1} < V_X < 0$, I_D increases according to

$$I_D = \frac{1}{2}\mu_n C_{ox}\frac{W}{L}\left[V_{GS} - V_{THO} - \gamma\left(\sqrt{2\Phi_F - V_X} - \sqrt{2\Phi_F}\right)\right]^2. \tag{2.24}$$

Fig. 2.23(b) shows the resulting behavior.

For body effect to manifest itself, the bulk potential, V_{sub}, need not change: if the source voltage varies with respect to V_{sub}, the same phenomenon occurs. For example, consider the circuit in Fig. 2.24(a), first ignoring body effect. We note that as V_{in} varies, V_{out} closely follows the input because the drain current remains equal to I_1. In fact, we can write

$$I_1 = \frac{1}{2}\mu_n C_{ox}\frac{W}{L}(V_{in} - V_{out} - V_{TH})^2, \tag{2.25}$$

concluding that $V_{in} - V_{out}$ is constant if I_1 is constant [Fig. 2.24(b)].

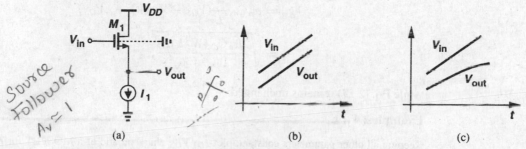

Figure 2.24 (a) A circuit in which the source-bulk voltage varies with input level, (b) input and output voltages with no body effect, (c) input and output voltages with body effect.

Now suppose the substrate is tied to ground and body effect is significant. Then, as V_{in} and hence V_{out} become more positive, the potential difference between the source and the bulk increases, raising the value of V_{TH}. Eq. (2.25) therefore implies that $V_{in} - V_{out}$ must increase so as to maintain I_D constant [Fig. 2.24(c)].

Body effect is usually undesirable. The change in the threshold voltage, e.g., as in Fig. 2.24(a), often complicates the design of analog (and even digital) circuits. Device technologists balance N_{sub} and C_{ox} to obtain a reasonable value for γ.

Channel-Length Modulation   In the analysis of channel pinch-off in Section 2.2, we noted that the actual length of the inverted channel gradually decreases as the potential difference between the gate and the drain increases. In other words, in (2.13), L' is in fact a function of V_{DS}. This effect is called "channel-length modulation." Writing $L' = L - \Delta L$, i.e., $1/L' \approx (1 + \Delta L/L)/L$, and assuming a first-order relationship between $\Delta L/L$ and V_{DS} such as $\Delta L/L = \lambda V_{DS}$, we have, in saturation,

$$I_D \approx \frac{1}{2}\mu_n C_{ox}\frac{W}{L}(V_{GS} - V_{TH})^2(1 + \lambda V_{DS}), \tag{2.26}$$

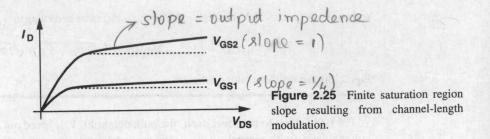

Figure 2.25 Finite saturation region slope resulting from channel-length modulation.

where λ is the channel-length modulation coefficient. Illustrated in Fig. 2.25, this phenomenon results in a nonzero slope in the I_D/V_{DS} characteristic and hence a nonideal current source between D and S in saturation. The parameter λ represents the *relative* variation in length for a given increment in V_{DS}. Thus, for longer channels, λ is smaller.

With channel-length modulation, some of the expressions derived for g_m must be modified. Equations (2.17) and (2.18) are respectively rewritten as

$$g_m = \mu_n C_{ox} \frac{W}{L}(V_{GS} - V_{TH})(1 + \lambda V_{DS}). \tag{2.27}$$

$$= \sqrt{\frac{2\mu_n C_{ox}(W/L)I_D}{1 + \lambda V_{DS}}}, \tag{2.28}$$

while Eq. (2.19) remains unchanged.

Example 2.4

Keeping all other parameters constant, plot I_D/V_{DS} characteristic of a MOSFET for $L = L_1$ and $L = 2L_1$.

Solution

Writing

$$I_D = \frac{1}{2}\mu_n C_{ox} \frac{W}{L}(V_{GS} - V_{TH})^2(1 + \lambda V_{DS}) \tag{2.29}$$

and $\lambda \propto 1/L$, we note that if the length is doubled, the slope of I_D vs. V_{DS} is divided by *four* because $\partial I_D/\partial V_{DS} \propto \lambda/L \propto 1/L^2$ (Fig. 2.26). For a given gate-source overdrive, a larger L gives a more

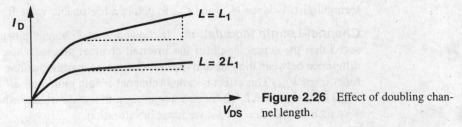

Figure 2.26 Effect of doubling channel length.

ideal current source while degrading the current capability of the device. Thus, W may need to be increased proportionally.

The linear approximation $\Delta L/L \propto V_{DS}$ becomes less accurate in short-channel transistors, resulting in a *variable* slope in the saturated I_D/V_{DS} characteristics. We return to this issue in Chapter 16.

The dependence of I_D upon V_{DS} in saturation may suggest that the bias current of a MOSFET can be defined by the proper choice of the drain-source voltage, allowing freedom in the choice of $V_{GS} - V_{TH}$. However, since the dependence on V_{DS} is much weaker, the drain-source voltage is not used to set the current. The effect of V_{DS} on I_D is usually considered an *error* and it is studied in Chapter 5.

Subthreshold Conduction   In our analysis of the MOSFET, we have assumed that the device turns off abruptly as V_{GS} drops below V_{TH}. In reality, for $V_{GS} \approx V_{TH}$, a "weak" inversion layer still exists and some current flows from D to S. Even for $V_{GS} < V_{TH}$, I_D is finite, but it exhibits an *exponential* dependence on V_{GS} [2, 3]. Called "subthreshold conduction," this effect can be formulated for V_{DS} greater than roughly 200 mV as

$$I_D = I_0 \exp \frac{V_{GS}}{\zeta V_T}, \tag{2.30}$$

where $\zeta > 1$ is a nonideality factor and $V_T = kT/q$. We also say the device operates in "week inversion." Except for ζ, (2.30) is similar to the exponential I_C/V_{BE} relationship in a bipolar transistor. The key point here is that as V_{GS} falls below V_{TH}, the drain current drops at a finite rate. With typical values of ζ, at room temperature V_{GS} must decrease by approximately 80 mV for I_D to decrease by one decade (Fig. 2.27). For example, if a

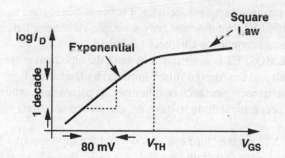

Figure 2.27   MOS subthreshold characteristics.

threshold of 0.3 V is chosen in a process to allow low-voltage operation, then when V_{GS} is reduced to zero, the drain current decreases by only a factor of $10^{3.75}$. Especially problematic in large circuits such as memories, subthreshold conduction can result in significant power dissipation (or loss of analog information).

It is appropriate at this point to return to the definition of the threshold voltage. One definition is to plot the inverse on-resistance of the device $R_{on}^{-1} = \mu C_{ox}(W/L)(V_{GS} - V_{TH})$ as a function of V_{GS} and extrapolate the result to zero, for which $V_{GS} = V_{TH}$. In rough calculations, we often view V_{TH} as the gate-source voltage yielding $I_D/W = 1\mu A/\mu m$ in saturation. For example, if a device with $W = 100~\mu m$ operates with $I_D = 100~\mu A$, it is in the vicinity of the subthreshold region. This view is nonetheless vague, especially as device length scales down in every technology generation.

We now re-examine Eq. (2.18) for the transconductance of a MOS device operating in the subthreshold region. Is it possible to achieve an arbitrarily high transconductance by increasing W while maintaining I_D constant? Is it possible to obtain a *higher* transconductance than that of a bipolar transistor (I_C/V_T) biased at the same current? Equation (2.18) was derived from the square-law characteristics $I_D = (1/2)\mu_n C_{ox}(W/L)(V_{GS} - V_{TH})^2$. However, if W increases while I_D remains constant, then $V_{GS} \to V_{TH}$ and the device enters the subthreshold region. As a result, the transconductance is calculated from (2.30) to be $g_m = I_D/(\zeta V_T)$, revealing that MOSFETs are inferior to bipolar transistors in this respect.

The exponential dependence of I_D upon V_{GS} in subthreshold operation may suggest the use of MOS devices in this regime so as to achieve a higher gain. However, since such conditions are met by only a large device width or low drain current, the speed of subthreshold circuits is severely limited.

Voltage Limitations   MOSFETs experience various breakdown effects if their terminal voltage differences exceed certain limits. At high gate-source voltages, the gate oxide breaks down irreversibly, damaging the transistor. In short-channel devices, an excessively large drain-source voltage widens the depletion region around the drain so much that it touches that around the source, creating a very large drain current. (This effect is called "punchthrough.") Other limitations relate to "hot electron effects" and are described in Chapter 16.

short channel, ↑↑ V_{DS}, depletion region around drain increases thereby increasing the depletion region around drain so much that it touches the ... squareanpheating ... very large I_D.

2.4 MOS Device Models

2.4.1 MOS Device Layout

For the developments in subsequent sections, it is beneficial to have some understanding of the layout of a MOSFET. We describe only a simple view here, deferring the fabrication details and structural subtleties to Chapters 17 and 18.

The layout of a MOSFET is determined by both the electrical properties required of the device in the circuit and the "design rules" imposed by the technology. For example, W/L is chosen to set the transconductance or other circuit parameters, while the minimum L is dictated by the process. In addition to the gate, the source and drain areas must be defined properly as well.

Shown in Fig. 2.28 are the "bird eye's view" and the top view of a MOSFET. The gate polysilicon and the source and drain terminals are typically tied to metal (aluminum) wires that serve as interconnects with low resistance and capacitance. To accomplish this, one or more "contact windows" must be opened in each region, filled with metal, and connected to the upper metal wires. Note that the gate poly extends beyond the channel area by some amount to ensure reliable definition of the "edge" of the transistor.

The source and drain junctions play an important role in the performance. To minimize the capacitance of S and D, the total area of each junction must be minimized. We see from Fig. 2.28 that one dimension of the junctions is equal to W. The other dimension must be large enough to accommodate the contact windows and is specified by the technology design rules.[7]

---

[7]This dimension is typically three to four times the minimum allowable channel length.

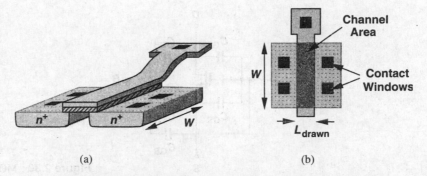

(a)   (b)

Figure 2.28   Bird's eye and vertical views of a MOS device.

Example 2.5

Draw the layout of the circuit shown in Fig. 2.29(a).

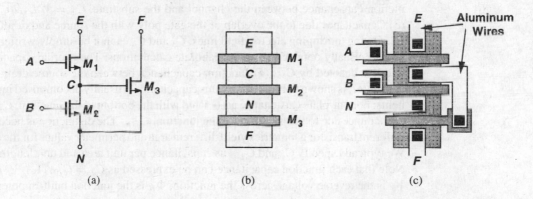

(a)   (b)   (c)

Figure 2.29

Solution

Noting that M_1 and M_2 share the same S/D junctions at node C and M_2 and M_3 also do so at node N, we surmise that the three transistors can be laid out as shown in Fig. 2.29(b). Connecting the remaining terminals, we obtain the layout in Fig. 2.29(c). Note that the gate polysilicon of M_3 cannot be directly tied to the source material of M_1, thus requiring a metal interconnect.

2.4.2 MOS Device Capacitances

The basic quadratic I/V relationships derived in the previous section along with corrections for body effect and channel-length modulation provide a reasonable model for understanding the "dc" behavior of CMOS circuits. In many analog circuits, however, the capacitances associated with the devices must also be taken into account so as to predict the "ac" behavior as well.

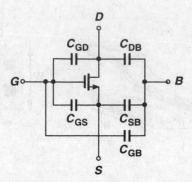

Figure 2.30 MOS capacitances.

We expect that a capacitance exists between every two of the four terminals of a MOSFET (Fig. 2.30).[8] Moreover, the value of each of these capacitances may depend on the bias conditions of the transistor. Considering the physical structure in Fig. 2.31(a), we identify the following. (1) Oxide capacitance between the gate and the channel, $C_1 = WLC_{ox}$; (2) Depletion capacitance between the channel and the substrate, $C_2 = WL\sqrt{q\epsilon_{si}N_{sub}/(4\Phi_F)}$; (3) Capacitance due to the overlap of the gate poly with the source and drain areas, C_3 and C_4. Owing to fringing electric field lines, C_3 and C_4 cannot be simply written as $WL_D C_{ox}$, and are usually obtained by more elaborate calculations. The overlap capacitance per unit width is denoted by C_{ov}; (4) Junction capacitance between the source/drain areas and the substrate. As shown in Fig. 2.31(b), this capacitance is usually decomposed into two components: bottom-plate capacitance associated with the bottom of the junction, C_j, and sidewall capacitance due to the perimeter of the junction, C_{jsw}. The distinction is necessary because different transistor geometries yield different area and perimeter values for the S/D junctions. We typically specify C_j and C_{jsw} as capacitance per unit area and unit length, respectively. Note that each junction capacitance can be expressed as $C_j = C_{j0}/[1 + V_R/\Phi_B]^m$, where V_R is the reverse voltage across the junction, Φ_B is the junction built-in potential, and m is a power typically in the range of 0.3 and 0.4.

---

[8]The capacitance between S and D is negligible.

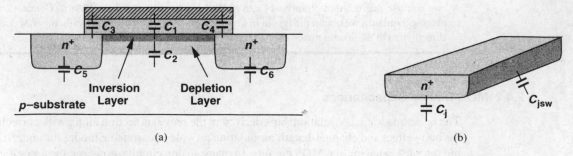

(a)   (b)

Figure 2.31 (a) MOS device capacitances, (b) decomposition of S/D junction capacitance into bottom-plate and sidewall components.

Example 2.6

Calculate the source and drain junction capacitances of the two structures shown in Fig. 2.32.

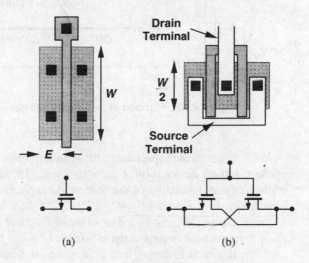

Figure 2.32

Solution

For the transistor in Fig. 2.32(a), we have

$$C_{DB} = C_{SB} = WEC_j + 2(W + E)C_{jsw}, \tag{2.31}$$

whereas for that in Fig. 2.32(b),

$$C_{DB} = \frac{W}{2}EC_j + 2\left(\frac{W}{2} + E\right)C_{jsw} \tag{2.32}$$

$$C_{SB} = 2\left[\frac{W}{2}EC_j + 2\left(\frac{W}{2} + E\right)C_{jsw}\right] \tag{2.33}$$

$$= WEC_j + 2(W + 2E)C_{jsw}. \tag{2.34}$$

Called a "folded" structure, the geometry in Fig. 2.32(b) exhibits substantially less drain junction capacitance than that in Fig. 2.32(a) while providing the same W/L.

In the above calaculations, we have assumed that the total source or drain perimeter, $2(W + E)$, is multiplied by C_{jsw}. In reality, the capacitance of the sidewall facing the channel may be less than that of the other three sidewalls because of the channel-stop implant (Chapter 17). Nonetheless, we typically assume all four sides have the same unit capacitance. The error resulting from this assumption is negligible because each node in a circuit is connected to a number of other device capacitances as well.

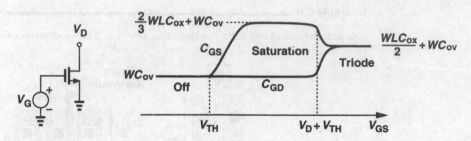

Figure 2.33 Variation of gate-source and gate-drain capacitances versus V_{GS}.

We now derive the capacitances between terminals of a MOSFET in different regions of operation. If the device is off, $C_{GD} = C_{GS} = C_{ov}W$, and the gate-bulk capacitance consists of the series combination of the gate oxide capacitance and the depletion region capacitance, i.e., $C_{GB} = (WLC_{ox})C_d/(WLC_{ox} + C_d)$, where L is the effective length and $C_d = WL\sqrt{q\epsilon_{si}N_{sub}/(4\Phi_F)}$. The value of C_{SB} and C_{DB} is a function of the source and drain voltages with respect to the substrate.

If the device is in deep triode region, i.e., if S and D have approximately equal voltages, then the gate-channel capacitance, WLC_{ox}, is divided equally between the gate and source terminals and the gate and drain terminals. This is because a change ΔV in the gate voltage draws equal amounts of charge from S and D. Thus, $C_{GD} = C_{GS} = WLC_{ox}/2 + WC_{ov}$.

If in saturation, a MOSFET exhibits a gate-drain capacitance of roughly WC_{ov}. The potential difference between the gate and the channel varies from V_{GS} at the source to $V_{GS} - V_{TH}$ at the pinch-off point, resulting in a nonuniform vertical electric field in the gate oxide along the channel. It can be proved that the equivalent capacitance of this structure excluding the gate-source overlap capacitance equals $2WLC_{ox}/3$ [1]. Thus, $C_{GS} = 2WL_{eff}C_{ox}/3 + WC_{ov}$. The behavior of C_{GD} and C_{GS} in different regions of operation is plotted in Fig. 2.33. Note that the above equations do not provide a smooth transition from one region of operation to another, creating convergence difficulties in simulation programs. This issue is revisited in Chapter 16.

The gate-bulk capacitance is usually neglected in the triode and saturation regions because the inversion layer acts as a "shield" between the gate and the bulk. In other words, if the gate voltage varies, the charge is supplied by the source and the drain rather than the bulk.

Example 2.7

Sketch the capacitances of M_1 in Fig. 2.34 as V_X varies from zero to 3 V. Assume $V_{TH} = 0.6$ V and $\lambda = \gamma = 0$.

Solution

To avoid confusion, we label the three terminals as shown in Fig. 2.34. For $V_X \approx 0$, M_1 is in the triode region, $C_{EN} \approx C_{EF} = (1/2)WLC_{ox} + WC_{ov}$, and C_{FB} is maximum. The value of C_{NB} is independent of V_X. As V_X exceeds 1 V, the role of the source and drain is exchanged [Fig. 2.35(a)],

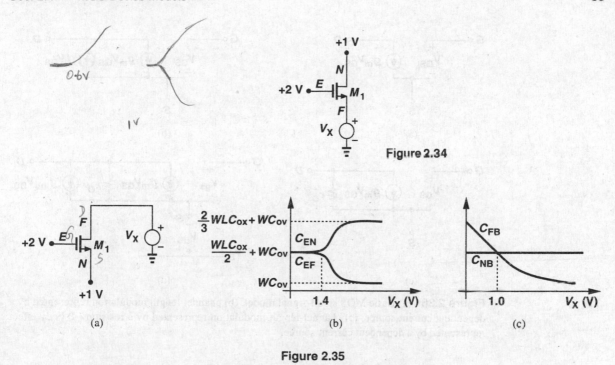

Figure 2.34

Figure 2.35

eventually bringing M_1 out of the triode region for $V_X \geq 2\,\text{V} - 0.6\,\text{V}$. The variation of the capacitances is plotted in Figs. 2.35(b) and (c).

2.4.3 MOS Small-Signal Model

The quadratic characteristics described by (2.8) and (2.9) along with the voltage-dependent capacitances derived above form the large-signal model of MOSFETs. Such a model proves essential in analyzing circuits in which the signal significantly disturbs the bias points, particularly if nonlinear effects are of concern. By contrast, if the perturbation in bias conditions is small, a small-signal model, i.e., an approximation of the large-signal model around the operating point, can be employed to simplify the calculations. Since in many analog circuits, MOSFETs are biased in the saturation region, we derive the corresponding small-signal model here. For transistors operating as switches, a linear resistor given by (2.11) together with device capacitances serves as a rough small-signal equivalent.

We derive the small-signal model by producing a small increment in a bias point and calculating the resulting increment in other bias parameters. Since the drain current is a function of the gate-source voltage, we incorporate a voltage-dependent current source equal to $g_m V_{GS}$ [Fig. 2.36(a)]. Note that the low-frequency impedance between G and S is very high. This is the small-signal model of an ideal MOSFET.

Owing to channel-length modulation, the drain current also varies with the drain-source voltage. This effect can also be modeled by a voltage-dependent current source [Fig. 2.36(b)], but a current source whose value linearly depends on the voltage across it is equivalent to

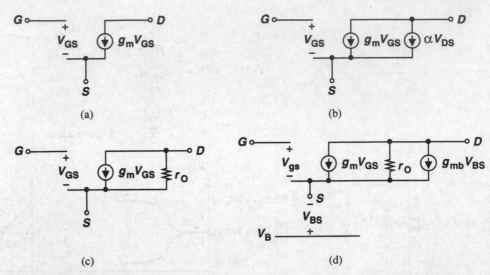

Figure 2.36 (a) Basic MOS small-signal model, (b) channel-length modulation represented by a dependent current source, (c) channel-length modulation represented by a resistor, (d) body effect represented by a dependent current source.

a linear resistor [Fig. 2.36(c)]. Tied between D and S, the resistor is given by

$$r_O = \frac{\partial V_{DS}}{\partial I_D} \tag{2.35}$$

$$= \frac{1}{\partial I_D / \partial V_{DS}}. \tag{2.36}$$

$$= \frac{1}{\frac{1}{2}\mu_n C_{ox}\frac{W}{L}(V_{GS} - V_{TH})^2 \cdot \lambda} \tag{2.37}$$

$$\approx \frac{1}{\lambda I_D}. \tag{2.38}$$

As seen throughout this book, the output resistance, r_O, impacts the performance of many analog circuits. For example, r_O limits the maximum voltage gain of most amplifiers.

Now recall that the bulk potential influences the threshold voltage and hence the gate-source overdrive. As demonstrated in Example 2.3, with all other terminals held at a constant voltage, the drain current is a function of the bulk voltage. That is, the bulk behaves as a second gate. Modeling this dependence by a current source connected between D and S [Fig. 2.36(d)], we write the value as $g_{mb}V_{bs}$, where $g_{mb} = \partial I_D / \partial V_{BS}$. In the saturation region, g_{mb} can be expressed as:

$$g_{mb} = \frac{\partial I_D}{\partial V_{BS}} \tag{2.39}$$

$$= \mu_n C_{ox} \frac{W}{L}(V_{GS} - V_{TH})\left(-\frac{\partial V_{TH}}{\partial V_{BS}}\right). \tag{2.40}$$

We also have

$$\frac{\partial V_{TH}}{\partial V_{BS}} = -\frac{\partial V_{TH}}{\partial V_{SB}} \tag{2.41}$$

$$= -\frac{\gamma}{2}(2\Phi_F + V_{SB})^{-1/2}. \tag{2.42}$$

Thus,

$$g_{mb} = g_m \frac{\gamma}{2\sqrt{2\Phi_F + V_{SB}}} \tag{2.43}$$

$$= \eta g_m, \tag{2.44}$$

where $\eta = g_{mb}/g_m$. As expected, g_{mb} is proportional to γ. Equation (2.43) also suggests that incremental body effect becomes less pronounced as V_{SB} increases. Note that $g_m V_{GS}$ and $g_{mb} V_{BS}$ have the same polarity, i.e., raising the gate voltage has the same effect as raising the bulk potential.

The model in Fig. 2.36(d) is adequate for most low-frequency small-signal analyses. In reality, each terminal of a MOSFET exhibits a finite ohmic resistance resulting from the resistivity of the material (and the contacts), but proper layout can minimize such resistances. For example, consider the two structures of Fig. 2.32, repeated in Fig. 2.37 along with the gate distributed resistance. We note that folding reduces the gate resistance by a factor of four.

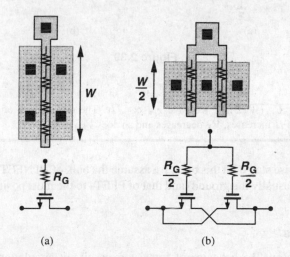

(a)   (b)

Figure 2.37 Reduction of gate resistance by folding.

Shown in Fig. 2.38, the complete small-signal model includes the device capacitances as well. The value of each capacitance is calculated according to the equations derived in Section 2.4.2. The reader may wonder how a complex circuit is analyzed intuitively if each transistor must be replaced by the model of Fig. 2.38. The first step is to determine

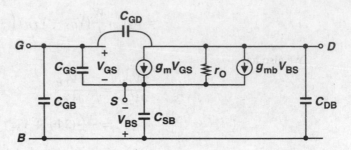

Figure 2.38 Complete MOS small-signal model.

the *simplest* device model that can represent the role of each transistor with reasonable accuracy. We provide some guidelines for this task at the end of Chapter 3.

Example 2.8 _____

Sketch g_m and g_{mb} of M_1 in Fig. 2.39 as a function of the bias current I_1.

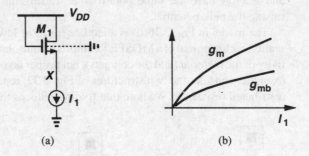

Figure 2.39

Solution

Since $g_m = \sqrt{2\mu_n C_{ox}(W/L)I_D}$, we have $g_m \propto \sqrt{I_1}$. The dependence of g_{mb} upon I_1 is less straightforward. As I_1 increases, V_X decreases and so does V_{SB}.

_____

Unless otherwise stated, in this book we assume the bulk of all NFETs is tied to the most negative supply (usually the ground) and that of PFETs to the most positive supply (usually V_{DD}).

2.4.4 MOS SPICE models

In order to represent the behavior of transistors in circuit simulations, SPICE requires an accurate model for each device. Over the last two decades, MOS modeling has made tremendous progress, reaching quite sophisticated levels so as to represent high-order effects in short-channel devices.

Table 2.1   Level 1 SPICE Models for NMOS and PMOS Devices.

NMOS Model

LEVEL = 1	VTO = 0.7	GAMMA = 0.45	PHI = 0.9
NSUB = 9e+14	LD = 0.08e−6	UO = 350	LAMBDA = 0.1
TOX = 9e−9	PB = 0.9	CJ = 0.56e−3	CJSW = 0.35e−11
MJ = 0.45	MJSW = 0.2	CGDO = 0.4e−9	JS = 1.0e−8

PMOS Model

LEVEL = 1	VTO = −0.8	GAMMA = 0.4	PHI = 0.8
NSUB = 5e+14	LD = 0.09e−6	UO = 100	LAMBDA = 0.2
TOX = 9o−9	PB = 0.9	CJ = 0.94e−3	CJSW = 0.32e−11
MJ = 0.5	MJSW = 0.3	CGDO = 0.3e−9	JS = 0.5e−8

In this section, we describe the simplest MOS SPICE model, known as "Level 1," and provide typical values for each parameter in the model corresponding to a 0.5-μm technology. Chapter 16 describes more accurate SPICE models. Table 2.1 shows the model parameters for NMOS and PMOS devices. The parameters are defined as below:

VTO: threshold voltage with zero V_{SB} (unit: V)

GAMMA: body effect coefficient (unit: $V^{1/2}$)

PHI: $2\Phi_F$ (unit: V)

TOX: gate oxide thickness (unit: m)

NSUB: substrate doping (unit: cm^{-3})

LD: source/drain side diffusion (unit: m)

UO: channel mobility (unit: $cm^2/V/s$)

LAMBDA: channel-length modulation coefficient (unit: V^{-1})

CJ: source/drain bottom-plate junction capacitance per unit area (unit: F/m^2)

CJSW: source/drain sidewall junction capacitance per unit length (unit: F/m)

PB: source/drain junction built-in potential (unit: V)

MJ: exponent in CJ equation (unitless)

MJSW: exponent in CJSW equation (unitless)

CGDO: gate-drain overlap capacitance per unit width (unit: F/m)

CGSO: gate-source overlap capacitance per unit width (unit: F/m)

JS: source/drain leakage current per unit area (unit: A/m^2)

2.4.5 NMOS versus PMOS Devices

In most CMOS technologies, PMOS devices are quite inferior to NMOS transistors. For example, due to the lower mobility of holes, $\mu_p C_{ox} \approx 0.25\mu_n C_{ox}$ in modern processes, yielding low current drive and transconductance. Moreover, for given dimensions and bias currents, NMOS transistors exhibit a higher output resistance, providing more ideal current sources and higher gain in amplifiers. For these reasons, it is preferred to incorporate NFETs rather than PFETs wherever possible.

2.4.6 Long-Channel versus Short-Channel Devices

In this chapter, we have employed a very simple view of MOSFETs so as to understand the basic principles of their operation. Most of our treatment is valid for "long-channel" devices, e.g., transistors having a minimum length of about 4 μm. Many of the relationships derived here must be reexamined and revised for short-channel MOSFETs. Furthermore, the SPICE models necessary for simulation of today's devices need to be much more sophisticated than the Level 1 model. For example, the intrinsic gain, $g_m r_O$, calculated from the device parameters in Table 2.1 is quite higher than actual values. These issues are studied in Chapter 16.

The reader may wonder why we begin with a simplistic view of devices if such a view does not lead to a high accuracy in predicting the performance of circuits. The key point is that the simple model provides a great deal of intuition that is necessary in analog design. As we will see throughout this book, we often encounter a trade-off between intuition and rigor, and our approach is to establish the intuition first and gradually complete our understanding so as to achieve rigor as well.

Appendix A: Behavior of MOS Device as a Capacitor

In this chapter, we have limited our treatment of MOS devices to a basic level. However, the behavior of a MOSFET as a capacitor merits some attention. Recall that if the source, drain, and bulk of an NFET are grounded and the gate voltage rises, an inversion layer begins to form for $V_{GS} \approx V_{TH}$. We also noted that for $0 < V_{GS} < V_{TH}$, the device operates in the subthreshold region.

Now consider the NFET of Fig. 2.40. The transistor can be considered a two-terminal

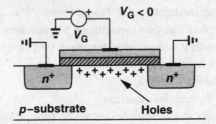

Figure 2.40 NMOS operating in accumulation mode.

device and hence its capacitance can be examined for different gate voltages. Let us begin with a very *negative* gate-source voltage. The negative potential on the gate attracts the holes in the substrate to the oxide interface. We say the MOSFET operates in the "accumulation" region. The two-terminal device can be viewed as a capacitor having a unit-area capacitance of C_{ox} because the two "plates" of the capacitor are separated by t_{ox}.

As V_{GS} rises, the density of holes at the interface falls, a depletion region begins to form under the oxide, and the device enters weak inversion. In this mode, the capacitance consists of the series combination of C_{ox} and C_{dep}. Finally, as V_{GS} exceeds V_{TH}, the oxide-silicon interface sustains a channel and the unit-area capacitance returns to C_{ox}. Figure 2.41 plots the behavior.

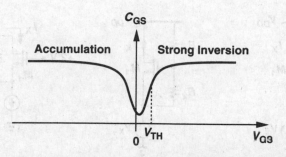

Figure 2.41 Capacitance-voltage characteristic of an NMOS device.

Problems

Unless otherwise stated, in the following problems, use the device data shown in Table 2.1 and assume $V_{DD} = 3$ V where necessary.

2.1. For $W/L = 50/0.5$, plot the drain current of an NFET and a PFET as a function of $|V_{GS}|$ as $|V_{GS}|$ varies from 0 to 3 V. Assume $|V_{DS}| = 3$ V.

2.2. For $W/L = 50/0.5$, and $|I_D| = 0.5$ mA, calculate the transconductance and output impedance of both NMOS and PMOS devices. Also, find the "intrinsic gain," defined as $g_m r_O$.

2.3. Derive expressions for $g_m r_O$ in terms of I_D and W/L. Plot $g_m r_O$ as a function of I_D with L as a parameter. Note that $\lambda \propto 1/L$.

2.4. Plot I_D versus V_{GS} for an MOS transistor (a) with V_{DS} as a parameter, (b) with V_{BS} as a parameter. Identify the break points in the characteristics.

2.5. Sketch I_X and the transconductance of the transistor as a function of V_X for each circuit in Fig. 2.42 as V_X varies from 0 to V_{DD}. For part (a), assume V_X varies from 0 to 1.5 V.

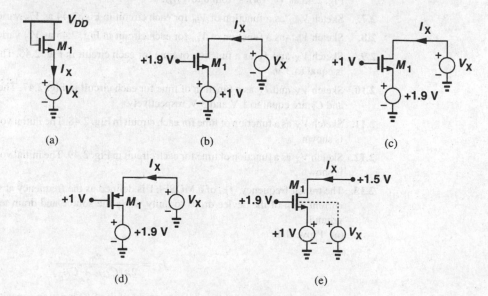

Figure 2.42

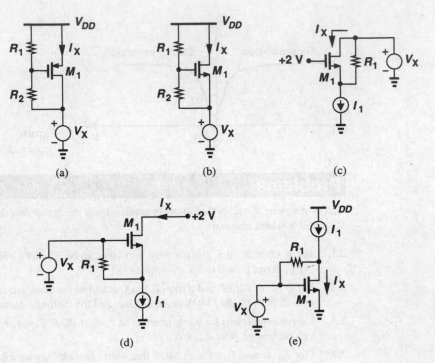

Figure 2.43

2.6. Sketch I_X and the transconductance of the transistor as a function of V_X for each circuit in Fig. 2.43 as V_X varies from 0 to V_{DD}.

2.7. Sketch V_{out} as a function of V_{in} for each circuit in Fig. 2.44 as V_{in} varies from 0 to V_{DD}.

2.8. Sketch V_{out} as a function of V_{in} for each circuit in Fig. 2.45 as V_{in} varies from 0 to V_{DD}.

2.9. Sketch V_X and I_X as a function of time for each circuit in Fig. 2.46. The initial voltage of C_1 is equal to 3 V.

2.10. Sketch V_X and I_X as a function of time for each circuit in Fig. 2.47. The initial voltages of C_1 and C_2 are equal to 1 V and 3 V, respectively.

2.11. Sketch V_X as a function of time for each circuit in Fig. 2.48. The initial voltage of each capacitor is shown.

2.12. Sketch V_X as a function of time for each circuit in Fig. 2.49. The initial voltage of each capacitor is shown.

2.13. The transit frequency, f_T, of a MOSFET is defined as the frequency at which the small-signal current gain of the device drops to unity while the source and drain terminals are held at ac ground.
(a) Prove that

$$f_T = \frac{g_m}{2\pi(C_{GD} + C_{GS})}.$$
(2.45)

Note that f_T does not include the effect of the S/D junction capacitance.

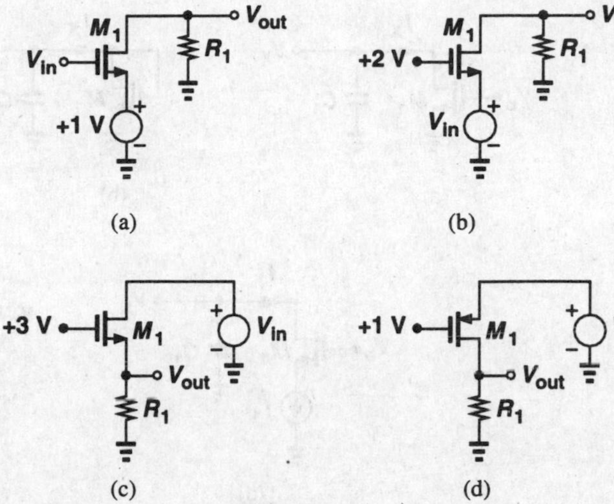

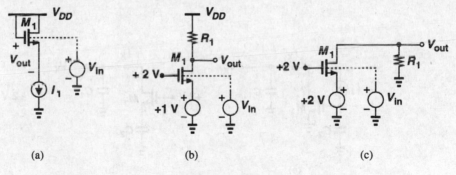

Figure 2.44

Figure 2.45

(b) Suppose the gate resistance, R_G, is significant and the device is modeled as a distributed set of n transistors each with a gate resistance equal to R_G/n. Prove that the f_T of the device is independent of R_G and still equal to the value given above.

(c) For a given bias current, the minimum allowable drain-source voltage for operation in saturation can be reduced only by increasing the width and hence the capacitances of the transistor. Using square-law characteristics, prove that

$$f_T = \frac{\mu_n}{2\pi} \frac{V_{GS} - V_{TH}}{L^2}. \tag{2.46}$$

This relation indicates how the speed is limited as a device is designed to operate with lower supply voltages.

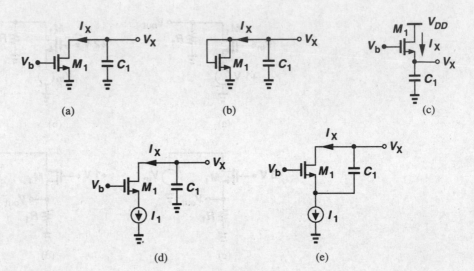

(a)   (b)   (c)

(d)   (e)

Figure 2.46

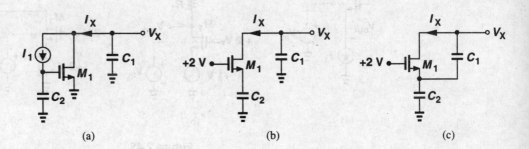

(a)   (b)   (c)

Figure 2.47

2.14. Calculate the f_T of a MOS device in the subthreshold region and compare the result with those obtained in Problem 2.13.

2.15. For a saturated NMOS device having $W = 50$ μm and $L = 0.5$ μm, calculate all of the capacitances. Assume the minimum (lateral) dimension of the S/D areas is 1.5 μm and the device is folded as shown in Fig. 2.32(b). What is the f_T if the drain current is 1 mA?

2.16. Consider the structure shown in Fig. 2.50. Determine I_D as a function of V_{GS} and V_{DS} and prove that the structure can be viewed as a single transistor having an aspect ratio $W/(2L)$. Assume $\lambda = \gamma = 0$.

2.17. For an NMOS device operating in saturation, plot W/L versus $V_{GS} - V_{TH}$ if (a) I_D is constant, (b) g_m is constant.

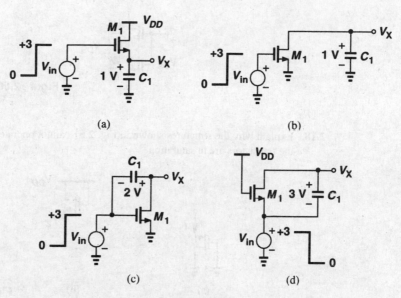

(a)

(b)

(c)

(d)

Figure 2.48

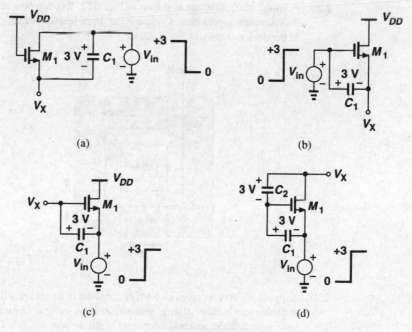

(a)

(b)

(c)

(d)

Figure 2.49

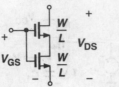

Figure 2.50

2.18. Explain why the structures shown in Fig. 2.51 cannot operate as current sources even though the transistors are in saturation.

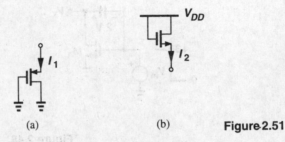

(a)   (b)   **Figure 2.51**

2.19. Considering the body effect as "backgate effect," explain intuitively why γ is directly proportional to $\sqrt{N_{sub}}$ and inversely proportional to C_{ox}.

2.20. A "ring" MOS structure is shown in Fig. 2.52. Explain how the device operates and estimate its equivalent aspect ratio. Compare the drain junction capacitance of this structure with that of the devices shown in Fig. 2.32.

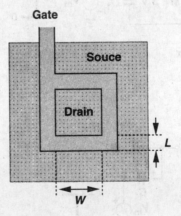

Figure 2.52

2.21. Suppose we have received an NMOS transistor in a package with four unmarked pins. Describe the minimum number of dc measurement steps using an ohmmeter necessary to determine the gate, source/drain, and bulk terminals of the device.

2.22. Repeat Problem 2.21 if the type of the device (NFET or PFET) is not known.

2.23. For an NMOS transistor, the threshold voltage is known but $\mu_n C_{ox}$ and W/L are not. Assume $\lambda = \gamma = 0$. If we cannot measure C_{ox} independently, is it possible to devise a sequence of dc measurement tests to determine $\mu_n C_{ox}$ and W/L? What if we have two transistors and we know one has twice the aspect ratio of the other?

2.24. Sketch I_X versus V_X for each of the composite structures shown in Fig. 2.53 with V_G as a parameter. Also, sketch the equivalent transconductance. Assume $\lambda = \gamma = 0$.

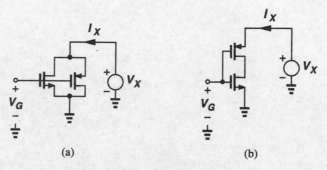

(a)   (b)

Figure 2.53

2.25. An NMOS current source with $I_D = 0.5$ mA must operate with drain-source voltages as low as 0.4 V. If the minimum required output impedance is 20 kΩ, determine the width and length of the device. Calculate the gate-source, gate-drain, and drain-substrate capacitance if the device is folded as in Fig. 2.32 and $E = 3\ \mu m$.

2.26. Consider the circuit shown in Fig. 2.54, where the initial voltage at node X is equal to V_{DD}. Assuming $\lambda = \gamma = 0$ and neglecting other capacitances, plot V_X and V_Y versus time if (a) V_{in} is a positive step with amplitude $V_0 > V_{TH}$, (b) V_{in} is a negative step with amplitude $V_0 = V_{TH}$.

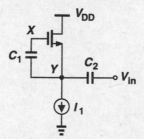

Figure 2.54

2.27. An NMOS device operating in the subthreshold region has a ζ of 1.5. What variation in V_{GS} results in a ten-fold change in I_D? If $I_D = 10\ \mu A$, what is g_m?

2.28. Consider an NMOS device with $V_G = 1.5$ V and $V_S = 0$. Explain what happens if we continually decrease V_D below zero or increase V_{sub} above zero.

References

1. R.S. Muller and T. I. Kamins, *Device Electronics for Integrated Circuits,* Second Ed., New York: Wiley, 1986.
2. Y. Tsividis, *Operation and Modeling of the MOS Transistor,* Second Ed., Boston: McGraw-Hill, 1999.
3. Y. Taur and T. H. Ning, *Fundamentals of Modern VLSI Devices,* New York: Cambridge University Press, 1998.

Chapter 3

Single-Stage Amplifiers

Amplification is an essential function in most analog (and many digital) circuits. We amplify an analog or digital signal because it may be too small to drive a load, overcome the noise of a subsequent stage, or provide logical levels to a digital circuit. Amplification also plays a critical role in feedback systems (Chapter 8).

In this chapter, we study the low-frequency behavior of single-stage CMOS amplifiers. Analyzing both the large-signal and the small-signal characteristics of each circuit, we develop intuitive techniques and models that prove useful in understanding more complex systems. An important part of a designer's job is to use proper approximations so as to create a simple mental picture of a complicated circuit. The intuition thus gained makes it possible to formulate the behavior of most circuits by inspection rather than by lengthy calculations.

Following a brief review of basic concepts, we describe in this chapter four types of amplifiers: common-source and common-gate topologies, source followers, and cascode configurations. In each case, we begin with a simple model and gradually add second-order phenomena such as channel-length modulation and body effect.

3.1 Basic Concepts

The input-output characteristic of an amplifier is generally a nonlinear function (Fig. 3.1) that can be approximated by a polynomial over some signal range:

$$y(t) \approx \alpha_0 + \alpha_1 x(t) + \alpha_2 x^2(t) + \cdots + \alpha_n x^n(t) \quad x_1 \leq x \leq x_2. \tag{3.1}$$

The input and output may be current or voltage quantities. For a sufficiently narrow range of x,

$$y(t) \approx \alpha_0 + \alpha_1 x(t), \tag{3.2}$$

where α_0 can be considered the operating (bias) point and α_1 the small-signal gain. So long as $\alpha_1 x(t) \ll \alpha_0$, the bias point is disturbed negligibly, (3.2) provides a reasonable

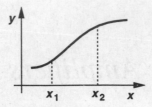

Figure 3.1 Input-output characteristic of a nonlinear system.

approximation, and higher order terms are insignificant. In other words, $\Delta y = \alpha_1 \Delta x$, indicating a linear relationship between the *increments* at the input and output. As $x(t)$ increases in magnitude, higher order terms manifest themselves, leading to nonlinearity and necessitating large-signal analysis. From another point of view, if the slope of the characteristic (the incremental gain) *varies* with the signal level, then the system is nonlinear. These concepts are described in detail in Chapter 13.

What aspects of the performance of an amplifier are important? In addition to gain and speed, such parameters as power dissipation, supply voltage, linearity, noise, or maximum voltage swings may be important. Furthermore, the input and output impedances determine how the circuit interacts with preceding and subsequent stages. In practice, most of these parameters trade with each other, making the design a multi-dimensional optimization problem. Illustrated in the "analog design octagon" of Fig. 3.2, such trade-offs present many challenges in the design of high-performance amplifiers, requiring intuition and experience to arrive at an acceptable compromise.

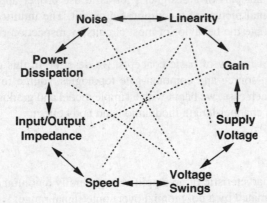

Figure 3.2 Analog design octagon.

3.2 Common-Source Stage

3.2.1 Common-Source Stage with Resistive Load

By virtue of its transconductance, a MOSFET converts variations in its gate-source voltage to a small-signal drain current, which can pass through a resistor to generate an output voltage. Shown in Fig. 3.3(a), the common-source (CS) stage performs such an operation.

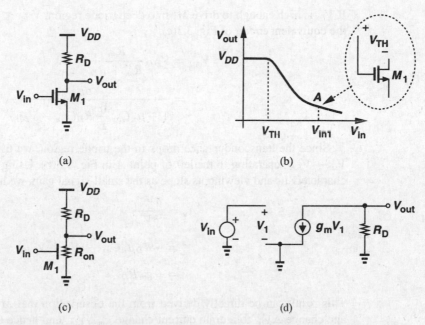

Figure 3.3   (a) Common-source stage, (b) input-output characteristic, (c) equivalent circuit in deep triode region, (d) small-signal model for the saturation region.

We study both the large-signal and the small-signal behavior of the circuit. Note that the input impedance of the circuit is very high at low frequencies.

If the input voltage increases from zero, M_1 is off and $V_{out} = V_{DD}$ [Fig. 3.3(b)]. As V_{in} approaches V_{TH}, M_1 begins to turn on, drawing current from R_D and lowering V_{out}. If V_{DD} is not excessively low, M_1 turns on in saturation, and we have

$$V_{out} = V_{DD} - R_D \frac{1}{2} \mu_n C_{ox} \frac{W}{L} (V_{in} - V_{TH})^2, \tag{3.3}$$

where channel-length modulation is neglected. With further increase in V_{in}, V_{out} drops more and the transistor continues to operate in saturation until V_{in} exceeds V_{out} by V_{TH} [point A in Fig. 3.3(b)]. At this point,

$$V_{in1} - V_{TH} = V_{DD} - R_D \frac{1}{2} \mu_n C_{ox} \frac{W}{L} (V_{in1} - V_{TH})^2, \tag{3.4}$$

from which $V_{in1} - V_{TH}$ and hence V_{out} can be calculated.

For $V_{in} > V_{in1}$, M_1 is in the triode region:

$$V_{out} = V_{DD} - R_D \frac{1}{2} \mu_n C_{ox} \frac{W}{L} \left[2(V_{in} - V_{TH})V_{out} - V_{out}^2 \right]. \tag{3.5}$$

If V_{in} is high enough to drive M_1 into deep triode region, $V_{out} \ll 2(V_{in} - V_{TH})$, and, from the equivalent circuit of Fig. 3.3(c),

$$V_{out} = V_{DD} \frac{R_{on}}{R_{on} + R_D} \tag{3.6}$$

$$= \frac{V_{DD}}{1 + \mu_n C_{ox} \dfrac{W}{L} R_D (V_{in} - V_{TH})}. \tag{3.7}$$

Since the transconductance drops in the triode region, we usually ensure that $V_{out} > V_{in} - V_{TH}$, operating to the left of point A in Fig. 3.3(b). Using (3.3) as the input-output characteristic and viewing its slope as the small-signal gain, we have:

$$A_v = \frac{\partial V_{out}}{\partial V_{in}} \tag{3.8}$$

$$= -R_D \mu_n C_{ox} \frac{W}{L} (V_{in} - V_{TH}) \tag{3.9}$$

$$= -g_m R_D. \tag{3.10}$$

This result can be directly derived from the observation that M_1 converts an input voltage change ΔV_{in} to a drain current change $g_m \Delta V_{in}$, and hence an output voltage change $-g_m R_D \Delta V_{in}$. The small-signal model of Fig. 3.3(d) yields the same result.

Even though derived for small-signal operation, the equation $A_v = -g_m R_D$ predicts certain effects if the circuit senses a *large* signal swing. Since g_m itself varies with the input signal according to $g_m = \mu_n C_{ox}(W/L)(V_{GS} - V_{TH})$, the gain of the circuit changes substantially if the signal is large. In other words, if the gain of the circuit *varies* significantly with the signal swing, then the circuit operates in the large-signal mode. The dependence of the gain upon the signal level leads to nonlinearity (Chapter 13), usually an undesirable effect.

A key result here is that to minimize the nonlinearity, the gain equation must be a weak function of signal-dependent parameters such as g_m. We present several examples of this concept in this chapter and in Chapter 13.

Example 3.1

Sketch the drain current and transconductance of M_1 in Fig. 3.3(a) as a function of the input voltage.

Solution

The drain current becomes significant for $V_{in} > V_{TH}$, eventually approaching V_{DD}/R_D if $R_{on1} \ll R_D$ [Fig. 3.4(a)]. Since in saturation, $g_m = \mu_n C_{ox}(W/L)(V_{in} - V_{TH})$, the transconductance begins to rise for $V_{in} > V_{TH}$. In the triode region, $g_m = \mu_n C_{ox}(W/L)V_{DS}$, falling as V_{in} exceeds V_{in1} [Fig. 3.4(b)].

How do we maximize the voltage gain of a common-source stage? Writing (3.10) as

$$A_v = -\sqrt{2\mu_n C_{ox} \frac{W}{L} I_D} \frac{V_{RD}}{I_D}, \tag{3.11}$$

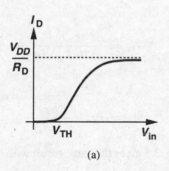
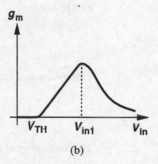

$$(a) \qquad\qquad\qquad (b)$$

Figure 3.4

where V_{RD} denotes the voltage drop across R_D, we have

$$A_v = -\sqrt{2\mu_n C_{ox} \frac{W}{L}} \frac{V_{RD}}{\sqrt{I_D}}. \tag{3.12}$$

Thus, the magnitude of A_v can be increased by increasing W/L or V_{RD} or decreasing I_D if other parameters are constant. It is important to understand the trade-offs resulting from this equation. A larger device size leads to greater device capacitances, and a higher V_{RD} limits the maximum voltage swings. For example, if $V_{DD} - V_{RD} = V_{in} - V_{TH}$, then M_1 is at the edge of the triode region, allowing only very small swings at the output (and input). If V_{RD} remains constant and I_D is reduced, then R_D must increase, thereby leading to a greater time constant at the output node. In other words, as noted in the analog design octagon, the circuit exhibits trade-offs between gain, bandwidth, and voltage swings. Lower supply voltages further tighten these trade-offs.

For large values of R_D, the effect of channel length modulation in M_1 becomes significant. Modifying (3.4) to include this effect,

$$V_{out} = V_{DD} - R_D \frac{1}{2} \mu_n C_{ox} \frac{W}{L} (V_{in} - V_{TH})^2 (1 + \lambda V_{out}), \tag{3.13}$$

we have

$$\frac{\partial V_{out}}{\partial V_{in}} = -R_D \mu_n C_{ox} \frac{W}{L} (V_{in} - V_{TH})(1 + \lambda V_{out})$$

$$- R_D \frac{1}{2} \mu_n C_{ox} \frac{W}{L} (V_{in} - V_{TH})^2 \lambda \frac{\partial V_{out}}{\partial V_{in}}. \tag{3.14}$$

Using the approximation $I_D \approx (1/2)\mu_n C_{ox}(W/L)(V_{in} - V_{TH})^2$, we obtain:

$$A_v = -R_D g_m - R_D I_D \lambda A_v \tag{3.15}$$

and hence

$$A_v = -\frac{g_m R_D}{1 + R_D \lambda I_D}. \tag{3.16}$$

Since $\lambda I_D = 1/r_O$,

$$A_v = -g_m \frac{r_O R_D}{r_O + R_D}. \tag{3.17}$$

The small-signal model of Fig. 3.5 gives the same result with much less effort. That is, since

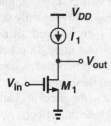

Figure 3.5 Small-signal model of CS stage including the transistor output resistance.

$g_m V_1(r_O \| R_D) = -V_{out}$ and $V_1 = V_{in}$, we have $V_{out}/V_{in} = -g_m(r_O \| R_D)$. Note that, as mentioned in Chapter 1, V_{in}, V_1, and V_{out} in this figure denote small-signal quantities.

Example 3.2 _____

Assuming M_1 in Fig. 3.6 is biased in saturation, calculate the small-signal voltage gain of the circuit.

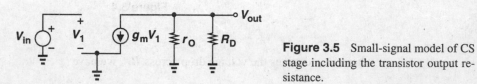

Figure 3.6

Solution

Since I_1 introduces an infinite impedance, the gain is limited by the output resistance of M_1:

$$A_v = -g_m r_O. \tag{3.18}$$

Called the "intrinsic gain" of a transistor, this quantity represents the maximum voltage gain that can be achieved using a single device. In today's CMOS technology, $g_m r_O$ of short-channel devices is between roughly 10 and 30. Thus, we usually assume $1/g_m \ll r_O$.

In Fig. 3.6, Kirchhoff's current law (KCL) requires that $I_{D1} = I_1$. Then, how can V_{in} change the current of M_1 if I_1 is constant? Writing the total drain current of M_1 as

$$I_{D1} = \frac{1}{2}\mu_n C_{ox}(V_{in} - V_{TH})^2(1 + \lambda V_{out}) \tag{3.19}$$

$$= I_1, \tag{3.20}$$

we note that V_{in} appears in the square term and V_{out} in the linear term. As V_{in} increases, V_{out} must decrease such that the product remains constant. We may nevertheless say "I_{D1} increases as V_{in} increases." This statement simply refers to the quadratic part of the equation.

3.2.2 CS Stage with Diode-Connected Load

In many CMOS technologies, it is difficult to fabricate resistors with tightly-controlled values or a reasonable physical size (Chapter 17). Consequently, it is desirable to replace R_D in Fig. 3.3(a) with a MOS transistor.

A MOSFET can operate as a small-signal resistor if its gate and drain are shorted [Fig. 3.7(a)]. Called a "diode-connected" device in analogy with its bipolar counterpart,

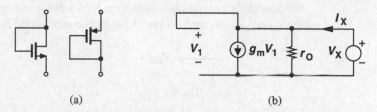

| (a) | (b) |

Figure 3.7 (a) Diode-connected NMOS and PMOS devices, (b) small-signal equivalent circuit.

this configuration exhibits a small-signal behavior similar to a two-terminal resistor. Note that the transistor is always in saturation because the drain and the gate have the same potential. Using the small-signal equivalent shown in Fig. 3.7(b) to obtain the impedance of the device, we write $V_1 = V_X$ and $I_X = V_X/r_O + g_m V_X$. That is, the impedance of the diode is simply equal to $(1/g_m)\|r_O \approx 1/g_m$. If body effect exists, we can use the circuit in Fig. 3.8 to write $V_1 = -V_X$, $V_{bs} = -V_X$ and

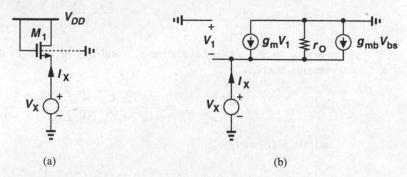

| (a) | (b) |

Figure 3.8 (a) Arrangement for measuring the equivalent resistance of a diode-connected MOSFET, (b) small-signal equivalent circuit.

$$(g_m + g_{mb})V_X + \frac{V_X}{r_O} = I_X. \qquad (3.21)$$

It follows that

$$\frac{V_X}{I_X} = \frac{1}{g_m + g_{mb} + r_O^{-1}} \tag{3.22}$$

$$= \frac{1}{g_m + g_{mb}} \| r_O \tag{3.23}$$

$$\approx \frac{1}{g_m + g_{mb}}. \tag{3.24}$$

Interestingly, the impedance seen at the source of M_1 is *lower* when body effect is included. Intuitive explanation of this effect is left as an exercise for the reader.

We now study a common-source stage with a diode-connected load (Fig. 3.9). For negligible channel-length modulation, (3.24) can be substituted in (3.10) for the load impedance,

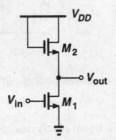

Figure 3.9 CS stage with diode-connected load.

yielding

$$A_v = -g_{m1} \frac{1}{g_{m2} + g_{mb2}} \tag{3.25}$$

$$= -\frac{g_{m1}}{g_{m2}} \frac{1}{1 + \eta}, \tag{3.26}$$

where $\eta = g_{mb2}/g_{m2}$. Expressing g_{m1} and g_{m2} in terms of device dimensions and bias currents, we have

$$A_v = -\frac{\sqrt{2\mu_n C_{ox}(W/L)_1 I_{D1}}}{\sqrt{2\mu_n C_{ox}(W/L)_2 I_{D2}}} \frac{1}{1 + \eta}, \tag{3.27}$$

and, since $I_{D1} = I_{D2}$,

$$A_v = -\sqrt{\frac{(W/L)_1}{(W/L)_2}} \frac{1}{1 + \eta}. \tag{3.28}$$

This equation reveals an interesting property: if the variation of η with the output voltage is neglected, the gain is independent of the bias currents and voltages (so long as M_1 stays in saturation). In other words, as the input and output signal levels vary, the gain remains relatively constant, indicating that the input-output characteristic is relatively linear.

The linear behavior of the circuit can also be confirmed by large-signal analysis. Neglecting channel-length modulation for simplicity, we have in Fig. 3.9

$$\frac{1}{2}\mu_n C_{ox} \left(\frac{W}{L}\right)_1 (V_{in} - V_{TH1})^2 = \frac{1}{2}\mu_n C_{ox} \left(\frac{W}{L}\right)_2 (V_{DD} - V_{out} - V_{TH2})^2, \qquad (3.29)$$

and hence

$$\sqrt{\left(\frac{W}{L}\right)_1} (V_{in} - V_{TH1}) = \sqrt{\left(\frac{W}{L}\right)_2} (V_{DD} - V_{out} - V_{TH2}). \qquad (3.30)$$

Thus, if the variation of V_{TH2} with V_{out} is small, the circuit exhibits a linear input-output characteristic. The small-signal gain can also be computed by differentiating both sides with respect to V_{in}:

$$\sqrt{\left(\frac{W}{L}\right)_1} = \sqrt{\left(\frac{W}{L}\right)_2} \left(-\frac{\partial V_{out}}{\partial V_{in}} - \frac{\partial V_{TH2}}{\partial V_{in}}\right), \qquad (3.31)$$

which, upon application of the chain rule $\partial V_{TH2}/\partial V_{in} = (\partial V_{TH2}/\partial V_{out})(\partial V_{out}/\partial V_{in}) = \eta(\partial V_{out}/\partial V_{in})$, reduces to

$$\frac{\partial V_{out}}{\partial V_{in}} = -\sqrt{\frac{(W/L)_1}{(W/L)_2}} \frac{1}{1+\eta}. \qquad (3.32)$$

It is instructive to study the overall large-signal characteristic of the circuit as well. But let us first consider the circuit shown in Fig. 3.10(a). What is the final value of V_{out} if I_1 drops to zero? As I_1 decreases, so does the overdrive of M_2. Thus, for small I_1, $V_{GS2} \approx V_{TH2}$ and $V_{out} \approx V_{DD} - V_{TH2}$. In reality, the subthreshold conduction in M_2 eventually brings V_{out} to V_{DD} if I_D approaches zero, but at very low current levels, the finite capacitance at the output node slows down the change from $V_{DD} - V_{TH2}$ to V_{DD}. This is illustrated in the time-domain waveforms of Fig. 3.10(b). For this reason, in circuits that have frequent switching activity, we assume V_{out} remains around $V_{DD} - V_{TH2}$ when I_1 falls to small values.

Now we return to the circuit of Fig. 3.9. Plotted in Fig. 3.11 versus V_{in}, the output voltage equals $V_{DD} - V_{TH2}$ if $V_{in} < V_{TH1}$. For $V_{in} > V_{TH1}$, Eq. (3.30) holds and V_{out} follows an approximately straight line. As V_{in} exceeds $V_{out} + V_{TH1}$ (beyond point A), M_1 enters the triode region, and the characteristic becomes nonlinear.

The diode-connected load of Fig. 3.9 can be implemented with a PMOS device as well. Shown in Fig. 3.12, the circuit is free from body effect, providing a small-signal voltage gain equal to

$$A_v = -\sqrt{\frac{\mu_n(W/L)_1}{\mu_p(W/L)_2}}, \qquad (3.33)$$

where channel-length modulation is neglected.

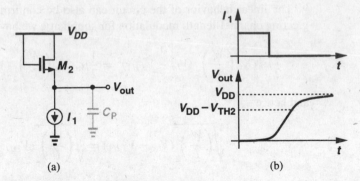

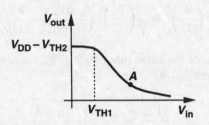

Figure 3.10   (a) Diode-connected device with stepped bias current, (b) variation of source voltage versus time.

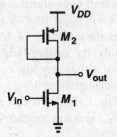

Figure 3.11   Input-output characteristic of a CS stage with diode-connected load.

Figure 3.12   CS stage with diode-connected PMOS device.

Equations (3.28) and (3.33) indicate that the gain of a common-source stage with diode-connected load is a relatively weak function of the device dimensions. For example, to achieve a gain of 10, $\mu_n(W/L)_1/[\mu_p(W/L)_2] = 100$, implying that, with $\mu_n \approx 2\mu_p$, we must have $(W/L)_1 \approx 50(W/L)_2$. In a sense, a high gain requires a "strong" input device and a "weak" load device. In addition to disproportionately wide or long transistors (and hence a large input or load capacitance), a high gain translates to another important limitation: reduction in allowable voltage swings. Specifically, since in Fig. 3.12, $I_{D1} = |I_{D2}|$,

$$\mu_n \left(\frac{W}{L}\right)_1 (V_{GS1} - V_{TH1})^2 \approx \mu_p \left(\frac{W}{L}\right)_2 (V_{GS2} - V_{TH2})^2, \tag{3.34}$$

revealing that

$$\frac{|V_{GS2} - V_{TH2}|}{V_{GS1} - V_{TH1}} \approx A_v. \tag{3.35}$$

In the above example, the overdrive voltage of M_2 must be 10 times that of M_1. For example, with $V_{GS1} - V_{TH1} = 200$ mV, and $|V_{TH2}| = 0.7$ V, we have $|V_{GS2}| = 2.7$ V, severely limiting the output swing. This is another example of the trade-offs suggested by the analog design octagon. Note that, with diode-connected loads, the swing is constrained by both the required overdrive voltage and the threshold voltage. That is, even with a small overdrive, the output level cannot exceed $V_{DD} - |V_{TH}|$.

An interesting paradox arises here if we write $g_m = \mu C_{ox}(W/L)|V_{GS} - V_{TH}|$. The voltage gain of the circuit is then given by

$$A_v = \frac{g_{m1}}{g_{m2}} \tag{3.36}$$

$$= \frac{\mu_n C_{ox}(W/L)_1(V_{GS1} - V_{TH1})}{\mu_p C_{ox}(W/L)_2|V_{GS2} - V_{TH2}|}. \tag{3.37}$$

Equation (3.37) implies that A_v is *inversely* proportional to $|V_{GS2} - V_{TH2}|$. It is left for the reader to resolve the seemingly opposite trends suggested by (3.35) and (3.37).

Example 3.3

In the circuit of Fig. 3.13, M_1 is biased in saturation with a drain current equal to I_1. The current source $I_S = 0.75I_1$ is added to the circuit. How is (3.35) modified for this case?

Solution

Since $|I_{D2}| = I_1/4$, we have

$$A_v \approx -\frac{g_{m1}}{g_{m2}} \tag{3.38}$$

$$= -\sqrt{\frac{4\mu_n(W/L)_1}{\mu_p(W/L)_2}}. \tag{3.39}$$

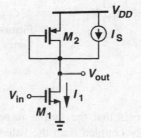

Figure 3.13

Moreover,

$$\mu_n \left(\frac{W}{L}\right)_1 (V_{GS1} - V_{TH1})^2 \approx 4\mu_p \left(\frac{W}{L}\right)_2 (V_{GS2} - V_{TH2})^2, \tag{3.40}$$

yielding

$$\frac{|V_{GS2} - V_{TH2}|}{V_{GS1} - V_{TH1}} \approx \frac{A_v}{4}. \tag{3.41}$$

Thus, for a gain of 10, the overdrive of M_2 need be only 2.5 times that of M_1. Alternatively, for a given overdrive voltage, this circuit achieves a gain four times that of the stage in Fig. 3.12. Intuitively, this is because for a given $|V_{GS2} - V_{TH2}|$, if the current decreases by a factor of 4, then $(W/L)_2$ must decrease proportionally, and $g_{m2} = \sqrt{2\mu_p C_{ox}(W/L)_2 I_{D2}}$ is lowered by the same factor.

We should also mention that in today's CMOS technology, channel-length modulation is quite significant and, more importantly, the behavior of transistors notably departs from the square law (Chapter 16). Thus, the gain of the stage in Fig. 3.9 must be expressed as

$$A_v = -g_{m1}\left(\frac{1}{g_{m2}}\|r_{O1}\|r_{O2}\right), \tag{3.42}$$

where g_{m1} and g_{m2} must be obtained as described in Chapter 16.

3.2.3 CS Stage with Current-Source Load

In applications requiring a large voltage gain in a single stage, the relationship $A_v = -g_m R_D$ suggests that we increase the load impedance of the CS stage. With a resistor or diode-connected load, however, increasing the load resistance limits the output voltage swing.

A more practical approach is to replace the load with a current source. Described briefly in Example 3.2, the resulting circuit is shown in Fig. 3.14, where both transistors operate in saturation. Since the total impedance seen at the output node is equal to $r_{O1}\|r_{O2}$, the gain is

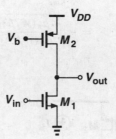

Figure 3.14 CS stage with current-source load.

$$A_v = -g_{m1}(r_{O1}\|r_{O2}). \tag{3.43}$$

The key point here is that the output impedance and the minimum required $|V_{DS}|$ of M_2 are less strongly coupled than the value and voltage drop of a resistor. The voltage

$|V_{DS2,min}| = |V_{GS2} - V_{TH2}|$ can be reduced to even a few hundred millivolts by simply increasing the width of M_2. If r_{O2} is not sufficiently high, the length and width of M_2 can be increased to achieve a smaller λ while maintaining the same overdrive voltage. The penalty is the large capacitance introduced by M_2 at the output node.

We should remark that the output bias voltage of the circuit in Fig. 3.14 is not well-defined. Thus, the stage is reliably biased only if a feedback loop forces V_{out} to a known value (Chapter 8). The large-signal analysis of the circuit is left as an exercise for the reader.

As explained in Chapter 2, the output impedance of MOSFETs at a given drain current can be scaled by changing the channel length, i.e., to the first order, $\lambda \propto 1/L$ and hence $r_O \propto L/I_D$. Since the gain of the stage shown in Fig. 3.14 is proportional to $r_{O1} \| r_{O2}$, we may surmise that longer transistors yield a higher voltage gain.

Let us consider M_1 and M_2 separately. If L_1 is scaled by a factor α (> 1), then W_1 may need to be scaled proportionally as well. This is because, for a given drain current, $V_{GS1} - V_{TH1} \propto 1/\sqrt{(W/L)_1}$, i.e., if W_1 is not scaled, the overdrive voltage increases, limiting the output voltage swing. Also, since $g_{m1} \propto \sqrt{(W/L)_1}$, scaling up only L_1 lowers g_{m1}.

In applications where these issues are unimportant, W_1 can remain constant while L_1 increases. Thus, the intrinsic gain of the transistor can be written as

$$g_{m1}r_{O1} = \sqrt{2 \left(\frac{W}{L} \right)_1 \mu_n C_{ox} I_D \frac{1}{\lambda I_D}}, \qquad (3.44)$$

indicating that the gain *increases* with L because λ depends more strongly on L than g_m does. Also, note that $g_m r_O$ *decreases* as I_D increases.

Increasing L_2 while keeping W_2 constant increases r_{O2} and hence the voltage gain, but at the cost of higher $|V_{DS2}|$ required to maintain M_2 in saturation.

3.2.4 CS Stage with Triode Load

A MOS device operating in deep triode region behaves as a resistor and can therefore serve as the load in a CS stage. Illustrated in Fig. 3.15, such a circuit biases the gate of M_2 at a sufficiently low level, ensuring the load is in deep triode region for all output voltage swings.

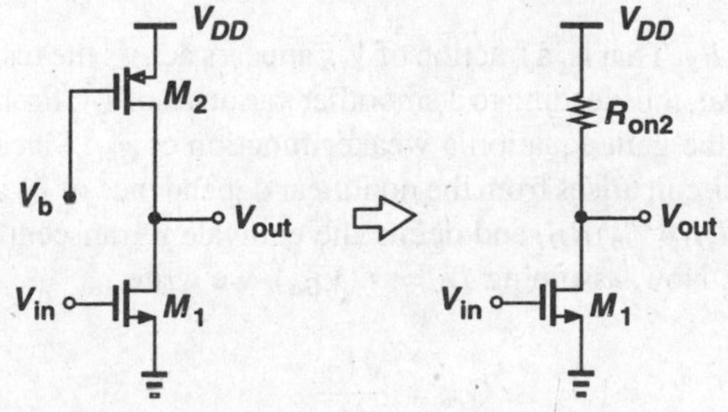

Figure 3.15 CS stage with triode load.

Since

$$R_{on2} = \frac{1}{\mu_p C_{ox}(W/L)_2(V_{DD} - V_b - |V_{THP}|)}, \tag{3.45}$$

the voltage gain can be readily calculated.

The principal drawback of this circuit stems from the dependence of R_{on2} upon $\mu_p C_{ox}$, V_b, and V_{THP}. Since $\mu_p C_{ox}$ and V_{THP} vary with process and temperature and since generating a precise value for V_b requires additional complexity, this circuit is difficult to use. Triode loads, however, consume less voltage headroom then do diode-connected devices because in Fig. 3.15 $V_{out,max} = V_{DD}$ whereas in Fig. 3.12, $V_{out,max} \approx V_{DD} - |V_{THP}|$.

3.2.5 CS Stage with Source Degeneration

In some applications, the square-law dependence of the drain current upon the overdrive voltage introduces excessive nonlinearity, making it desirable to "soften" the device characteristic. In Section 3.2.2, we noted the linear behavior of a CS stage using a diode-connected load. Alternatively, as depicted in Fig. 3.16, this can be accomplished by placing a "degeneration" resistor in series with the source terminal. Here, as V_{in} increases, so do I_D and the

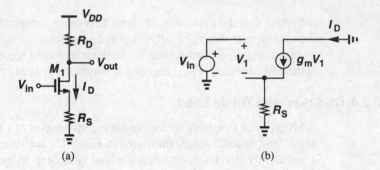

Figure 3.16 CS stage with source degeneration.

voltage drop across R_S. That is, a fraction of V_{in} appears across the resistor rather than as the gate-source overdrive, thus leading to a smoother variation of I_D. From another perspective, we intend to make the gain equation a weaker function of g_m. Since $V_{out} = -I_D R_D$, the nonlinearity of the circuit arises from the nonlinear dependence of I_D upon V_{in}. We note that $\partial V_{out}/\partial V_{in} = -(\partial I_D/\partial V_{in})R_D$, and define the equivalent transconductance of the circuit as $G_m = \partial I_D/\partial V_{in}$. Now, assuming $I_D = f(V_{GS})$, we write

$$G_m = \frac{\partial I_D}{\partial V_{in}} \tag{3.46}$$

$$= \frac{\partial f}{\partial V_{GS}} \frac{\partial V_{GS}}{\partial V_{in}}. \tag{3.47}$$

209.147.153.70

Since $V_{GS} = V_{in} - I_D R_S$, we have $\partial V_{GS}/\partial V_{in} = 1 - R_S \partial I_D/\partial V_{in}$, obtaining

$$G_m = \left(1 - R_S \frac{\partial I_D}{\partial V_{in}}\right) \frac{\partial f}{\partial V_{GS}}. \tag{3.48}$$

But, $\partial f/\partial V_{GS}$ is the transconductance of M_1, and

$$G_m = \frac{g_m}{1 + g_m R_S}. \tag{3.49}$$

The small-signal voltage gain is thus equal to

$$A_v = -G_m R_D \tag{3.50}$$

$$= \frac{-g_m R_D}{1 + g_m R_S}. \tag{3.51}$$

The same result can be derived using the small-signal model of Fig. 3.16(b). Equation (3.49) implies that as R_S increases, G_m becomes a weaker function of g_m and hence the drain current. In fact, for $R_S \gg 1/g_m$, we have $G_m \approx 1/R_S$, i.e., $\Delta I_D \approx \Delta V_{in}/R_S$, indicating that most of the change in V_{in} appears across R_S. We say the drain current is a "linearized" function of the input voltage. The linearization is obtained at the cost of lower gain [and higher noise (Chapter 7)].

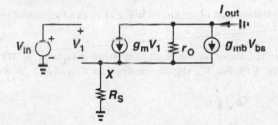

Figure 3.17 Small-signal equivalent circuit of a degenerated CS stage.

For our subsequent calculations, it is useful to determine G_m in the presence of body effect and channel-length modulation. With the aid of the equivalent circuit shown in Fig. 3.17, we recognize that the current through R_S equals I_{out} and, therefore, $V_{in} = V_1 + I_{out} R_S$. Summing the currents at node X, we have

$$I_{out} = g_m V_1 - g_{mb} V_X - \frac{I_{out} R_S}{r_O} \tag{3.52}$$

$$= g_m(V_{in} - I_{out} R_S) + g_{mb}(-I_{out} R_S) - \frac{I_{out} R_S}{r_O}. \tag{3.53}$$

It follows that

$$G_m = \frac{I_{out}}{V_{in}} \tag{3.54}$$

$$= \frac{g_m r_O}{R_S + [1 + (g_m + g_{mb})R_S]r_O}. \tag{3.55}$$

Let us now examine the large-signal behavior of the CS stage with $R_S = 0$ and $R_S \neq 0$. For $R_S = 0$, our derivations in Chapter 2 indicate that I_D and g_m vary as shown in Fig. 3.18(a). For $R_S \neq 0$, the turn-on behavior is similar to that in Fig. 3.18(a) because,

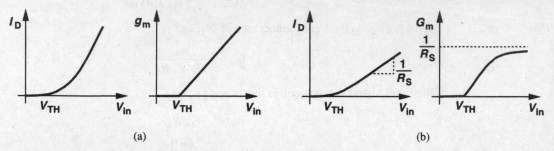

Figure 3.18 Drain current and transconductance of a CS device (a) without and (b) with source degeneration.

at low current levels, $1/g_m \gg R_S$ and hence $G_m \approx g_m$ [Fig. 3.18(b)]. As the overdrive and therefore g_m increase, the effect of degeneration, $1 + g_m R_S$ in (3.49), becomes more significant. For large values of V_{in} (if M_1 is still saturated), I_D is approximately linear and G_m approaches $1/R_S$.

Example 3.4 _____

Plot the small-signal voltage gain of the circuit in Fig. 3.16 as a function of the input bias voltage.

Solution

Using the results derived above for the equivalent transconductance of M_1 and R_S, we arrive at the plot shown in Fig. 3.19. For V_{in} slightly greater than V_{TH}, $1/g_m \gg R_S$ and $A_v \approx -g_m R_D$.

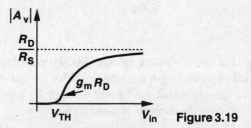

Figure 3.19

As V_{in} increases, degeneration becomes more significant and $A_v = -g_m R_D/(1 + g_m R_S)$. For large values of V_{in}, $G_m \approx 1/R_S$ and $A_v = -R_D/R_S$. However, if $V_{in} > V_{out} + V_{TH}$, that is, if $R_D I_D > V_{TH} + V_{DD} - V_{in}$, M_1 enters the triode region and A_v drops.

_____

Equation (3.51) can be rewritten as

$$A_v = -\frac{R_D}{\dfrac{1}{g_m} + R_S}. \tag{3.56}$$

This result allows formulating the gain by inspection. First, let us examine the denominator of (3.56). The expression is equal to the *series* combination of the inverse transconductance of the device and the explicit resistance seen from the source to ground. We call the denominator "the resistance seen in the source path" because if, as shown in Fig. 3.20, we disconnect the bottom terminal of R_S from ground and calculate the resistance seen "looking up" (while setting the input to zero), we obtain $R_S + 1/g_m$.

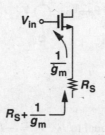

Figure 3.20   Resistance seen in the source path.

Noting that the numerator of (3.56) is the resistance seen at the drain, we view the magnitude of the gain as the resistance seen at the drain node divided by the total resistance in the source path. This method greatly simplifies the analysis of more complex circuits.

Example 3.5

Assuming $\lambda = \gamma = 0$, calculate the small-signal gain of the circuit shown in Fig. 3.21(a).

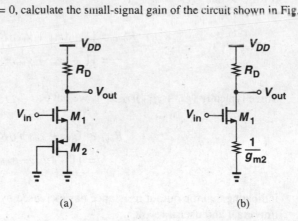

(a)   (b)

Figure 3.21

Solution

Noting that M_2 is a diode-connected device and simplifying the circuit to that shown in Fig. 3.21(b), we use the above rule to write

$$A_v = -\frac{R_D}{\dfrac{1}{g_{m1}} + \dfrac{1}{g_{m2}}}. \tag{3.57}$$

Another important consequence of source degeneration is the increase in the output resistance of the stage. We calculate the output resistance first with the aid of the equivalent circuit shown in Fig. 3.22. Note that body effect is also included to arrive at a general result.

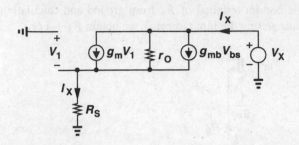

Figure 3.22 Equivalent circuit for calculating the output resistance of a degenerated CS stage.

Since the current through R_S is equal to I_X, $V_1 = -I_X R_S$ and the current flowing through r_O is given by $I_X - (g_m + g_{mb})V_1 = I_X + (g_m + g_{mb})R_S I_X$. Adding the voltage drops across r_O and R_S, we obtain

$$r_O[I_X + (g_m + g_{mb})R_S I_X] + I_X R_S = V_X. \tag{3.58}$$

It follows that

$$R_{out} = [1 + (g_m + g_{mb})R_S]r_O + R_S \tag{3.59}$$

$$= [1 + (g_m + g_{mb})r_O]R_S + r_O. \tag{3.60}$$

Since typically $(g_m + g_{mb})r_O \gg 1$, we have

$$R_{out} \approx (g_m + g_{mb})r_O R_S + r_O \tag{3.61}$$

$$= [1 + (g_m + g_{mb})R_S]r_O, \tag{3.62}$$

indicating that the output resistance has increased by a factor $1 + (g_m + g_{mb})R_S$. This is an important and useful result.

To gain more insight, let us consider the circuit of Fig. 3.22 with $R_S = 0$ and $R_S > 0$. If $R_S = 0$, then $g_m V_1 = g_{mb} V_{bs} = 0$ and $I_X = V_X/r_O$. On the other hand, if $R_S > 0$, we have $I_X R_S > 0$ and $V_1 < 0$, obtaining *negative* $g_m V_1$ and $g_{mb} V_{bs}$. Thus, the current supplied by V_X is less than V_X/r_O.

The relationships in (3.60) and (3.62) can also be derived by inspection. As shown in Fig. 3.23(a), we apply a voltage to the output node, change its value by ΔV, and measure the resulting change, ΔI, in the output current. Since the current through R_S must change by ΔI, we first compute the voltage change across R_S. To this end, we draw the circuit as shown in Fig. 3.23(b) and note that the resistance seen looking into the source of M_1 is equal to $1/(g_m + g_{mb})$ [Eq. (3.24)], thus arriving at the equivalent circuit in Fig. 3.23(c).

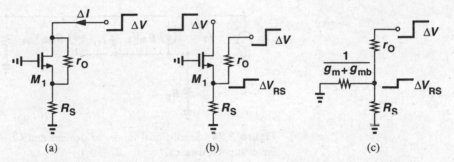

Figure 3.23 (a) Change in drain current in response to change in applied voltage to drain, (b) equivalent of (a), (c) small-signal model.

The voltage change across R_S is therefore equal to

$$\Delta V_{RS} = \Delta V \frac{\dfrac{1}{g_m + g_{mb}} \| R_S}{\dfrac{1}{g_m + g_{mb}} \| R_S + r_O}. \tag{3.63}$$

The change in the current is

$$\Delta I = \frac{\Delta V_{RS}}{R_S} \tag{3.64}$$

$$= \Delta V \frac{1}{[1 + (g_m + g_{mb})R_S]r_O + R_S}, \tag{3.65}$$

that is,

$$\frac{\Delta V}{\Delta I} = [1 + (g_m + g_{mb})R_S]r_O + R_S. \tag{3.66}$$

With the foregoing developments, we can now compute the gain of a degenerated CS stage in the general case, taking into account both body effect and channel-length modulation. In the equivalent circuit depicted in Fig. 3.24, the current through R_S must equal that through R_D, i.e., $-V_{out}/R_D$. Thus, the source voltage with respect to ground (and the bulk) is equal to $-V_{out}R_S/R_D$ and hence $V_1 = V_{in} + V_{out}R_S/R_D$. The current through r_O can therefore be written as

$$I_{ro} = -\frac{V_{out}}{R_D} - (g_m V_1 + g_{mb} V_{bs}) \tag{3.67}$$

$$= -\frac{V_{out}}{R_D} - \left[g_m \left(V_{in} + V_{out} \frac{R_S}{R_D} \right) + g_{mb} V_{out} \frac{R_S}{R_D} \right]. \tag{3.68}$$

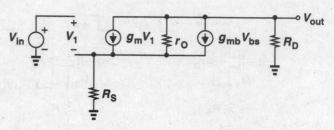

Figure 3.24 Small-signal model of degenerated CS stage with finite output resistance.

Since the voltage drop across r_O and R_S must add up to V_{out}, we have

$$V_{out} = I_{r_O} r_O - \frac{V_{out}}{R_D} R_S \tag{3.69}$$

$$= -\frac{V_{out}}{R_D} r_O - \left[g_m \left(V_{in} + V_{out} \frac{R_S}{R_D} \right) + g_{mb} V_{out} \frac{R_S}{R_D} \right] r_O - V_{out} \frac{R_S}{R_D}. \tag{3.70}$$

It follows that

$$\frac{V_{out}}{V_{in}} = \frac{-g_m r_O R_D}{R_D + R_S + r_O + (g_m + g_{mb}) R_S r_O}. \tag{3.71}$$

To gain more insight into this result, we recognize that the last three terms in the denominator, namely, $R_S + r_O + (g_m + g_{mb}) R_S r_O$, represent the output resistance of a MOS device degenerated by a resistor R_S, as originally derived in (3.60). Let us now rewrite (3.71) as

$$A_v = \frac{-g_m r_O R_D [R_S + r_O + (g_m + g_{mb}) R_S r_O]}{R_D + R_S + r_O + (g_m + g_{mb}) R_S r_O} \cdot \frac{1}{R_S + r_O + (g_m + g_{mb}) R_S r_O} \tag{3.72}$$

$$= -\frac{g_m r_O}{R_S + r_O + (g_m + g_{mb}) R_S r_O} \cdot \frac{R_D [R_S + r_O + (g_m + g_{mb}) R_S r_O]}{R_D + R_S + r_O + (g_m + g_{mb}) R_S r_O}. \tag{3.73}$$

The two fractions in (3.73) represent two important parameters of the circuit: the first is identical to that in (3.55), i.e., the equivalent transconductance of a degenerated MOSFET; and the second denotes the parallel combination of R_D and $R_S + r_O + (g_m + g_{mb}) R_S r_O$, i.e., the overall output resistance of the circuit.

The above discussion suggests that in some circuits it may be easier to calculate the voltage gain by exploiting the following lemma.

Lemma.   In a linear circuit, the voltage gain is equal to $-G_m R_{out}$, where G_m denotes the transconductance of the circuit when the output is shorted to ground and R_{out} represents the output resistance of the circuit when the input voltage is set to zero [Fig. 3.25(a)].

The lemma can be proved with the aid of Fig. 3.25 by noting that the output port of a linear circuit can be modeled by a Norton equivalent. That is, the output voltage is equal to $-I_{out} R_{out}$, and I_{out} can be obtained by measuring the short-circuit current at the output.

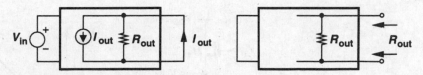

Figure 3.25   Modeling output port of an amplifier by a Norton equivalent.

Defining $G_m = I_{out}/V_{in}$, we have $V_{out} = -G_m V_{in} R_{out}$. This lemma proves useful if G_m and R_{out} can be determined by inspection.

Example 3.6

Calculate the voltage gain of the circuit shown in Fig. 3.26. Assume I_0 is ideal.

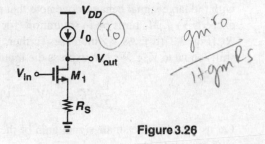

Figure 3.26

Solution

The transconductance and output resistance of the stage are given by Eqs. (3.55) and (3.60), respectively. Thus,

$$A_v = -\frac{g_m r_O}{R_S + [1 + (g_m + g_{mb})R_S]r_O}\{[1 + (g_m + g_{mb})r_O]R_S + r_O\} \tag{3.74}$$

$$= -g_m r_O. \tag{3.75}$$

Interestingly, the voltage gain is equal to the intrinsic gain of the transistor and independent of R_S. This is because, if I_0 is ideal, the current through R_S cannot change and hence the small-signal voltage drop across R_S is zero—as if R_S were zero itself.

3.3 Source Follower

Our analysis of the common-source stage indicates that, to achieve a high voltage gain with limited supply voltage, the load impedance must be as large as possible. If such a stage is to drive a low-impedance load, then a "buffer" must be placed after the amplifier so as to drive the load with negligible loss of the signal level. The source follower (also called the "common-drain" stage) can operate as a voltage buffer.

Illustrated in Fig. 3.27(a), the source follower senses the signal at the gate and drives

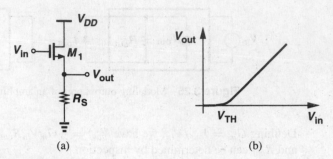

Figure 3.27 (a) Source follower, and (b) its input-output characteristic.

the load at the source, allowing the source potential to "follow" the gate voltage. Beginning with the large-signal behavior, we note that for $V_{in} < V_{TH}$, M_1 is off and $V_{out} = 0$. As V_{in} exceeds V_{TH}, M_1 turns on in saturation (for typical values of V_{DD}) and I_{D1} flows through R_S [Fig. 3.27(b)]. As V_{in} increases further, V_{out} follows the input with a difference (level shift) equal to V_{GS}. We can express the input-output characteristic as:

$$\frac{1}{2}\mu_n C_{ox}\frac{W}{L}(V_{in} - V_{TH} - V_{out})^2 R_S = V_{out}. \tag{3.76}$$

Let us calculate the small-signal gain of the circuit by differentiating both sides of (3.76) with respect to V_{in}:

$$\frac{1}{2}\mu_n C_{ox}\frac{W}{L}2(V_{in} - V_{TH} - V_{out})\left(1 - \frac{\partial V_{TH}}{\partial V_{in}} - \frac{\partial V_{out}}{\partial V_{in}}\right)R_S = \frac{\partial V_{out}}{\partial V_{in}}. \tag{3.77}$$

Since $\partial V_{TH}/\partial V_{in} = \eta\, \partial V_{out}/\partial V_{in}$,

$$\frac{\partial V_{out}}{\partial V_{in}} = \frac{\mu_n C_{ox}\dfrac{W}{L}(V_{in} - V_{TH} - V_{out})R_S}{1 + \mu_n C_{ox}\dfrac{W}{L}(V_{in} - V_{TH} - V_{out})R_S(1 + \eta)}. \tag{3.78}$$

Also, note that

$$g_m = \mu_n C_{ox}\frac{W}{L}(V_{in} - V_{TH} - V_{out}). \tag{3.79}$$

Consequently,

$$A_v = \frac{g_m R_S}{1 + (g_m + g_{mb})R_S}. \tag{3.80}$$

The same result is more easily obtained with the aid of a small-signal equivalent circuit. From Fig. 3.28, we have $V_{in} - V_1 = V_{out}$, $V_{bs} = -V_{out}$, and $g_m V_1 - g_{mb}V_{out} = V_{out}/R_S$.

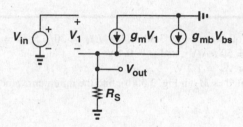

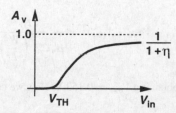

Figure 3.28 Small-signal equivalent circuit of source follower.

Figure 3.29 Voltage gain of source follower versus input voltage.

Thus, $V_{out}/V_{in} = g_m R_S / [1 + (g_m + g_{mb})R_S]$.

Sketched in Fig. 3.29 vs. V_{in}, the voltage gain begins from zero for $V_{in} \approx V_{TH}$ (that is, $g_m \approx 0$) and monotonically increases. As the drain current and g_m increase, A_v approaches $g_m/(g_m + g_{mb}) = 1/(1 + \eta)$. Since η itself slowly decreases with V_{out}, A_v would eventually become equal to unity, but for typical allowable source-bulk voltages, η remains greater than roughly 0.2.

An important result of (3.80) is that even if $R_S = \infty$, the voltage gain of a source follower is not equal to one. We return to this point later. Note that M_1 in Fig. 3.27 does not enter the triode region if V_{in} remains below V_{DD}.

In the source follower of Fig. 3.27, the drain current of M_1 heavily depends on the input dc level. For example, if V_{in} changes from 1.5 V to 2 V, I_D may increase by a factor of 2 and hence $V_{GS} - V_{TH}$ by $\sqrt{2}$, thereby introducing substantial nonlinearity in the input-output characteristic. To alleviate this issue, the resistor can be replaced by a current source as shown in Fig. 3.30(a). The current source itself is implemented as an NMOS transistor operating in the saturation region [Fig. 3.30(b)].

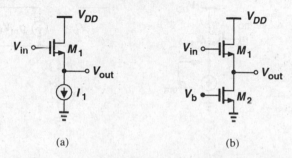

(a)   (b)

Figure 3.30 Source follower using an NMOS transistor as current source.

$$\frac{V_{out}}{V_{in}} =$$

Example 3.7

Suppose in the source follower of Fig. 3.30(a), $(W/L)_1 = 20/0.5$, $I_1 = 200 \ \mu A$, $V_{TH0} = 0.6$ V, $2\Phi_F = 0.7$ V , $\mu_n C_{ox} = 50 \ \mu A/V^2$, and $\gamma = 0.4$ V^2.

(a) Calculate V_{out} for $V_{in} = 1.2$ V.

(b) If I_1 is implemented as M_2 in Fig. 3.30(b), find the minimum value of $(W/L)_2$ for which M_2 remains saturated.

Solution

(a) Since the threshold voltage of M_1 depends on V_{out}, we perform a simple iteration. Noting that

$$(V_{in} - V_{TH} - V_{out})^2 = \frac{2I_D}{\mu_n C_{ox} \left(\dfrac{W}{L}\right)_1}, \tag{3.81}$$

we first assume $V_{TH} \approx 0.6$ V, obtaining $V_{out} = 0.153$ V. Now we calculate a new V_{TH} as

$$V_{TH} = V_{TH0} + \gamma(\sqrt{2\Phi_F + V_{SB}} - \sqrt{2\Phi_F}) \tag{3.82}$$

$$= 0.635 \text{ V}. \tag{3.83}$$

This indicates that V_{out} is approximately 35 mV less than that calculated above, i.e., $V_{out} \approx 0.119$ V.

(b) Since the drain-source voltage of M_2 is equal to 0.119 V, the device is saturated only if $(V_{GS} - V_{TH})_2 \leq 0.119$ V. With $I_D = 200 \ \mu A$, this gives $(W/L)_2 \geq 283/0.5$. Note the substantial drain junction and overlap capacitance contributed by M_2 to the output node.

To gain a better understanding of source followers, let us calculate the small-signal output resistance of the circuit in Fig. 3.31(a). Using the equivalent circuit of Fig. 3.31(b) and noting that $V_1 = -V_X$, we write

$$I_X - g_m V_X - g_{mb} V_X = 0. \tag{3.84}$$

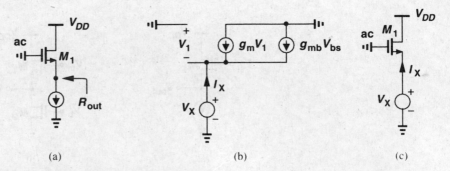

(a)   (b)   (c)

Figure 3.31   Calculation of the output impedance of a source follower.

It follows that

$$R_{out} = \frac{1}{g_m + g_{mb}}.$$ (3.85)

Interestingly, body effect decreases the output resistance of source followers. To understand why, suppose in Fig. 3.31(c), V_X decreases by ΔV so that the drain current increases. With no body effect, only the gate-source voltage of M_1 would increase by ΔV. With body effect, on the other hand, the threshold voltage of the device decreases as well. Thus, in $(V_{GS} - V_{TH})^2$ the first term increases and the second decreases, resulting in a greater change in the drain current and hence a lower output impedance.

The above phenomenon can also be studied with the aid of the small-signal model shown in Fig. 3.32(a). It is important to note that the magnitude of the current source $g_{mb}V_{bs}$ is

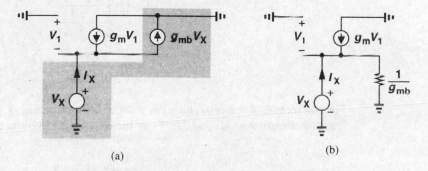

(a)   (b)

Figure 3.32   Source follower including body effect.

linearly proportional to the voltage across it. Such behavior is that of a simple resistor equal to $1/g_{mb}$, yielding the small-signal model shown in Fig. 3.32(b). The equivalent resistor simply appears in parallel with the output, thereby lowering the overall output resistance. The reader can show that, without $1/g_{mb}$, the output resistance equals $1/g_m$, concluding that

$$R_{out} = \frac{1}{g_m} \parallel \frac{1}{g_{mb}}$$ (3.86)

$$= \frac{1}{g_m + g_{mb}}.$$ (3.87)

Modeling the effect of g_{mb} by a resistor—which is only valid for source followers—also helps explain the less-than-unity voltage gain implied by (3.80) for $R_S = \infty$. As shown in

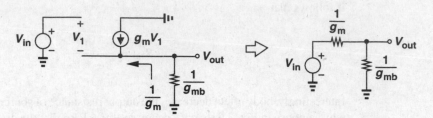

Figure 3.33   Representation of intrinsic source follower by a Thevenin equivalent.

the Thevenin equivalent of Fig. 3.33,

$$A_v = \frac{\dfrac{1}{g_{mb}}}{\dfrac{1}{g_m} + \dfrac{1}{g_{mb}}} \tag{3.88}$$

$$= \frac{g_m}{g_m + g_{mb}}. \tag{3.89}$$

For completeness, we also study the source follower of Fig. 3.34(a) with finite channel-length modulation in M_1 and M_2. From the equivalent circuit in Fig. 3.34(b), we have

$$A_v = \frac{\dfrac{1}{g_{mb}} \| r_{O1} \| r_{o2} \| R_L}{\dfrac{1}{g_{mb}} \| r_{O1} \| r_{o2} \| R_L + \dfrac{1}{g_m}}. \tag{3.90}$$

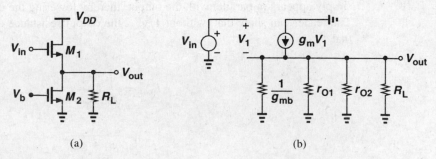

(a)   (b)

Figure 3.34   (a) Source follower driving load resistance, (b) small-signal equivalent circuit.

Example 3.8

Calculate the voltage gain of the circuit shown in Fig. 3.35.

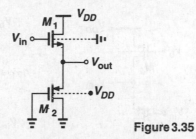

Figure 3.35

Solution

The impedance seen looking into the source of M_2 is equal to $[1/(g_{m2} + g_{mb2})]\|r_{O2}$. Thus,

$$A_v = \frac{\dfrac{1}{g_{m2} + g_{mb2}}\|r_{O2}\|r_{O1}\|\dfrac{1}{g_{mb1}}}{\dfrac{1}{g_{m2} + g_{mb2}}\|r_{O2}\|r_{O1}\|\dfrac{1}{g_{mb1}} + \dfrac{1}{g_{m1}}}. \tag{3.91}$$

Source followers exhibit a high input impedance and a moderate output impedance, but at the cost of two drawbacks: nonlinearity and voltage headroom limitation. We consider these issues in detail.

As mentioned in relation to Fig. 3.27(a), even if a source follower is biased by an ideal current source, its input-output characteristic displays some nonlinearity due to the nonlinear dependence of V_{TH} upon the source potential. In submicron technologies, r_O of the transistor also changes substantially with V_{DS}, thus introducing additional variation in the small-signal gain of the circuit (Chapter 16). For this reason, typical source followers suffer from several percent of nonlinearity.

The nonlinearity due to body effect can be eliminated if the bulk is tied to the source. This is usually possible only for PFETs because all NFETs share the same substrate. Fig. 3.36 shows a PMOS source follower employing two separate n-wells so as to eliminate the body effect of M_1. The lower mobility of PFETs, however, yields a higher output impedance in this case than that available in an NMOS counterpart.

Source followers also shift the dc level of the signal by V_{GS}, thereby consuming voltage headroom and limiting the voltage swings. To understand this point, consider the example illustrated in Fig. 3.37, a cascade of a common-source stage and a source follower. Without the source follower, the minimum allowable value of V_X would be equal to $V_{GS1} - V_{TH1}$ (for M_1 to remain in saturation). With the source follower, on the other hand, V_X must be greater than $V_{GS2} + (V_{GS3} - V_{TH3})$ so that M_3 is saturated. For comparable overdrive voltages in M_1 and M_3, this means the allowable swing at X is reduced by V_{GS2}, a substantial amount.

It is also instructive to compare the gain of source followers and common-source stages when the load impedance is relatively low. A practical example is the need to drive an external 50-Ω termination in a high-frequency setup. As shown in Fig. 3.38(a), the load can

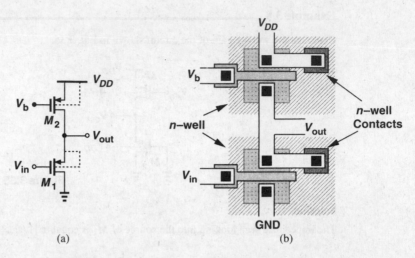

Figure 3.36   PMOS source follower with no body effect.

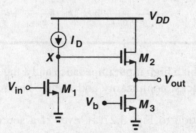

Figure 3.37   Cascade of source follower and CS stage.

be driven by a source follower with an overall voltage gain of

$$\frac{V_{out}}{V_{in}}\Big|_{SF} \approx \frac{R_L}{R_L + 1/g_{m1}}. \tag{3.92}$$

On the other hand, as depicted in Fig. 3.38(b), the load can be included as part of a common-

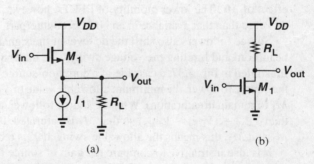

Figure 3.38   (a) Source follower and (b) CS stage driving a load resistance.

[handwritten margin notes: 1.86 / 7.27 / 1092 / 312 / 412 / 10 / 11.3412 / 7.27 × 156 / 100 / 156 + 11.34 / 167.34]

source stage, providing a gain of

$$\frac{V_{out}}{V_{in}}\Big|_{CS} \approx -g_{m1}R_L. \tag{3.93}$$

The key difference between these two topologies is the achievable voltage gain for a given bias current. For example, if $1/g_{m1} \approx R_L$, then the source follower exhibits a gain of at most 0.5 whereas the common-source stage provides a gain close to unity. Thus, source followers are not necessarily efficient drivers.

The drawbacks of source followers, namely, nonlinearity due to body effect, voltage headroom consumption due to level shift, and poor driving capability, limit the use of this topology. Perhaps the most common application of source followers is in performing voltage level shift.

Example 3.9

(a) In the circuit of Fig. 3.39(a), calculate the voltage gain if C_1 acts as an ac short at the frequency of interest. What is the maximum dc level of the input signal for which M_1 remains saturated?

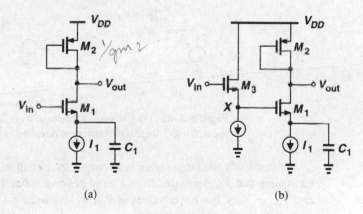

Figure 3.39

(b) To accommodate an input dc level close to V_{DD}, the circuit is modified as shown in Fig. 3.39(b). What relationship among the gate-source voltages of M_1-M_3 guarantees that M_1 is saturated?

Solution

(a) The gain is given by

$$A_v = -g_{m1}[r_{O1}\|r_{O2}\|(1/g_{m2})]. \tag{3.94}$$

Since $V_{out} = V_{DD} - |V_{GS2}|$, the maximum allowable dc level of V_{in} is equal to $V_{DD} - |V_{GS2}| + V_{TH1}$.

(b) If $V_{in} = V_{DD}$, then $V_X = V_{DD} - V_{GS3}$. For M_1 to be saturated, $V_{DD} - V_{GS3} - V_{TH1} \leq V_{DD} - |V_{GS2}|$ and hence $V_{GS3} + V_{TH1} \geq |V_{GS2}|$.

As explained in Chapter 7, source followers also introduce substantial noise. For this reason, the circuit of Fig. 3.39(b) is ill-suited to low-noise applications.

3.4 Common-Gate Stage

In common-source amplifiers and source followers, the input signal is applied to the gate of a MOSFET. It is also possible to apply the signal to the source terminal. Shown in Fig. 3.40(a), a common-gate (CG) stage senses the input at the source and produces the output at the drain. The gate is connected to a dc voltage to establish proper operating conditions. Note that the bias current of M_1 flows through the input signal source. Alternatively, as depicted in Fig. 3.40(b), M_1 can be biased by a constant current source, with the signal capacitively coupled to the circuit.

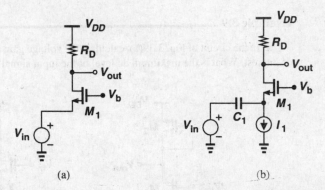

Figure 3.40   (a) Common-gate stage with direct coupling at input, (b) CG stage with capacitive coupling at input.

We first study the large-signal behavior of the circuit in Fig. 3.40(a). For simplicity, let us assume that V_{in} decreases from a large positive value. For $V_{in} \geq V_b - V_{TH}$, M_1 is off and $V_{out} = V_{DD}$. For lower values of V_{in}, we can write

$$I_D = \frac{1}{2}\mu_n C_{ox}\frac{W}{L}(V_b - V_{in} - V_{TH})^2, \tag{3.95}$$

if M_1 is in saturation. As V_{in} decreases, so does V_{out}, eventually driving M_1 into the triode region if

$$V_{DD} - \frac{1}{2}\mu_n C_{ox}\frac{W}{L}(V_b - V_{in} - V_{TH})^2 R_D = V_b - V_{TH}. \tag{3.96}$$

The input-output characteristic is shown in Fig. 3.41. If M_1 is saturated, we can express the output voltage as

$$V_{out} = V_{DD} - \frac{1}{2}\mu_n C_{ox}\frac{W}{L}(V_b - V_{in} - V_{TH})^2 R_D, \tag{3.97}$$

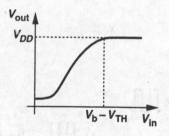

Figure 3.41 Common-gate input-output characteristic.

obtaining a small-signal gain of

$$\frac{\partial V_{out}}{\partial V_{in}} = -\mu_n C_{ox} \frac{W}{L} (V_b - V_{in} - V_{TH}) \left(-1 - \frac{\partial V_{TH}}{\partial V_{in}} \right) R_D. \tag{3.98}$$

Since $\partial V_{TH}/\partial V_{in} = \partial V_{TH}/\partial V_{SB} = \eta$, we have

$$\frac{\partial V_{out}}{\partial V_{in}} = \mu_n C_{ox} \frac{W}{L} R_D (V_b - V_{in} - V_{TH})(1 + \eta) \tag{3.99}$$

$$= g_m (1 + \eta) R_D. \tag{3.100}$$

Note that the gain is positive. Interestingly, body effect increases the equivalent transconductance of the stage.

The input impedance of the circuit is also important. We note that, for $\lambda = 0$, the impedance seen at the source of M_1 in Fig. 3.40(a) is the same as that at the source of M_1 in Fig. 3.31, namely, $1/(g_m + g_{mb}) = 1/[g_m(1 + \eta)]$. Thus, the body effect decreases the input impedance of the common-gate stage. The relatively low input impedance of the common-gate stage proves useful in some applications.

Example 3.10

In Fig. 3.42, transistor M_1 senses ΔV and delivers a proportional current to a 50-Ω transmission line. The other end of the line is terminated by a 50-Ω resistor in Fig. 3.42(a) and a common-gate stage in Fig. 3.42(b). Assume $\lambda = \gamma = 0$.

(a) Calculate V_{out}/V_{in} at low frequencies for both arrangements.

(b) What condition is necessary to minimize wave reflection at node X?

Solution

(a) For small signals applied to the gate of M_1, the drain current experiences a change equal to $g_{m1} \Delta V_X$. This current is drawn from R_D in Fig. 3.42(a) and M_2 in Fig. 3.42(b), producing an output voltage swing equal to $-g_{m1} \Delta V_X R_D$. Thus, $A_v = -g_m R_D$ for both cases.

(b) To minimize reflection at node X, the resistance seen at the source of M_2 must equal 50 Ω and the reactance must be small. Thus, $1/(g_m + g_{mb}) = 50 \Omega$, which can be ensured by proper sizing and biasing of M_2. To minimize the capacitances of the transistor, it is desirable to use a small device biased at a large current. (Recall that $g_m = \sqrt{2\mu_n C_{ox}(W/L)I_D}$.) In addition to higher power dissipation, this remedy also requires a large V_{GS} for M_2.

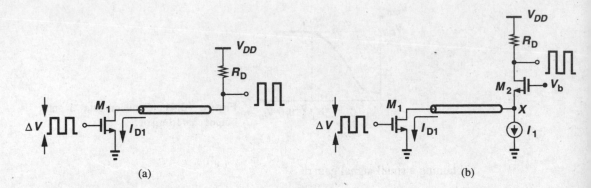

(a)   (b)

Figure 3.42

The key point in this example is that, while the overall voltage gain in both arrangements equals $-g_{m1}R_D$, the value of R_D in Fig. 3.42(b) can be much greater than 50 Ω without introducing reflections at point X. Thus, the common-gate circuit can provide a much higher voltage gain than that in Fig. 3.42(a).

Now let us study the common-gate topology in a more general case, taking into account both the output impedance of the transistor and the impedance of the signal source. Depicted in Fig. 3.43(a), the circuit can be analyzed with the aid of its equivalent shown

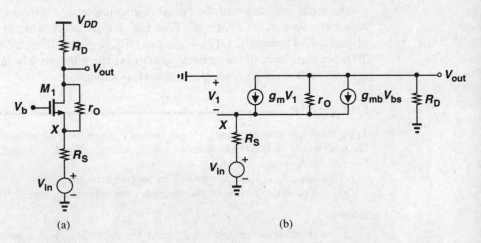

(a)   (b)

Figure 3.43   (a) CG stage with finite output resistance, (b) small-signal equivalent circuit.

in Fig. 3.43(b). Noting that the current flowing through R_S is equal to $-V_{out}/R_D$, we have:

$$V_1 - \frac{V_{out}}{R_D}R_S + V_{in} = 0. \tag{3.101}$$

Moreover, since the current through r_O is equal to $-V_{out}/R_D - g_m V_1 - g_{mb} V_1$, we can write

$$r_O \left(\frac{-V_{out}}{R_D} - g_m V_1 - g_{mb} V_1 \right) - \frac{V_{out}}{R_D} R_S + V_{in} = V_{out}. \tag{3.102}$$

Upon substitution for V_1 from (3.102), (3.101) reduces to

$$r_O \left[\frac{-V_{out}}{R_D} - (g_m + g_{mb}) \left(V_{out} \frac{R_S}{R_D} - V_{in} \right) \right] - \frac{V_{out} R_S}{R_D} + V_{in} = V_{out}. \tag{3.103}$$

It follows that

$$\frac{V_{out}}{V_{in}} = \frac{(g_m + g_{mb})r_O + 1}{r_O + (g_m + g_{mb})r_O R_S + R_S + R_D} R_D. \tag{3.104}$$

Note the similarity between (3.104) and (3.71). The gain of the common-gate stage is slightly higher due to body effect.

Example 3.11

Calculate the voltage gain of the circuit shown in Fig. 3.44(a) if $\lambda \neq 0$ and $\gamma \neq 0$.

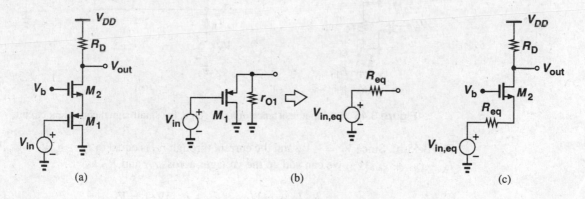

Figure 3.44

Solution

We first find the Thevenin equivalent of M_1. As shown in Fig. 3.44(b), M_1 operates as a source follower and the equivalent Thevenin voltage is given by

$$V_{in,eq} = \frac{r_{O1} \left\| \dfrac{1}{g_{mb1}} \right.}{r_{O1} \left\| \dfrac{1}{g_{mb1}} + \dfrac{1}{g_{m1}} \right.} V_{in}, \tag{3.105}$$

and the equivalent Thevenin resistance is

$$R_{eq} = r_{O1} \left\| \frac{1}{g_{mb1}} \right\| \frac{1}{g_{m1}}. \tag{3.106}$$

Redrawing the circuit as in Fig. 3.44(c), we use (3.104) to write

$$\frac{V_{out}}{V_{in}} = \frac{(g_{m2} + g_{mb2})r_{O2} + 1}{r_{O2} + [1 + (g_{m2} + g_{mb2})r_{O2}]\left(r_{O1} \left\| \frac{1}{g_{mb1}} \right\| \frac{1}{g_{m1}}\right) + R_D} R_D \frac{r_{O1} \left\| \frac{1}{g_{mb1}}\right.}{r_{O1} \left\| \frac{1}{g_{mb1}} + \frac{1}{g_{m2}}\right.}. \tag{3.107}$$

The input and output impedances of the common-gate topology are also of interest. To obtain the impedance seen at the source [Fig. 3.45(a)], we use the equivalent circuit in

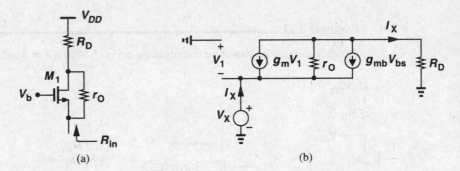

(a)   (b)

Figure 3.45   (a) Input resistance of a CG stage, (b) small-signal equivalent circuit.

Fig. 3.45(b). Since $V_1 = -V_X$ and the current through r_O is equal to $I_X + g_m V_1 + g_{mb} V_1 = I_X - (g_m + g_{mb})V_X$, we can add up the voltages across r_O and R_D as

$$R_D I_X + r_O[I_X - (g_m + g_{mb})V_X] = V_X. \tag{3.108}$$

Thus,

$$\frac{V_X}{I_X} = \frac{R_D + r_O}{1 + (g_m + g_{mb})r_O} \tag{3.109}$$

$$\approx \frac{R_D}{(g_m + g_{mb})r_O} + \frac{1}{g_m + g_{mb}}, \tag{3.110}$$

if $(g_m + g_{mb})r_O \gg 1$. This result reveals that the drain impedance is divided by $(g_m + g_{mb})r_O$ when seen at the source. This is particularly important in short-channel devices because of their low intrinsic gain. Two special cases of (3.109) are worth studying. First, suppose

$R_D = 0$. Then,

$$\frac{V_X}{I_X} = \frac{r_O}{1 + (g_m + g_{mb})r_O} \qquad (3.111)$$

$$= \frac{1}{\dfrac{1}{r_O} + g_m + g_{mb}}, \qquad (3.112)$$

which is simply the impedance seen at the source of a source follower, a predictable result because if $R_D = 0$, the circuit configuration is the same as in Fig. 3.31(a).

Second, let us replace R_D with an ideal current source. Equation (3.110) predicts that the input impedance approaches *infinity*. While somewhat surprising, this result can be explained with the aid of Fig. 3.46. Since the total current through the transistor is fixed and equal to I_1, a change in the source potential cannot change the device current, and hence $I_X = 0$. In other words, the input impedance of a common-gate stage is relatively low *only* if the load impedance connected to the drain is small.

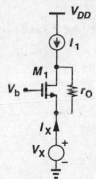

Figure 3.46 Input resistance of a CG stage with ideal current source load.

Example 3.12

Calculate the voltage gain of a common-gate stage with a current-source load [Fig. 3.47(a)].

Solution

Letting R_D approach infinity in (3.104), we have

$$A_v = (g_m + g_{mb})r_O + 1. \qquad (3.113)$$

Interestingly, the gain does not depend on R_S. From our foregoing discussion, we recognize that if $R_D \to \infty$, so does the impedance seen at the source of M_1, and the small-signal voltage at node X becomes *equal* to V_{in}. We can therefore simplify the circuit as shown in Fig. 3.47(b), readily arriving at (3.113).

In order to calculate the output impedance of the common-gate stage, we use the circuit

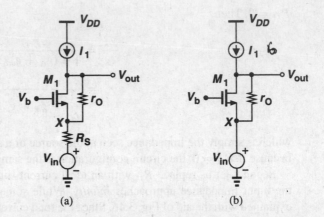

Figure 3.47

in Fig. 3.48. We note that the result is similar to that in Fig. 3.22 and hence

$$R_{out} = \{[1 + (g_m + g_{mb})r_O]R_S + r_O\}\|R_D. \tag{3.114}$$

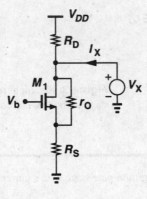

Figure 3.48   Calculation of output resistance of a CG stage.

Example 3.13

As seen in Example 3.10 the input signal of a common-gate stage may be a current rather than a voltage. Shown in Fig. 3.49 is such an arrangement. Calculate V_{out}/I_{in} and the output impedance of the circuit if the input current source exhibits an output impedance equal to R_P.

Solution

To find V_{out}/I_{in}, we replace I_{in} and R_P with a Thevenin equivalent and use (3.104) to write

$$\frac{V_{out}}{I_{in}} = \frac{(g_m + g_{mb})r_O + 1}{r_O + (g_m + g_{mb})r_O R_P + R_P + R_D} R_D R_P. \tag{3.115}$$

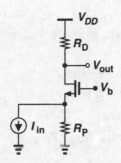

Figure 3.49

The output impedance is simply equal to

$$R_{out} = \{[1 + (g_m + g_{mb})r_O]R_P + r_O\}\|R_D. \tag{3.116}$$

3.5 Cascode Stage

As mentioned in Example 3.10 the input signal of a common-gate stage may be a current. We also know that a transistor in a common-source arrangement converts a voltage signal to a current signal. The cascade of a CS stage and a CG stage is called a "cascode"[1] topology, providing many useful properties. Fig. 3.50 shows the basic configuration: M_1 generates a small-signal drain current proportional to V_{in} and M_2 simply routes the current to R_D.

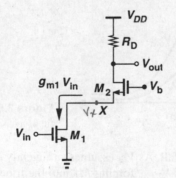

Figure 3.50 Cascode stage.

We call M_1 the input device and M_2 the cascode device. Note that in this example, M_1 and M_2 carry equal currents. As we describe the attributes of the circuit in this section, many advantages of the cascode topology over a simple common-source stage become evident.

First, let us study the bias conditions of the cascode. For M_1 to operate in saturation, $V_X \geq V_{in} - V_{TH1}$. If M_1 and M_2 are both in saturation, then V_X is determined primarily by

---

[1]The term *cascode* is believed to be the acronym for "cascaded triodes," possibly invented in vacuum tube days.

V_b: $V_X = V_b - V_{GS2}$. Thus, $V_b - V_{GS2} \geq V_{in} - V_{TH1}$ and hence $V_b > V_{in} + V_{GS2} - V_{TH1}$ (Fig. 3.51). For M_2 to be saturated, $V_{out} \geq V_b - V_{TH2}$, that is, $V_{out} \geq V_{in} - V_{TH1} + V_{GS2} -$

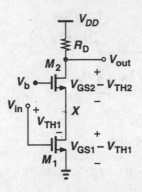

Figure 3.51 Allowable voltages in cascode stage.

V_{TH2} if V_b is chosen to place M_1 at the edge of saturation. Consequently, the minimum output level for which both transistors operate in saturation is equal to the overdrive voltage of M_1 plus that of M_2. In other words, addition of M_2 to the circuit reduces the output voltage swing by at least the overdrive voltage of M_2. We also say M_2 is "stacked" on top of M_1.

We now analyze the large-signal behavior of the cascode stage shown in Fig. 3.50 as V_{in} goes from zero to V_{DD}. For $V_{in} \leq V_{TH1}$, M_1 and M_2 are off, $V_{out} = V_{DD}$, and $V_X \approx V_b - V_{TH2}$ (if subthreshold conduction is neglected) (Fig. 3.52). As V_{in} exceeds V_{TH1}, M_1 begins to draw current, and V_{out} drops. Since I_{D2} increases, V_{GS2} must increase

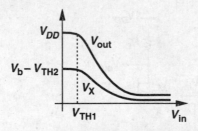

Figure 3.52 Input-output characteristic of a cascode stage.

as well, causing V_X to fall. As V_{in} assumes sufficiently large values, two effects occur: (1) V_X drops below V_{in} by V_{TH1}, forcing M_1 into the triode region; (2) V_{out} drops below V_b by V_{TH2}, driving M_2 into the triode region. Depending on the device dimensions and the values of R_D and V_b, one effect may occur before the other. For example, if V_b is relatively low, M_1 may enter the triode region first. Note that if M_2 goes into deep triode region, V_X and V_{out} become nearly equal.

Let us now consider the small-signal characteristics of a cascode stage, assuming both transistors operate in saturation. If $\lambda = 0$, the voltage gain is equal to that of a common-source stage because the drain current produced by the input device must flow through the cascode device. Illustrated in the equivalent circuit of Fig. 3.53, this result is independent of the transconductance and body effect of M_2.

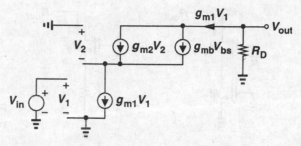

Figure 3.53 Small-signal equivalent circuit of cascode stage.

Example 3.14

Calculate the voltage gain of the circuit shown in Fig. 3.54 if $\lambda = 0$.

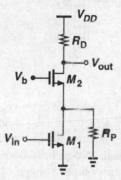

Figure 3.54

Solution

The small-signal drain current of M_1, $g_{m1} V_{in}$, is divided between R_P and the impedance seen looking into the source of M_2, $1/(g_{m2} + g_{mb2})$. Thus, the current flowing through M_2 is

$$I_{D2} = g_{m1} V_{in} \frac{(g_{m2} + g_{mb2}) R_P}{1 + (g_{m2} + g_{mb2}) R_P}. \qquad (3.117)$$

The voltage gain is therefore given by

$$A_v = -\frac{g_{m1}(g_{m2} + g_{mb2}) R_P R_D}{1 + (g_{m2} + g_{mb2}) R_P}. \qquad (3.118)$$

An important property of the cascode structure is its high output impedance. As illustrated in Fig. 3.55, for calculation of R_{out}, the circuit can be viewed as a common-source stage with a degeneration resistor equal to r_{O1}. Thus, from (3.60),

$$R_{out} = [1 + (g_{m2} + g_{mb2}) r_{O2}] r_{O1} + r_{O2}. \qquad (3.119)$$

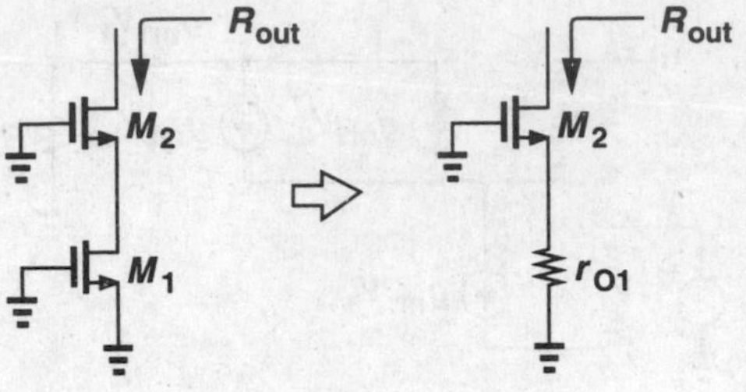

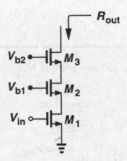

Figure 3.55 Calculation of output resistance of cascode stage.

Figure 3.56 Triple cascode.

Assuming $g_m r_O \gg 1$, we have $R_{out} \approx (g_{m2} + g_{mb2}) r_{O2} r_{O1}$. That is, M_2 boosts the output impedance of M_1 by a factor of $(g_{m2} + g_{mb2}) r_{O2}$. As shown in Fig. 3.56, cascoding can be extended to three or more stacked devices to achieve a higher output impedance, but the required additional voltage headroom makes such configurations less attractive. For example, the minimum output voltage of a triple cascode is equal to the sum of three overdrive voltages.

To appreciate the usefulness of a high output impedance, recall from the lemma in Section 3.2.3 that the voltage gain can be written as $G_m R_{out}$. Since G_m is typically determined by the transconductance of a transistor, e.g., M_1 in Fig. 3.50, and hence bears trade-offs with the bias current and device capacitances, it is desirable to increase the voltage gain by maximizing R_{out}. Shown in Fig. 3.57 is an example. If both M_1 and M_2 operate in saturation,

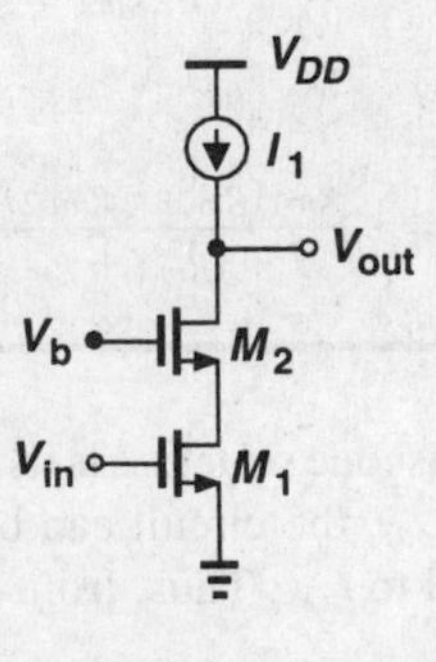

Figure 3.57 Cascode stage with current-source load.

then $G_m \approx g_{m1}$ and $R_{out} \approx (g_{m2} + g_{mb2}) r_{O2} r_{O1}$, yielding $A_v = (g_{m2} + g_{mb2}) r_{O2} g_{m1} r_{O1}$.

Thus, the maximum voltage gain is roughly equal to the *square* of the intrinsic gain of the transistors.

Example 3.15

Calculate the exact voltage gain of the circuit shown in Fig. 3.57.

Solution

The actual G_m of the stage is slightly less than g_{m1} because a fraction of the small-signal current produced by M_1 is shunted to ground by r_{O1}. As depicted in Fig. 3.58:

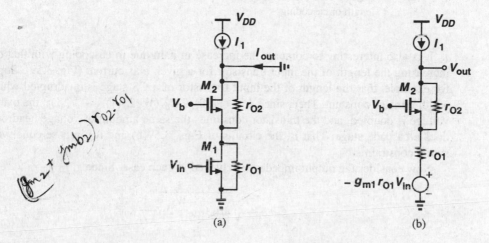

$(g_{m2} + g_{mb2})r_{O2} \cdot r_{O1}$

(a)   (b)

Figure 3.58

$$I_{out} = g_{m1}V_{in}\frac{r_{O1}}{r_{O1} + \dfrac{1}{g_{m2} + g_{mb2}} \| r_{O2}}. \tag{3.120}$$

It follows that the overall transconductance is equal to

$$G_m = \frac{g_{m1}r_{O1}[r_{O2}(g_{m2} + g_{mb2}) + 1]}{r_{O1}r_{O2}(g_{m2} + g_{mb2}) + r_{O1} + r_{O2}}, \tag{3.121}$$

and hence the voltage gain is given by

$$|A_v| = G_m R_{out} \tag{3.122}$$

$$= g_{m1}r_{O1}[(g_{m2} + g_{mb2})r_{O2} + 1]. \tag{3.123}$$

If we had assumed $G_m \approx g_m$, then $|A_v| \approx g_{m1}\{[1 + (g_{m2} + g_{mb2})r_{O2}]r_{O1} + r_{O2}\}$.

Another approach to calculating the voltage gain is to replace V_{in} and M_1 by a Thevenin equivalent, reducing the circuit to a common-gate stage. Illustrated in Fig. 3.58(b), this method in conjunction with (3.104) gives the same result as (3.123).

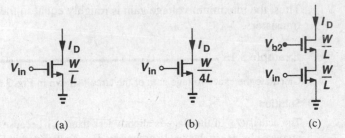

(a)   (b)   (c)

Figure 3.59 Increasing output impedance by increasing the device length or cascoding.

It is also interesting to compare the increase in gain due to cascoding with that due to increasing the length of the input transistor for a given bias current (Fig. 3.59). Suppose, for example, that the length of the input transistor of a CS stage is quadrupled while the width remains constant. Then, since $I_D = (1/2)\mu_n C_{ox}(W/L)(V_{GS} - V_{TH})^2$, the overdrive voltage is doubled, and the transistor consumes the same amount of voltage headroom as does a cascode stage. That is, the circuits of Figs. 3.59(b) and (c) impose equal voltage swing constraints.

Now consider the output impedance achieved in each case. Since

$$g_m r_O = \sqrt{2\mu_n C_{ox}\frac{W}{L}I_D\frac{1}{\lambda I_D}}, \tag{3.124}$$

and $\lambda \propto 1/L$, quadrupling L only doubles the value of $g_m r_O$ while cascoding results in an output impedance of roughly $(g_m r_O)^2$. Note that the transconductance of M_1 in Fig. 3.59(b) is half that in Fig. 3.59(c), leading to higher noise (Chapter 7).

A cascode structure need not operate as an amplifier. Another popular application of this topology is in building constant current sources. The high output impedance yields a current source closer to the ideal, but at the cost of voltage headroom. For example, current source I_1 in Fig. 3.57 can be implemented with a PMOS cascode (Fig. 3.60), exhibiting an impedance equal to $[1 + (g_{m3} + g_{mb3})r_{O3}]r_{O4} + r_{O3}$. If the gate bias voltages are chosen

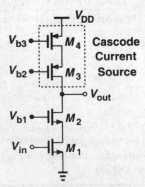

Figure 3.60 NMOS cascode amplifier with PMOS cascode load.

properly, the maximum output swing is equal to $V_{DD} - (V_{GS1} - V_{TH1}) - (V_{GS2} - V_{TH2}) - |V_{GS3} - V_{TH3}| - |V_{GS4} - V_{TH4}|$.

We calculate the voltage gain with the aid of the lemma illustrated in Fig. 3.25. Writing $G_m \approx g_{m1}$ and

$$R_{out} = \{[1 + (g_{m2} + g_{mb2})r_{O2}]r_{O1} + r_{O2}\} \| \{[1 + (g_{m3} + g_{mb3})r_{O3}]r_{O4} + r_{O3}\}, \quad (3.125)$$

we have $|A_v| \approx g_{m1}R_{out}$. For typical values, we approximate the voltage gain as

$$|A_v| \approx g_{m1}[(g_{m2}r_{O2}r_{O1}) \| (g_{m3}r_{O3}r_{O4})]. \quad (3.126)$$

Shielding Property   Recall from Fig. 3.23 that the high output impedance arises from the fact that if the output node voltage is changed by ΔV, the resulting change at the source of the cascode device is much less. In a sense, the cascode transistor "shields" the input device from voltage variations at the output. The shielding property of cascodes proves useful in many circuits.

Example 3.16 ───

Two identical NMOS transistors are used as constant current sources in a system [Fig. 3.61(a)]. However, due to internal circuitry of the system, V_X is higher than V_Y by ΔV.

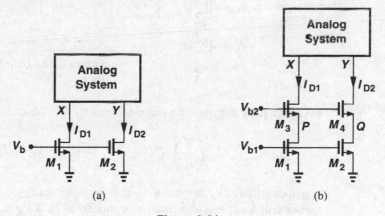

(a)   (b)

Figure 3.61

(a) Calculate the resulting difference between I_{D1} and I_{D2} if $\lambda \neq 0$.
(b) Add cascode devices to M_1 and M_2 and repeat part (a).

Solution

(a) We have

$$I_{D1} - I_{D2} = \frac{1}{2}\mu_n C_{ox}\frac{W}{L}(V_b - V_{TH})^2(\lambda V_{DS1} - \lambda V_{DS2}) \quad (3.127)$$

$$= \frac{1}{2}\mu_n C_{ox}\frac{W}{L}(V_b - V_{TH})^2(\lambda \Delta V). \quad (3.128)$$

(b) As shown in Fig. 3.61(b), cascoding reduces the effect of V_X and V_Y upon I_{D1} and I_{D2}, respectively. As depicted in Fig. 3.23 and implied by Eq. (3.63), a difference ΔV between V_X and V_Y translates to a difference ΔV_{PQ} between P and Q equal to

$$\Delta V_{PQ} = \Delta V \frac{r_{O1}}{[1 + (g_{m3} + g_{mb3})r_{O3}]r_{O1} + r_{O3}} \tag{3.129}$$

$$\approx \frac{\Delta V}{(g_{m3} + g_{mb3})r_{O3}}. \tag{3.130}$$

Thus,

$$I_{D1} - I_{D2} = \frac{1}{2}\mu_n C_{ox} \frac{W}{L}(V_b - V_{TH})^2 \frac{\lambda \Delta V}{(g_{m3} + g_{mb3})r_{O3}}. \tag{3.131}$$

In other words, cascoding reduces the mismatch between I_{D1} and I_{D2} by $(g_{m3} + g_{mb3})r_{O3}$.

The shielding property of cascodes diminishes if the cascode device enters the triode region. To understand why, let us consider the circuit in Fig. 3.62, assuming V_X decreases from a large positive value. As V_X falls below $V_{b2} - V_{TH2}$, M_2 requires a greater gate-source

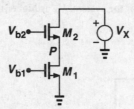

Figure 3.62   Output swing of cascode stage.

overdrive so as to sustain the current drawn by M_1. We can write

$$I_{D2} = \frac{1}{2}\mu_n C_{ox}\left(\frac{W}{L}\right)_2 [2(V_{b2} - V_P - V_{TH2})(V_X - V_P) - (V_X - V_P)^2], \tag{3.132}$$

concluding that as V_X decreases, V_P also drops so that I_{D2} remains constant. In other words, variation of V_X is less attenuated as it appears at P. If V_X falls sufficiently, V_P goes below $V_{b1} - V_{TH1}$, driving M_1 into the triode region.

3.5.1 Folded Cascode

The idea behind the cascode structure is to convert the input voltage to a current and apply the result to a common-gate stage. However, the input device and the cascode device need not be of the same type. For example, as depicted in Fig. 3.63(a), a PMOS-NMOS combination performs the same function. In order to bias M_1 and M_2, a current source must be added as in Fig. 3.63(b). The small-signal operation is as follows. If V_{in} becomes more positive, $|I_{D1}|$ decreases, forcing I_{D2} to increase and hence V_{out} to drop. The voltage gain and output impedance of the circuit can be obtained as calculated for the NMOS-NMOS

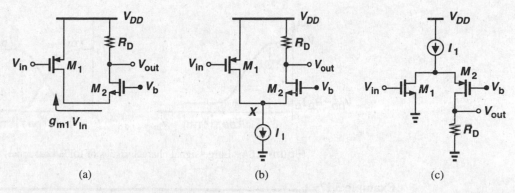

Figure 3.63 (a) Simple folded cascode, (b) folded cascode with proper biasing, (c) folded cascode with NMOS input.

cascode of Fig. 3.50. Shown in Fig. 3.63(c) is an NMOS-PMOS cascode. The advantages and disadvantages of these types will be explained later.

The structures of Figs. 3.63(b) and (c) are called "folded cascode" stages because the small-signal current is "folded" up [in Fig. 3.63(b)] or down [in Fig. 3.63(c)]. Note that the total bias current in this case must be higher than that in Fig. 3.50 to achieve comparable performance.

It is instructive to examine the large-signal behavior of a folded-cascode stage. Suppose in Fig. 3.63(b), V_{in} decreases from V_{DD} to zero. For $V_{in} > V_{DD} - |V_{TH1}|$, M_1 is off and M_2 carries all of I_1,[2] yielding $V_{out} = V_{DD} - I_1 R_D$. For $V_{in} < V_{DD} - |V_{TH1}|$, M_1 turns on in saturation, giving

$$I_{D2} = I_1 - \frac{1}{2}\mu_p C_{ox} \left(\frac{W}{L}\right)_1 (V_{DD} - V_{in} - |V_{TH1}|)^2. \qquad (3.133)$$

As V_{in} drops, I_{D2} decreases further, falling to zero if $I_{D1} = I_1$. For this to occur:

$$\frac{1}{2}\mu_p C_{ox} \left(\frac{W}{L}\right)_1 (V_{DD} - V_{in1} - |V_{TH1}|)^2 = I_1. \qquad (3.134)$$

Thus,

$$V_{in1} = V_{DD} - \sqrt{\frac{2I_1}{\mu_p C_{ox}(W/L)_1}} - |V_{TH1}|. \qquad (3.135)$$

If V_{in} falls below this level, I_{D1} tends to be greater than I_1 and M_1 enters the triode region so as to allow $I_{D1} = I_1$. The result is plotted in Fig. 3.64.

What happens to V_X in the above test? As I_{D2} drops, V_X rises, reaching $V_b - V_{TH2}$ for $I_{D2} = 0$. As M_1 enters the triode region, V_X approaches V_{DD}.

---

[2]If I_1 is excessively large, M_2 may enter deep triode region, possibly driving I_1 into the triode region as well.

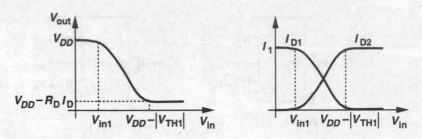

Figure 3.64   Large-signal characteristics of folded cascode.

Example 3.17

Calculate the output impedance of the folded cascode shown in Fig. 3.65 where M_3 operates as a current source.

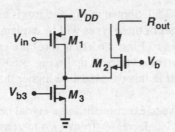

Figure 3.65

Solution

Using (3.60), we have

$$R_{out} = [1 + (g_{m2} + g_{mb2})r_{O2}](r_{O1} \| r_{O3}) + r_{O2}. \tag{3.136}$$

Thus, the circuit exhibits an output impedance lower than that of a nonfolded cascode.

In order to achieve a high voltage gain, the load of a folded cascode can be implemented as a cascode itself (Fig. 3.66). This structure is studied more extensively in Chapter 9.

Throughout this chapter, we have attempted to *increase* the output resistance of voltage amplifiers so as to obtain a high gain. This may seem to make the speed of the circuit quite susceptible to the load capacitance. However, as explained in Chapter 8, a high output impedance per se does not pose a serious issue if the amplifier is placed in a proper feedback loop.

3.6 Choice of Device Models

In this chapter, we have developed various expressions for the properties of single-stage amplifiers. For example, the voltage gain of a degenerated common-source stage can be as simple as $-R_D/(R_S + g_m^{-1})$ or as complex as Eq. (3.71). How does one choose a sufficiently accurate device model or expression?

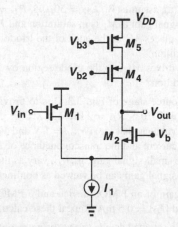

Figure 3.66 Folded cascode with cascode load.

The proper choice is not always straightforward and it is a skill gained by practice, experience, and intuition. However, some general principles in choosing the model for each transistor can be followed. First, break the circuit down into a number of familiar topologics. Next, concentrate on each subcircuit and use the simplest transistor model (a single voltage-dependent current source for FETs operating in saturation) for all transistors. If the drain of a device is connected to a high impedance (e.g., the drain of another), then add r_O to its model. At this point, the basic properties of most circuits can be determined by inspection. In a second, more accurate iteration, the body effect of devices whose source or bulk is not at ac ground can be included as well.

For bias calculations, it is usually adequate to neglect channel-length modulation and body effect in the first pass. These effects do introduce some error but they can be included in the next iteration step—after the basic properties are understood.

In today's analog design, simulation of circuits is essential because the behavior of short-channel MOSFETs cannot be predicted accurately by hand calculations. Nonetheless, if the designer avoids a simple and intuitive analysis of the circuit and hence skips the task of gaining insight, then he/she cannot interpret the simulation results intelligently. For this reason, we say, "Don't let the computer think for you."

Problems

Unless otherwise stated, in the following problems, use the device data shown in Table 2.1 and assume $V_{DD} = 3$ V where necessary. All device dimensions are effective values and in microns.

3.1. For the circuit of Fig. 3.9, calculate the small-signal voltage gain if $(W/L)_1 = 50/0.5$, $(W/L)_2 = 10/0.5$, and $I_{D1} = I_{D2} = 0.5$ mA. What is the gain if M_2 is implemented as a diode-connected PMOS device (Fig. 3.12)?

3.2. In the circuit of Fig. 3.14, assume $(W/L)_1 = 50/0.5$, $(W/L)_2 = 50/2$, and $I_{D1} = I_{D2} = 0.5$ mA when both devices are in saturation. Recall that $\lambda \propto 1/L$.
(a) Calculate the small-signal voltage gain.
(b) Calculate the maximum output voltage swing while both devices are saturated.

3.3. In the circuit of Fig. 3.3(a), assume $(W/L)_1 = 50/0.5$, $R_D = 2$ kΩ, and $\lambda = 0$.
   (a) What is the small-signal gain if M_1 is in saturation and $I_D = 1$ mA?
   (b) What input voltage places M_1 at the edge of the triode region? What is the small-signal gain under this condition?
   (c) What input voltage drives M_1 into the triode region by 50 mV? What is the small-signal gain under this condition?

3.4. Suppose the common-source stage of Fig. 3.3(a) is to provide an output swing from 1 V to 2.5 V. Assume $(W/L)_1 = 50/0.5$, $R_D = 2$ kΩ, and $\lambda = 0$.
   (a) Calculate the input voltages that yield $V_{out} = 1$ V and $V_{out} = 2.5$ V.
   (b) Calculate the drain current and the transconductance of M_1 for both cases.
   (c) How much does the small-signal gain, $g_m R_D$, vary as the output goes from 1 V to 2.5 V? (Variation of small-signal gain can be viewed as nonlinearity.)

3.5. Calculate the intrinsic gain of an NMOS device and a PMOS device operating in saturation with $W/L = 50/0.5$ and $|I_D| = 0.5$ mA. Repeat these calculations if $W/L = 100/1$.

3.6. Plot the intrinsic gain of a satuated device versus the gate-source voltage if **(a)** the drain current is constant, **(b)** W and L are constant.

3.7. Plot the intrinsic gain of a saturated device versus W/L if **(a)** the gate-source voltage is constant, **(b)** the drain current is constant.

3.8. An NMOS transistor with $W/L = 50/0.5$ is biased with $V_G = +1.2$ V and $V_S = 0$. The drain voltage is varied from 0 to 3 V.
   (a) Assuming the bulk voltage is zero, plot the intrinsic gain versus V_{DS}.
   (b) Repeat part **(a)** for a bulk voltage of -1 V.

3.9. For an NMOS device operating in saturation, plot g_m, r_O, and $g_m r_O$ as the bulk voltage goes from 0 to $-\infty$ while other terminal voltages remain constant.

3.10. Consider the circuit of Fig. 3.9 with $(W/L)_1 = 50/0.5$ and $(W/L)_2 = 10/0.5$. Assume $\lambda = \gamma = 0$.
   (a) At what input voltage is M_1 at the edge of the triode region? What is the small-signal gain under this condition?
   (b) What input voltage drives M_1 into the triode region by 50 mV? What is the small-signal gain under this condition?

3.11. Repeat Problem 3.10 if body effect is not neglected.

3.12. In the circuit of Fig. 3.13, $(W/L)_1 = 20/0.5$, $I_1 = 1$ mA, and $I_S = 0.75$ mA. Assuming $\lambda = 0$, calculate $(W/L)_2$ such that M_1 is at the edge of the triode region. What is the small-signal voltage gain under this condition?

3.13. Plot the small-signal gain of the circuit shown in Fig. 3.13 as I_S goes from 0 to $0.75I_1$. Assume M_1 is always saturated and neglect channel-length modulation and body effect.

3.14. The circuit of Fig. 3.14 is designed to provide an output voltage swing of 2.2 V with a bias current of 1 mA and a small-signal voltage gain of 100. Calculate the dimensions of M_1 and M_2.

3.15. Sketch V_{out} versus V_{in} for the circuits of Fig. 3.67 as V_{in} varies from 0 to V_{DD}. Identify important transition points.

3.16. Sketch V_{out} versus V_{in} for the circuits of Fig. 3.68 as V_{in} varies from 0 to V_{DD}. Identify important transition points.

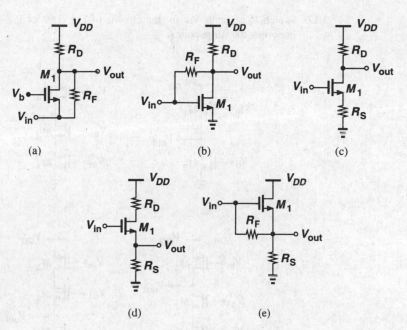

(a)   (b)   (c)

(d)   (e)

Figure 3.67

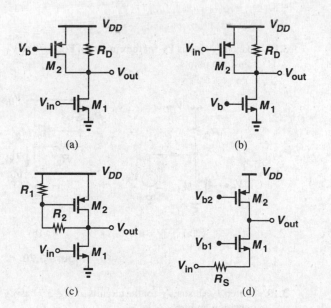

(a)   (b)

(c)   (d)

Figure 3.68

3.17. Sketch V_{out} versus V_{in} for the circuits of Fig. 3.69 as V_{in} varies from 0 to V_{DD}. Identify important transition points.

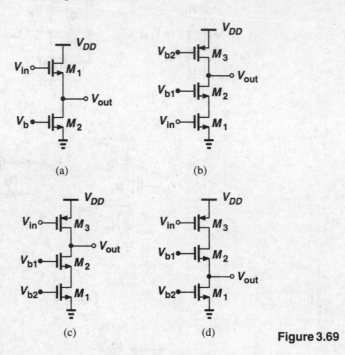

(a)   (b)

(c)   (d)

Figure 3.69

3.18. Sketch I_X versus V_X for the circuits of Fig. 3.70 as V_X varies from 0 to V_{DD}. Identify important transition points.

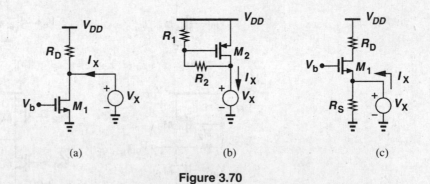

(a)   (b)   (c)

Figure 3.70

3.19. Sketch I_X versus V_X for the circuits of Fig. 3.71 as V_X varies from 0 to V_{DD}. Identify important transition points.

3.20. Assuming all MOSFETs are in saturation, calculate the small-signal voltage gain of each circuit in Fig. 3.72 ($\lambda \neq 0$, $\gamma = 0$).

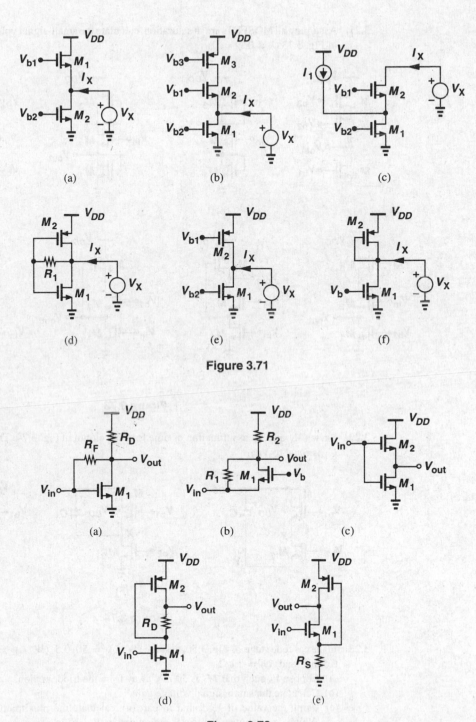

Figure 3.71

Figure 3.72

3.21. Assuming all MOSFETs are in saturation, calculate the small-signal voltage gain of each circuit in Fig. 3.73 ($\lambda \neq 0$, $\gamma = 0$).

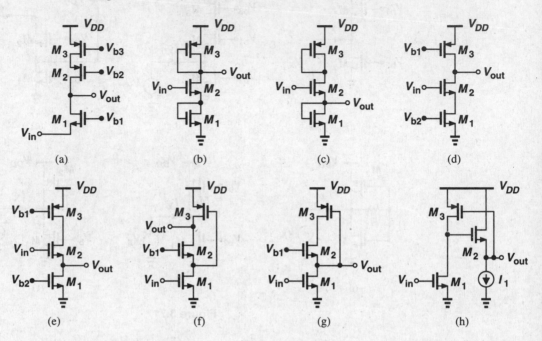

Figure 3.73

3.22. Sketch V_X and V_Y as a function of time for each circuit in Fig. 3.74. The initial voltage across C_1 is equal to V_{DD}.

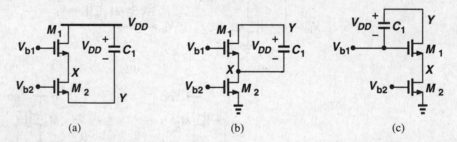

Figure 3.74

3.23. In the cascode stage of Fig. 3.50, assume $(W/L)_1 = 50/0.5$, $(W/L)_2 = 10/0.5$, $I_{D1} = I_{D2} = 0.5$ mA, and $R_D = 1$ kΩ.
  (a) Choose V_b such that M_1 is 50 mV away from the triode region.
  (b) Calculate the small-signal voltage gain.
  (c) Using the value of V_b found in part (a), calculate the maximum output voltage swing. Which device enters the triode region first as V_{out} falls?
  (d) Calculate the swing at node X for the maximum output swing obtained above.

3.24. Consider the circuit of Fig. 3.16 with $(W/L)_1 = 50/0.5$, $R_D = 2 \, k\Omega$, and $R_S = 200 \, \Omega$.
  (a) Calculate the small-signal voltage gain if $I_D = 0.5 \, mA$.
  (b) Assuming $\lambda = \gamma = 0$, calculate the input voltage that places M_1 at the edge of the triode region. What is the gain under this condition?

3.25. Suppose the circuit of Fig. 3.15 is designed for a voltage gain of 5. If $(W/L)_1 = 20/0.5$, $I_{D1} = 0.5 \, mA$, and $V_b = 0 \, V$.
  (a) Calculate the aspect ratio of M_2.
  (b) What input level places M_1 at the edge of the triode region. What is the small-signal gain under this condition?
  (c) What input level places M_2 at the edge of the saturation region? What is the small-signal gain under this condition?

3.26. Sketch the small-signal voltage gain of the circuit shown in Fig. 3.15 as V_b varies from 0 to V_{DD}. Consider two cases: (a) M_1 enters the triode region before M_2 is saturated; (b) M_1 enters the triode region after M_2 is saturated.

3.27. A source follower can operate as a level shifter. Suppose the circuit of Fig. 3.30(b) is designed to shift the voltage level by 1 V, i.e., $V_{in} - V_{out} = 1 \, V$.
  (a) Calculate the dimensions of M_1 and M_2 if $I_{D1} = I_{D2} = 0.5 \, mA$, $V_{GS2} - V_{GS1} = 0.5 \, V$, and $\lambda = \gamma = 0$.
  (b) Repeat part (a) if $\gamma = 0.45 \, V^{-1}$ and $V_{in} = 2.5 \, V$. What is the minimum input voltage for which M_2 remains saturated?

3.28. Sketch the small-signal gain, V_{out}/V_{in}, of the cascode stage shown in Fig. 3.50 as V_b goes from 0 to V_{DD}. Assume $\lambda = \gamma = 0$.

3.29. The cascode of Fig. 3.60 is designed to provide an output swing of 1.9 V with a bias current of 0.5 mA. If $\gamma = 0$ and $(W/L)_{1-4} = W/L$, calculate V_{b1}, V_{b2}, and W/L. What is the voltage gain if $L = 0.5 \, \mu m$?

Chapter 4

Differential Amplifiers

The differential amplifier is among the most important circuit inventions, dating back to the vacuum tube era. Offering many useful properties, differential operation has become the dominant choice in today's high-performance analog and mixed-signal circuits.

This chapter deals with the analysis and design of CMOS differential amplifiers. Following a review of single-ended and differential operation, we describe the basic differential pair, and analyze both the large-signal and the small-signal behavior. Next, we introduce the concept of common-mode rejection and formulate it for differential amplifiers. We then study differential pairs with diode-connected and current-source loads as well as differential cascode stages. Finally, we describe the Gibert cell.

4.1 Single-Ended and Differential Operation

A single-ended signal is defined as one that is measured with respect to a fixed potential, usually the ground. A differential signal is defined as one that is measured between two nodes that have *equal* and *opposite* signal excursions around a fixed potential. In the strict sense, the two nodes must also exhibit equal impedances to that potential. Fig. 4.1 illustrates the two types of signals conceptually. The "center" potential in differential signaling is called the "common-mode" (CM) level.

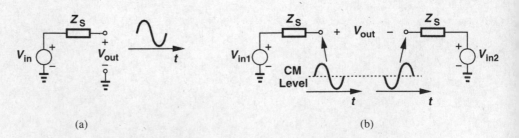

(a)   (b)

Figure 4.1 (a) Single-ended and (b) differential signals.

An important advantage of differential operation over single-ended signaling is higher immunity to "environmental" noise. Consider the example depicted in Fig. 4.2, where two adjacent lines in a circuit carry a small, sensitive signal and a large clock waveform. Due to capacitive coupling between the lines, transitions on line L_2 corrupt the signal on line L_1. Now suppose, as shown in Fig. 4.2(b), the sensitive signal is distributed as two equal and opposite phases. If the clock line is placed midway between the two, the transitions disturb the differential phases by equal amounts, leaving the *difference* intact. Since the common-mode level of the two phases is disturbed but the differential output is not corrupted, we say this arrangement "rejects" common-mode noise.

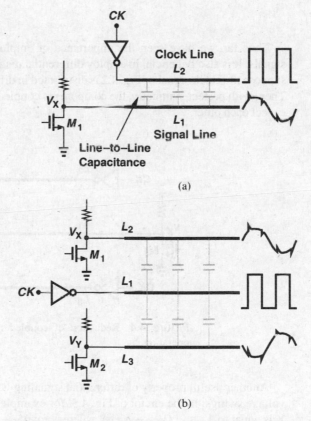

(a)

(b)

Figure 4.2 (a) Corruption of a signal due to coupling,
(b) reduction of coupling by differential operation.

Another example of common-mode rejection occurs with noisy supply voltages. In Fig. 4.3(a), if V_{DD} varies by ΔV, then V_{out} changes by approximately the same amount, i.e., the output is quite susceptible to noise on V_{DD}. Now consider the circuit in Fig. 4.3(b). Here, if the circuit is symmetric, noise on V_{DD} affects V_X and V_Y but not $V_X - V_Y = V_{out}$. Thus, the circuit of Fig. 4.3(b) is much more robust to supply noise.

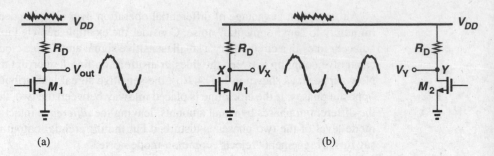

Figure 4.3   Effect of supply noise on (a) a single-ended circuit, (b) a differential circuit.

Thus far, we have seen the importance of employing differential paths for sensitive signals. It is also beneficial to employ differential distribution for *noisy lines*. For example, suppose the clock signal of Fig. 4.2 is distributed in differential form on two lines (Fig. 4.4). Then, with perfect symmetry, the components coupled from CK and $\overline{CK}$ to the signal line cancel each other.

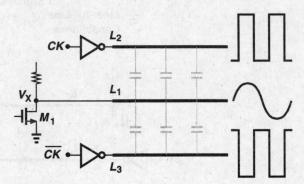

Figure 4.4   Reduction of coupled noise by differential operation.

Another useful property of differential signaling is the increase in maximum achievable voltage swings. In the circuit of Fig. 4.3, for example, the maximum output swing at X or Y is equal to $V_{DD} - (V_{GS} - V_{TH})$, whereas for $V_X - V_Y$, the peak-to-peak swing is equal to $2[V_{DD} - (V_{GS} - V_{TH})]$.

Other advantages of differential circuits over single-ended counterparts include simpler biasing and higher linearity (Chapter 13).

While it may seem that differential circuits occupy twice as much area as single-ended alternatives, in practice this is a minor drawback. Also, the suppression of nonideal effects by differential operation often results in a *smaller* area than that of a brute-force single-ended design. Furthermore, the numerous advantages of differential operation by far outweigh the possible increase in the area.

4.2 Basic Differential Pair

How do we amplify a differential signal? As suggested by the observations in the previous section, we may incorporate two identical single-ended signal paths to process the two phases [Fig. 4.5(a)]. Such a circuit indeed offers some of the advantages of differential

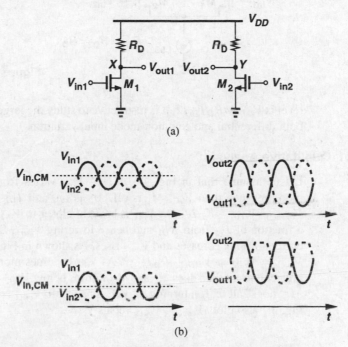

Figure 4.5   (a) Simple differential circuit, (b) illustration of sensitivity to the input common-mode level.

signaling: high rejection of supply noise, higher output swings, etc. But what happens if V_{in1} and V_{in2} experience a large common-mode disturbance or simply do not have a well-defined common-mode dc level? As the input CM level, $V_{in,CM}$, changes, so do the bias currents of M_1 and M_2, thus varying both the transconductance of the devices and the output CM level. The variation of the transconductance in turn leads to a change in the small-signal gain while the departure of the output CM level from its ideal value lowers the maximum allowable output swings. For example, as shown in Fig. 4.5(b), if the input CM level is excessively low, the minimum values of V_{in1} and V_{in2} may in fact turn off M_1 and M_2, leading to severe clipping at the output. Thus, it is important that the bias currents of the devices have minimal dependence on the input CM level.

A simple modification can resolve the above issue. Shown in Fig. 4.6, the "differential pair"[1] employs a current source I_{SS} to make $I_{D1} + I_{D2}$ independent of $V_{in,CM}$. Thus, if $V_{in1} = V_{in2}$, the bias current of each transistor equals $I_{SS}/2$ and the output common-mode

---

[1]Also called a source-coupled pair or (in the British literature) a long-tailed pair.

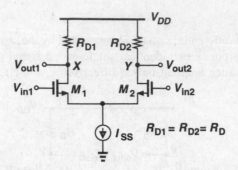

Figure 4.6 Basic differential pair.

level is $V_{DD} - R_D I_{SS}/2$. It is instructive to study the large-signal behavior of the circuit for both differential and common-mode input variations.

4.2.1 Qualitative Analysis

Let us assume that in Fig. 4.6, $V_{in1} - V_{in2}$ varies from $-\infty$ to $+\infty$. If V_{in1} is much more negative than V_{in2}, M_1 is off, M_2 is on, and $I_{D2} = I_{SS}$. Thus, $V_{out1} = V_{DD}$ and $V_{out2} = V_{DD} - R_D I_{SS}$. As V_{in1} is brought closer to V_{in2}, M_1 gradually turns on, drawing a fraction of I_{SS} from R_{D1} and hence lowering V_{out1}. Since $I_{D1} + I_{D2} = I_{SS}$, the drain current of M_2 decreases and V_{out2} rises. As shown in Fig. 4.7(a), for $V_{in1} = V_{in2}$, we have $V_{out1} = V_{out2} = V_{DD} - R_D I_{SS}/2$. As V_{in1} becomes more positive than V_{in2}, M_1 carries a greater current than does M_2 and V_{out1} drops below V_{out2}. For sufficiently large $V_{in1} - V_{in2}$, M_1 "hogs" all of I_{SS}, turning M_2 off. As a result, $V_{out1} = V_{DD} - R_D I_{SS}$ and $V_{out2} = V_{DD}$. Fig. 4.7 also plots $V_{out1} - V_{out2}$ versus $V_{in1} - V_{in2}$.

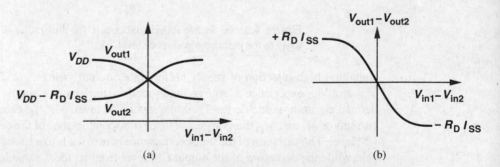

Figure 4.7 Input-output characteristics of a differential pair.

The foregoing analysis reveals two important attributes of the differential pair. First, the maximum and minimum levels at the output are well-defined (V_{DD} and $V_{DD} - R_D I_{SS}$, respectively) and independent of the input CM level. Second, the small-signal gain (the slope of $V_{out1} - V_{out2}$ versus $V_{in1} - V_{in2}$) is maximum for $V_{in1} = V_{in2}$, gradually falling to zero as $|V_{in1} - V_{in2}|$ increases. In other words, the circuit becomes more nonlinear as the input voltage swing increases. For $V_{in1} = V_{in2}$, we say the circuit is in equilibrium.

Now let us consider the common-mode behavior of the circuit. As mentioned earlier, the role of the tail current source is to suppress the effect of input CM level variations on the operation of M_1 and M_2 and the output level. Does this mean that $V_{in,CM}$ can assume arbitrarily low or high values? To answer this question, we set $V_{in1} = V_{in2} = V_{in,CM}$ and vary $V_{in,CM}$ from 0 to V_{DD}. Fig. 4.8(a) shows the circuit with I_{SS} implemented by an NFET. Note that the symmetry of the pair requires that $V_{out1} = V_{out2}$.

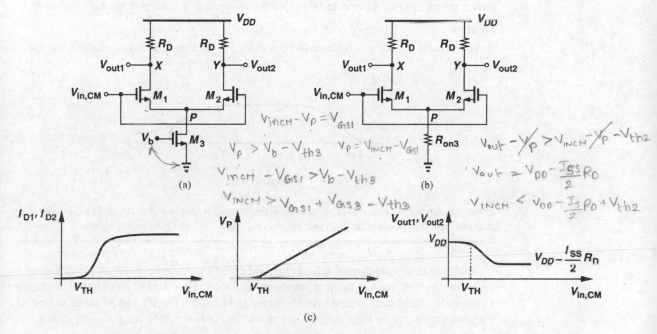

The following handwritten annotations appear on the figure:

$$V_{INCM} - V_P = V_{GS1}$$
$$V_P > V_b - V_{th3}$$
$$V_{INCM} - V_{GS1} > V_b - V_{th3}$$
$$V_{INCM} > V_{GS1} + V_{GS3} - V_{th3}$$

$$V_P = V_{INCM} - V_{GS1}$$

$$V_{out} - V_P > V_{INCM} - V_P - V_{th2}$$
$$V_{out} = V_{DD} - \frac{I_{SS}}{2}R_D$$
$$V_{INCM} < V_{DD} - \frac{I_S}{2}R_D + V_{th2}$$

(a)   (b)

(c)

Figure 4.8 (a) Differential pair sensing an input common-mode change, (b) equivalent circuit if M_3 operates in deep triode region, (c) common-mode input-output characteristics.

What happens if $V_{in,CM} = 0$? Since the gate potential of M_1 and M_2 is not more positive than their source potential, both devices are off, yielding $I_{D3} = 0$. This indicates that M_3 is in deep triode region because V_b is high enough to create an inversion layer in the transistor. With $I_{D1} = I_{D2} = 0$, the circuit is incapable of signal amplification, and $V_{out1} = V_{out2} = V_{DD}$.

Now suppose $V_{in,CM}$ becomes more positive. Modeling M_3 by a resistor as in Fig. 4.8(b), we note that M_1 and M_2 turn on if $V_{in,CM} \geq V_{TH}$. Beyond this point, I_{D1} and I_{D2} continue to increase and V_P also rises [Fig. 4.8(c)]. In a sense, M_1 and M_2 constitute a source follower, forcing V_P to track $V_{in,CM}$. For a sufficiently high $V_{in,CM}$, the drain-source voltage of M_3 exceeds $V_{GS3} - V_{TH3}$, allowing the device to operate in saturation. The total current through M_1 and M_2 then remains constant. We conclude that for proper operation, $V_{in,CM} \geq V_{GS1} + (V_{GS3} - V_{TH3})$.

What happens if $V_{in,CM}$ rises further? Since V_{out1} and V_{out2} are relatively constant, we expect that M_1 and M_2 enter the triode region if $V_{in,CM} > V_{out1} + V_{TH} = V_{DD} - R_D I_{SS}/2 + V_{TH}$. This sets an upper limit on the input CM level. In summary, the allowable value of

$V_{in,CM}$ is bounded as follows:

$$V_{GS1} + (V_{GS3} - V_{TH3}) \le V_{in,CM} \le \min\left[V_{DD} - R_D\frac{I_{SS}}{2} + V_{TH}, \; V_{DD}\right]. \qquad (4.1)$$

Example 4.1

Sketch the small-signal differential gain of a differential pair as a function of the input CM level.

Solution

As shown in Fig. 4.9, the gain begins to increase as $V_{in,CM}$ exceeds V_{TH}. After the tail current source

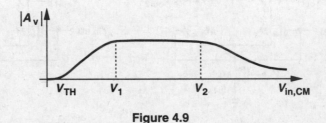

Figure 4.9

enters saturation ($V_{in,CM} = V_1$), the gain remains relatively constant. Finally, if $V_{in,CM}$ is so high that the input transistors enter the triode region ($V_{in,CM} = V_2$), the gain begins to fall.

With our understanding of differential and common-mode behavior of the differential pair, we can now answer another important question: How large can the output voltage swings of a differential pair be? As illustrated in Fig. 4.10, for M_1 and M_2 to be saturated, each output can go as high as V_{DD} but as low as approximately $V_{in,CM} - V_{TH}$. In other

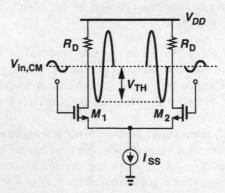

Figure 4.10 Maximum allowable output swings in a differential pair.

words, the higher the input CM level, the smaller the allowable output swings. For this reason, it is desirable to choose a relatively low $V_{in,CM}$, but the preceding stage may not provide such a level easily.

An interesting trade-off exists in the circuit of Fig. 4.10 between the maximum value of $V_{in,CM}$ and the differential *gain*. Similar to a simple common-source stage (Chapter 3), the gain of a differential pair is a function of the dc drop across the load resistors. Thus, if $R_D I_{SS}/2$ is large, $V_{in,CM}$ must remain close to ground potential.

4.2.2 Quantitative Analysis

We now quantify the behavior of a MOS differential pair as a function of the input differential voltage. We begin with large-signal analysis to arrive at an expression for the plots shown in Fig. 4.7.

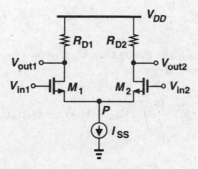

Figure 4.11   Differential pair.

For the differential pair in Fig. 4.11, we have $V_{out1} = V_{DD} - R_{D1} I_{D1}$ and $V_{out2} = V_{DD} - R_{D2} I_{D2}$, i.e., $V_{out1} - V_{out2} = R_{D2} I_{D2} - R_{D1} I_{D1} = R_D(I_{D2} - I_{D1})$ if $R_{D1} = R_{D2} = R_D$. Thus, we simply calculate I_{D1} and I_{D2} in terms of V_{in1} and V_{in2}, assuming the circuit is symmetric, M_1 and M_2 are saturated, and $\lambda = 0$. Since the voltage at node P is equal to $V_{in1} - V_{GS1}$ and $V_{in2} - V_{GS2}$,

$$V_{in1} - V_{in2} = V_{GS1} - V_{GS2}. \qquad V_P = V_{in1} - V_{GS1} \qquad (4.2)$$
$$= V_{in2} - V_{GS2}$$

For a square-law device, we have:

$$(V_{GS} - V_{TH})^2 = \frac{I_D}{\frac{1}{2}\mu_n C_{ox}\frac{W}{L}}, \qquad (4.3)$$

and, therefore,

$$V_{GS} = \sqrt{\frac{2I_D}{\mu_n C_{ox}\frac{W}{L}}} + V_{TH}. \qquad (4.4)$$

It follows from (4.2) and (4.4) that

$$V_{in1} - V_{in2} = \sqrt{\frac{2I_{D1}}{\mu_n C_{ox}\frac{W}{L}}} - \sqrt{\frac{2I_{D2}}{\mu_n C_{ox}\frac{W}{L}}}. \qquad (4.5)$$

Our objective is to calculate the differential output current, $I_{D1} - I_{D2}$. Squaring the two sides of (4.5) and recognizing that $I_{D1} + I_{D2} = I_{SS}$, we obtain

$$(V_{in1} - V_{in2})^2 = \frac{2}{\mu_n C_{ox} \dfrac{W}{L}} (I_{SS} - 2\sqrt{I_{D1} I_{D2}}). \tag{4.6}$$

That is,

$$\frac{1}{2}\mu_n C_{ox} \frac{W}{L}(V_{in1} - V_{in2})^2 - I_{SS} = -2\sqrt{I_{D1} I_{D2}}. \tag{4.7}$$

Squaring the two sides again and noting that $4I_{D1}I_{D2} = (I_{D1} + I_{D2})^2 - (I_{D1} - I_{D2})^2 = I_{SS}^2 - (I_{D1} - I_{D2})^2$, we arrive at

$$(I_{D1} - I_{D2})^2 = -\frac{1}{4}\left(\mu_n C_{ox} \frac{W}{L}\right)^2 (V_{in1} - V_{in2})^4 + I_{SS}\mu_n C_{ox} \frac{W}{L}(V_{in1} - V_{in2})^2. \tag{4.8}$$

Thus,

$$I_{D1} - I_{D2} = \frac{1}{2}\mu_n C_{ox} \frac{W}{L}(V_{in1} - V_{in2})\sqrt{\frac{4I_{SS}}{\mu_n C_{ox} \dfrac{W}{L}} - (V_{in1} - V_{in2})^2}. \tag{4.9}$$

As expected, $I_{D1} - I_{D2}$ is an odd function of $V_{in1} - V_{in2}$, falling to zero for $V_{in1} = V_{in2}$. As $|V_{in1} - V_{in2}|$ increases from zero, $|I_{D1} - I_{D2}|$ also increases because the factor preceding the square root rises more rapidly than the argument in the square root drops.[2]

Before examining (4.9) further, it is instructive to calculate the slope of the characteristic, i.e., the equivalent G_m of M_1 and M_2. Denoting $I_{D1} - I_{D2}$ and $V_{in1} - V_{in2}$ by ΔI_D and ΔV_{in}, respectively, the reader can show that

$$\frac{\partial \Delta I_D}{\partial \Delta V_{in}} = \frac{1}{2}\mu_n C_{ox} \frac{W}{L} \cdot \frac{\dfrac{4I_{SS}}{\mu_n C_{ox} W/L} - 2\Delta V_{in}^2}{\sqrt{\dfrac{4I_{SS}}{\mu_n C_{ox} W/L} - \Delta V_{in}^2}}. \tag{4.10}$$

For $\Delta V_{in} = 0$, $G_m = \sqrt{\mu_n C_{ox}(W/L)I_{SS}}$. Moreover, since $V_{out1} - V_{out2} = R_D \Delta I = R_D G_m \Delta V_{in}$, we can write the small-signal differential voltage gain of the circuit in the equilibrium condition as

$$|A_v| = \sqrt{\mu_n C_{ox} \frac{W}{L} I_{SS}} R_D. \tag{4.11}$$

---

[2]It is interesting to note that, even though I_{D1} and I_{D2} are *even* functions of their respective gate-source voltages, $I_{D1} - I_{D2}$ is an odd function of $V_{in1} - V_{in2}$. This effect is studied in Chapter 13.

Equation (4.10) also suggests that G_m falls to zero for $\Delta V_{in} = \sqrt{2I_{SS}/(\mu_n C_{ox} W/L)}$. As we will see below, this value of ΔV_{in} plays an important role in the operation of the circuit.

Let us now examine Eq. (4.9) more closely. It appears that the argument in the square root drops to zero for $\Delta V_{in} = \sqrt{4I_{SS}/(\mu_n C_{ox} W/L)}$, implying that ΔI_D crosses zero at *two* different values of ΔV_{in}. This was not predicted in our qualitative analysis in Fig. 4.7. This conclusion, however, is incorrect. To understand why, recall that (4.9) was derived with the assumption that both M_1 and M_2 are on. In reality, as ΔV_{in} exceeds a limit, one transistor carries the entire I_{SS}, turning off the other.[3] Denoting this value by ΔV_{in1}, we have $I_{D1} = I_{SS}$ and $\Delta V_{in1} = V_{GS1} - V_{TH}$ because M_2 is nearly off. It follows that

$$\Delta V_{in1} = \sqrt{\frac{2I_{SS}}{\mu_n C_{ox} \dfrac{W}{L}}}. \qquad (4.12)$$

For $\Delta V_{in} > \Delta V_{in1}$, M_2 is off and (4.9) does not hold. As mentioned above, G_m falls to zero for $\Delta V_{in} = \Delta V_{in1}$. Figure 4.12 plots the behavior.

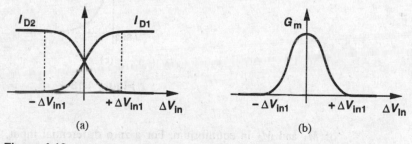

(a)   (b)

Figure 4.12 Variation of drain currents and overall transconductance of a differential pair versus input voltage.

Example 4.2

Plot the input-output characteristic of a differential pair as the device width and the tail current vary.

Solution

Consider the characteristic shown in Fig. 4.13(a). As W/L increases, ΔV_{in1} decreases, narrowing the input range across which both devices are on [Fig. 4.13(b)]. As I_{SS} increases, both the input range and the output current swing increase [Fig. 4.13(c)]. Intuitively, we expect the circuit to become more linear as I_{SS} increases or W/L decreases.

The value of ΔV_{in1} given by (4.12) in essence represents the maximum differential input that the circuit can "handle." It is possible to relate ΔV_{in1} to the overdrive voltage

---

[3] We neglect subthreshold conduction here.

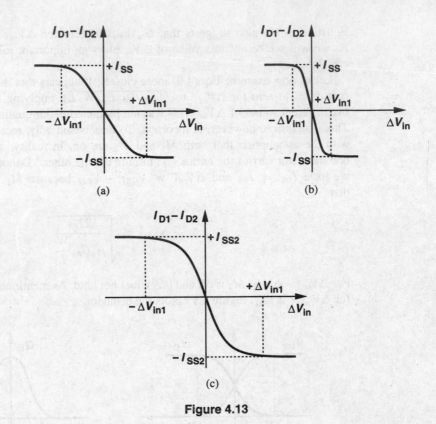

Figure 4.13

of M_1 and M_2 in equilibrium. For a zero differential input, $I_{D1} = I_{D2} = I_{SS}/2$, and hence

$$(V_{GS} - V_{TH})_{1,2} = \sqrt{\frac{I_{SS}}{\mu_n C_{ox} \dfrac{W}{L}}}. \qquad (4.13)$$

Thus, the equilibrium overdrive is equal to $\Delta V_{in1}/\sqrt{2}$. The point is that increasing ΔV_{in1} to make the circuit more linear inevitably increases the overdrive voltage of M_1 and M_2. For a given I_{SS}, this is accomplished only by reducing W/L and hence the transconductance of the transistors.

We now study the small-signal behavior of differential pairs. As depicted in Fig. 4.14, we apply small signals V_{in1} and V_{in2} and assume M_1 and M_2 are saturated. What is the differential voltage gain, $V_{out}/(V_{in1} - V_{in2})$? Recall from Eq. (4.11) that this quantity equals $\sqrt{\mu_n C_{ox} I_{SS} W/L} R_D$. Since in the vicinity of equilibrium, each transistor carries approximately $I_{SS}/2$, this expression reduces to $g_m R_D$, where g_m denotes the transconductance of M_1 and M_2. To arrive at the same result by small-signal analysis, we employ two different methods, each providing insight into the circuit's operation. We assume $R_{D1} = R_{D2} = R_D$.

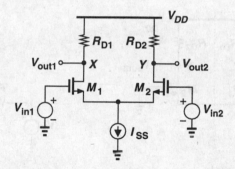

Figure 4.14 Differential pair with small-signal inputs.

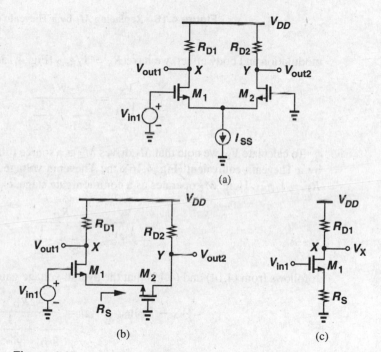

Figure 4.15 (a) Differential pair sensing one input signal, (b) circuit of (a) viewed as a CS stage degenerated by M_2, (c) equivalent circuit of (b).

Method I The circuit of Fig. 4.14 is driven by two independent signals. Thus, the output can be computed by superposition.

Let us set V_{in2} to zero and find the effect of V_{in1} at X and Y [Fig. 4.15(a)]. To obtain V_X, we note that M_1 forms a common-source stage with a degeneration resistance equal to the impedance seen looking into the source of M_2 [Fig. 4.15(b)]. Neglecting channel-length

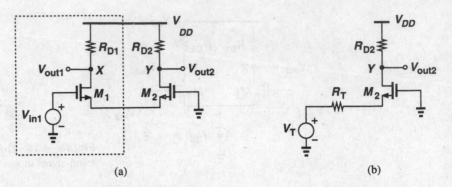

Figure 4.16   Replacing M_1 by a Thevenin equivalent.

modulation and body effect, we have $R_S = 1/g_{m2}$ [Fig. 4.15(c)] and

$$\frac{V_X}{V_{in1}} = \frac{-R_D}{\dfrac{1}{g_{m1}} + \dfrac{1}{g_{m2}}}. \tag{4.14}$$

To calculate V_Y, we note that M_1 drives M_2 as a source follower and replace V_{in1} and M_1 by a Thevenin equivalent (Fig. 4.16): the Thevenin voltage $V_T = V_{in1}$ and the resistance $R_T = 1/g_{m1}$. Here, M_2 operates as a common-gate stage, exhibiting a gain equal to

$$\frac{V_Y}{V_{in1}} = \frac{R_D}{\dfrac{1}{g_{m2}} + \dfrac{1}{g_{m1}}}. \tag{4.15}$$

It follows from (4.14) and (4.15) that the overall voltage gain for V_{in1} is

$$(V_X - V_Y)|_{\text{Due to } V_{in1}} = \frac{-2R_D}{\dfrac{1}{g_{m1}} + \dfrac{1}{g_{m2}}} V_{in1}, \tag{4.16}$$

which, for $g_{m1} = g_{m2} = g_m$ reduces to

$$(V_X - V_Y)|_{\text{Due to } V_{in1}} = -g_m R_D V_{in1}. \tag{4.17}$$

By virtue of symmetry, the effect of V_{in2} at X and Y is identical to that of V_{in1} except for a change in the polarities:

$$(V_X - V_Y)|_{\text{Due to } V_{in2}} = g_m R_D V_{in2}. \tag{4.18}$$

Adding the two sides of (4.17) and (4.18) to perform superposition, we have

$$\frac{(V_X - V_Y)_{tot}}{V_{in1} - V_{in2}} = -g_m R_D. \tag{4.19}$$

Comparison of (4.17), (4.18), and (4.19) indicates that the magnitude of the differential gain is equal to $g_m R_D$ regardless of how the inputs are applied: in Figs. 4.15 and 4.16, the input is applied to only one side whereas in Fig. 4.14 the input is the *difference* between two sources. It is also important to recognize that if the output is single-ended, i.e., it is sensed between X or Y and ground, the gain is halved.

Example 4.3

In the circuit of Fig. 4.17, M_2 is twice as wide as M_1. Calculate the small-signal gain if the bias values of V_{in1} and V_{in2} are equal.

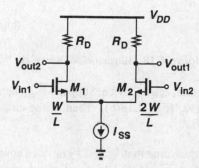

Figure 4.17

Solution

If the gates of M_1 and M_2 are at the same dc potential, then $V_{GS1} = V_{GS2}$ and $I_{D2} = 2I_{D1} = 2I_{SS}/3$. Thus, $g_{m1} = \sqrt{2\mu_n C_{ox}(W/L)I_{SS}/3}$ and $g_{m2} = \sqrt{2\mu_n C_{ox}(2W/L)2I_{SS}/3} = 2g_{m1}$. Following the same procedure as above, the reader can show that

$$|A_v| = \frac{2R_D}{\dfrac{1}{g_{m1}} + \dfrac{1}{2g_{m1}}} \tag{4.20}$$

$$= \frac{4}{3} g_{m1} R_D. \tag{4.21}$$

Note that, for a given I_{SS}, this value is lower than the gain of a symmetric differential pair (with $2W/L$ for each device) [Eq. (4.19)] because g_{m1} is smaller.

How does the gain of a differential pair compare with that of a common-source stage? For a given *total* bias current, the value of g_m in (4.19) is $1/\sqrt{2}$ times that of a single transistor biased at I_{SS} with the same dimensions. Thus, the total gain is proportionally less. Equivalently, for given device dimensions and load impedance, a differential pair achieves the same gain as a CS stage at the cost of twice the bias current.

Method II   If a fully-symmetric differential pair senses differential inputs (i.e., the two inputs change by equal and opposite amounts from the equilibrium condition), then the concept of "half circuit" can be applied. We first prove a lemma.

Lemma.   Consider the symmetric circuit shown in Fig. 4.18(a), where D_1 and D_2 represent

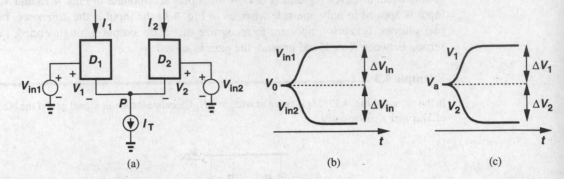

Figure 4.18   Illustration of why node P is a virtual ground.

any three-terminal active device. Suppose V_{in1} changes from V_0 to $V_0 + \Delta V_{in}$ and V_{in2} from V_0 to $V_0 - \Delta V_{in}$ [Fig. 4.18(b)]. Then, if the circuit remains linear, V_P does not change. Assume $\lambda = 0$.

Proof.   Let us assume that V_1 and V_2 have an equilibrium value of V_a and change by ΔV_1 and ΔV_2, respectively [Fig. 4.18(c)]. The output currents therefore change by $g_m \Delta V_1$ and $g_m \Delta V_2$. Since $I_1 + I_2 = I_T$, we have $g_m \Delta V_1 + g_m \Delta V_2 = 0$, i.e., $\Delta V_1 = -\Delta V_2$. We also know $V_{in1} - V_1 = V_{in2} - V_2$, and hence $V_0 + \Delta V_{in} - (V_a + \Delta V_1) = V_0 - \Delta V_{in} - (V_a + \Delta V_2)$. Consequently, $2\Delta V_{in} = \Delta V_1 - \Delta V_2 = 2\Delta V_1$. In other words, if V_{in1} and V_{in2} change by $+\Delta V_{in}$ and $-\Delta V_{in}$, respectively, then V_1 and V_2 change by the same values, i.e., a differential change in the inputs is simply "absorbed" by V_1 and V_2. In fact, since $V_P = V_{in1} - V_1$, and since V_1 exhibits the same change as V_{in1}, V_P does not change.   ❑

The proof of the foregoing lemma can also be invoked from symmetry. As long as the operation remains linear so that the difference between the bias currents of D_1 and D_2 is negligible, the circuit is symmetric. Thus, V_P cannot "favor" the change at one input and "ignore" the other.

From yet another point of view, the effect of D_1 and D_2 at node P can be represented by Thevenin equivalents (Fig. 4.19). If V_{T1} and V_{T2} change by equal and opposite amounts and R_{T1} and R_{T2} are equal, then V_P remains constant. We emphasize that this is valid if the changes are small such that we can assume $R_{T1} = R_{T2}$.[4]

The above lemma greatly simplifies the small-signal analysis of differential amplifiers. As shown in Fig. 4.20, since V_P experiences no change, node P can be considered "ac ground" and the circuit can be decomposed into two separate halves, hence the term "half-circuit concept" [1]. We can write $V_X / V_{in1} = -g_m R_D$ and $V_Y / (-V_{in1}) = -g_m R_D$, where V_{in1} and $-V_{in1}$ denote the voltage *change* on each side. Thus, $(V_X - V_Y)/(2V_{in1}) = -g_m R_D$.

---

[4]It is also possible to derive an expression for the large-signal behavior of V_P and prove that for small $V_{in1} - V_{in2}$, V_P remains constant. We defer this calculation to Chapter 14.

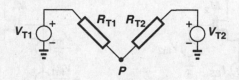

Figure 4.19 Replacing each half of a differential pair by a Thevenin equivalent.

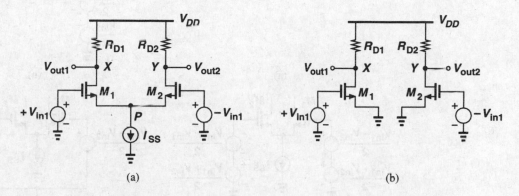

(a)   (b)

Figure 4.20 Application of the half-circuit concept.

Example 4.4

Calculate the differential gain of the circuit of Fig. 4.20(a) if $\lambda \neq 0$.

Solution

Applying the half-circuit concept as illustrated in Fig. 4.21, we have $V_X/V_{in1} = -g_m(R_D \| r_{O1})$ and $V_Y/(-V_{in1}) = -g_m(R_D \| r_{O2})$, thus arriving at $(V_X - V_Y)/(2V_{in1}) = -g_m(R_D \| r_O)$, where $r_O = r_{O1} = r_{O2}$. Note that Method I would require lengthy calculations here.

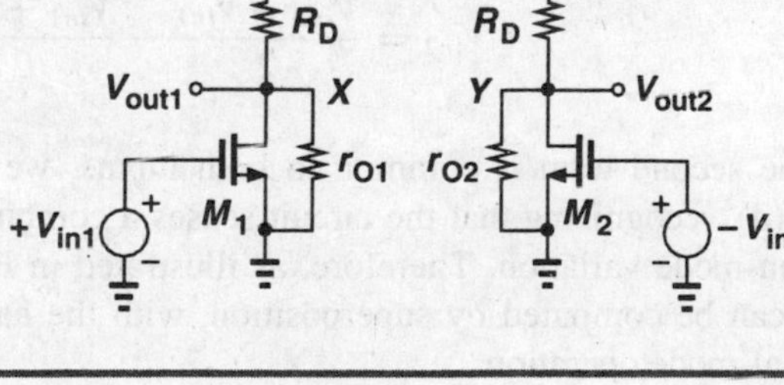

Figure 4.21

The half-circuit concept provides a powerful technique for analyzing symmetric differential pairs with fully differential inputs. But what happens if the two inputs are not fully

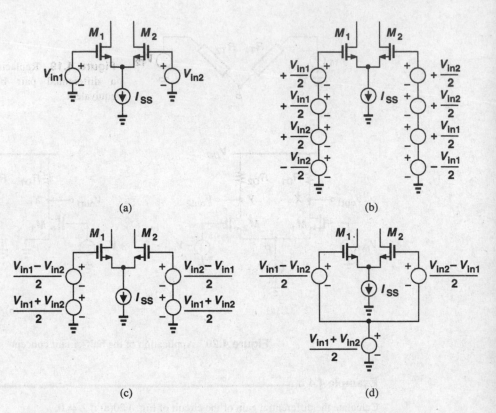

Figure 4.22 Conversion of arbitrary inputs to differential and common-mode components.

differential [Fig. 4.22(a)]? As depicted in Figs. 4.22(b) and (c), the two inputs V_{in1} and V_{in2} can be viewed as

$$V_{in1} = \frac{V_{in1} - V_{in2}}{2} + \frac{V_{in1} + V_{in2}}{2} \tag{4.22}$$

$$V_{in2} = \frac{V_{in2} - V_{in1}}{2} + \frac{V_{in1} + V_{in2}}{2}. \tag{4.23}$$

Since the second term is common to both inputs, we obtain the equivalent circuit in Fig. 4.22(d), recognizing that the circuit senses a combination of a differential input and a common-mode variation. Therefore, as illustrated in Fig. 4.23, the effect of each type of input can be computed by superposition, with the half-circuit concept applied to the differential-mode operation.

Example 4.5 ───

In the circuit of Fig. 4.20(a), calculate V_X and V_Y if $V_{in1} \neq -V_{in2}$ and $\lambda \neq 0$.

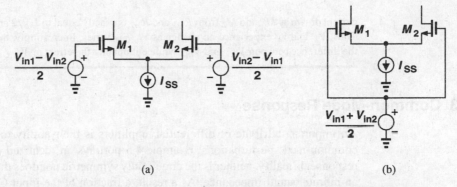

(a)   (b)

Figure 4.23 Superposition for differential and common-mode signals.

Solution

For differential-mode operation, we have from Fig. 4.24(a)

$$V_X = -g_m(R_D \| r_{O1})\frac{V_{in1} - V_{in2}}{2} \tag{4.24}$$

$$V_Y = -g_m(R_D \| r_{O2})\frac{V_{in2} - V_{in1}}{2}. \tag{4.25}$$

That is,

$$V_X - V_Y = -g_m(R_D \| r_O)(V_{in1} - V_{in2}), \tag{4.26}$$

which is to be expected.

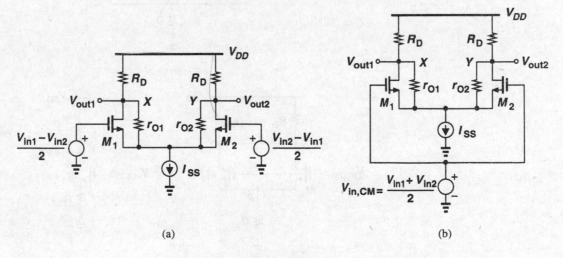

(a)   (b)

Figure 4.24

For common-mode operation, the circuit reduces to that in Fig. 4.24(b). How much do V_X and V_Y change as $V_{in,CM}$ changes? If the circuit is fully symmetric and I_{SS} an ideal current source, the current drawn by M_1 and M_2 from R_{D1} and R_{D2} is exactly equal to $I_{SS}/2$ and independent of $V_{in,CM}$. Thus, V_X and V_Y experience no change as $V_{in,CM}$ varies. Interestingly, the circuit simply amplifies the difference between V_{in1} and V_{in2} while eliminating the effect of $V_{in,CM}$.

4.3 Common-Mode Response

An important attribute of differential amplifiers is their ability to suppress the effect of common-mode perturbations. Example 4.5 portrays an idealized case of common-mode response. In reality, neither is the circuit fully symmetric nor does the current source exhibit an infinite output impedance. As a result, a fraction of the input CM variation appears at the output.

We first assume the circuit is symmetric but the current source has a finite output impedance, R_{SS} [Fig. 4.25(a)]. As $V_{in,CM}$ changes, so does V_P, thereby increasing the drain currents of M_1 and M_2 and lowering both V_X and V_Y. Owing to symmetry, V_X remains equal to V_Y and, as depicted in Fig. 4.25(b), the two nodes can be shorted together. Since M_1 and M_2 are now "in parallel," i.e., they share all of their respective terminals, the

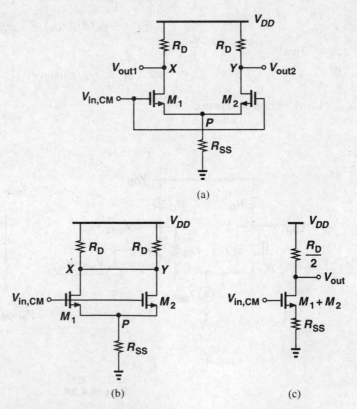

Figure 4.25 (a) Differential pair sensing CM input, (b) simplified version of (a), (c) equivalent circuit of (b).

circuit can be reduced to that in Fig. 4.25(c). Note that the compound device, $M_1 + M_2$, has twice the width and the bias current of each of M_1 and M_2 and, therefore, twice their transconductance. The CM gain of the circuit is thus equal to

$$A_{v,CM} = \frac{V_{out}}{V_{in,CM}} \tag{4.27}$$

$$= -\frac{R_D/2}{1/(2g_m) + R_{SS}}, \tag{4.28}$$

where g_m denotes the transconductance of each of M_1 and M_2 and $\lambda = \gamma = 0$.

What is the significance of this calculation? In a symmetric circuit, input CM variations disturb the bias points, altering the small-signal gain and possibly limiting the output voltage swings. This can be illustrated by an example.

Example 4.6 ——————————————————————————————————

The circuit of Fig. 4.26 uses a resistor rather than a current source to define a tail current of 1 mA.

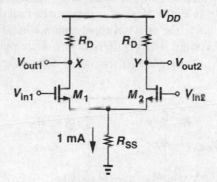

Figure 4.26

Assume $(W/L)_{1,2} = 25/0.5$, $\mu_n C_{ox} = 50\ \mu A/V^2$, $V_{TH} = 0.6$ V, $\lambda = \gamma = 0$, and $V_{DD} = 3$ V.

(a) What is the required input CM for which R_{SS} sustains 0.5 V?

(b) Calculate R_D for a differential gain of 5.

(c) What happens at the output if the input CM level is 50 mV higher than the value calculated in (a)?

Solution

(a) Since $I_{D1} = I_{D2} = 0.5$ mA, we have

$$V_{GS1} = V_{GS2} = \sqrt{\frac{2I_{D1}}{\mu_n C_{ox}\dfrac{W}{L}}} + V_{TH} \tag{4.29}$$

$$= 1.23\ V. \tag{4.30}$$

Thus, $V_{in,CM} = V_{GS1} + 0.5$ V= 1.73 V. Note that $R_{SS} = 500\ \Omega$.

(b) The transconductance of each device is $g_m = \sqrt{2\mu_n C_{ox}(W/L)I_{D1}} = 1/(632\ \Omega)$, requiring $R_D = 3.16$ kΩ for a gain of 5.

Note that the output bias level is equal to $V_{DD} - I_{D1}R_D = 1.42$ V. Since $V_{in,CM} = 1.73$ V and $V_{TH} = 0.6$ V, the transistors are 290 mV away from the triode region.

(c) If $V_{in,CM}$ increases by 50 mV, the equivalent circuit of Fig. 4.25(c) suggests that V_X and V_Y drop by

$$|\Delta V_{X,Y}| = \Delta V_{in,CM}\frac{R_D/2}{R_{SS} + 1/(2g_m)} \tag{4.31}$$

$$= 50\,\text{mV} \times 1.94 \tag{4.32}$$

$$= 96.8\,\text{mV}. \tag{4.33}$$

Now, M_1 and M_2 are only 143 mV away from the triode region because the input CM level has increased by 50 mV and the output CM level has decreased by 96.8 mV.

The foregoing discussion indicates that the finite output impedance of the tail current source results in some common-mode gain in a symmetric differential pair. Nonetheless, this is usually a minor concern. More troublesome is the variation of the *differential* output as a result of a change in $V_{in,CM}$, an effect that occurs because in reality the circuit is not fully symmetric, i.e., the two sides suffer from slight mismatches during manufacturing. For example, in Fig. 4.25(a), R_{D1} may not be exactly equal to R_{D2}.

We now study the effect of input common-mode variation if the circuit is asymmetric and the tail current source suffers from a finite output impedance. Suppose, as shown in Fig. 4.27, $R_{D1} = R_D$ and $R_{D2} = R_D + \Delta R_D$, where ΔR_D denotes a small mismatch and

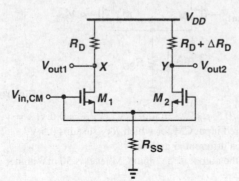

Figure 4.27 Common-mode response in the presence of resistor mismatch.

the circuit is otherwise symmetric. What happens to V_X and V_Y as $V_{in,CM}$ increases? Since M_1 and M_2 are identical, I_{D1} and I_{D2} increase by $[g_m/(1+2g_mR_{SS})]\Delta V_{in,CM}$, but V_X and V_Y change by different amounts:

$$\Delta V_X = -\Delta V_{in,CM}\frac{g_m}{1 + 2g_mR_{SS}}R_D \tag{4.34}$$

$$\Delta V_Y = -\Delta V_{in,CM}\frac{g_m}{1 + 2g_mR_{SS}}(R_D + \Delta R_D). \tag{4.35}$$

Thus, a common-mode change at the input introduces a *differential* component at the output. We say the circuit exhibits common-mode to differential conversion. This is a critical problem because if the input of a differential pair includes both a differential signal and common-mode noise, the circuit corrupts the amplified differential signal by the input CM change. The effect is illustrated in Fig. 4.28.

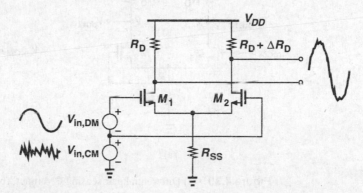

Figure 4.28   Effect of CM noise in the presence of resistor mismatch.

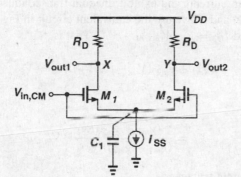

Figure 4.29   CM response with finite tail capacitance.

In summary, the common-mode response of differential pairs depends on the output impedance of the tail current source and asymmetries in the circuit, manifesting itself through two effects: variation of the output CM level (in the absence of mismatches) and conversion of input common-mode variations to differential components at the output. In analog circuits, the latter effect is much more severe than the former. For this reason, the common-mode response should usually be studied with mismatches taken into account.

How significant is common-mode to differential conversion? We make two observations. First, as the *frequency* of the CM disturbance increases, the total capacitance shunting the tail current source introduces larger tail current variations. Thus, even if the output *resistance* of the current source is high, common-mode to differential conversion becomes significant at high frequencies. Shown in Fig. 4.29, this capacitance arises from the parasitics of the current source itself as well as the source-bulk junctions of M_1 and M_2. Second, the asymmetry in the circuit stems from both the load resistors and the input transistors, the latter contributing a typically much greater mismatch.

Let us now study the asymmetry resulting from mismatches between M_1 and M_2 in Fig. 4.30(a). Owing to dimension and threshold voltage mismatches, the two transistors

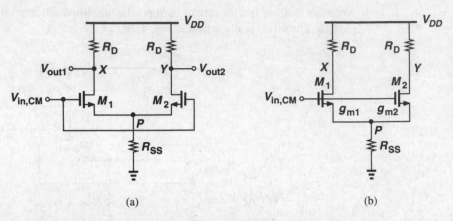

(a)

Figure 4.30 (a) Differential pair sensing CM input, (b) equivalent circuit of (a).

carry slightly different currents and exhibit unequal transconductances. To calculate the gain from $V_{in,CM}$ to X and Y, we use the equivalent circuit in Fig. 4.30(b), writing $I_{D1} = g_{m1}(V_{in,CM} - V_P)$ and $I_{D2} = g_{m2}(V_{in,CM} - V_P)$. That is,

$$(g_{m1} + g_{m2})(V_{in,CM} - V_P)R_{SS} = V_P, \tag{4.36}$$

and

$$V_P = \frac{(g_{m1} + g_{m2})R_{SS}}{(g_{m1} + g_{m2})R_{SS} + 1} V_{in,CM}. \tag{4.37}$$

We now obtain the output voltages as

$$V_X = -g_{m1}(V_{in,CM} - V_P)R_D \tag{4.38}$$

$$= \frac{-g_{m1}}{(g_{m1} + g_{m2})R_{SS} + 1} R_D V_{in,CM} \tag{4.39}$$

and

$$V_Y = -g_{m2}(V_{in,CM} - V_P)R_D \tag{4.40}$$

$$= \frac{-g_{m2}}{(g_{m1} + g_{m2})R_{SS} + 1} R_D V_{in,CM}. \tag{4.41}$$

The differential component at the output is therefore given by

$$V_X - V_Y = -\frac{g_{m1} - g_{m2}}{(g_{m1} + g_{m2})R_{SS} + 1} R_D V_{in,CM}. \tag{4.42}$$

In other words, the circuit converts input CM variations to a differential error by a factor

equal to

$$A_{CM-DM} = -\frac{\Delta g_m R_D}{(g_{m1} + g_{m2})R_{SS} + 1},$$

(4.43)

where A_{CM-DM} denotes common-mode to differential-mode conversion and $\Delta g_m = g_{m1} - g_{m2}$.

Example 4.7

Two differential pairs are cascaded as shown in Fig. 4.31. Transistors M_3 and M_4 suffer from a g_m

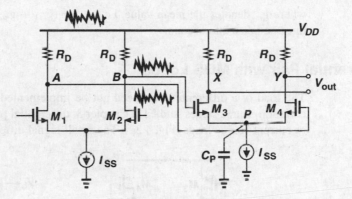

Figure 4.31

mismatch of Δg_m, the total parasitic capacitance at node P is represented by C_P, and the circuit is otherwise symmetric. What fraction of the supply noise appears as a differential component at the output? Assume $\lambda = \gamma = 0$.

Solution

Neglecting the capacitance at nodes A and B, we note that the supply noise appears at these nodes with no attenuation. Substituting $1/(C_P s)$ for R_{SS} in (4.43) and taking the magnitude, we have

$$|A_{CM-DM}| = \frac{\Delta g_m R_D}{\sqrt{1 + (g_{m3} + g_{m4})^2 \left|\dfrac{1}{C_P \omega}\right|^2}}.$$

(4.44)

The key point is that the effect becomes more noticeable as the supply noise frequency, ω, increases.

For meaningful comparison of differential circuits, the undesirable differential component produced by CM variations must be normalized to the wanted differential output resulting from amplification. We define the "common-mode rejection ratio" (CMRR) as

$$CMRR = \left|\frac{A_{DM}}{A_{CM-DM}}\right|.$$

(4.45)

If only g_m mismatch is considered, the reader can show from the analysis of Fig. 4.15 that

$$|A_{DM}| = \frac{R_D}{2} \frac{g_{m1} + g_{m2} + 4g_{m1}g_{m2}R_{SS}}{1 + (g_{m1} + g_{m2})R_{SS}}, \tag{4.46}$$

where it is assumed $V_{in1} = -V_{in2}$, and hence

$$CMRR = \frac{g_{m1} + g_{m2} + 4g_{m1}g_{m2}R_{SS}}{2\Delta g_m} \tag{4.47}$$

$$\approx \frac{g_m}{\Delta g_m}(1 + 2g_m R_{SS}), \tag{4.48}$$

where g_m denotes the mean value, i.e., $g_m = (g_{m1} + g_{m2})/2$. In practice, all mismatches must be taken into account.

4.4 Differential Pair with MOS Loads

The load of a differential pair need not be implemented by linear resistors. As with the common-source stages studied in Chapter 3, differential pairs can employ diode-connected or current-source loads (Fig. 4.32). The small-signal differential gain can be derived using

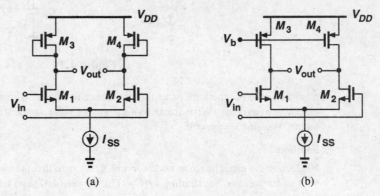

Figure 4.32 Differential pair with (a) diode-connected and (b) current-source loads.

the half-circuit concept. For Fig. 4.32(a),

$$A_v = -g_{mN}\left(g_{mP}^{-1}\|r_{ON}\|r_{OP}\right) \tag{4.49}$$

$$\approx -\frac{g_{mN}}{g_{mP}}, \tag{4.50}$$

where subscripts N and P denote NMOS and PMOS, respectively. Expressing g_{mN} and g_{mP} in terms of device dimensions, we have

$$A_v \approx -\sqrt{\frac{\mu_n(W/L)_N}{\mu_p(W/L)_P}}. \tag{4.51}$$

For Fig. 4.32(b), we have

$$A_v = -g_{mN}(r_{ON} \| r_{OP}).\qquad(4.52)$$

In the circuit of Fig. 4.32(a), the diode-connected loads consume voltage headroom, thus creating a trade-off between the output voltage swings, the voltage gain, and the input CM range. Recall from Eq. (3.35) that, for given bias current and input device dimensions, the circuit's gain and the PMOS overdrive voltage scale together. To achieve a higher gain, $(W/L)_P$ must decrease, thereby increasing $|V_{GSP} - V_{THP}|$ and lowering the CM level at nodes X and Y.

In order to alleviate the above difficulty, part of the bias currents of the input transistors can be provided by PMOS current sources. Illustrated in Fig. 4.33, the idea is to lower the g_m of the load devices by reducing their current rather than their aspect ratio. For example,

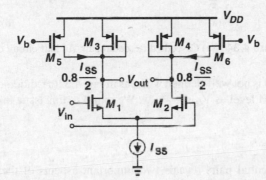

Figure 4.33 Addition of current sources to increase the voltage gain.

if M_5 and M_6 carry 80% of the drain current of M_1 and M_2, the current through M_3 and M_4 is reduced by a factor of five. For a given $|V_{GSP} - V_{THP}|$, this translates to a factor of five reduction in the transconductance of M_3 and M_4 because the aspect ratio of the devices can be lowered by the same factor. Thus, the differential gain is now approximately five times that of the case with no PMOS current sources.

The small-signal gain of the differential pair with current-source loads is relatively low— in the range of 10 to 20 in submicron technologies. How do we increase the voltage gain? Borrowing ideas from the amplifiers in Chapter 3, we increase the output impedance of both PMOS and NMOS devices by cascoding, in essence creating a differential version of the cascode stage introduced in Chapter 3. The result is depicted in Fig. 4.34(a). To calculate the gain, we construct the half circuit of Fig. 4.34(b), which is similar to the cascode stage of Fig. 3.60. Thus,

$$|A_v| \approx g_{m1}[(g_{m3}r_{O3}r_{O1})\|(g_{m5}r_{O5}r_{O7})].\qquad(4.53)$$

Cascoding therefore increases the differential gain substantially but at the cost of consuming more voltage headroom.

As a final note, we should mention that high-gain fully differential amplifiers require a means of defining the output common-mode level. For example, in Fig. 4.32(b), the output

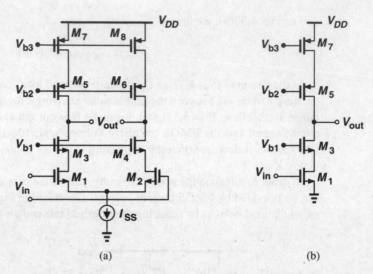

Figure 4.34   (a) Cascode differential pair, (b) half circuit of (a).

common-mode level is not well-defined whereas in Fig. 4.32(a), diode-connected transistors define the output CM level as $V_{DD} - V_{GSP}$. We return to this issue in Chapter 9.

4.5 Gilbert Cell

Our study of differential pairs reveals two important aspects of their operation: (1) the small-signal gain of the circuit is a function of the tail current and (2) the two transistors in a differential pair provide a simple means of steering the tail current to one of two destinations. By combining these two properties, we can develop a versatile building block.

Suppose we wish to construct a differential pair whose gain is varied by a control voltage. This can be accomplished as depicted in Fig. 4.35(a), where the control voltage defines the

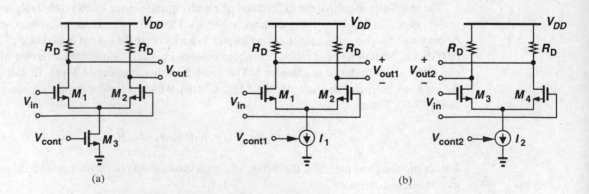

Figure 4.35   (a) Simple VGA, (b) two stages providing variable gain.

tail current and hence the gain. In this topology, $A_v = V_{out}/V_{in}$ varies from zero (if $I_{D3} = 0$) to a maximum value given by voltage headroom limitations and device dimensions. This circuit is a simple example of a "variable-gain amplifier" (VGA). VGAs find application in systems where the signal amplitude may experience large variations and hence requires inverse changes in the gain.

Now suppose we seek an amplifier whose gain can be continuously varied from a negative value to a positive value. Consider two differential pairs that amplify the input by opposite gains [Fig. 4.35(b)]. We now have $V_{out1}/V_{in} = -g_m R_D$ and $V_{out2}/V_{in} = +g_m R_D$, where g_m denotes the transconductance of each transistor in equilibrium. If I_1 and I_2 vary in opposite directions, so do $|V_{out1}/V_{in}|$ and $|V_{out2}/V_{in}|$.

But how should V_{out1} and V_{out2} be combined into a single output? As illustrated in Fig. 4.36(a), the two voltages can be summed, producing $V_{out} = V_{out1} + V_{out2} = A_1 V_{in} + A_2 V_{in}$,

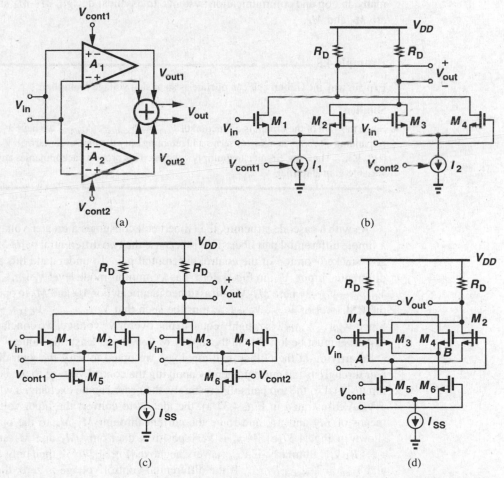

Figure 4.36 (a) Summation of the output voltages of two amplifiers, (b) summation in the current domain, (c) use of M_5-M_6 to control the gain, (d) Gilbert cell.

where A_1 and A_2 are controlled by V_{cont1} and V_{cont2}, respectively. The actual implementation is in fact quite simple: since $V_{out1} = R_D I_{D1} - R_D I_{D2}$ and $V_{out2} = R_D I_{D4} - R_D I_{D3}$, we have $V_{out1} + V_{out2} = R_D(I_{D1} + I_{D4}) - R_D(I_{D2} + I_{D3})$. Thus, rather than add V_{out1} and V_{out2}, we simply short the corresponding drain terminals to sum the currents and subsequently generate the output voltage [Fig. 4.36(b)]. Note that if $I_1 = 0$, then $V_{out} = +g_m R_D V_{in}$ and if $I_2 = 0$, then $V_{out} = -g_m R_D V_{in}$. For $I_1 = I_2$, the gain drops to zero.

In the circuit of Fig. 4.36(b), V_{cont1} and V_{cont2} must vary I_1 and I_2 in opposite directions such that the gain of the amplifier changes monotonically. What circuit can vary two currents in opposite directions? A differential pair provides such a characteristic, leading to the topology of Fig. 4.36(c). Note that for a large $|V_{cont1} - V_{cont2}|$, all of the tail current is steered to one of the top differential pairs and the gain from V_{in} to V_{out} is at its most positive or most negative value. For $V_{cont1} = V_{cont2}$, the gain is zero. For simplicity, we redraw the circuit as shown in Fig. 4.36(d). Called the "Gilbert cell" [2], this circuit is widely used in many analog and communication systems. In a typical design, M_1-M_4 are identical and so are M_5 and M_6.

Example 4.8

Explain why the Gilbert cell can operate as an analog voltage multiplier.

Solution

Since the gain of the circuit is a function of $V_{cont} = V_{cont1} - V_{cont2}$, we have $V_{out} = V_{in} \cdot f(V_{cont})$. Expanding $f(V_{cont})$ in a Taylor series and retaining only the first-order term, αV_{cont}, we have $V_{out} = \alpha V_{in} V_{cont}$. Thus, the circuit can multiply voltages. This property accompanies any voltage-controlled variable-gain amplifier.

As with a cascode structure, the Gilbert cell consumes a greater voltage headroom than a simple differential pair does. This is because the two differential pairs M_1-M_2 and M_3-M_4 are "stacked" on top of the control differential pair. To understand this point, suppose the differential input, V_{in}, in Fig. 4.36(d) has a common-mode level $V_{CM,in}$. Then, $V_A = V_B = V_{CM,in} - V_{GS1}$, where M_1-M_4 are assumed identical. For M_5 and M_6 to operate in saturation, the CM level of V_{cont}, $V_{CM,cont}$, must be such that $V_{CM,cont} \leq V_{CM,in} - V_{GS1} + V_{TH5,6}$. Since $V_{GS1} - V_{TH5,6}$ is roughly equal to one overdrive voltage, we conclude that the control CM level must be lower than the input CM level by at least this value.

In arriving at the Gilbert cell topology, we opted to vary the gain of each differential pair through its tail current, thereby applying the control voltage to the bottom pair and the input signal to the top pairs. Interestingly, the order can be exchanged while still obtaining a VGA. Illustrated in Fig. 4.37(a), the idea is to convert the input voltage to current by means of M_5 and M_6 and route the current through M_1-M_4 to the output nodes. If, as shown in Fig. 4.37(b), V_{cont} is very positive, then only M_1 and M_2 are on and $V_{out} = g_{m5,6} R_D V_{in}$. Similarly, if V_{cont} is very negative [Fig. 4.37(c)], then only M_3 and M_4 are on and $V_{out} = -g_{m5,6} R_D V_{in}$. If the differential control voltage is zero, then $V_{out} = 0$. The input differential pair may incorporate degeneration to provide a linear voltage-to-current conversion.

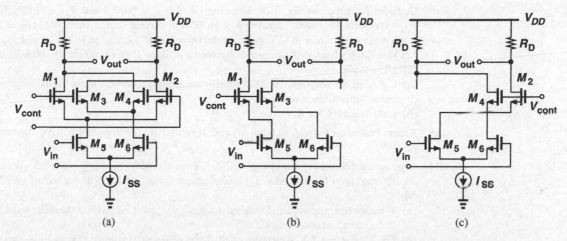

Figure 4.37 (a) Gilbert cell sensing the input voltage by the bottom differential pair, (b) signal path for very positive V_{cont}, (c) signal path for very negative V_{cont}.

Problems

Unless otherwise stated, in the following problems, use the device data shown in Table 2.1 and assume $V_{DD} = 3$ V where necessary. All device dimensions are effective values and in microns.

4.1. Suppose the total capacitance between adjacent lines in Fig. 4.2 is 10 fF and the capacitance from the drains of M_1 and M_2 to ground is 100 fF.
    (a) What is the amplitude of the glitches in the analog output in Fig. 4.2(a) for a clock swing of 3 V?
    (b) If in Fig. 4.2(b), the capacitance between L_1 and L_2 is 10% less than that between L_1 and L_3, what is the amplitude of the glitches in the differential analog output for a clock swing of 3 V?

4.2. Sketch the small-signal differential voltage gain of the circuit shown in Fig. 4.8(a) if V_{DD} varies from 0 to 3 V. Assume $(W/L)_{1-3} = 50/0.5$, $V_{in,CM} = 1.3$ V, and $V_b = 1$ V.

4.3. Construct the plots of Fig. 4.8(c) for a differential pair using PMOS transistors.

4.4. In the circuit of Fig. 4.10, $(W/L)_{1,2} = 50/0.5$ and $I_{SS} = 0.5$ mA.
    (a) What is the maximum allowable output voltage swing if $V_{in,CM} = 1.2$ V?
    (b) What is the voltage gain under this condition?

4.5. A differential pair uses input NMOS devices with $W/L = 50/0.5$ and a tail current of 1 mA.
    (a) What is the equilibrium overdrive voltage of each transistor?
    (b) How is the tail current shared between the two sides if $V_{in1} - V_{in2} = 50$ mA?
    (c) What is the equivalent G_m under this condition?
    (d) For what value of $V_{in1} - V_{in2}$ does the G_m drop by 10%? By 90%?

4.6. Repeat Problem 4.5 with $W/L = 25/0.5$ and compare the results.

4.7. Repeat Problem 4.5 with a tail current of 2 mA and compare the results.

4.8. Sketch I_{D1} and I_{D2} in Fig. 4.17 versus $V_{in1} - V_{in2}$. For what value of $V_{in1} - V_{in2}$ are the two currents equal?

4.9. Consider the circuit of Fig. 4.28, assuming $(W/L)_{1,2} = 50/0.5$ and $R_D = 2\ k\Omega$. Suppose R_{SS} represents the output impedance of an NMOS current source with $(W/L)_{SS} = 50/0.5$ and a drain current of 1 mA. The input signal consists of $V_{in,DM} = 10\ mV_{pp}$ and $V_{in,CM} = 1.5\ V + V_n(t)$, where $V_n(t)$ denotes noise with a peak-to-peak amplitude of 100 mV. Assume $\Delta R/R = 0.5\%$.
  (a) Calculate the output differential signal-to-noise ratio, defined as the signal amplitude divided by the noise amplitude.
  (b) Calculate the CMRR.

4.10. Repeat Problem 4.9 if $\Delta R = 0$ but M_1 and M_2 suffer from a threshold voltage mismatch of 1 mV.

4.11. Suppose the differential pair of Fig. 4.32(a) is designed with $(W/L)_{1,2} = 50/0.5$, $(W/L)_{3,4} = 10/0.5$, and $I_{SS} = 0.5$ mA. Also, I_{SS} is implemented with an NMOS device having $(W/L)_{SS} = 50/0.5$.
  (a) What are the minimum and maximum allowable input CM levels if the differential swings at the input and output are small?
  (b) For $V_{in,CM} = 1.2$ V, sketch the small-signal differential voltage gain as V_{DD} goes from 0 to 3 V.

4.12. In Problem 4.11, suppose M_1 and M_2 have a threshold voltage mismatch of 1 mV. What is the CMRR?

4.13. In Problem 4.11, suppose $W_3 = 10\ \mu$m but $W_4 = 11\ \mu$m. Calculate the CMRR.

4.14. For the differential pairs of Fig. 4.32(a) and (b), calculate the differential voltage gain if $I_{SS} = 1$ mA, $(W/L)_{1,2} = 50/0.5$, and $(W/L)_{3,4} = 50/1$. What is the minimum allowable input CM level if I_{SS} requires at least 0.4 V across it? Using this value for $V_{in,CM}$, calculate the maximum output voltage swing in each case.

4.15. In the circuit of Fig. 4.33, assume $I_{SS} = 1$ mA and $W/L = 50/0.5$ for all of the transistors.
  (a) Determine the voltage gain.
  (b) Calculate V_b such that $I_{D5} = I_{D6} = 0.8(I_{SS}/2)$.
  (c) If I_{SS} requires a minimum voltage of 0.4 V, what is the maximum differential output swing?

4.16. Assuming all of the circuits shown in Fig. 4.38 are symmetric, sketch V_{out} as (a) V_{in1} and V_{in2} vary differentially from zero to V_{DD}, and (b) V_{in1} and V_{in2} are equal and they vary from zero to V_{DD}.

4.17. Assuming all of the circuits shown in Fig. 4.39 are symmetric, sketch V_{out} as (a) V_{in1} and V_{in2} vary differentially from zero to V_{DD}, and (b) V_{in1} and V_{in2} are equal and they vary from zero to V_{DD}.

4.18. Assuming all of the transistors in the circuits of Figs. 4.38 and 4.39 are saturated and $\lambda \neq 0$, calculate the small-signal differential voltage gain of each circuit.

4.19. Consider the circuit shown in Fig. 4.40.
  (a) Sketch V_{out} as V_{in1} and V_{in2} vary differentially from zero to V_{DD}.
  (b) If $\lambda = 0$, obtain an expression for the voltage gain. What is the voltage gain if $W_{3,4} = 0.8W_{5,6}$?

4.20. For the circuit shown in Fig. 4.41,
  (a) Sketch V_{out}, V_X, and V_Y as V_{in1} and V_{in2} vary differentially from zero to V_{DD}.
  (b) Calculate the small-signal differential voltage gain.

4.21. Assuming no symmetry in the circuit of Fig. 4.42 and using no equivalent circuits, calculate the small-signal voltage gain $(V_{out})/(V_{in1} - V_{in2})$ if $\lambda = 0$ and $\gamma \neq 0$.

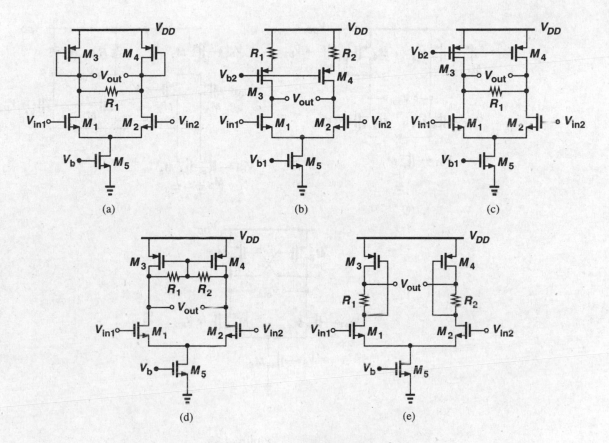

Figure 4.38

4.22. Due to a manufacturing defect, a large parasitic resistance has appeared between the drain and source terminals of M_1 in Fig. 4.43. Assuming $\lambda = \gamma = 0$, calculate the small-signal gain, common-mode gain, and CMRR.

4.23. Due to a manufacturing defect, a large parasitic resistance has appeared between the drains of M_1 and M_4 in the circuit of Fig. 4.44. Assuming $\lambda = \gamma = 0$, calculate the small-signal gain, common-mode gain, and CMRR.

4.24. In the circuit of Fig. 4.45, all of the transistors have a W/L of 50/0.5 and M_3 and M_4 are to operate in deep triode region with an on-resistance of 2 kΩ. Assuming $I_{D5} = 20$ μA and $\lambda = \gamma = 0$, calculate the input common-mode level that yields such resistance. Sketch V_{out1} and V_{out2} as V_{in1} and V_{in2} vary differentially from 0 to V_{DD}.

4.25. In the circuit of Fig. 4.32(b), $(W/L)_{1-4} = 50/0.5$ and $I_{SS} = 1$ mA.
(a) What is the small-signal differential gain?
(b) For $V_{in,CM} = 1.5$ V, what is the maximum allowable output voltage swing?

4.26. In the circuit of Fig. 4.33, assume M_5 and M_6 have a small threshold voltage mismatch of ΔV and I_{SS} has an output impedance R_{SS}. Calculate the CMRR.

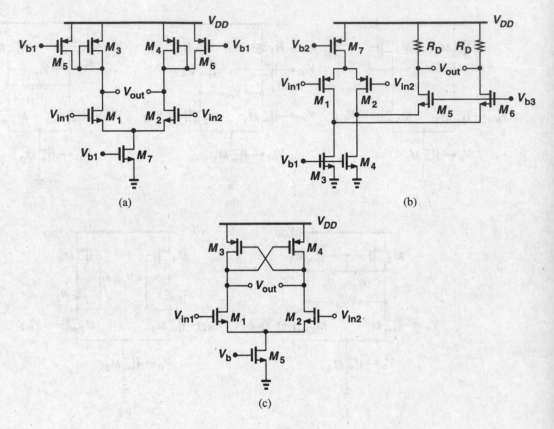

(a)

(b)

(c)

Figure 4.39

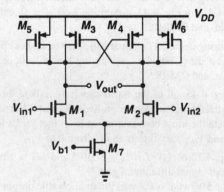

Figure 4.40

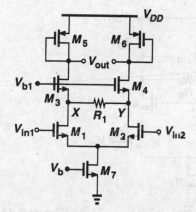

Figure 4.41

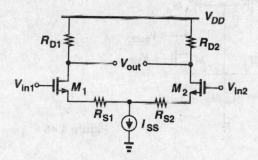

Figure 4.42

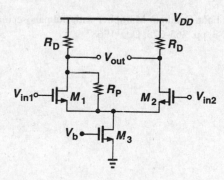

Figure 4.43

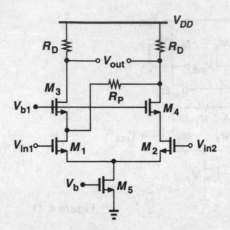

Figure 4.44

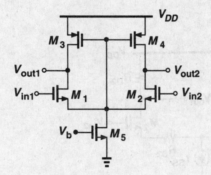

Figure 4.45

References

1. P. R. Gray and R. G. Meyer, *Analysis and Design of Analog Integrated Circuits*, Third Ed., New York: Wiley, 1993.
2. B. Gilbert, "A Precise Four-Quadrant Multiplier with Subnanosecond Response," *IEEE J. Solid-State Circuits*, vol. SC-3, pp. 365–373, Dec. 1968.

Chapter 5

Passive and Active Current Mirrors

Our study of single-stage and differential amplifiers in Chapters 3 and 4 points to the wide usage of current sources. In these circuits current sources act as a large resistor without consuming excessive voltage headroom. We also noted that MOS devices operating in saturation can act as a current source.

Current sources find other applications in analog design as well. For example, some digital-to-analog (D/A) converters employ an array of current sources to produce an analog output proportional to the digital input. Also, current sources, in conjunction with "current mirrors," can perform useful functions on analog signals.

This chapter deals with the design of current mirrors as both bias elements and signal processing components. Following a review of basic current mirrors, we study cascode mirror operation. Next, we analyze active current mirrors and describe the properties of differential pairs using such circuits as loads.

5.1 Basic Current Mirrors

Fig. 5.1 illustrates two examples where a current source proves useful. From our study in Chapter 2, recall that the output resistance and capacitance and the voltage headroom of a current source trade with the magnitude of the output current. In addition to these issues, several other aspects of current sources are important: supply, process, and temperature dependence, output noise current, and matching with other current sources. We postpone noise and matching considerations to Chapters 7 and 13, respectively.

How should a MOSFET be biased so as to operate as a stable current source? To gain a better view of the issues, let us consider the simple resistive biasing shown in Fig. 5.2. Assuming M_1 is in saturation, we can write

$$I_{out} \approx \frac{1}{2}\mu_n C_{ox}\frac{W}{L}\left(\frac{R_2}{R_1+R_2}V_{DD}-V_{TH}\right)^2. \tag{5.1}$$

This expression reveals various dependencies of I_{out} upon the supply, process, and temperature. The overdrive voltage is a function of V_{DD} and V_{TH}; the threshold voltage may

135

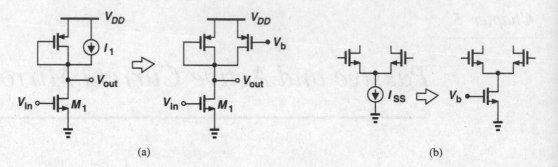

(a)   (b)

Figure 5.1 Applications of current sources.

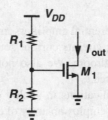

Figure 5.2 Definition of current by resistive divider.

vary by 100 mV from wafer to wafer. Furthermore, both μ_n and V_{TH} exhibit temperature dependence. Thus, I_{out} is poorly defined. The issue becomes more severe as the device is biased with a smaller overdrive voltage, e.g., to consume less headroom. With a nominal overdrive of, say, 200 mV, a 50-mV error in V_{TH} results in a 44% error in the output current.

It is important to note that the above process and temperature dependencies exist even if the gate voltage is not a function of the supply voltage. In other words, if the gate-source *voltage* of a MOSFET is precisely defined, then its drain *current* is not! For this reason, we must seek other methods of biasing MOS current sources.

The design of current sources in analog circuits is based on "copying" currents from a reference, with the assumption that *one* precisely-defined current source is already available. While this method may appear to entail an endless cycle, it is carried out as illustrated in Fig. 5.3. A relatively complex circuit—sometimes requiring external adjustments—is used

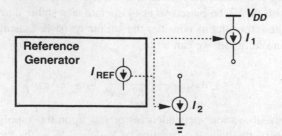

Figure 5.3 Use of a reference to generate various currents.

to generate a stable reference current, I_{REF}, which is then copied to many current sources in the system. We study the copying operation here and the reference generator circuit in Chapter 11.

How do we generate copies of a reference current? For example, in Fig. 5.4, how do we guarantee $I_{out} = I_{REF}$? For a MOSFET, if $I_D = f(V_{GS})$, where $f(\cdot)$ denotes the

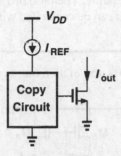

Figure 5.4 Conceptual means of copying currents.

functionality of I_D versus V_{GS}, then $V_{GS} = f^{-1}(I_D)$. That is, if a transistor is biased at I_{REF}, then it produces $V_{GS} = f^{-1}(I_{REF})$ [Fig. 5.5(a)]. Thus, if this voltage is applied to the gate and source terminals of a second MOSFET, the resulting current is $I_{out} = ff^{-1}(I_{REF}) = I_{REF}$ [Fig. 5.5(b)]. From another point of view, two identical MOS devices that have equal gate–source voltages and operate in saturation carry equal currents (if $\lambda = 0$).

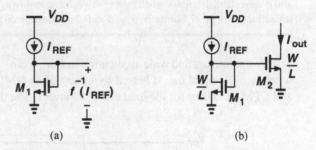

(a)   (b)

Figure 5.5 (a) Diode-connected device providing inverse function, (b) basic current mirror.

The structure consisting of M_1 and M_2 in Fig. 5.5(b) is called a "current mirror." In the general case, the devices need not be identical. Neglecting channel-length modulation, we can write

$$I_{REF} = \frac{1}{2}\mu_n C_{ox} \left(\frac{W}{L}\right)_1 (V_{GS} - V_{TH})^2 \tag{5.2}$$

$$I_{out} = \frac{1}{2}\mu_n C_{ox} \left(\frac{W}{L}\right)_2 (V_{GS} - V_{TH})^2, \tag{5.3}$$

obtaining

$$I_{out} = \frac{(W/L)_2}{(W/L)_1} I_{REF}.$$ (5.4)

The key property of this topology is that it allows precise copying of the current with no dependence on process and temperature. The ratio of I_{out} and I_{REF} is given by the *ratio* of device dimensions, a quantity that can be controlled with reasonable accuracy.

Example 5.1

In Fig. 5.6, find the drain current of M_4 if all of the transistors are in saturation.

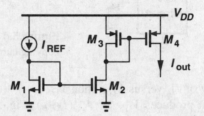

Figure 5.6

Solution

We have $I_{D2} = I_{REF}[(W/L)_2/(W/L)_1]$. Also, $|I_{D3}| = |I_{D2}|$ and $I_{D4} = I_{D3}[(W/L)_4/(W/L)_3]$. Thus, $|I_{D4}| = \alpha\beta I_{REF}$, where $\alpha = (W/L)_2/(W/L)_1$ and $\beta = (W/L)_4/(W/L)_3$. Proper choice of α and β can establish large or small ratios between I_{D4} and I_{REF}. For example, $\alpha = \beta = 5$ yields a magnification factor of 25. Similarly, $\alpha = \beta = 0.2$ can be utilized to generate a small, well-defined current.

Current mirrors find wide application in analog circuits. Fig. 5.7 illustrates a typical case, where a differential pair is biased by means of an NMOS mirror for the tail current source and a PMOS mirror for the load current sources. The device dimensions shown establish a

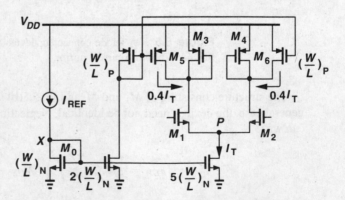

Figure 5.7   Current mirrors used to bias a differential amplifier.

drain current of $0.4I_T$ in M_5 and M_6, reducing the drain current of M_3 and M_4 and hence increasing the gain.

Current mirrors usually employ the same *length* for all of the transistors so as to minimize errors due to the side-diffusion of the source and drain areas (L_D). For example, in Fig. 5.7, the NMOS current sources must have the same channel length as M_0. This is because if, L_{drawn} is, say, doubled, then $L_{eff} = L_{drawn} - 2L_D$ is not. Furthermore, the threshold voltage of short-channel devices exhibits some dependence on the channel length (Chapter 16). Thus, current ratioing is achieved by only scaling the width of transistors.[1]

We should also mention that current mirrors can process *signals* as well. In Fig. 5.5(b), for example, if I_{REF} increases by ΔI, then I_{out} increases by $\Delta I(W/L)_2/(W/L)_1$. That is, the circuit *amplifies* the small-signal current if $(W/L)_2/(W/L)_1 > 1$ (but at the cost of proportional multiplication of the bias current).

Example 5.2 ────────────────────────────────

Calculate the small-signal voltage gain of the circuit shown in Fig. 5.8.

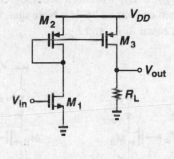

Figure 5.8

Solution

The small-signal drain current of M_1 is equal to $g_{m1}V_{in}$. Since $I_{D2} = I_{D1}$ and $I_{D3} = I_{D2}(W/L)_3/(W/L)_2$, the small-signal drain current of M_3 is equal to $g_{m1}V_{in}(W/L)_3/(W/L)_2$, yielding a voltage gain of $g_{m1}R_L(W/L)_3/(W/L)_2$.

5.2 Cascode Current Mirrors

In our discussion of current mirrors thus far, we have neglected channel length modulation. In practice, this effect results in significant error in copying currents, especially if minimum-length transistors are used so as to minimize the width and hence the output capacitance of the current source. For the simple mirror of Fig. 5.5(b), we can write

$$I_{D1} = \frac{1}{2}\mu_n C_{ox} \left(\frac{W}{L}\right)_1 (V_{GS} - V_{TH})^2(1 + \lambda V_{DS1}) \tag{5.5}$$

────────────────

[1] As explained in Chapter 18, the widths are actually scaled by placing multiple unit transistors in parallel rather than making a device wider.

$$I_{D2} = \frac{1}{2}\mu_n C_{ox}\left(\frac{W}{L}\right)_2 (V_{GS} - V_{TH})^2(1 + \lambda V_{DS2}), \tag{5.6}$$

and hence

$$\frac{I_{D2}}{I_{D1}} = \frac{(W/L)_2}{(W/L)_1} \cdot \frac{1 + \lambda V_{DS2}}{1 + \lambda V_{DS1}}. \tag{5.7}$$

While $V_{DS1} = V_{GS1} = V_{GS2}$, V_{DS2} may not equal V_{GS2} because of the circuitry fed by M_2. For example, in Fig. 5.7, the potential at node P is determined by the input common-mode level and the gate-source voltage of M_1 and M_2, and it may not equal V_X.

In order to suppress the effect of channel-length modulation, a cascode current source can be used. As shown in Fig. 5.9(a), if V_b is chosen such that $V_Y = V_X$, then I_{out} closely tracks I_{REF}. This is because, as described in conjunction with Fig. 3.61, the cascode device "shields" the bottom transistor from variations in V_P. With the aid of Fig. 3.23, the reader can prove that $\Delta V_Y \approx \Delta V_P/[(g_{m3} + g_{mb3})r_{O3}]$. Thus, we say that V_Y remains close to V_X and hence $I_{D2} \approx I_{D1}$ with high accuracy. Such accuracy is obtained at the cost of the voltage headroom consumed by M_3. Note that, while L_1 must be equal to L_2, the length of M_3 need not be equal to L_1 and L_2.

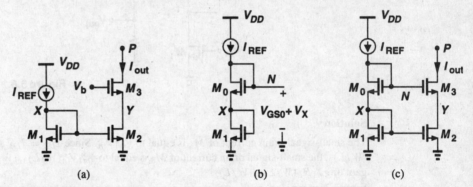

Figure 5.9   (a) Cascode current source, (b) modification of mirror circuit to generate the cascode bias voltage, (c) cascode current mirror.

How do we generate V_b in Fig. 5.9(a)? Since the objective is to ensure $V_Y = V_X$, we must guarantee $V_b - V_{GS3} = V_X$ or $V_b = V_{GS3} + V_X$. This result suggests that if a gate-source voltage is added to V_X, the required value of V_b can be obtained. Depicted in Fig. 5.9(b), the idea is to place another diode-connected device, M_0, in series with M_1, thereby generating a voltage $V_N = V_{GS0} + V_X$. Proper choice of the dimensions of M_0 with respect to those of M_3 yields $V_{GS0} = V_{GS3}$. Connecting node N to the gate of M_3 as shown in Fig. 5.9(c), we have $V_{GS0} + V_X = V_{GS3} + V_Y$. Thus, if $(W/L)_3/(W/L)_0 = (W/L)_2/(W/L)_1$, then $V_{GS3} = V_{GS0}$ and $V_X = V_Y$. Note that this result holds even if M_0 and M_3 suffer from body effect.

Example 5.3

In Fig. 5.10, sketch V_X and V_Y as a function of I_{REF}. If I_{REF} requires 0.5 V to operate as a current source, what is its maximum value?

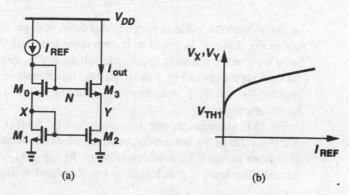

Figure 5.10

Solution

Since M_2 and M_3 are properly ratioed with respect to M_1 and M_0, we have $V_Y = V_X \approx \sqrt{2I_{REF}/[\mu_n C_{ox}(W/L)_1]} + V_{TH1}$. The behavior is plotted in Fig. 5.10(b).

To find the maximum value of I_{REF}, we note that

$$V_N = V_{GS0} + V_{GS1} \tag{5.8}$$

$$= \sqrt{\frac{2I_{REF}}{\mu_n C_{ox}}}\left[\sqrt{\left(\frac{L}{W}\right)_0} + \sqrt{\left(\frac{L}{W}\right)_1}\right] + V_{TH0} + V_{TH1}. \tag{5.9}$$

Thus,

$$V_{DD} - \sqrt{\frac{2I_{REF}}{\mu_n C_{ox}}}\left[\sqrt{\left(\frac{L}{W}\right)_0} + \sqrt{\left(\frac{L}{W}\right)_1}\right] - V_{TH0} - V_{TH1} = 0.5 \text{ V}. \tag{5.10}$$

and hence

$$I_{REF,max} = \frac{\mu_n C_{ox}}{2} \frac{(V_{DD} - 0.5 \text{ V} - V_{TH0} - V_{TH1})^2}{(\sqrt{(L/W)_0} + \sqrt{(L/W)_1})^2}. \tag{5.11}$$

While operating as a current source with high output impedance and accurate value, the topology of Fig. 5.9(c) nonetheless consumes substantial voltage headroom. For simplicity, let us neglect the body effect and assume all of the transistors are identical. Then, the minimum allowable voltage at node P is equal to

$$V_N - V_{TH} = V_{GS0} + V_{GS1} - V_{TH} \tag{5.12}$$

$$= (V_{GS0} - V_{TH}) + (V_{GS1} - V_{TH}) + V_{TH}, \tag{5.13}$$

i.e., two overdrive voltages plus one threshold voltage. How does this value compare with that in Fig. 5.9(a) if V_b could be chosen more arbitrarily? As shown in Fig. 3.51, V_b could be so low that the minimum allowable voltage at P is merely two overdrive voltages. Thus, the cascode mirror of Fig. 5.9(c) "wastes" one threshold voltage in the headroom. This is because $V_{DS2} = V_{GS2}$, whereas V_{DS2} could be as low as $V_{GS2} - V_{TH}$ while maintaining M_2 in saturation.

Fig. 5.11 summarizes our discussion. In Fig. 5.11(a), V_b is chosen to allow the lowest possible value of V_P but the output current does not accurately track I_{REF} because M_1 and M_2 sustain unequal drain-source voltages. In Fig. 5.11(b), higher accuracy is achieved but the minimum level at P is higher by one threshold voltage.

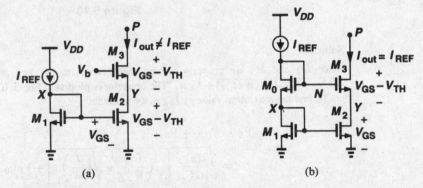

Figure 5.11   (a) Cascode current source with minimum headroom voltage, (b) headroom consumed by a cascode mirror.

Before resolving this issue, it is instructive to examine the large-signal behavior of a cascode current source.

Example 5.4

In Fig. 5.12(a), assuming all of the transistors are identical, sketch I_X and V_B as V_X drops from a large positive value.

Solution

For $V_X \geq V_N - V_{TH}$, both M_2 and M_3 are in saturation, $I_X = I_{REF}$ and $V_B = V_A$. As V_X drops, which transistor enters the triode region first, M_3 or M_2? Suppose M_2 enters the triode region before M_3 does. For this to occur, V_{DS2} must drop and, since V_{GS2} is constant, so must I_{D2}. This means V_{GS3} increases while I_{D3} decreases, which is not possible if M_3 is still in saturation. Thus, M_3 enters the triode region first.

As V_X falls below $V_N - V_{TH}$, M_3 enters the triode region, requiring a greater gate-source overdrive to carry the same current. Thus, as shown in Fig. 5.12(b), V_B begins to drop, causing I_{D2} and hence

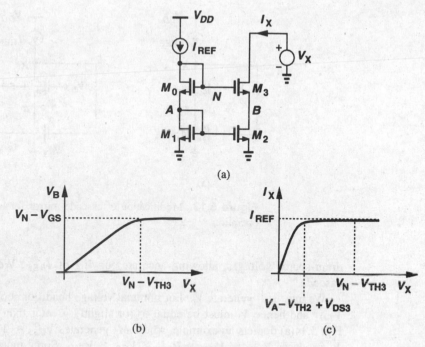

Figure 5.12

I_X to decrease slightly. As V_X and V_B decrease further, eventually we have $V_B < V_A - V_{TH}$, and M_2 enters the triode region. At this point, I_{D2} begins to drop sharply. For $V_X = 0$, $I_X = 0$, and M_2 and M_3 operate in deep triode region. Note that as V_X drops below $V_N - V_{TH3}$, the output impedance of the cascode falls rapidly because g_{m3} degrades in the triode region.

In order to eliminate the accuracy-headroom trade-off described above, we first study the modification depicted in Fig. 5.13(a). Note that this circuit is in fact a cascode topology with its output shorted to its input. How can we choose V_b so that both M_1 and M_2 are in saturation? We must have $V_b - V_{TH2} \leq V_X(= V_{GS1})$ for M_2 to be saturated and $V_{GS1} - V_{TH1} \leq V_A(= V_b - V_{GS2})$ for M_1 to be saturated. Thus,

$$V_{GS2} + (V_{GS1} - V_{TH1}) \leq V_b \leq V_{GS1} + V_{TH2}. \qquad (5.14)$$

A solution exists if $V_{GS2} + (V_{GS1} - V_{TH1}) \leq V_{GS1} + V_{TH2}$, i.e., if $V_{GS2} - V_{TH2} \leq V_{TH1}$. We must therefore size M_2 such that its overdrive voltage remains less than one threshold voltage.

Now consider the circuit shown in Fig. 5.13(b), where all of the transistors are in saturation and proper ratioing ensures that $V_{GS2} = V_{GS4}$. If $V_b = V_{GS2} + (V_{GS1} - V_{TH1}) = V_{GS4} + (V_{GS3} - V_{TH3})$, then the cascode current source M_3-M_4 consumes minimum headroom (the overdrive of M_3 plus that of M_4) while M_1 and M_3 sustain equal

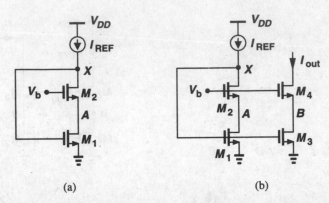

Figure 5.13 Modification of cascode mirror for low-voltage operation.

drain-source voltages, allowing accurate copying of I_{REF}. We call this a "low-voltage cascode."

We must still generate V_b. For minimal voltage headroom consumption, $V_A = V_{GS1} - V_{TH1}$ and hence V_b must be equal to (or slightly greater than) $V_{GS2} + (V_{GS1} - V_{TH1})$. Fig. 5.14(a) depicts an example, where M_5 generates $V_{GS5} \approx V_{GS2}$ and M_6 together with R_b produces $V_{DS6} = V_{GS6} - R_b I_1 \approx V_{GS1} - V_{TH1}$. Some inaccuracy nevertheless arises because M_5 does not suffer from body effect whereas M_2 does. Also, the magnitude of $R_b I_1$ is not well-controlled.

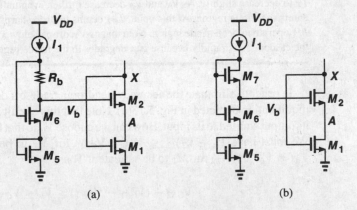

Figure 5.14 Generation of gate voltage V_b for cascode mirrors.

An alternative circuit is shown in Fig. 5.14(b), where the diode-connected transistor M_7 has a large W/L so that $V_{GS7} \approx V_{TH7}$. That is, $V_{DS6} \approx V_{GS6} - V_{TH7}$ and hence $V_b = V_{GS5} + V_{GS6} - V_{TH7}$. While requiring no resistors, this circuit nonetheless suffers from similar errors due to body effect. Some margin is therefore necessary to ensure M_1 and M_2 remain in saturation.

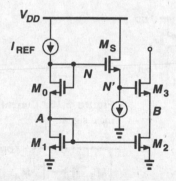

Figure 5.15 Low-voltage cascode using a source follower level shifter.

We should mention that low-voltage cascodes can also be biased using source followers. Shown in Fig. 5.15, the idea is to shift the gate voltage of M_3 down with respect to V_N by interposing a source follower. If M_S is biased at a very low current density, $I_D/(W/L)$, then its gate–source voltage is approximately equal to V_{TH3}, i.e., $V_{N'} \approx V_N - V_{TH3}$, and

$$V_B = V_{GS1} + V_{GS0} - V_{TH3} - V_{GS3} \qquad (5.15)$$

$$= V_{GS1} - V_{TH3}, \qquad (5.16)$$

implying that M_2 is at the edge of the triode region. In this topology, however, $V_{DS2} \neq V_{DS1}$, introducing substantial mismatch. Also, if the body effect is considered for M_0, M_S, and M_3, it is difficult to guarantee that M_2 operates in saturation. We should mention that, in addition to reducing the systematic mismatch due to channel-length modulation, the cascode structure also provides a high output impedance.

5.3 Active Current Mirrors

As mentioned earlier and exemplified by the circuit of Fig. 5.8, current mirrors can also process signals, i.e., operate as "active" elements. Particularly useful is a type of mirror topology used in conjunction with differential pairs. In this section, we study this circuit and its properties.

First, let us examine the circuit shown in Fig. 5.16, where M_1 and M_2 are identical. Neglecting channel-length modulation, we have $I_{out} = I_{in}$, i.e., with the direction shown for I_{in} and I_{out}, the circuit performs no *inversion*. From the small-signal point of view, if I_{in} increases by ΔI, so does I_{out}.

Now consider the differential amplifier of Fig. 5.17(a), where a current source in a mirror arrangement serves as the load and the output is single-ended. What is the small-signal gain, $A_v = V_{out}/V_{in}$, of this circuit? We calculate A_v using two different approaches,[2] assuming $\gamma = 0$ for simplicity.

---

[2] Note that, owing to the lack of symmetry, the half-circuit concept cannot be applied here.

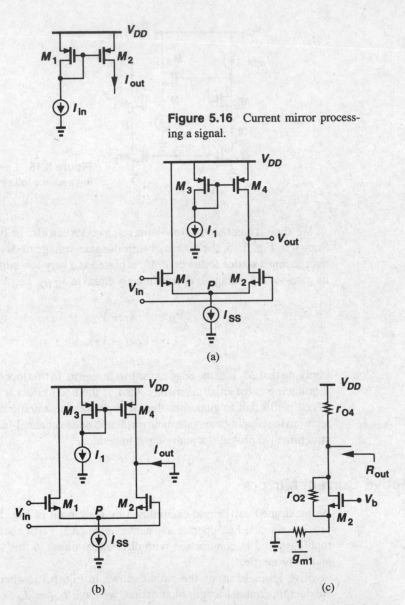

Figure 5.16 Current mirror processing a signal.

Figure 5.17 (a) Differential pair with current-source load, (b) circuit for calculation of G_m, (c) circuit for calculation of R_{out}.

Writing $|A_v| = G_m R_{out}$ and recognizing from Fig. 5.17(b) that $G_m = I_{out}/V_{in} = (g_{m1}V_{in}/2)/V_{in} = g_{m1}/2$, we simply need to compute R_{out}. As illustrated in Fig. 5.17(c), for this calculation, M_2 is degenerated by the source output impedance, $1/g_{m1}$, of M_1, thereby exhibiting an output impedance equal to $(1 + g_{m2}r_{O2})(1/g_{m1,2}) + r_{O2} = 2r_{O2} + 1/g_{m1} \approx$

$2r_{O2}$. Thus, $R_{out} \approx (2r_{O2}) \| r_{O4}$, and

$$|A_v| \approx \frac{g_{m1}}{2}[(2r_{O2}) \| r_{O4}].   \qquad (5.17)$$

Interestingly, if $r_{O4} \to \infty$, then $A_v \to g_{m1}r_{O2}$. This can be explained by the second approach.

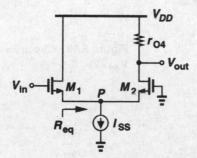

Figure 5.18   Circuit for calculation of V_P/V_{in}.

In our second approach, we calculate V_P/V_{in} and V_{out}/V_P and multiply the results to obtain V_{out}/V_{in}. With the aid of Fig. 5.18,

$$\frac{V_P}{V_{in}} = \frac{R_{eq}}{R_{eq} + \dfrac{1}{g_{m1}}},   \qquad (5.18)$$

where R_{eq} denotes the resistance seen looking into the source of M_2. Since the drain of M_2 is terminated by a relatively large resistance, r_{O4}, the value of R_{eq} must be obtained from Eq. (3.110):

$$R_{eq} \approx \frac{1}{g_{m2}} + \frac{r_{O4}}{g_{m2}r_{O2}}   \qquad (5.19)$$

$$= \frac{1}{g_{m2}}\left(1 + \frac{r_{O4}}{r_{O2}}\right).   \qquad (5.20)$$

It follows that

$$\frac{V_P}{V_{in}} = \frac{1 + \dfrac{r_{O4}}{r_{O2}}}{2 + \dfrac{r_{O4}}{r_{O2}}}.   \qquad (5.21)$$

Note that if $r_{O4} \to 0$, $V_P/V_{in} \to 1/2$ and if $r_{O4} \to \infty$, then $V_P/V_{in} \to 1$.

We now calculate V_{out}/V_P while taking r_{O2} into account. From Fig. 5.19,

$$\frac{V_{out}}{V_P} = \frac{1 + g_{m2}r_{O2}}{1 + \dfrac{r_{O2}}{r_{O4}}}   \qquad (5.22)$$

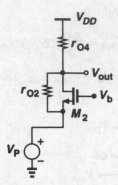

Figure 5.19 Circuit for calculation of V_{out}/V_P.

$$\approx \frac{g_{m2}r_{O2}}{1+\dfrac{r_{O2}}{r_{O4}}}. \tag{5.23}$$

From (5.21) and (5.23), we have

$$\frac{V_{out}}{V_{in}} = \frac{1+\dfrac{r_{O4}}{r_{O2}}}{2+\dfrac{r_{O4}}{r_{O2}}} \cdot \frac{g_{m2}r_{O2}}{1+\dfrac{r_{O2}}{r_{O4}}} \tag{5.24}$$

$$= \frac{g_{m2}r_{O2}r_{O4}}{2r_{O2}+r_{O4}} \tag{5.25}$$

$$= \frac{g_{m2}}{2}[(2r_{O2})\|r_{O4}]. \tag{5.26}$$

In the circuit of Fig. 5.17, the small-signal drain current of M_1 is "wasted." As conceptually shown in Fig. 5.20(a), it is desirable to utilize this current with proper polarity at the output. This can be accomplished as depicted in Fig. 5.20(b), where M_3 and M_4 are identical. To see how M_3 enhances the gain, suppose the gate voltage of M_1 increases by a small amount, increasing I_{D1} by ΔI and decreasing I_{D2} by ΔI. Since $|I_{D3}|$ and hence $|I_{D4}|$ also increase by ΔI, we observe that the output voltage tends to increase through two mechanisms: the drain current of M_2 drops *and* the drain current of M_4 rises.[3] In contrast to the circuit of Fig. 5.17, here M_4 assists M_2 with the voltage change at the output. This configuration is called a differential pair with active current mirror.[4] An important property of this circuit is that it converts a differential input to a single-ended output.

---

[3]The reader may wonder how this is possible if KCL requires that $I_{D2} = |I_{D4}|$. The explanation in Example 3.2 clarifies this issue.

[4]It is also called a differential pair with active load.

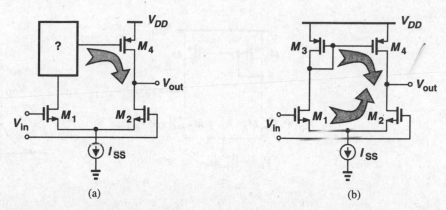

Figure 5.20   (a) Concept of combining the drain currents of M_1 and M_2, (b) realization of (a).

5.3.1 Large-Signal Analysis

Let us study the large-signal behavior of the circuit. To this end, we replace the ideal tail current source by a MOSFET as shown in Fig. 5.21(a). If V_{in1} is much more negative than V_{in2}, M_1 is off and so are M_3 and M_4. Since no current can flow from V_{DD}, both M_2 and M_5 operate in deep triode region, carrying zero current. Thus, $V_{out} = 0.$[5] As V_{in1} approaches V_{in2}, M_1 turns on, drawing part of I_{D5} from M_3 and turning M_4 on. The output voltage then depends on the difference between I_{D4} and I_{D2}. For a small difference between V_{in1} and V_{in2}, both M_2 and M_4 are saturated, providing a high gain [Fig. 5.21(b)]. As V_{in1} becomes more positive than V_{in2}, I_{D1}, $|I_{D3}|$, and $|I_{D4}|$ increase and I_{D2} decreases, eventually driving M_4 into the triode region. If $V_{in1} - V_{in2}$ is sufficiently large, M_2 turns off, M_4 operates in deep triode region with zero current, and $V_{out} = V_{DD}$. Note that if $V_{in1} > V_F + V_{TH}$, then M_1 enters the triode region.

The choice of the input common-mode voltage of the circuit is also important. For M_2 to be saturated, the output voltage cannot be less than $V_{in,CM} - V_{TH}$. Thus, to allow maximum output swings, the input CM level must be as low as possible, with the minimum given by $V_{GS1,2} + V_{DS5,min}$. The direct relationship between the input CM level and the output swing in this circuit is a critical drawback.

What is the output voltage of the circuit when $V_{in1} = V_{in2}$? With perfect symmetry, $V_{out} = V_F = V_{DD} - |V_{GS3}|$. This can be proved by contradiction as well. Suppose, for example, that $V_{out} < V_F$. Then, due to channel-length modulation, M_1 must carry a greater current than M_2 (and M_4 a greater current than M_3). In other words, the total current through M_1 is greater than half of I_{SS}. But this means that the total current through M_3 also exceeds $I_{SS}/2$, violating the assumption that M_4 carries more current than M_3. In reality, however, asymmetries in the circuit may result in a large deviation in V_{out}, possibly driving M_2 or M_4 into the triode region. For example, if the threshold voltage of M_2 is slightly smaller

---

[5]If V_{in1} is greater than one threshold voltage with respect to ground, M_5 may draw a small current from M_1, raising V_{out} slightly.

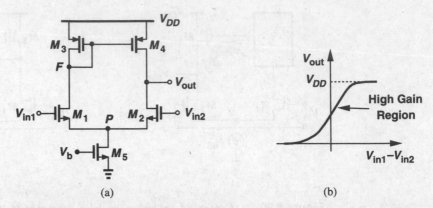

(a)   (b)

Figure 5.21 (a) Differential pair with active current mirror and realistic current source, (b) large-signal input-output characteristic.

than that of M_1, the former carries a greater current than the latter even with $V_{in1} = V_{in2}$, causing V_{out} to drop significantly. For this reason, the circuit is rarely used in an open-loop configuration to amplify small signals.

Example 5.5 ──

Assuming perfect symmetry, sketch the output voltage of the circuit in Fig. 5.22(a) as V_{DD} varies from 3 V to zero. Assume that for $V_{DD} = 3$ V all of the devices are saturated.

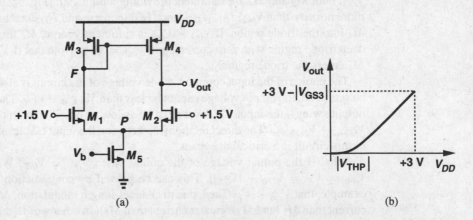

(a)   (b)

Figure 5.22

Solution

For $V_{DD} = 3$ V, symmetry requires that $V_{out} = V_F$. As V_{DD} drops, so do V_F and V_{out} with a slope close to unity [Fig. 5.22(b)]. As V_F and V_{out} fall below $+1.5$ V $-V_{THN}$, M_1 and M_2 enter the triode region, but their drain currents are constant if M_5 is saturated. Further decrease in V_{DD} and hence V_F and V_{out} causes V_{GS1} and V_{GS2} to increase, eventually driving M_5 into the triode region.

Thereafter, the bias current of all of the transistors drops, lowering the rate at which V_{out} decreases. For $V_{DD} < |V_{THP}|$, we have $V_{out} = 0$.

5.3.2 Small-Signal Analysis

We now analyze the small-signal properties of the circuit shown in Fig. 5.21(a), assuming $\gamma = 0$ for simplicity. Can we apply the half-circuit concept to calculate the differential gain here? As illustrated in Fig. 5.23, with small differential inputs, the voltage swings at nodes

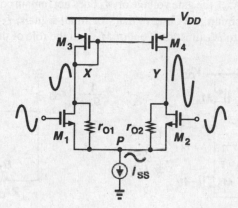

Figure 5.23 Asymmetric swings in a differential pair with active current mirror.

X and Y are vastly different. This is because the diode-connected device M_3 yields a much lower voltage gain from the input to node X than that from the input to node Y. As a result, the effects of V_X and V_Y at node P (through r_{O1} and r_{O2}, respectively) do not cancel each other and this node cannot necessarily be considered a virtual ground. We compute the gain using two different approaches.

In the first approach, we write $|A_v| = G_m R_{out}$ and obtain G_m and R_{out} separately. For the calculation of G_m, consider Fig. 5.24(a). The circuit is not quite symmetric but

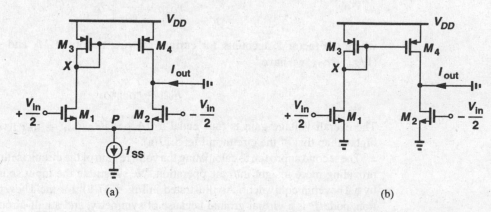

(a)   (b)

Figure 5.24   (a) Circuit for calculation of G_m, (b) circuit of (a) with node P grounded.

because the impedance seen at node X is relatively low and the swing at this node small, the current returning from X to P through r_{O1} is negligible and node P can be viewed as a virtual ground [Fig. 5.24(b)]. Thus, $I_{D1} = |I_{D3}| = |I_{D4}| = g_{m1,2}V_{in}/2$ and $I_{D2} = -g_{m1,2}V_{in}/2$, yielding $I_{out} = -g_{m1,2}V_{in}$ and hence $|G_m| = g_{m1,2}$. Note that, by virtue of active current mirror operation, this value is twice the transconductance of the circuit of Fig. 5.17(b).

Calculation of R_{out} is less straightforward. We may surmise that the output resistance of this circuit is equal to that of the circuit in Fig. 5.17(c), namely, $(2r_{O2})\|r_{O4}$. In reality, however, the active mirror operation yields a different value because when a voltage is applied to the output to measure R_{out}, the gate voltage of M_4 does not remain constant. Rather than draw the entire equivalent circuit, we observe that, for small signals, I_{SS} is open [Fig. 5.25(a)], any current flowing into M_1 must flow out of M_2, and the role of the two transistors can be

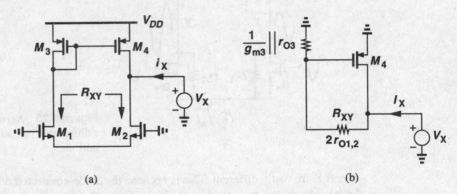

(a)   (b)

Figure 5.25 (a) Circuit for calculating R_{out}, (b) substitution of M_1 and M_2 by a resistor.

represented by a resistor $R_{XY} = 2r_{O1,2}$ [Fig. 5.25(b)]. As a consequence, the current drawn from V_X by R_{XY} is mirrored by M_3 into M_4 with unity gain. We can therefore write:

$$I_X = 2\frac{V_X}{2r_{O1,2} + \dfrac{1}{g_{m3}} \| r_{O3}} + \frac{V_X}{r_{O4}}, \tag{5.27}$$

where the factor 2 accounts for current copying action of M_3 and M_4. For $2r_{O1,2} \gg (1/g_{m3})\|r_{O3}$, we have

$$R_{out} \approx r_{O2}\|r_{O4}. \tag{5.28}$$

The overall voltage gain is thus equal to $|A_v| = G_m R_{out} = g_{m1,2}(r_{O2}\|r_{O4})$, somewhat higher than that of the circuit in Fig. 5.17(a).

The second approach to calculating the voltage gain of the circuit is illustrated in Fig. 5.26, providing more insight into the operation. We substitute the input source and M_1 and M_2 by a Thevenin equivalent. As illustrated in Fig. 5.27(a), for the Thevenin voltage calculation, node P is a virtual ground because of symmetry, and a half-circuit equivalent yields $V_{eq} = g_{m1,2}r_{O1,2}V_{in}$. Moreover, the output resistance is $R_{eq} = 2r_{O1,2}$. From Fig. 5.27(b),

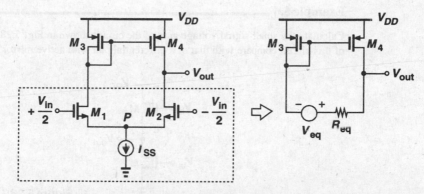

Figure 5.26 Substitution of the input differential pair by a Thevenin equivalent.

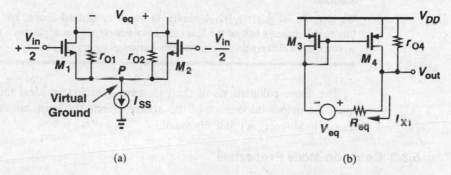

(a)   (b)

Figure 5.27 (a) Calculation of the Thevenin equivalent voltage, (b) simplified circuit.

we note that the current through R_{eq} is

$$I_{X1} = \frac{V_{out} - g_{m1,2}r_{O1,2}V_{in}}{2r_{O1,2} + \dfrac{1}{g_{m3}} \Big\| r_{O3}}. \tag{5.29}$$

The fraction of this current that flows through $1/g_{m3}$ is mirrored into M_4 with unity gain. That is,

$$2\frac{V_{out} - g_{m1,2}r_{O1,2}V_{in}}{2r_{O1,2} + \dfrac{1}{g_{m3}} \Big\| r_{O3}} \cdot \frac{r_{O3}}{r_{O3} + 1/g_{m3}} = -\frac{V_{out}}{r_{O4}}. \tag{5.30}$$

Assuming $2r_{O1,2} \gg (1/g_{m3,4}) \| r_{O3,4}$, we obtain

$$\frac{V_{out}}{V_{in}} = \frac{g_{m1,2}r_{O3,4}r_{O1,2}}{r_{O1,2} + r_{O3,4}} \tag{5.31}$$

$$= g_{m1,2}(r_{O1,2} \| r_{O3,4}). \tag{5.32}$$

Example 5.6 ⎯⎯⎯⎯⎯⎯⎯⎯⎯⎯⎯⎯⎯⎯⎯⎯⎯⎯⎯⎯⎯⎯⎯⎯⎯⎯⎯⎯⎯⎯

Calculate the small-signal voltage gain of the circuit shown in Fig. 5.28. How does the performance of this circuit compare with that of a differential pair with active mirror?

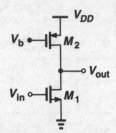

Figure 5.28

Solution

We have $A_v = g_{m1}(r_{O1} \| r_{O2})$, similar to the value derived above. For given device dimensions, this circuit requires half of the bias current to achieve the same gain as a differential pair. However, advantages of differential operation often outweigh the power penalty.

⎯⎯⎯

The above calculations of the gain have assumed an ideal tail current source. In reality, the output impedance of this source affects the gain, but the error with respect to $g_{m1,2}(r_{O1,2} \| r_{O3,4})$ is relatively small.

5.3.3 Common-Mode Properties

Let us now study the common-mode properties of the differential pair with active current mirror. We assume $\gamma = 0$ for simplicity and leave a more general analysis including body effect for the reader. Our objective is to predict the consequences of a finite output impedance in the tail current source. As depicted in Fig. 5.29, a change in the input CM level leads to

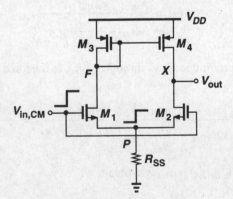

Figure 5.29 Differential pair with active current mirror sensing a common-mode change.

a change in the bias current of all of the transistors. How do we define the common-mode gain here? Recall from Chapter 4 that the CM gain represents the *corruption* of the output signal of interest due to variations of the input CM level. In the circuits of Chapter 3, the output signal was sensed differentially and hence the CM gain was defined in terms of the output differential component generated by the input CM change. In the circuit of Fig. 5.29, on the other hand, the output signal of interest is sensed with respect to ground. Thus, we define the CM gain in terms of the single-ended output component produced by the input CM change:

$$A_{CM} = \frac{\Delta V_{out}}{\Delta V_{in,CM}}. \tag{5.33}$$

To determine A_{CM}, we observe that if the circuit is symmetric, $V_{out} = V_F$ for any input CM level. For example, as $V_{in,CM}$ increases, V_F drops and so does V_{out}. In other words, nodes F and X can be shorted [Fig. 5.30(a)], resulting in the equivalent circuit shown

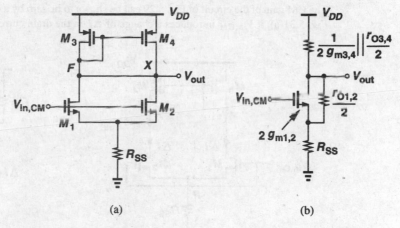

(a)   (b)

Figure 5.30   (a) Simplified circuit of Fig. 5.29, (b) equivalent circuit of (a).

in Fig. 5.30(b). Here, M_1 and M_2 appear in parallel and so do M_3 and M_4. It follows that

$$A_{CM} \approx \frac{-\dfrac{1}{2g_{m3,4}} \left\| \dfrac{r_{O3,4}}{2}\right.}{\dfrac{1}{2g_{m1,2}} + R_{SS}} \tag{5.34}$$

$$= \frac{-1}{1 + 2g_{m1,2}R_{SS}} \frac{g_{m1,2}}{g_{m3,4}}, \tag{5.35}$$

where we have assumed $1/(2g_{m3,4}) \ll r_{O3,4}$ and neglected the effect of $r_{O1,2}/2$. The CMRR is then given by

$$CMRR = \left| \frac{A_{DM}}{A_{CM}} \right| \tag{5.36}$$

$$= g_{m1,2}(r_{O1,2} \| r_{O3,4}) \frac{g_{m3,4}(1 + 2g_{m1,2}R_{SS})}{g_{m1,2}} \tag{5.37}$$

$$= (1 + 2g_{m1,2}R_{SS})g_{m3,4}(r_{O1,2} \| r_{O3,4}). \tag{5.38}$$

Equation (5.35) indicates that, even with perfect symmetry, the output signal is corrupted by input CM variations, a drawback that does not exist in the fully differential circuits of Chapter 3. High-frequency common-mode noise therefore degrades the performance considerably as the capacitance shunting the tail current source exhibits a lower impedance.

Example 5.7

The CM gain of the circuit of Fig. 5.29 can be shown to be *zero* by a (flawed) argument. As shown in Fig. 5.31(a), if $V_{in,CM}$ introduces a change of ΔI in the drain current of each input transistor, then

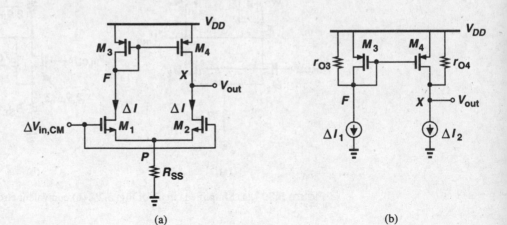

(a)   (b)

Figure 5.31

I_{D3} also experiences the same change and so does I_{D4}. Thus, M_4 seemingly provides the additional current required by M_2, and the output voltage need not change, i.e., $A_{CM} = 0$. Explain the flaw in this proof.

Solution

The assumption that ΔI_{D4} completely cancels the effect of ΔI_{D2} is incorrect. Consider the equivalent circuit shown in Fig. 5.31(b). Since

$$\Delta V_F = \Delta I_1 \left(\frac{1}{g_{m3}} \middle\| r_{O3} \right), \tag{5.39}$$

we have

$$|\Delta I_{D4}| = g_{m4}\Delta V_F \tag{5.40}$$

$$= g_{m4}\Delta I_1 \frac{r_{O3}}{1 + g_{m3}r_{O3}}. \tag{5.41}$$

This current and ΔI_2 $(= \Delta I_1 = \Delta I)$ give a net voltage change equal to

$$\Delta V_{out} = (\Delta I_1 g_{m4}\frac{r_{O3}}{1 + g_{m3}r_{O3}} - \Delta I_2)r_{O4} \tag{5.42}$$

$$= -\Delta I \frac{1}{g_{m3}r_{O3} + 1}r_{O4}, \tag{5.43}$$

which is equal to the voltage change at node F.

It is also instructive to calculate the common-mode gain in the presence of mismatches. As an example, we consider the case where the input transistors exhibit slightly different transconductances [Fig. 5.32(a)]. How does V_{out} depend on $V_{in,CM}$? Since the change at

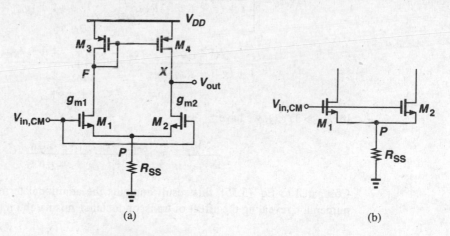

(a)   (b)

Figure 5.32   Differential pair with g_m mismatch.

nodes F and X is relatively small, we can compute the change in I_{D1} and I_{D2} while neglecting the effect of r_{O1} and r_{O2}. As shown in Fig. 5.32(b), the voltage change at P can be obtained by considering M_1 and M_2 as a single transistor (in a source follower configuration) with a transconductance equal to $g_{m1} + g_{m2}$, i.e.,

$$\Delta V_P = \Delta V_{in,CM}\frac{R_{SS}}{R_{SS} + \dfrac{1}{g_{m1} + g_{m2}}}, \tag{5.44}$$

where body effect is neglected. The changes in the drain currents of M_1 and M_2 are therefore given by

$$\Delta I_{D1} = g_{m1}(\Delta V_{in,CM} - \Delta V_P) \tag{5.45}$$

$$= \frac{\Delta V_{in,CM}}{R_{SS} + \dfrac{1}{g_{m1} + g_{m2}}} \frac{g_{m1}}{g_{m1} + g_{m2}} \tag{5.46}$$

$$\Delta I_{D2} = g_{m2}(\Delta V_{in,CM} - \Delta V_P) \tag{5.47}$$

$$= \frac{\Delta V_{in,CM}}{R_{SS} + \dfrac{1}{g_{m1} + g_{m2}}} \frac{g_{m2}}{g_{m1} + g_{m2}}. \tag{5.48}$$

The change ΔI_{D1} multiplied by $(1/g_{m3}) \| r_{O3}$ yields $|\Delta I_{D4}| = g_{m4}[(1/g_{m3}) \| r_{O3}]\Delta I_{D1}$. The difference between this current and ΔI_{D2} flows through the output impedance of the circuit, which is equal to r_{O4} because we have neglected the effect of r_{O1} and r_{O2}:

$$\Delta V_{out} = \left[\frac{g_{m1}\Delta V_{in,CM}}{1 + (g_{m1} + g_{m2})R_{SS}} \frac{r_{O3}}{r_{O3} + \dfrac{1}{g_{m3}}} - \frac{g_{m2}\Delta V_{in,CM}}{1 + (g_{m1} + g_{m2})R_{SS}} \right] r_{O4} \tag{5.49}$$

$$= \frac{\Delta V_{in,CM}}{1 + (g_{m1} + g_{m2})R_{SS}} \frac{(g_{m1} - g_{m2})r_{O3} - g_{m2}/g_{m3}}{r_{O3} + \dfrac{1}{g_{m3}}} r_{O4}. \tag{5.50}$$

If $r_{O3} \gg 1/g_{m3}$, we have

$$\frac{\Delta V_{out}}{\Delta V_{in,CM}} \approx \frac{(g_{m1} - g_{m2})r_{O3} - g_{m2}/g_{m3}}{1 + (g_{m1} + g_{m2})R_{SS}}. \tag{5.51}$$

Compared to Eq. (5.35), this result contains the additional term $(g_{m1} - g_{m2})r_{O3}$ in the numerator, revealing the effect of transconductance mismatch on the common-mode gain.

Problems

Unless otherwise stated, in the following problems, use the device data shown in Table 2.1 and assume $V_{DD} = 3$ V where necessary. All device dimensions are effective values and in microns.

5.1. In Fig. 5.2, assume $(W/L)_1 = 50/0.5$, $\lambda = 0$, $I_{out} = 0.5$ mA, and M_1 is saturated.
  (a) Determine R_2/R_1.
  (b) Calculate the sensitivity of I_{out} to V_{DD}, defined as $\partial I_{out}/\partial V_{DD}$ and normalized to I_{out}.
  (c) How much does I_{out} change if V_{TH} changes by 50 mV?
  (d) If the temperature dependence of μ_n is expressed as $\mu_n \propto T^{-3/2}$ but V_{TH} is independent of temperature, how much does I_{out} vary if T changes from $300°$K to $370°$K?

(e) What is the worst-case change in I_{out} if V_{DD} changes by 10%, V_{TH} by 50 mV, and T from 300°K to 370°K?

5.2. Consider the circuit of Fig. 5.6. Assuming I_{REF} is *ideal*, sketch I_{out} versus V_{DD} as V_{DD} varies from 0 to 3 V.

5.3. In the circuit of Fig. 5.7, $(W/L)_N = 10/0.5$, $(W/L)_P = 10/0.5$, and $I_{REF} = 100 \ \mu A$. The input CM level applied to the gates of M_1 and M_2 is equal to 1.3 V.
   (a) Assuming $\lambda = 0$, calculate V_P and the drain voltage of the PMOS diode-connected transistors.
   (b) Now take channel-length modulation into account to determine I_T and the drain current of the PMOS diode-connected transistors more accurately.

5.4. Consider the circuit of Fig. 5.8; sketch V_{out} versus V_{DD} as V_{DD} varies from 0 to 3 V.

5.5. Consider the circuit of Fig. 5.9(a), assuming $(W/L)_{1-3} = 40/0.5$, $I_{REF} = 0.3 \ mA$, and $\gamma = 0$.
   (a) Determine V_b such that $V_X = V_Y$.
   (b) If V_b deviates from the value calculated in part (a) by 100 mV, what is the mismatch between I_{out} and I_{REF}?
   (c) If the circuit fed by the cascode current source changes V_P by 1 V, how much does V_Y change?

5.6. The circuit of Fig. 5.13 is designed with $(W/L)_{1,2} = 20/0.5$, $(W/L)_{3,4} = 60/0.5$, and $I_{REF} = 100 \ \mu A$.
   (a) Determine V_X and the acceptable range of V_b.
   (b) Estimate the deviation of I_{out} from 300 μA if the drain voltage of M_4 is higher than V_X by 1 V.

5.7. The circuit of Fig. 5.17(a) is designed with $(W/L)_{1-4} = 50/0.5$ and $I_{SS} = 2I_1 = 0.5 \ mA$.
   (a) Calculate the small-signal voltage gain.
   (b) Determine the maximum output voltage swing if the input CM level is 1.3 V.

5.8. Consider the circuit of Fig. 5.22(a) with $(W/L)_{1-5} = 50/0.5$ and $I_{D5} = 0.5 \ mA$.
   (a) Calculate the deviation of V_{out} from V_F if $|V_{TH3}|$ is 1 mV less than $|V_{TH4}|$.
   (b) Determine the CMRR of the amplifier.

5.9. Sketch V_X and V_Y as a function of V_{DD} for each circuit in Fig. 5.33. Assume the transistors in each circuit are identical.

5.10. Sketch V_X and V_Y as a function of V_{DD} for each circuit in Fig. 5.34. Assume the transistors in each circuit are identical.

5.11. For each circuit in Fig. 5.35, sketch V_X and V_Y as a function of V_1 for $0 < V_1 < V_{DD}$. Assume the transistors in each circuit are identical.

5.12. For each circuit in Fig. 5.36, sketch V_X and V_Y as a function of V_1 for $0 < V_1 < V_{DD}$. Assume the transistors in each circuit are identical.

5.13. For each circuit in Fig. 5.37, sketch V_X and V_Y as a function of I_{REF}.

5.14. For the circuit of Fig. 5.38, sketch I_{out}, V_X, V_A, and V_B as a function of (a) I_{REF}, (b) V_b.

5.15. In the circuit shown in Fig. 5.39, a source follower using a wide transistor and a small bias current is inserted in series with the gate of M_3 so as to bias M_2 at the edge of saturation. Assuming M_0-M_3 are identical and $\lambda \neq 0$, estimate the mismatch between I_{out} and I_{REF} if (a) $\gamma = 0$, (b) $\gamma \neq 0$.

5.16. Sketch V_X and V_Y as a function of time for each circuit in Fig. 5.40. Assume the transistors in each circuit are identical.

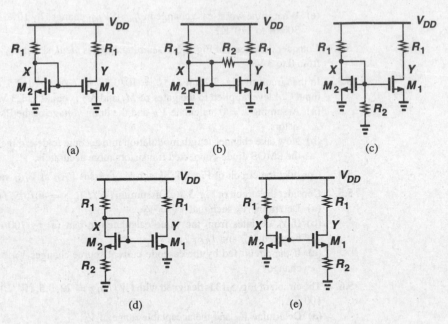

Figure 5.33

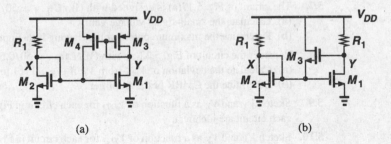

Figure 5.34

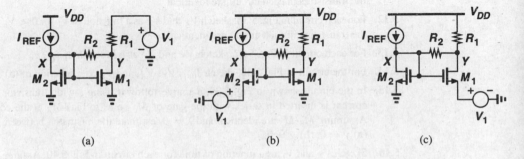

Figure 5.35

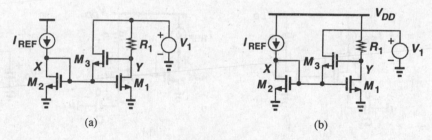

(a)   (b)

Figure 5.36

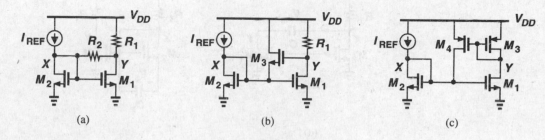

(a)   (b)   (c)

Figure 5.37

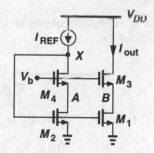

Figure 5.38

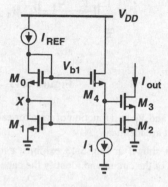

Figure 5.39

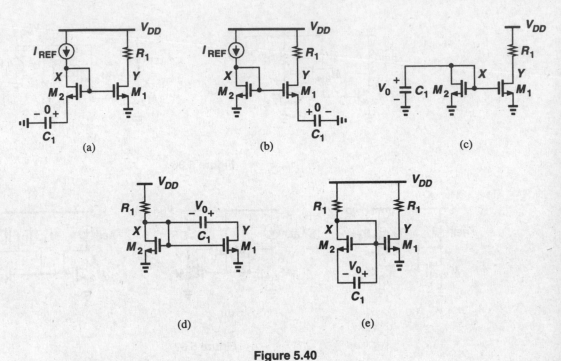

Figure 5.40

5.17. Sketch V_X and V_Y as a function of time for each circuit in Fig. 5.41. Assume the transistors in each circuit are identical.

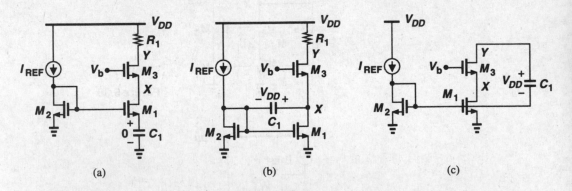

Figure 5.41

5.18. Sketch V_X and V_Y as a function of time for each circuit in Fig. 5.42. Assume the transistors in each circuit are identical.

5.19. The circuit shown in Fig. 5.43 exhibits a *negative* input capacitance. Calculate the input impedance of the circuit and identify the capacitive component.

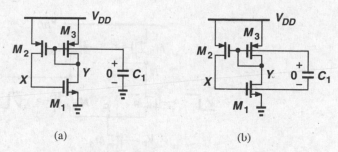

Figure 5.42

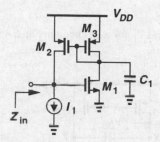

Figure 5.43

5.20. Due to a manufacturing defect, a large parasitic resistance, R_1, has appeared in the circuits of Fig. 5.44. Calculate the gain of each circuit.

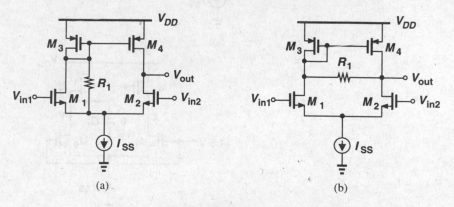

Figure 5.44

5.21. In digital circuits such as memories, a differential pair with active current mirror is used to convert a small differential signal to a large single-ended swing (Fig. 5.45). In such applications, it is desirable that the output levels be as close to the supply rails as possible. Assuming moderate differential input swings (e.g., $\Delta V = 0.1$ V) around a common-mode level $V_{in,CM}$ and a high gain in the circuit, explain why V_{min} depends on $V_{in,CM}$.

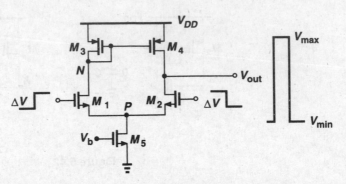

Figure 5.45

5.22. Sketch V_X and V_Y for each circuit in Fig. 5.46 as a function of time. The initial voltage across C_1 is shown.

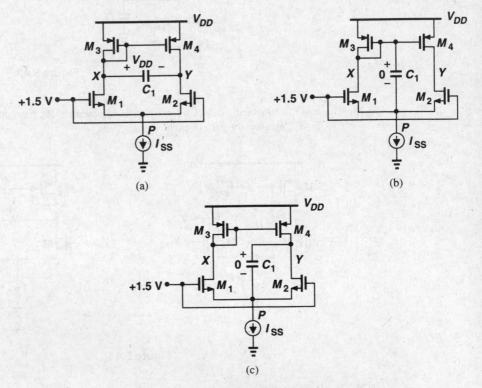

Figure 5.46

5.23. If in Fig. 5.47, ΔV is small enough that all of the transistors remain in saturation, determine the time constant and the initial and final values of V_{out}.

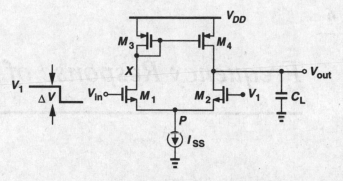

Figure 5.47

Chapter 6

Frequency Response of Amplifiers

Our analysis of simple amplifiers has thus far focussed on low-frequency characteristics, neglecting the effect of device and load capacitances. In most analog circuits, however, the speed trades with many other parameters such as gain, power dissipation, and noise. It is therefore necessary to understand the frequency response limitations of each circuit.

In this chapter, we study the response of single-stage and differential amplifiers in the frequency domain. Following a review of basic concepts, we analyze the high-frequency behavior of common-source and common-gate stages and source followers. Next, we deal with cascode and differential amplifiers. Finally, we consider the effect of active current mirrors on the frequency response of differential pairs.

6.1 General Considerations

6.1.1 Miller Effect

An important phenomenon that occurs in many analog (and digital) circuits is related to "Miller Effect," as described by Miller in a theorem.

Miller's Theorem. If the circuit of Fig. 6.1(a) can be converted to that of Fig. 6.1(b), then $Z_1 = Z/(1 - A_v)$ and $Z_2 = Z/(1 - A_v^{-1})$, where $A_v = V_Y/V_X$.

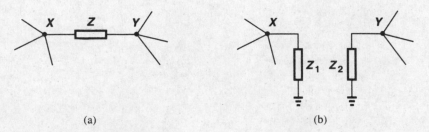

(a)   (b)

Figure 6.1 Application of Miller effect to a floating impedance.

Proof.   The current flowing through Z from X to Y is equal to $(V_X - V_Y)/Z$. For the two circuits to be equivalent, the same current must flow through Z_1. Thus,

$$\frac{V_X - V_Y}{Z} = \frac{V_X}{Z_1}, \tag{6.1}$$

that is,

$$Z_1 = \frac{Z}{1 - \dfrac{V_Y}{V_X}}. \tag{6.2}$$

Similarly,

$$Z_2 = \frac{Z}{1 - \dfrac{V_X}{V_Y}}. \tag{6.3}$$

□

Example 6.1 ──

Consider the circuit shown in Fig. 6.2(a), where the voltage amplifier has a negative gain equal to $-A$ and is otherwise ideal. Calculate the input capacitance of the circuit.

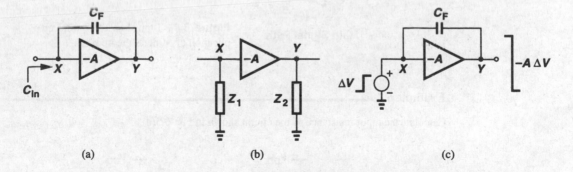

(a)   (b)   (c)

Figure 6.2

Solution

Using Miller's theorem to convert the circuit to that shown in Fig. 6.2(b), we have $Z = 1/(C_F s)$ and $Z_1 = [1/(C_F s)]/(1 + A)$. That is, the input capacitance is equal to $C_F(1 + A)$.

Why is C_F multiplied by $1 + A$? Suppose, as depicted in Fig. 6.2(c), we measure the input capacitance by applying a voltage step at the input and calculating the charge supplied by the voltage source. A step equal to ΔV at X results in a change of $-A\Delta V$ at Y, yielding a total change of $(1+A)\Delta V$ in the voltage across C_F. Thus, the charge drawn by C_F from V_{in} is equal to $(1 + A)C_F\Delta V$ and the equivalent input capacitance equal to $(1 + A)C_F$.

It is important to understand that (6.2) and (6.3) hold *if* we know a priori that the circuit of Fig. 6.1(a) can be converted to that of Fig. 6.1(b). That is, Miller's theorem does not stipulate the conditions under which this conversion is valid. If the impedance Z forms the only signal path between X and Y, then the conversion is often invalid. Illustrated in Fig. 6.3 for a simple resistive divider, the theorem gives a correct input impedance but an incorrect

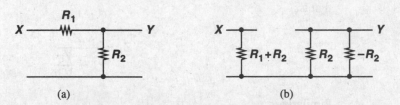

(a)   (b)

Figure 6.3   Improper application of Miller's theorem.

gain. Nevertheless, Miller's theorem proves useful in cases where the impedance Z appears in parallel with the main signal (Fig. 6.4).

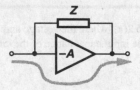

Main Signal Path

Figure 6.4   Typical case for valid application of Miller's theorem.

Example 6.2 _____

Calculate the input resistance of the circuit shown in Fig. 6.5(a).

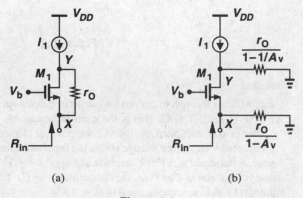

(a)   (b)

Figure 6.5

Solution

The reader can prove that the voltage gain from X to Y is equal to $1 + (g_m + g_{mb})r_O$. As shown in Fig. 6.5(b), the input resistance is given by the parallel combination of $r_O/(1 - A_v)$ and $1/(g_m + g_{mb})$. Since A_v is usually greater than unity, $r_O/(1 - A_v)$ is a *negative* resistance. We therefore have

$$R_{in} = \frac{r_O}{1 - [1 + (g_m + g_{mb})r_O]} \left\| \frac{1}{g_m + g_{mb}} \right. \tag{6.4}$$

$$= \frac{-1}{g_m + g_{mb}} \left\| \frac{1}{g_m + g_{mb}} \right. \tag{6.5}$$

$$= \infty. \tag{6.6}$$

This is the same result as obtained in Chapter 3 (Fig. 3.46) by direct calculation.

We should also mention that, strictly speaking, the value of $A_v = V_Y/V_X$ in (6.2) and (6.3) must be calculated at the frequency of interest, complicating the algebra significantly. However, in many cases we use the low-frequency value of A_v to gain insight into the behavior of the circuit.

If applied to obtain the input-output transfer function, Miller's theorem cannot be used simultaneously to calculate the output impedance. To derive the transfer function, we apply a voltage source to the *input* of the circuit, obtaining a value for V_Y/V_X in Fig. 6.1(a). On the other hand, to determine the output impedance, we apply a voltage source to the *output* of the circuit, obtaining a value for V_X/V_Y that may not be equal to the inverse of the V_Y/V_X measured in the first test. For example, the circuit of Fig. 6.5(b) may suggest that the output impedance is equal to

$$R_{out} = \frac{r_O}{1 - 1/A_v} \tag{6.7}$$

$$= \frac{r_O}{1 - [1 + (g_m + g_{mb})r_O]^{-1}} \tag{6.8}$$

$$= \frac{1}{g_m + g_{mb}} + r_O, \tag{6.9}$$

whereas the actual value is equal to r_O (if X is grounded). Other subtleties of Miller's theorem are decribed in the appendix.

6.1.2 Association of Poles with Nodes

Consider the simple cascade of amplifiers depicted in Fig. 6.6. Here, A_1 and A_2 are ideal voltage amplifiers, R_1 and R_2 model the output resistance of each stage, C_{in} and C_N represent the input capacitance of each stage, and C_P denotes the load capacitance. The overall transfer function can be written as

$$\frac{V_{out}}{V_{in}}(s) = \frac{A_1}{1 + R_S C_{in}s} \cdot \frac{A_2}{1 + R_1 C_N s} \cdot \frac{1}{1 + R_2 C_P s}. \tag{6.10}$$

The circuit exhibits three poles, each of which is determined by the total capacitance seen from each node to ground multiplied by the total resistance seen at the node to ground. We can therefore associate each pole with one node of the circuit, i.e., $\omega_j = \tau_j^{-1}$, where τ_j is the product of the capacitance and resistance seen at node j to ground. From this perspective, we may say "each node in the circuit contributes one pole to the transfer function."

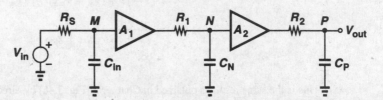

Figure 6.6 Cascade of amplifiers.

The above statement is not valid in general. For example, in the circuit of Fig. 6.7, the location of the poles is difficult to calculate because R_3 and C_3 create interaction between

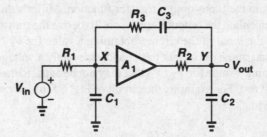

Figure 6.7 Example of interaction between nodes.

X and Y. Nevertheless, in many circuits association of one pole with each node provides an intuitive approach to estimating the transfer function: we simply multiply the total equivalent capacitance by the total incremental resistance (both from the node of interest to ground), thus obtaining an equivalent time constant and hence a pole frequency.

Example 6.3 _____

In Fig. 6.8, calculate the pole associated with node X.

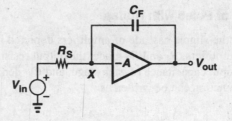

Figure 6.8

Solution

From Fig. 6.2(b), the total equivalent capacitance seen from X to ground equals $(1 + A)C_F$. Since this capacitance is driven by R_S, the pole frequency is equal to $1/[R_S(1 + A)C_F]$ (in rad/s). We call this the "input pole."

The above approach does suffer from some limitations. In particular, the simplification of the circuit through the use of Miller effect often discards the *zeros* of the transfer function. However, the utility of the method becomes apparent in more complex topologies, as described in the following example.

Example 6.4

Neglecting channel-length modulation, compute the transfer function of the common-gate stage shown in Fig. 6.9.

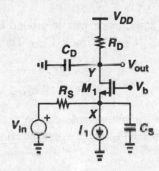

Figure 6.9 Common-gate stage with parasitic capacitances.

Solution

In this circuit, the capacitances contributed by M_1 are connected from the input and output nodes to ground. At node X, $C_S = C_{GS1} + C_{SB1}$, giving a pole frequency

$$\omega_{in} = \left[(C_{GS1} + C_{SB1}) \left(R_S \left\| \frac{1}{g_{m1} + g_{mb1}} \right. \right) \right]^{-1}$$ (6.11)

Similarly, at node Y, $C_D = C_{DG} + C_{DB}$, yielding a pole frequency

$$\omega_{out} = [(C_{DG} + C_{DB})R_D]^{-1}.$$ (6.12)

The overall transfer function is thus given by

$$\frac{V_{out}}{V_{in}}(s) = \frac{(g_m + g_{mb})R_D}{1 + (g_m + g_{mb})R_S} \cdot \frac{1}{\left(1 + \dfrac{s}{\omega_{in}}\right)\left(1 + \dfrac{s}{\omega_{out}}\right)},$$ (6.13)

where the first fraction represents the low-frequency gain of the circuit. Note that if we do not neglect r_{O1}, the input and output nodes interact, making it difficult to calculate the poles.

6.2 Common-Source Stage

The common-source topology exhibits a relatively high input impedance while providing voltage gain and requiring a minimal voltage headroom. As such, it finds wide application in analog circuits and its frequency response is of interest.

Shown in Fig. 6.10 is a common-source stage driven by a finite source resistance, R_S. We identify all of the capacitances in the circuit, noting that C_{GS} and C_{DB} are "grounded" capacitances while C_{GD} appears between the input and the output. Assuming that $\lambda = 0$ and M_1 operates in saturation, let us first estimate the transfer function by associating one pole with each node. The total capacitance seen from X to ground is equal to C_{GS} plus the Miller multiplication of C_{GD}: $C_{GS} + (1 - A_v)C_{GD}$, where $A_v = -g_m R_D$. The magnitude of the input pole is therefore given by

$$\omega_{in} = \frac{1}{R_S[C_{GS} + (1 + g_m R_D)C_{GD}]}. \tag{6.14}$$

At the output node, the total capacitance seen to ground is equal to C_{DB} plus the Miller effect of C_{GD}: $C_{DB} + (1 - A_v^{-1})C_{GD} \approx C_{DB} + C_{GD}$. Thus,

$$\omega_{out} = \frac{1}{R_D(C_{DB} + C_{GD})}. \tag{6.15}$$

Another approximation of the output pole can be obtained if R_S is relatively large. Simplifying the circuit as shown in Fig. 6.11, where the effect of R_S is neglected, the reader can prove that

$$Z_X = \frac{1}{C_{eq}s} \left\| \left(\frac{C_{GD} + C_{GS}}{C_{GD}} \cdot \frac{1}{g_{m1}} \right), \right. \tag{6.16}$$

where $C_{eq} = C_{GD}C_{GS}/(C_{GD} + C_{GS})$. Thus, the output pole is roughly equal to

$$\omega_{out} = \frac{1}{\left[R_D \left\| \left(\frac{C_{GD} + C_{GS}}{C_{GD}} \cdot \frac{1}{g_{m1}} \right) \right] (C_{eq} + C_{DB})}. \tag{6.17}$$

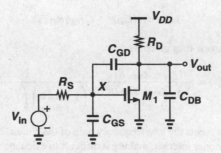

Figure 6.10 High-frequency model of a common-source stage.

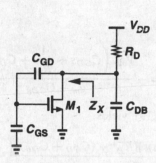

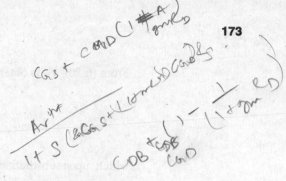

Figure 6.11 Model for calculation of output impedance.

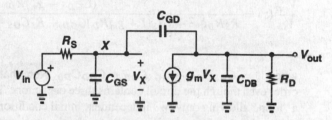

Figure 6.12 Equivalent circuit of Fig. 6.10.

We then surmise that the transfer function is

$$\frac{V_{out}}{V_{in}}(s) = \frac{-g_m R_D}{\left(1 + \dfrac{s}{\omega_{in}}\right)\left(1 + \dfrac{s}{\omega_{out}}\right)}. \tag{6.18}$$

Note that r_{O1} and any load capacitance can easily be included here.

The primary error in this estimation is that we have not considered the existence of zeros in the circuit. Another concern stems from approximating the gain of the amplifier by $-g_m R_D$ whereas in reality the gain varies with frequency (for example, due to the capacitance at the output node).

We now obtain the exact transfer function, investigating the validity of the above approach. Using the equivalent circuit depicted in Fig. 6.12, we can sum the currents at each node:

$$\frac{V_X - V_{in}}{R_S} + V_X C_{GS}s + (V_X - V_{out})C_{GD}s = 0 \tag{6.19}$$

$$(V_{out} - V_X)C_{GD}s + g_m V_X + V_{out}\left(\frac{1}{R_D} + C_{DB}s\right) = 0. \tag{6.20}$$

From (6.20), V_X is obtained as

$$V_X = -\frac{V_{out}\left(C_{GD}s + \dfrac{1}{R_D} + C_{DB}s\right)}{g_m - C_{GD}s}, \tag{6.21}$$

which, upon substitution in (6.19), yields

$$-V_{out}\frac{[R_S^{-1} + (C_{GS} + C_{GD})s][R_D^{-1} + (C_{GD} + C_{DB})s]}{g_m - C_{GD}s} - V_{out}C_{GD}s = \frac{V_{in}}{R_S}. \tag{6.22}$$

That is,

$$\frac{V_{out}}{V_{in}}(s) = \frac{(C_{GD}s - g_m)R_D}{R_S R_D \xi s^2 + [R_S(1 + g_m R_D)C_{GD} + R_S C_{GS} + R_D(C_{GD} + C_{DB})]s + 1}, \tag{6.23}$$

where $\xi = C_{GS}C_{GD} + C_{GS}C_{DB} + C_{GD}C_{DB}$. Note that the transfer function is of second order even though the circuit contains three capacitors. This is because the capacitors form a "loop," allowing only *two* independent initial conditions in the circuit and hence yielding a second-order differential equation for the time response.

If manipulated judiciously, Eq. (6.23) reveals several interesting points about the circuit. While the denominator appears rather complicated, it can yield intuitive expressions for the two poles, ω_{p1} and ω_{p2}, if we assume $|\omega_{p1}| \ll |\omega_{p2}|$ [1]. Writing the denominator as

$$D = \left(\frac{s}{\omega_{p1}} + 1\right)\left(\frac{s}{\omega_{p2}} + 1\right) \tag{6.24}$$

$$= \frac{s^2}{\omega_{p1}\omega_{p2}} + \left(\frac{1}{\omega_{p1}} + \frac{1}{\omega_{p2}}\right)s + 1, \tag{6.25}$$

we recognize that the coefficient of s is approximately equal to $1/\omega_{p1}$ if ω_{p2} is much farther from the origin. It follows from (6.23) that

$$\omega_{p1} = \frac{1}{R_S(1 + g_m R_D)C_{GD} + R_S C_{GS} + R_D(C_{GD} + C_{DB})}. \tag{6.26}$$

How does this compare with the "input" pole given by (6.14)? The only difference results from the term $R_D(C_{GD} + C_{DB})$, which may be negligible in some cases. The key point here is that the intuitive approach of associating a pole with the input node provides a rough estimate with much less effort. We also note that the Miller multiplication of C_{GD} by the low-frequency gain of the amplifier is relatively accurate in this case.

Example 6.5

For the circuit shown in Fig. 6.13, calculate the transfer function (with $\lambda = 0$) and explain why Miller effect vanishes as C_{DB} increases.

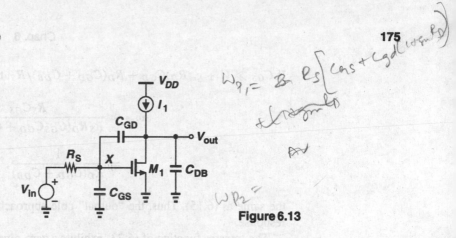

Figure 6.13

Handwritten notes (top right):
$$\omega_{p_1} = \frac{1}{R_S[C_{GS} + C_{gd}(1 + g_m R_D)]}$$

A_V

$\omega_{p_2} =$

Solution

Using (6.23) and letting R_D approach infinity, we have

$$\frac{V_{out}}{V_{in}}(s) = \frac{C_{GD}s - g_m}{R_S \xi s^2 + [g_m R_S C_{GD} + (C_{GD} + C_{DB})]s} \tag{6.27}$$

$$= \frac{C_{GD}s - g_m}{s[R_S(C_{GS}C_{GD} + C_{GS}C_{DB} + C_{GD}C_{DB})s + (g_m R_S + 1)C_{GD} + C_{DB}]}. \tag{6.28}$$

As expected, the circuit exhibits two poles—one at the origin because the dc gain is infinity. The magnitude of the other pole is given by

$$\omega_2 \approx \frac{(1 + g_m R_S)C_{GD} + C_{DB}}{R_S(C_{GD}C_{GS} + C_{GS}C_{DB} + C_{GD}C_{DB})}. \tag{6.29}$$

For large C_{DB}, this expression reduces to

$$\omega_2 \approx \frac{1}{R_S(C_{GS} + C_{GD})}, \tag{6.30}$$

indicating that C_{GD} experiences no Miller multiplication. This can be explained by noting that, for large C_{DB}, the voltage gain from node X to the output begins to drop even at low frequencies. As a result, for frequencies close to $[R_S(C_{GS} + C_{GD})]^{-1}$, the effective gain is quite small and $C_{GD}(1 - A_v) \approx C_{GD}$. Such a case is an example where the application of Miller effect using low-frequency gain does not provide a reasonable estimate.

From (6.23), we can also estimate the second pole of the CS stage of Fig. 6.10. Since the coefficient of s^2 is equal to $(\omega_{p1}\omega_{p2})^{-1}$, we have

$$\omega_{p2} = \frac{1}{\omega_{p1}} \cdot \frac{1}{R_S R_D(C_{GS}C_{GD} + C_{GS}C_{DB} + C_{GD}C_{DB})} \tag{6.31}$$

$$= \frac{R_S(1 + g_m R_D)C_{GD} + R_S C_{GS} + R_D(C_{GD} + C_{DB})}{R_S R_D(C_{GS}C_{GD} + C_{GS}C_{DB} + C_{GD}C_{DB})}. \tag{6.32}$$

If $C_{GS} \gg (1 + g_m R_D)C_{GD} + R_D(C_{GD} + C_{DB})/R_S$, then

$$\omega_{p2} \approx \frac{R_S C_{GS}}{R_S R_D(C_{GS}C_{GD} + C_{GS}C_{DB})} \tag{6.33}$$

$$= \frac{1}{R_D(C_{GD} + C_{DB})}, \tag{6.34}$$

the same as (6.15). Thus, the "output" pole approach is valid only if C_{GS} dominates the response.

The transfer function of (6.23) exhibits a zero given by $\omega_z = +g_m/C_{GD}$, an effect not predicted by the simple approach leading to (6.18). Located in the *right* half plane, the zero arises from direct coupling of the input to the output through C_{GD}. As illustrated in Fig. 6.14, C_{GD} provides a feedforward path that conducts the input signal to the output at very high frequencies, resulting in a slope in the frequency response that is less negative than -40 dB/dec. As explained in Chapter 10, a zero in the right half plane introduces stability issues in feedback amplifiers.

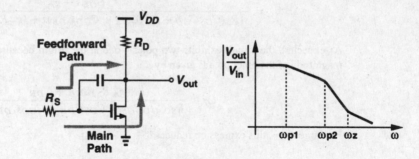

Figure 6.14 Feedforward path through C_{GD} (log-log scale).

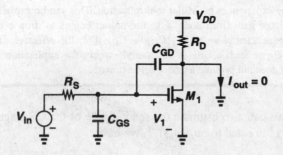

Figure 6.15 Calculation of the zero in a CS stage.

The zero, s_z, can also be computed by noting that the transfer function $V_{out}(s)/V_{in}(s)$ must drop to zero for $s = s_z$. For a finite V_{in}, this means that $V_{out}(s_z) = 0$ and hence the output can be *shorted* to ground at this (possibly complex) frequency with no current flowing through the short (Fig. 6.15). Therefore, the currents through C_{GD} and M_1 are equal and opposite:

$$V_1 C_{GD} s_z = g_m V_1. \tag{6.35}$$

That is, $s_z = +g_m/C_{GD}$.[1]

In high-speed applications, the input impedance of the common-source stage is also important. As a first-order approximation, we have from Fig. 6.16(a)

$$Z_{in} = \frac{1}{[C_{GS} + (1 + g_m R_D)C_{GD}]s}. \tag{6.36}$$

But at high frequencies, the effect of the output node must be taken into account. Ignoring C_{GS} for the moment and using the circuit of Fig. 6.16(b), we write

$$(I_X - g_m V_X)\frac{R_D}{1 + R_D C_{DBS}} + \frac{I_X}{C_{GDS}} = V_X, \tag{6.37}$$

and hence

$$\frac{V_X}{I_X} = \frac{1 + R_D(C_{GD} + C_{DB})s}{C_{GDS}(1 + g_m R_D + R_D C_{DBS})}. \tag{6.38}$$

The actual input impedance consists of the parallel combination of (6.38) and $1/(C_{GSS})$.

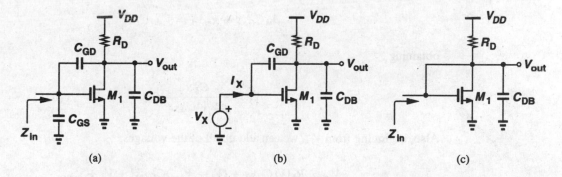

Figure 6.16   Calculation of input impedance of a CS stage.

At frequencies where $|R_D(C_{GD} + C_{DB})s| \ll 1$ and $|R_D C_{DBS}| \ll 1 + g_m R_D$, (6.38) reduces to $[(1 + g_m R_D)C_{GDS}]^{-1}$ (as expected), indicating that the input impedance is primarily capacitive. At higher frequencies, however, (6.38) contains both real and imaginary parts. In fact, if C_{GD} is large, it provides a low-impedance path between the gate and drain of M_1, yielding the equivalent circuit of Fig. 6.16(c) and suggesting that $1/g_{m1}$ and R_D appear in parallel with the input.

---

[1] This approach is similar to expressing the transfer function as $G_m Z_{out}$ and finding the zeros of G_m and Z_{out}.

6.3 Source Followers

Source followers are occasionally employed as level shifters or buffers, impacting the overall frequency response. Consider the circuit depicted in Fig. 6.17(a), where C_L represents the total capacitance seen at the output node to ground, including C_{SB1}. The strong inter-

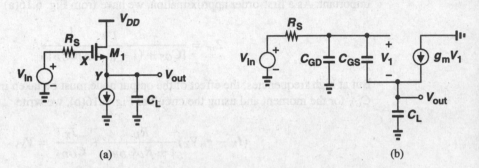

Figure 6.17 (a) Source follower, (b) high-frequency equivalent circuit.

action between nodes X and Y through C_{GS} in Fig. 6.17(a) makes it difficult to associate a pole with each node in a source follower. Neglecting body effect for simplicity and using the equivalent circuit shown in Fig. 6.17(b), we can sum the currents at the output node:

$$V_1 C_{GS}s + g_m V_1 = V_{out}C_L s, \qquad (6.39)$$

obtaining

$$V_1 = \frac{C_L s}{g_m + C_{GS}s} V_{out}. \qquad (6.40)$$

Also, beginning from V_{in}, we can add up all of the voltages:

$$V_{in} = R_S[V_1 C_{GS}s + (V_1 + V_{out})C_{GD}s] + V_1 + V_{out}. \qquad (6.41)$$

Substituting for V_1 from (6.40), we have

$$\frac{V_{out}}{V_{in}}(s) = \frac{g_m + C_{GS}s}{R_S(C_{GS}C_L + C_{GS}C_{GD} + C_{GD}C_L)s^2 + (g_m R_S C_{GD} + C_L + C_{GS})s + g_m}. \qquad (6.42)$$

Interestingly, the transfer function contains a zero in the *left* half plane. This is because the signal conducted by C_{GS} at high frequencies adds with the same polarity to the signal produced by the intrinsic transistor.

If the two poles of (6.42) are assumed far apart, then the more significant one has a magnitude of

$$\omega_{p1} \approx \frac{g_m}{g_m R_S C_{GD} + C_L + C_{GS}} \tag{6.43}$$

$$= \frac{1}{R_S C_{GD} + \dfrac{C_L + C_{GS}}{g_m}}. \tag{6.44}$$

Also, if $R_S = 0$, then $\omega_{p1} = g_m/(C_L + C_{GS})$.

Let us now calculate the input impedance of the circuit, noting that C_{GD} simply shunts the input and can be ignored initially. From the equivalent shown in Fig. 6.18, the small-signal

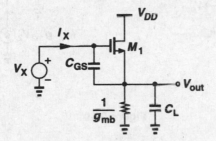

Figure 6.18   Calculation of source follower input impedance.

gate-source voltage of M_1 is equal to $I_X/(C_{GS}s)$, giving a source current of $g_m I_X/(C_{GS}s)$. Starting from the input and adding the voltages, we have

$$V_X = \frac{I_X}{C_{GS}s} + \left(I_X + \frac{g_m I_X}{C_{GS}s}\right)\left(\frac{1}{g_{mb}}\bigg\|\frac{1}{C_L s}\right), \tag{6.45}$$

that is,

$$Z_{in} = \frac{1}{C_{GS}s} + \left(1 + \frac{g_m}{C_{GS}s}\right)\frac{1}{g_{mb} + C_L s}. \tag{6.46}$$

At relatively low frequencies, $g_{mb} \gg |C_L s|$ and

$$Z_{in} \approx \frac{1}{C_{GS}s}\left(1 + \frac{g_m}{g_{mb}}\right) + \frac{1}{g_{mb}}, \tag{6.47}$$

indicating that the equivalent input capacitance is equal to $C_{GS}g_{mb}/(g_m + g_{mb})$. This result can also be obtained by Miller approximation. Since the low-frequency gain from the input to the output equals $g_m/(g_m + g_{mb})$, the effect of C_{GS} at the input can be expressed as $C_{GS}[1 - g_m/(g_m + g_{mb})] = C_{GS}g_{mb}(g_m + g_{mb})$. In other words, the overall input capacitance is equal to C_{GD} plus a *fraction* of C_{GS}.

At high frequencies, $g_{mb} \ll |C_L s|$ and

$$Z_{in} \approx \frac{1}{C_{GS}s} + \frac{1}{C_L s} + \frac{g_m}{C_{GS}C_L s^2}. \tag{6.48}$$

For a given $s = j\omega$, the input impedance consists of the series combination of capacitors C_{GS} and C_L and a *negative* resistance equal to $-g_m/(C_{GS}C_L\omega^2)$. The negative resistance property can be utilized in oscillators [2].

Example 6.6

Calculate the transfer function of the circuit shown in Fig. 6.19(a).

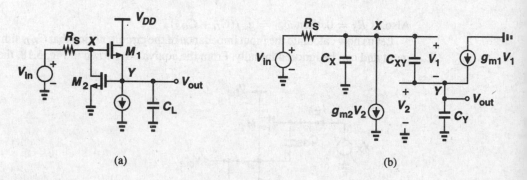

(a)   (b)

Figure 6.19

Solution

Let us first identify all of the capacitances in the circuit. At node X, C_{GD1} and C_{DB2} are connected to ground and C_{GS1} and C_{GD2} to Y. At node Y, C_{SB1}, C_{GS2}, and C_L are connected to ground. Similar to the source follower of Fig. 6.17(b), this circuit has three capacitances in a loop and hence a second-order transfer function. Using the equivalent circuit shown in Fig. 6.19(b), where $C_X = C_{GD1}+C_{DB2}$, $C_{XY} = C_{GS1} + C_{GD2}$, and $C_Y = C_{SB1} + C_{GS2} + C_L$, we have $V_1C_{XY}s + g_{m1}V_1 = V_{out}C_Ys$ and hence $V_1 = V_{out}C_Ys/(C_{XY}s + g_{m1})$. Also, since $V_2 = V_{out}$, the summation of currents at node X gives

$$(V_1 + V_{out})C_Xs + g_{m2}V_{out} + V_1C_{XY}s = \frac{V_{in} - V_1 - V_{out}}{R_S}. \tag{6.49}$$

Substituting for V_1 and simplifying the result, we obtain

$$\frac{V_{out}}{V_{in}}(s) = \frac{g_{m1} + C_{XY}s}{R_S\xi s^2 + [C_Y + g_{m1}R_SC_X + (1 + g_{m2}R_S)C_{XY}]s + g_{m1}(1 + g_{m2}R_S)}, \tag{6.50}$$

where $\xi = C_XC_Y + C_XC_{XY} + C_YC_{XY}$. As expected, (6.50) reduces to a form similar to (6.42) for $g_{m2} = 0$.

The output impedance of source followers is also of interest. In Fig. 6.17(a), the body effect and C_{SB} simply yield an impedance in parallel with the output. Ignoring this impedance and neglecting C_{GD}, we note from the equivalent circuit of Fig. 6.20(a) that $V_1C_{GS}s +$

$g_m V_1 = -I_X$. Also, $V_1 C_{GS} s R_S + V_1 = -V_X$. Dividing both sides of these equations gives

$$Z_{out} = \frac{V_X}{I_X} \tag{6.51}$$

$$= \frac{R_S C_{GS} s + 1}{g_m + C_{GS} s}. \tag{6.52}$$

It is instructive to examine the magnitude of this impedance as a function of frequency. At low frequencies, $Z_{out} \approx 1/g_m$, as expected. At very high frequencies, $Z_{out} \approx R_S$ (because C_{GS} shorts the gate and the source). We therefore surmise that $|Z_{out}|$ varies as shown in Figs. 6.20(b) or (c). Which one of these variations is more realistic? Operating as buffers, source followers must lower the output impedance, i.e., $1/g_m < R_S$. For this reason, the characteristic shown in Fig. 6.20(c) occurs more commonly than that in Fig. 6.20(b).

The behavior illustrated in Fig. 6.20(c) reveals an important attribute of source followers. Since the output impedance *increases* with frequency, we postulate that it contains an *inductive* component. To confirm this guess, we represent Z_{out} by a first-order passive network, noting that Z_{out} equals $1/g_m$ at $\omega = 0$ and R_S at $\omega = \infty$. The network can therefore be assumed as shown in Fig. 6.21 because Z_1 equals R_2 at $\omega = 0$ and $R_1 + R_2$ at $\omega = \infty$. In other words, $Z_1 = Z_{out}$ if $R_2 = 1/g_m$, $R_1 = R_S - 1/g_m$, and L is chosen properly.

To calculate L, we can simply obtain an expression for Z_1 in terms of the three components in Fig. 6.21 and equate the result to Z_{out} found above. Alternatively, since R_2 is a series component of Z_1, we can subtract its value from Z_{out}, thereby obtaining an expression

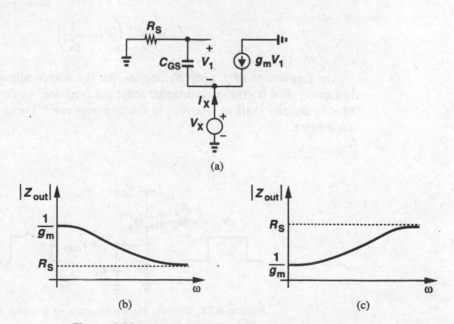

(a)

(b)   (c)

Figure 6.20 Calculation of source follower output impedance.

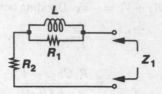

Figure 6.21 Equivalent output impedance of a source follower.

for the parallel combination of R_1 and L:

$$Z_{out} - \frac{1}{g_m} = \frac{C_{GS}s\left(R_S - \frac{1}{g_m}\right)}{g_m + C_{GS}s}. \tag{6.53}$$

Inverting the result to obtain the admittance of the parallel circuit, we have

$$\frac{1}{Z_{out} - \frac{1}{g_m}} = \frac{1}{R_S - \frac{1}{g_m}} + \frac{1}{\frac{C_{GS}s}{g_m}\left(R_S - \frac{1}{g_m}\right)}. \tag{6.54}$$

We can thus identify the first term on the right hand side as the inverse of R_1 and the second term as the inverse of an impedance equal to $(C_{GS}s/g_m)(R_S - 1/g_m)$, i.e., an inductor with the value

$$L = \frac{C_{GS}}{g_m}\left(R_S - \frac{1}{g_m}\right). \tag{6.55}$$

The dependence of L upon R_S implies that if a source follower is driven by a large resistance, then it exhibits substantial inductive behavior. As depicted in Fig. 6.22, this effect manifests itself as "ringing" in the step response if the circuit drives a large load capacitance.

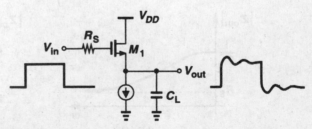

Figure 6.22 Ringing in step response of a source follower with heavy capacitive load.

6.4 Common-Gate Stage

As explained in Example 6.4, in a common-gate stage the input and output nodes are "isolated" if channel-length modulation is neglected. For a common-gate stage such as that in Fig. 6.23, the calculation of Example 6.4 suggested a transfer function

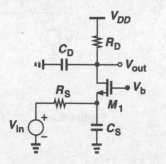

Figure 6.23 Common-gate stage at high frequencies.

$$\frac{V_{out}}{V_{in}}(s) = \frac{(g_m + g_{mb})R_D}{1 + (g_m + g_{mb})R_S} \frac{1}{\left(1 + \dfrac{C_S}{g_m + g_{mb} + R_S^{-1}}s\right)(1 + R_D C_D s)}. \tag{6.56}$$

An important property of this circuit is that it exhibits no Miller multiplication of capacitances, potentially achieving a wide band. Note, however, that the low input impedance may load the preceding stage. Furthermore, since the voltage drop across R_D is typically maximized to obtain a reasonable gain, the dc level of the input signal must be quite low.

If channel-length modulation is not negligible, the calculations become quite complex. Recall from Chapter 3 that the input impedance of a common-gate topology does depend on the drain load if $\lambda \neq 0$. From Eq. (3.110), we can express the impedance seen looking into the source of M_1 in Fig. 6.23 as

$$Z_{in} \approx \frac{Z_L}{(g_m + g_{mb})r_O} + \frac{1}{g_m + g_{mb}}, \tag{6.57}$$

where $Z_L = R_D \| [1/(C_D s)]$. Since Z_{in} now depends on Z_L, it is difficult to associate a pole with the input node.

Example 6.7

For the common-gate stage shown in Fig. 6.24(a), calculate the transfer function and the input impedance, Z_{in}. Explain why Z_{in} becomes independent of C_L as this capacitance increases.

Solution

Using the equivalent circuit shown in Fig. 6.24(b), we can write the current through R_S as $-V_{out}C_L s + V_1 C_{in} s$. Noting that the voltage across R_S plus V_{in} must equal $-V_1$, we have

$$(-V_{out}C_L s + V_1 C_{in} s)R_S + V_{in} = -V_1, \tag{6.58}$$

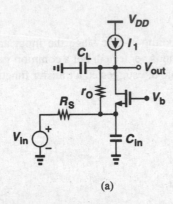

(a)

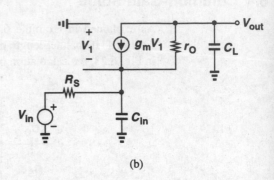

(b)

Figure 6.24

that is,

$$V_1 = -\frac{-V_{out}C_L s R_S + V_{in}}{1 + C_{in}R_S s}. \tag{6.59}$$

We also observe that the voltage across r_O minus V_1 equals V_{out}:

$$r_O(-V_{out}C_L s - g_m V_1) - V_1 = V_{out}. \tag{6.60}$$

Substituting for V_1 from (6.59), we obtain the transfer function:

$$\frac{V_{out}}{V_{in}}(s) = \frac{1 + g_m r_O}{r_O C_L C_{in} R_S s^2 + [r_O C_L + C_{in}R_S + (1 + g_m r_O)C_L R_S]s + 1}. \tag{6.61}$$

The reader can prove that body effect can be included by simply replacing g_m with $g_m + g_{mb}$. As expected, the gain at very low frequencies is equal to $1 + g_m r_O$. For Z_{in}, we can use (6.57) by replacing Z_L with $1/(C_L s)$, obtaining

$$Z_{in} = \frac{1}{g_m + g_{mb}} + \frac{1}{C_L s} \cdot \frac{1}{(g_m + g_{mb})r_O}. \tag{6.62}$$

We note that as C_L or s increases, Z_{in} approaches $1/(g_m + g_{mb})$ and hence the input pole can be defined as

$$\omega_{p,in} = \frac{1}{\left(R_S \left\|\dfrac{1}{g_m + g_{mb}}\right.\right)C_{in}}. \tag{6.63}$$

Why does Z_{in} become independent of C_L at high frequencies? This is because C_L lowers the voltage gain of the circuit, thereby suppressing the effect of the negative resistance introduced by Miller effect through r_O (Fig. 6.5). In the limit, C_L shorts the output node to ground, and r_O affects the input impedance negligibly.

If a common-gate stage is driven by a relatively large source impedance, then the output impedance of the circuit drops at high frequencies. This effect is better described in the context of cascode circuits.

6.5 Cascode Stage

As explained in Chapter 3, cascoding proves beneficial in increasing the voltage gain of amplifiers and the output impedance of current sources while providing shielding as well. The invention of the cascode (in the vacuum tube era), however, was motivated by the need for high-frequency amplifiers with relatively high input impedance. Viewed as a cascade of a common-source stage and a common-gate stage, a cascode circuit offers the speed of the latter—by suppressing the Miller effect—and the input impedance of the former.

Let us consider the cascode shown in Fig. 6.25, first identifying all of the device capacitances. At node A, C_{GS1} is connected to ground and C_{GD1} to node X. At node X, C_{DB1}, C_{SB2}, and C_{GS2} are tied to ground, and at node Y, C_{DB2}, C_{GD2}, and C_L are connected to ground. The Miller effect of C_{GD1} is determined by the gain from A to X. As an approximation, we use the low-frequency value of this gain, which for low values of R_D (or negligible channel-length modulation) is equal to $-g_{m1}/(g_{m2} + g_{mb2})$. Thus, if M_1 and M_2 have roughly equal dimensions, C_{GD1} is multiplied by approximately 2 rather than the large voltage gain in a simple common-source stage. We therefore say Miller effect is less significant in cascode amplifiers than in common-source stages. The pole associated with node A is estimated as

$$\omega_{p,A} = \frac{1}{R_S\left[C_{GS1} + \left(1 + \frac{g_{m1}}{g_{m2} + g_{mb2}}\right)C_{GD1}\right]}. \tag{6.64}$$

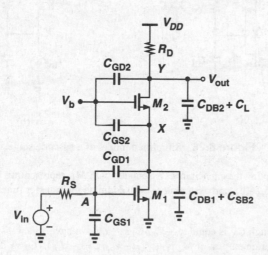

Figure 6.25 High-frequency model of a cascode stage.

We can also attribute a pole to node X. The total capacitance at this node is roughly equal to $2C_{GD1} + C_{DB1} + C_{SB2} + C_{GS2}$, giving a pole

$$\omega_{p,X} = \frac{g_{m2} + g_{mb2}}{2C_{GD1} + C_{DB1} + C_{SB2} + C_{GS2}}. \tag{6.65}$$

Finally, the output node yields a third pole:

$$\omega_{p,Y} = \frac{1}{R_D(C_{DB2} + C_L + C_{GD2})}. \tag{6.66}$$

The relative magnitudes of the three poles in a cascode circuit depend on the actual design parameters, but $\omega_{p,X}$ is typically chosen to be farther from the origin than the other two. As explained in Chapter 10, this choice plays an important role in the stability of op amps.

But what if R_D in Fig. 6.25 is replaced by a current source so as to achieve a higher dc gain? We know from Chapter 3 that the impedance seen at node X reaches high values if the load impedance at the drain of M_2 is large. For example, Eq. (3.110) predicts that the pole at node X may be quite lower than $(g_{m2} + g_{mb2})/C_X$ if R_D itself is the output impedance of a PMOS cascode current source. Interestingly, however, the overall transfer function is negligibly affected by this phenomenon. This can be better seen by an example.

Example 6.8

Consider the cascode stage shown in Fig. 6.26(a), where the load resistor is replaced by an ideal

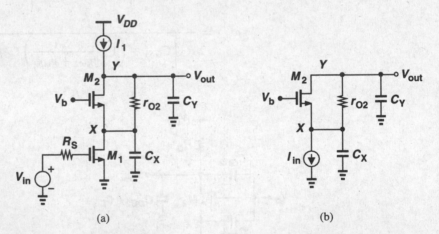

Figure 6.26 Simplified model of a cascode stage.

current source. Neglecting the capacitances associated with M_1, representing V_{in} and M_1 by a Norton equivalent as in Fig. 6.26(b), and assuming $\gamma = 0$, compute the transfer function.

Solution

Since the current through C_X is equal to $-V_{out}C_Ys - I_{in}$, we have $V_X = -(V_{out}C_Ys + I_{in})/(C_Xs)$, and the small-signal drain current of M_2 is $-g_{m2}(-V_{out}C_Ys - I_{in})/(C_Xs)$. The current through r_{O2}

is then equal to $-V_{out}C_Y s - g_{m2}(V_{out}C_Y s + I_{in})/(C_X s)$. Noting that V_X plus the voltage drop across r_{O2} is equal to V_{out}, we write

$$-r_{O2}\left[(V_{out}C_Y s + I_{in})\frac{g_{m2}}{C_X s} + V_{out}C_Y s\right] - (V_{out}C_Y s + I_{in})\frac{1}{C_X s} = V_{out}. \qquad (6.67)$$

That is,

$$\frac{V_{out}}{I_{in}} = -\frac{g_{m2}r_{O2} + 1}{C_X s} \cdot \frac{1}{1 + (1 + g_{m2}r_{O2})\dfrac{C_Y}{C_X} + C_Y r_{O2}s}, \qquad (6.68)$$

which, for $g_{m2}r_{O2} \gg 1$ and $g_{m2}r_{O2}C_Y/C_X \gg 1$ (i.e., $C_Y > C_X$), reduces to

$$\frac{V_{out}}{I_{in}} \approx -\frac{g_{m2}}{C_X s}\frac{1}{\dfrac{C_Y}{C_X}g_{m2} + C_Y s}. \qquad (6.69)$$

and hence

$$\frac{V_{out}}{V_{in}} = -\frac{g_{m1}g_{m2}}{C_Y C_X s}\frac{1}{g_{m2}/C_X + s}. \qquad (6.70)$$

The magnitude of the pole at node X is still given by g_{m2}/C_X. This is because at high frequencies (as we approach this pole) C_Y shunts the output node, dropping the gain and suppressing the Miller effect of r_{O2}.

If a cascode structure is used as a current source, then the variation of its output impedance with frequency is of interest. Neglecting C_{GD1} and C_Y in Fig. 6.26(a), we have

$$Z_{out} = (1 + g_{m2}r_{O2})Z_X + r_{O2}, \qquad (6.71)$$

where $Z_X = r_{O1}\|(C_X s)^{-1}$. Thus, Z_{out} contains a pole at $(r_{O1}C_X)^{-1}$ and falls at frequencies higher than this value.

6.6 Differential Pair

The versatility of differential pairs and their extensive use in analog systems motivate us to characterize their frequency response for both differential and common-mode signals.

Consider the simple differential pair shown in Fig. 6.27(a), with the differential half circuit and the common-mode equivalent circuit depicted in Figs. 6.27(b) and (c), respectively. For differential signals, the response is identical to that of a common-source stage, exhibiting Miller multiplication of C_{GD}. Note that since $+V_{in2}/2$ and $-V_{in2}/2$ are multiplied by the same transfer function, the number of poles in V_{out}/V_{in} is equal to that of each path (rather than the sum of the number of the poles in the two paths).

For common-mode signals, the total capacitance at node P in Fig. 6.27(c) determines the high-frequency gain. Arising from C_{GD3}, C_{DB3}, C_{SB1}, and C_{SB2}, this capacitance can be

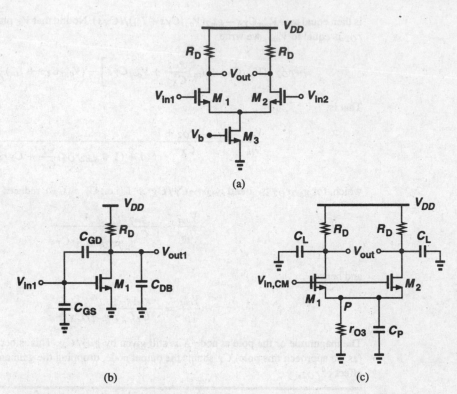

Figure 6.27 (a) Differential pair, (b) half-circuit equivalent, (c) equivalent circuit for common-mode inputs.

quite substantial if M_1-M_3 are wide transistors. For example, limited voltage headroom often necessitates that W_3 be so large that M_3 does not require a large drain-source voltage for operating in the saturation region. If only the mismatch between M_1 and M_2 is considered, the high-frequency common-mode gain can be calculated with the aid of Eq. (4.43). We replace r_{O3} with $r_{O3}\|[1/(C_P s)]$ and R_D by $R_D\|[1/(C_L s)]$, where C_L denotes the total capacitance seen at each output node. Thus,

$$A_{v,CM} = -\frac{\Delta g_m \left[R_D \left\| \left(\dfrac{1}{C_L s} \right) \right. \right]}{(g_{m1} + g_{m2}) \left[r_{O3} \left\| \left(\dfrac{1}{C_P s} \right) \right. \right] + 1}, \tag{6.72}$$

where other capacitances in the circuit are neglected.

This result suggests that, if the output pole is much farther from the origin than is the pole at node P, the common-mode rejection of the circuit degrades considerably at high frequencies. For example, as illustrated in Fig. 6.28, if the supply voltage contains high-frequency noise and the circuit exhibits mismatches, the resulting common-mode disturbance at node P leads to a differential noise component at the output.

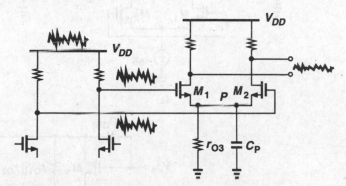

Figure 6.28   Effect of high-frequency supply noise in differential pairs.

We should emphasize that the circuit of Fig. 6.27(a) suffers from a trade-off between voltage headroom and $CMRR$. To minimize the headroom consumed by M_3, its width is maximized, introducing substantial capacitance at the sources of M_1 and M_2 and degrading the high-frequency $CMRR$. The issue becomes more serious at low supply voltages.

We now study the frequency response of differential pairs with high-impedance loads. Shown in Fig. 6.29(a) is a fully differential implementation. As with the topology of Fig. 6.27, this circuit can be analyzed for differential and common-mode signals separately. Note that here C_L includes the drain junction capacitance and the gate-drain overlap capacitance of each PMOS transistor as well. Also, as depicted in Fig. 6.29(b) for differential output signals, C_{GD3} and C_{GD4} conduct equal and opposite currents to node G, making this node an ac ground. (In practice, node G is nonetheless bypassed to ground by means of a capacitor.)

The differential half circuit is depicted in Fig. 6.29(c), with the output resistance of M_1 and M_3 shown explicitly. This topology implies that Eq. (6.23) can be applied to this circuit if R_L is replaced by $r_{O1} \| r_{O3}$. In practice, the relatively high value of this resistance makes the output pole, given by $[(r_{O1} \| r_{O3})C_L]^{-1}$, the "dominant" pole. We return to this observation in Chapter 10. The common-mode behavior of the circuit is similar to that of Fig. 6.27(c).

Let us now consider a differential pair with active current mirror (Fig. 6.30). How many poles does this circuit have? In contrast to the fully differential configuration of Fig. 6.29(a), this topology contains two signal paths with *different* transfer functions. The path consisting of M_3 and M_4 includes a pole at node E, approximately given by g_{m3}/C_E, where C_E denotes the total capacitance at E to ground. This capacitance arises from C_{GS3}, C_{GS4}, C_{DB3}, C_{DB1}, and the Miller effect of C_{GD1} and C_{GD4}. Even if only C_{GS3} and C_{GS4} are considered, the severe trade-off between g_m and C_{GS} of PMOS devices results in a pole that greatly impacts

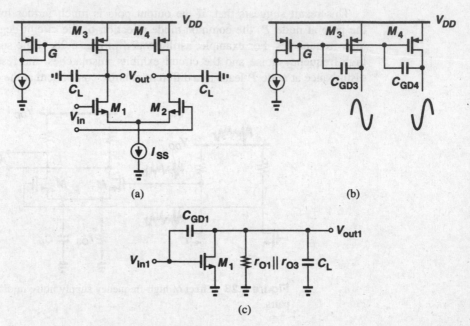

(a)   (b)

(c)

Figure 6.29 (a) Differential pair with current-source loads, (b) effect of differential swings at node G, (c) half-circuit equivalent.

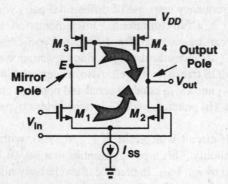

Figure 6.30 High-frequency behavior of differential pair with active current mirror.

the performance of the circuit. The pole associated with node E is called a "mirror pole." Note that, as with the circuit of Fig. 6.29(a), both signal paths shown in Fig. 6.30 contain a pole at the output node.

In order to estimate the frequency response of the differential pair with active current mirror, we construct the simplified model depicted in Fig. 6.31(a), where all other capacitances are neglected. Replacing V_{in}, M_1, and M_2 by a Thevenin equivalent, we arrive at the circuit of Fig. 6.31(b), where, from the analysis of Fig. 5.26, $V_X = g_{mN}r_{ON}V_{in}$ and $R_X = 2r_{ON}$. Here, the subscripts P and N refer to PMOS and NMOS devices, respectively,

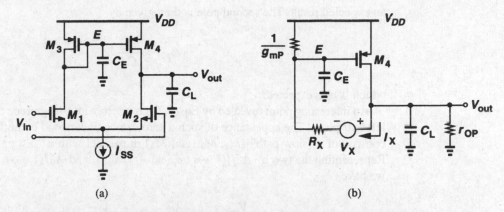

Figure 6.31 (a) Simplified high-frequency model of differential pair with active current mirror, (b) circuit of (a) with a Thevenin equivalent.

and we have assumed $1/g_{mP} \ll r_{OP}$. The small-signal voltage at E is equal to

$$V_E = (V_{out} - V_X) \frac{\dfrac{1}{C_E s + g_{mP}}}{\dfrac{1}{C_E s + g_{mP}} + R_X}, \tag{6.73}$$

and the small-signal drain current of M_4 is $g_{m4}V_E$. Noting that $-g_{m4}V_E - I_X = V_{out}(C_L s + r_{OP}^{-1})$, we have

$$\frac{V_{out}}{V_{in}}$$

$$= \frac{g_{mN}r_{ON}(2g_{mP} + C_E s)}{2r_{OP}r_{ON}C_E C_L s^2 + [(2r_{ON} + r_{OP})C_E + r_{OP}(1 + 2g_{mP}r_{ON})C_L]s + 2g_{mP}(r_{ON} + r_{OP})}. \tag{6.74}$$

Since the mirror pole is typically quite higher in magnitude than the output pole, we can utilize the results of Eq. (6.25) to write

$$\omega_{p1} \approx \frac{2g_{mP}(r_{ON} + r_{OP})}{(2r_{ON} + r_{OP})C_E + r_{OP}(1 + 2g_{mP}r_{ON})C_L}. \tag{6.75}$$

Neglecting the first term in the denominator and assuming $2g_{mP}r_{ON} \gg 1$, we have

$$\omega_{p1} \approx \frac{1}{(r_{ON} \| r_{OP})C_L}, \tag{6.76}$$

an expected result. The second pole is then given by

$$\omega_{p2} \approx \frac{g_{mP}}{C_E}, \tag{6.77}$$

which is also expected.

An interesting point revealed by Eq. (6.74) is a zero with a magnitude of $2g_{mP}/C_E$ in the left half plane. The appearance of such a zero can be understood by noting that the circuit consists of a "slow path" (M_1, M_3, and M_4) in parallel with a "fast path" (M_1 and M_2). Representing the two by $A_0/[(1 + s/\omega_{p1})(1 + s/\omega_{p2})]$ and $A_0/(1 + s/\omega_{p1})$, respectively, we have

$$\frac{V_{out}}{V_{in}} = \frac{A_0}{1 + s/\omega_{p1}} \left(\frac{1}{1 + s/\omega_{p2}} + 1 \right) \tag{6.78}$$

$$= \frac{A_0(2 + s/\omega_{p2})}{(1 + s/\omega_{p1})(1 + s/\omega_{p2})}. \tag{6.79}$$

That is, the system exibits a zero at $2\omega_{p2}$. The zero can also be obtained by the method of Fig. 6.15 (Problem 6.15).

Comparing the circuits of Figs. 6.29(a) and 6.30, we conclude that the former entails no mirror pole, another advantage of fully differential circuits over single-ended topologies.

Example 6.9

Not all fully differential circuits are free from mirror poles. Fig. 6.32(a) illustrates an example, where current mirrors M_3-M_5 and M_4-M_6 "fold" the signal current. Estimate the low-frequency gain and the transfer function of this circuit.

Solution

Neglecting channel-length modulation and using the differential half-circuit shown in Fig. 6.32(b), we observe that M_5 multiplies the drain current of M_3 by K, yielding an overall low-frequency voltage gain $A_v = g_{m1} K R_D$.

To obtain the transfer function, we utilize the equivalent circuit depicted in Fig. 6.32(c), including a source resistance R_S for completeness. To simplify calculations, we assume $R_D C_L$ is relatively small so that the Miller multiplication of C_{GD5} can be approximated as $C_{GD5}(1 + g_{m5}R_D)$. The circuit thus reduces to that in Fig. 6.32(d), where $C_X \approx C_{GS3} + C_{GS5} + C_{DB3} + C_{GD5}(1 + g_{m5}R_D) + C_{DB1}$. The overall transfer function is then equal to V_X/V_{in1} multiplied by V_{out1}/V_X. The former is readily obtained from (6.23) by replacing R_D with $1/g_{m3}$ and C_{DB} with C_X, while the latter is

$$\frac{V_{out1}}{V_X}(s) = -g_{m5}R_D \frac{1}{1 + R_D C_L s}. \tag{6.80}$$

Note that we have neglected the zero due to C_{GD5}.

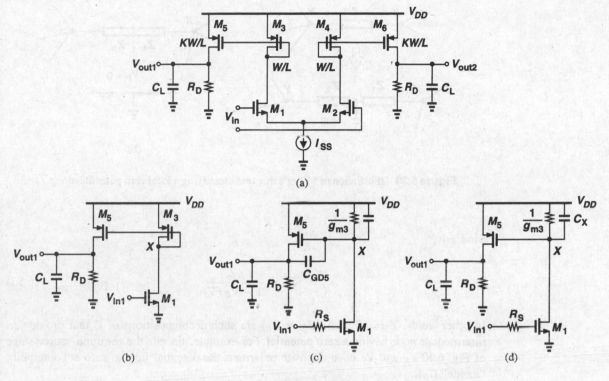

Figure 6.32

Appendix A: Dual of Miller's Theorem

In the Miller's theorem (Fig. 6.1), we readily observe that $Z_1 + Z_2 = Z$. This is no coincidence and it has interesting implications.

Redrawing Fig. 6.1 as shown in Fig. 6.33(a), we surmise that since the point between Z_1 and Z_2 can be grounded, then if we "walk" from X towards Y along the impedance Z, the local potential drops to zero at some intermediate point [Fig. 6.33(b)]. Indeed, for $V_P = 0$, we have

$$\frac{Z_a}{Z_a + Z_b}(V_Y - V_X) + V_X = 0, \tag{6.81}$$

and, since $Z_a + Z_b = Z$,

$$Z_a = \frac{Z}{1 - V_Y/V_X}. \tag{6.82}$$

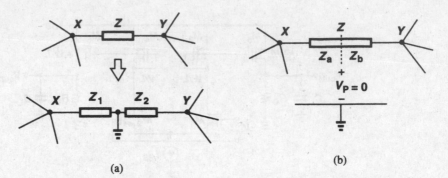

Figure 6.33 Illustration of Miller's theorem identifying a local zero potential along Z.

Similarly,

$$Z_b = \frac{Z}{1 - V_X/V_Y}. \tag{6.83}$$

In other words, $Z_1(= Z_a)$ and $Z_2(= Z_b)$ are such decompositions of Z that provide an intermediate node having a zero potential. For example, since in the common-source stage of Fig. 6.10 V_X and V_Y have opposite polarities, the potential falls to zero at some point "inside" C_{GD}.

The above observation explains the difficulty with the transformation depicted in Fig. 6.3. Drawing Fig. 6.33(b) for this case as in Fig. 6.34(a), we recognize that the circuit is

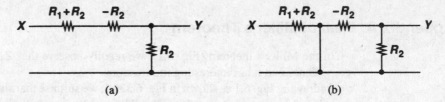

Figure 6.34 Resistive divider with decomposition of R_1.

still valid before node P is grounded because the current through $R_1 + R_2$ must equal that through $-R_2$. However, if, as shown in Fig. 6.34(b), node P is tied to ground, then the only current path between X and Y vanishes.

The concept of a zero local potential along the floating impedance Z also allows us to develop the "dual" of Miller's theorem, i.e., decomposition in terms of admittances and current ratios. Suppose two loops carrying currents I_1 and I_2 share an admittance Y [Fig. 6.35(a)]. Then, if Y is properly decomposed into two *parallel* admittances Y_1 and Y_2, the *current* flowing between the two is zero [Fig. 6.35(b)] and the connection can be broken [Fig. 6.35(c)]. In Fig. 6.35(a), the voltage across Y is equal to $(I_1 - I_2)/Y$ and in Fig. 6.35(c),

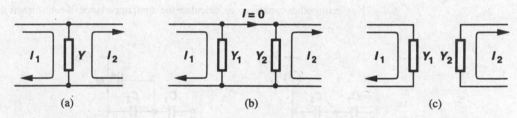

Figure 6.35 (a) Two loops sharing admittance Y, (b) decomposition of Y into Y_1 and Y_2 such that $I = 0$, (c) equivalent circuit

the voltage across Y_1 is I_1/Y_1. For the two circuits to be equivalent,

$$\frac{I_1 - I_2}{Y} = \frac{I_1}{Y1}, \tag{6.84}$$

and

$$Y_1 = \frac{Y}{1 - I_2/I_1}, \tag{6.85}$$

Note the duality between this expression and $Z_1 = (1 - V_Y/V_X)Z$. We also have

$$Y_2 = \frac{Y}{1 - I_1/I_2}. \tag{6.86}$$

Problems

Unless otherwise stated, in the following problems, use the device data shown in Table 2.1 and assume $V_{DD} = 3$ V where necessary. Also, assume all transistors are in saturation. All device dimensions are effective values and in microns.

6.1. In the circuit of Fig. 6.2(c), suppose the amplifier has a finite output resistance R_{out}.
   (a) Explain why the output jumps *up* by ΔV before it begins to go down. This indicates the existence of a zero in the transfer function.
   (b) Determine the transfer function and the step response without using Miller's theorem.

6.2. Repeat Problem 6.1 if the amplifier has an output resistance R_{out} and the circuit drives a load capacitance C_L.

6.3. The CS stage of Fig. 6.10 is designed with $(W/L)_1 = 50/0.5$, $R_S = 1$ kΩ and $R_D = 2$ kΩ. If $I_{D1} = 1$ mA, determine the poles and the zero of the circuit.

6.4. Consider the CS stage of Fig. 6.13, where I_1 is realized by a PMOS device operating in saturation. Assume $(W/L)_1 = 50/0.5$, $I_{D1} = 1$ mA, and $R_S = 1$ kΩ.
   (a) Determine the aspect ratio of the PMOS transistor such that the maximum allowable output level is 2.6 V. What is the maximum peak-to-peak swing?
   (b) Determine the poles and the zero.

6.5. A source follower employing an NFET with $W/L = 50/0.5$ and a bias current of 1 mA is driven by a source impedance of 10 kΩ. Calculate the equivalent inductance seen at the output.

6.6.   Neglecting other capacitances, calculate the input impedance of each circuit shown in Fig. 6.36.

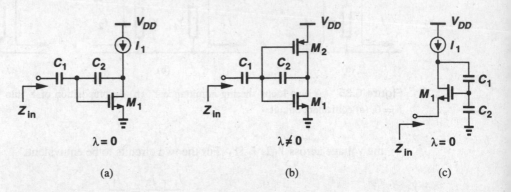

(a)   (b)   (c)

Figure 6.36

6.7.   Estimate the poles of each circuit in Fig. 6.37.

6.8.   Calculate the input impedance and the transfer function of each circuit in Fig. 6.38.

6.9.   Calculate the gain of each circuit in Fig. 6.39 at very low and very high frequencies. Neglect all other capacitances and assume $\lambda = 0$ for circuits (a) and (b) and $\gamma = 0$ for all of the circuits.

6.10.   Calculate the gain of each circuit in Fig. 6.40 at very low and very high frequencies. Neglect all other capacitances and assume $\lambda = \gamma = 0$.

6.11.   Consider the cascode stage shown in Fig. 6.41. In our analysis of the frequency response of a cascode stage, we assumed that the gate-drain overlap capacitance of M_1 is multiplied by $g_{m1}/(g_{m2} + g_{mb2})$. Recall from Chapter 3, however, that with a high resistance loading the drain of M_2, the resistance seen looking into the source of M_2 can be quite high, suggesting a much higher Miller multiplication factor for C_{GD1}. Explain why C_{GD1} is still multiplied by $1 + g_{m1}/(g_{m2} + g_{mb2})$ if C_L is relatively large.

6.12.   Neglecting other capacitances, calculate Z_X in the circuits of Fig. 6.42. Sketch $|Z_X|$ versus frequency.

6.13.   The common-gate stage of Fig. 6.23 is designed with $(W/L)_1 = 50/0.5$, $I_{D1} = 1$ mA, $R_D = 2$ kΩ, and $R_S = 1$ kΩ. Assuming $\lambda = 0$, determine the poles and the low-frequency gain. How do these results compare with those obtained in Problem 6.9?

6.14.   Suppose in the cascode stage of Fig. 6.25, a resistor R_G appears in series with the gate of M_2. Including only C_{GS2}, neglecting other capacitances, and assuming $\lambda = \gamma = 0$, determine the transfer function.

6.15.   Apply the method of Fig. 6.15 to the circuit of Fig. 6.31(b) to determine the zero of the transfer function.

6.16.   The circuit of Fig. 6.32(a) is designed with $(W/L)_{1,2} = 50/0.5$ and $(W/L)_{3,4} = 10/0.5$. If $I_{SS} = 100$ μA, $K = 2$, $C_L = 0$, and R_D is implemented by an NFET having $W/L = 50/0.5$, estimate the poles and zeros of the circuit. Assume the amplifier is driven by an ideal voltage source.

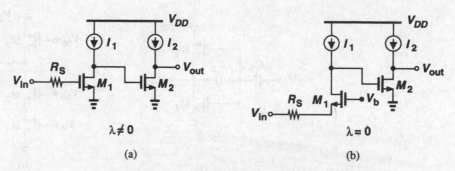

Figure 6.37

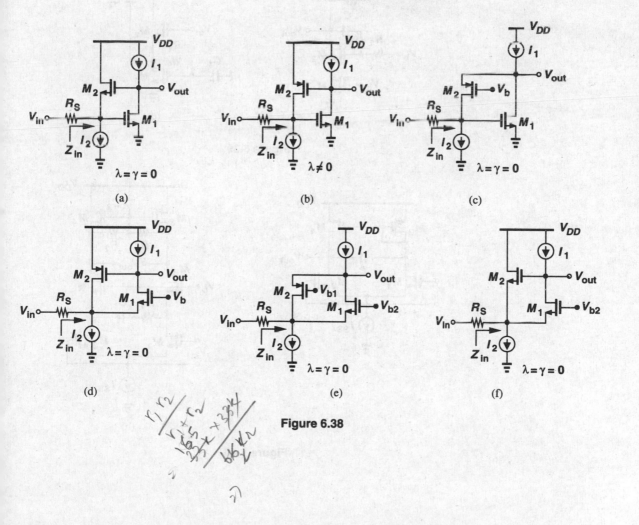

Figure 6.38

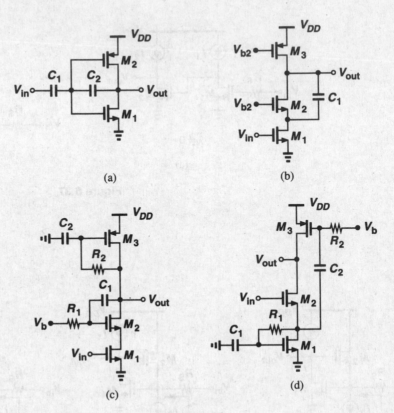

Figure 6.39

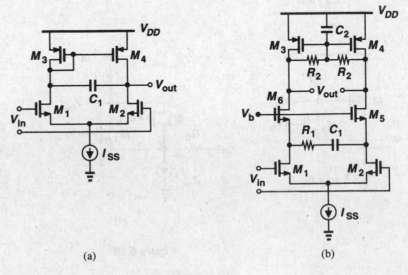

Figure 6.40

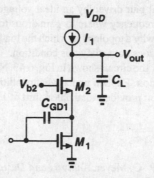

Figure 6.41

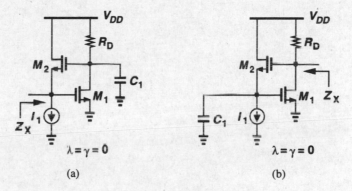

$\lambda = \gamma = 0$

(a)

$\lambda = \gamma = 0$

(b)

Figure 6.42

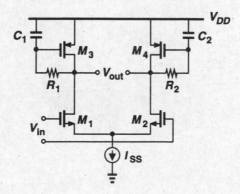

Figure 6.43

6.17. A differential pair driven by an ideal voltage source is required to have a total phase shift of 135° at the frequency where its gain drops to unity.

    (a) Explain why a topology in which the load is realized by diode-connected devices or current sources does not satisfy this condition.

    (b) Consider the circuit shown in Fig. 6.43. Neglecting other capacitances, determine the transfer function. Explain under what conditions the load exhibits an inductive behavior. Can this circuit provide a total phase shift of 135° at the frequency where its gain drops to unity?

References

1. P. R. Gray and R. G. Meyer, *Analysis and Design of Analog Integrated Circuits*, Third Ed., New York: Wiley, 1993.
2. B. Razavi, *RF Microelectronics*, Upper Saddle River, NJ: Prentice Hall, 1998.

Chapter 7

Noise

Noise limits the minimum signal level that a circuit can process with acceptable quality. Today's analog designers constantly deal with the problem of noise because it trades with power dissipation, speed, and linearity.

In this chapter, we describe the phenomenon of noise and its effect on analog circuits. The objective is to provide sufficient understanding of the problem so that further developments of analog circuits in the following chapters take noise into account as readily as other circuit parameters such as gain, input and output impedance, etc. Seemingly a complex subject, noise is introduced at this early stage so as to accompany the reader for the remainder of the book and become more intuitive through various examples.

Following a general description of noise characteristics in the frequency and time domains, we introduce thermal, shot, and flicker noise. Next, we consider methods of representing noise in circuits. Finally, we describe the effect of noise in single-stage and differential amplifiers along with trade-offs with other performance parameters.

7.1 Statistical Characteristics of Noise

Noise is a random process. For our purposes in this book, this statement means the value of noise cannot be predicted at any time even if the past values are known. Compare the output of a sinewave generator with that of a microphone picking up the sound of water flow in a river (Fig. 7.1). While the value of $x_1(t)$ at $t = t_1$ can be predicted from the observed waveform, the value of $x_2(t)$ at $t = t_2$ cannot. This is the principal difference between deterministic and random phenomena.

If the instantaneous value of noise in the time domain cannot be predicted, how can we incorporate noise in circuit analysis? This is accomplished by observing the noise for a long time and using the measured results to construct a "statistical model" for the noise. While the instantaneous *amplitude* of noise cannot be predicted, a statistical model provides knowledge about some other important properties of the noise that prove useful and adequate in circuit analysis.

Which properties of noise *can* be predicted? In many cases, the average power of noise is predictable. For example, if a microphone picking up the sound of a river is brought

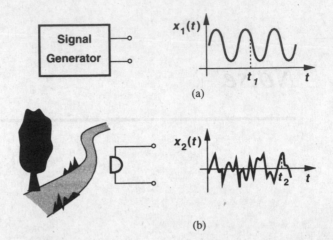

(a)

(b)

Figure 7.1   Output of a generator and the sound of a river.

closer to the river, the resulting electrical signal displays, on the average, larger excursions and hence higher power (Fig. 7.2). The reader may wonder if a random process can be so random that even its average power is unpredictable. Such processes do exist, but we are fortunate that most sources of noise in circuits exhibit a constant average power.

The concept of average power proves essential in our analysis and must be defined carefully. Recall from basic circuit theory that the average power delivered by a periodic

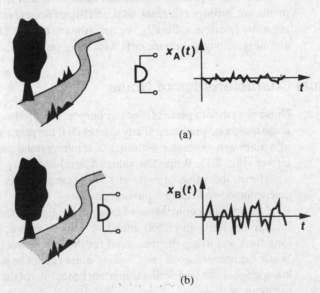

(a)

(b)

Figure 7.2   Illustration of the average power of a random signal.

voltage $v(t)$ to a load resistance R_L is given by

$$P_{av} = \frac{1}{T} \int_{-T/2}^{+T/2} \frac{v^2(t)}{R_L} dt, \tag{7.1}$$

where T denotes the period.[1] This quantity can be visualized as the average heat produced in R_L by $v(t)$.

How do we define P_{av} for a random signal? In the example of Fig. 7.2, we expect that $x_B(t)$ generates more heat than $x_A(t)$ if the microphone drives a resistive load. However, since the signals are not periodic, the measurement must be carried out over a long time:

$$P_{av} = \lim_{T\to\infty} \frac{1}{T} \int_{-T/2}^{+T/2} \frac{x^2(t)}{R_L} dt, \tag{7.2}$$

where $x(t)$ is a voltage quantity. Figure 7.3 illustrates the operation on $x_A(t)$ and $x_B(t)$; each signal is squared, the area under the resulting waveform is calculated for a long time T, and the average power is obtained by normalizing the area to T.[2]

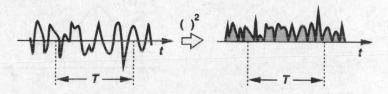

Figure 7.3   Average noise power.

To simplify calculations, we write the definition of P_{av} as

$$P_{av} = \lim_{T\to\infty} \frac{1}{T} \int_{-T/2}^{+T/2} x^2(t)dt, \tag{7.3}$$

where P_{av} is expressed in V^2 rather than W. The idea is that if we know P_{av} from (7.3), then the actual power delivered to a load R_L can be readily calculated as P_{av}/R_L. In analogy with deterministic signals, we can also define a root-mean-square (rms) voltage for noise as $\sqrt{P_{av}}$ where P_{av} is given by (7.3).

7.1.1 Noise Spectrum

The concept of average power becomes more versatile if defined with regard to the *frequency content* of noise. The noise made by a group of men contains weaker high-frequency components than that made by a group of women, a difference observable from the "spectrum"

---

[1]To be more rigorous, $v^2(t)$ should be replaced by $v(t) \cdot v^*(t)$, where $v^*(t)$ is the complex conjugate waveform.

[2]Strictly speaking, this definition holds only for "stationary" processes [1].

of each type of noise. Also called the "power spectral density" (PSD), the spectrum shows how much power the signal carries at each frequency. More specifically, the PSD, $S_x(f)$, of a noise waveform $x(t)$ is defined as the average power carried by $x(t)$ in a one-hertz bandwidth around f. That is, as illustrated in Fig. 7.4(a), we apply $x(t)$ to a bandpass filter

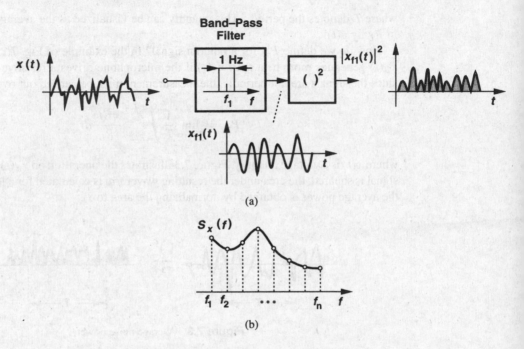

(a)

(b)

Figure 7.4 Calculation of noise spectrum.

with center frequency f_1 and 1-Hz bandwidth, square the output, and calculate the average over a long time to obtain $S_X(f_1)$. Repeating the procedure with bandpass filters having different center frequencies, we arrive at the overall shape of $S_X(f)$ [Fig. 7.4(b)].[3] While it is possible that the PSD of a random process is random itself, most of the noise sources of interest to us exhibit a predictable spectrum.

As with the definition of P_{av} in (7.3), it is customary to eliminate R_L from $S_X(f)$. Thus, since each value on the plot in Fig. 7.4(b) is measured for a 1-Hz bandwidth, $S_X(f)$ is expressed in V^2/Hz rather than W/Hz. It is also common to take the square root of $S_X(f)$, expressing the result in $V/\sqrt{Hz}$. For example, we say the input noise voltage of an amplifier at 100 MHz is equal to 3 $nV/\sqrt{Hz}$, simply to mean that the average power in a 1-Hz bandwidth at 100 MHz is equal to $(3 \times 10^{-9})^2 \, V^2$.

An example of a common type of noise PSD is the "white spectrum," also called white noise. Shown in Fig. 7.5, such a PSD displays the same value at all frequencies (similar

---

[3]In signal processing theory, the PSD is defined as the Fourier transform of the autocorrelation function of the noise. The two definitions are equivalent in most cases of interest to us.

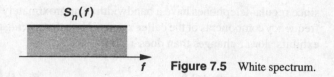

Figure 7.5   White spectrum.

to white light). Strictly speaking, we note that white noise does not exist because the total area under the power spectral density, i.e., the total power carried by the noise, is infinite. In practice, however, any noise spectrum that is flat *in the band of interest* is usually called white.

The PSD is a powerful tool in analyzing the effect of noise in circuits, especially in conjunction with the following theorem.

Theorem   If a signal with spectrum $S_X(f)$ is applied to a linear time-invariant system with transfer function $H(s)$, then the output spectrum is given by

$$S_Y(f) = S_X(f)|H(f)|^2, \qquad (7.4)$$

where $H(f) = H(s = 2\pi jf)$. The proof can be found in textbooks on signal processing or communications, e.g., [1].

This theorem agrees with our intuition that the spectrum of the signal should be "shaped" by the transfer function of the system (Fig. 7.6). For example, as illustrated in Fig. 7.7,

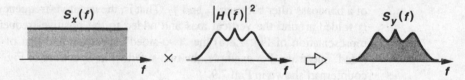

Figure 7.6   Noise shaping by a transfer function.

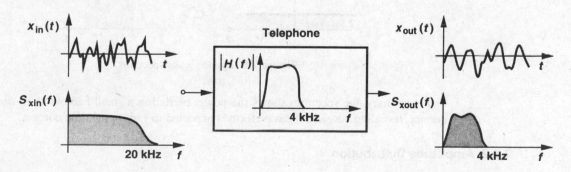

Figure 7.7   Spectral shaping by telephone bandwidth.

since regular telephones have a bandwidth of approximately 4 kHz, they suppress the high-frequency components of the caller's voice. Note that, owing to its limited bandwidth, $x_{out}(t)$ exhibits slower changes than does $x_{in}(t)$.

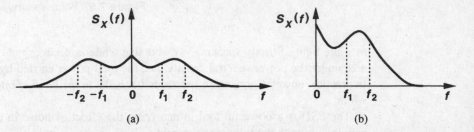

Figure 7.8   (a) Two-sided and (b) one-sided noise spectra.

Since $S_X(f)$ is an even function of f for real $x(t)$ [1], as depicted in Fig. 7.8, the total power carried by $x(t)$ in the frequency range $[f_1 \ f_2]$ is equal to

$$P_{f1,f2} = \int_{-f_2}^{-f_1} S_X(f)df + \int_{+f_1}^{+f_2} S_X(f)df \tag{7.5}$$

$$= \int_{+f_1}^{+f_2} 2S_X(f)df. \tag{7.6}$$

In fact, the integral in (7.6) is the quantity measured by a power meter sensing the output of a bandpass filter between f_1 and f_2. That is, the negative-frequency part of the spectrum is folded around the vertical axis and added to the positive-frequency part. We call the representation of Fig. 7.8(a) the "two-sided" spectrum and that of Fig. 7.8(b) the "one-sided" spectrum. For example, the two-sided white spectrum of Fig. 7.5 has the one-sided counterpart shown in Fig. 7.9.

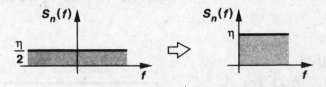

Figure 7.9   Folded white spectrum.

In summary, the spectrum shows the power carried in a small bandwidth at each frequency, revealing how *fast* the waveform is expected to vary in the time domain.

7.1.2 Amplitude Distribution

As mentioned earlier, the instantaneous amplitude of noise is usually unpredictable. However, by observing the noise waveform for a long time, we can construct a "distribution" of the amplitude, indicating how *often* each value occurs. Also called the "probability density function" (PDF), the distribution of $x(t)$ is defined as

$$p_X(x)dx = \text{probability of } x < X < x + dx, \tag{7.7}$$

where X is the measured value of $x(t)$ at some point in time.

As illustrated in Fig. 7.10, to estimate the distribution, we sample $x(t)$ at many points, construct bins of small width, choose the bin height equal to the number of samples whose value falls between the two edges of the bin, and normalize the bin heights to the total number of samples. Note that the PDF provides no information as to how fast $x(t)$ varies in the time domain. For example, the sound generated by a violin may have the same amplitude distribution as that produced by a drum even though their frequency contents are vastly different.

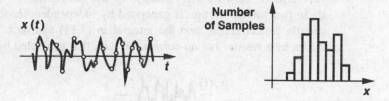

$x(t)$

Number of Samples

x

Figure 7.10 Amplitude distribution of noise.

An important example of PDFs is the Gaussian (or normal) distribution. The central limit theorem states that if many independent random processes with arbitrary PDFs are added, the PDF of the sum approaches a Gaussian distribution [1]. It is therefore not surprising that many natural phenomena exhibit Gaussian statistics. For example, since the noise of a resistor results from random "walk" of a very large number of electrons, each having relatively independent statistics, the overall amplitude follows a Gaussian PDF.

In this book, we employ the spectrum and average power of noise to a much greater extent than the amplitude distribution. For completeness, however, we note that the Gaussian PDF is defined as

$$p_X(x) = \frac{1}{\sigma\sqrt{2\pi}} \exp \frac{-(x-m)^2}{2\sigma^2}, \tag{7.8}$$

where σ and m are the standard deviation and mean of the distribution, respectively.

7.1.3 Correlated and Uncorrelated Sources

In analyzing circuits, we often need to add the effect of several sources of noise to obtain the total noise. While for deterministic voltages and currents, we simply use the superposition principle, the procedure is somewhat different for random signals. Since in noise analysis, ultimately the average noise *power* is of interest, we add two noise waveforms and take the average of the resulting power:

$$P_{av} = \lim_{T \to \infty} \frac{1}{T} \int_{-T/2}^{+T/2} [x_1(t) + x_2(t)]^2 dt \tag{7.9}$$

$$= \lim_{T \to \infty} \frac{1}{T} \int_{-T/2}^{+T/2} x_1^2(t)dt + \lim_{T \to \infty} \frac{1}{T} \int_{-T/2}^{+T/2} x_2^2(t)dt$$

$$+ \lim_{T \to \infty} \frac{1}{T} \int_{-T/2}^{+T/2} 2x_1(t)x_2(t)dt \tag{7.10}$$

$$= P_{av1} + P_{av2} + \lim_{T \to \infty} \frac{1}{T} \int_{-T/2}^{+T/2} 2x_1(t)x_2(t)dt, \tag{7.11}$$

where P_{av1} and P_{av2} denote the average power of $x_1(t)$ and $x_2(t)$, respectively. Called the "correlation" between $x_1(t)$ and $x_2(t)$,[4] the third term in (7.11) indicates how "similar" these two waveforms are. If generated by independent devices, the noise waveforms are usually "uncorrelated" and the integral in (7.11) vanishes. For example, the noise produced by a resistor has no correlation with that generated by a transistor. In such a case,

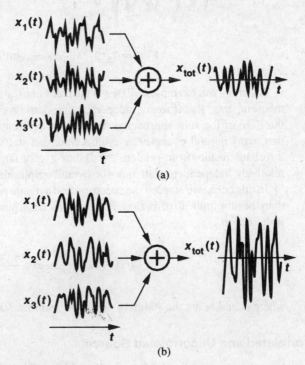

Figure 7.11 (a) Uncorrelated noise and (b) correlated noise generated in a stadium.

---

[4]This terminology applies only to stationary signals.

$P_{av} = P_{av1} + P_{av2}$. From this observation, we say superposition holds for the *power* of uncorrelated noise sources.

A familiar analogy is that of the spectators in a sports stadium. Before the game begins, many conversations are in progress, generating uncorrelated noise components [Fig. 7.11(a)]. During the game, the spectators applaud (or scream) simultaneously, producing correlated noise at much higher power level [Fig. 7.11(b)].

In most cases studied in this book, the noise sources are uncorrelated. One exception is studied in Section 7.3.

7.2 Types of Noise

Analog signals processed by integrated circuits are corrupted by two different types of noise: device electronic noise and "environmental" noise. The latter refers to (seemingly) random disturbances that a circuit experiences through the supply or ground lines or through the substrate. We focus on device electronic noise here and defer the study of environmental noise to Chapter 18.

7.2.1 Thermal Noise

Resistor Thermal Noise   The random motion of electrons in a conductor introduces fluctuations in the voltage measured across the conductor even if the average current is zero. Thus, the spectrum of thermal noise is proportional to the absolute temperature.

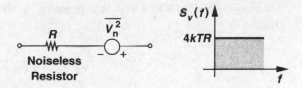

Figure 7.12   Thermal noise of a resistor.

As shown in Fig. 7.12, the thermal noise of a resistor R can be modeled by a series voltage source, with the one-sided spectral density

$$S_v(f) = 4kTR, \quad f \geq 0, \tag{7.12}$$

where $k = 1.38 \times 10^{-23}$ J/K is the Boltzmann constant. Note that $S_v(f)$ is expressed in V^2/Hz. Thus, we write $\overline{V_n^2} = 4kTR$, where the overline indicates averaging.[5] We may even say the noise "voltage" is given by $4kTR$ even though this quantity is in fact the noise voltage squared. For example, a 50-Ω resistor held at $T = 300°$K exhibits 8.28×10^{-19} V^2/Hz of thermal noise. To convert this number to a more familiar voltage quantity, we take the square root, obtaining 0.91 nV/$\sqrt{\text{Hz}}$. While the square root of hertz may appear strange,

---

[5]Some books write $\overline{V_n^2} = 4kTR\Delta f$ to emphasize that $4kTR$ is the noise power per unit bandwidth. To simplify the notation, we assume $\Delta f = 1$ Hz, unless otherwise stated.

it is helpful to remember that 0.91 nV/$\sqrt{\text{Hz}}$ has little significance per se and simply means that the power in a 1-Hz bandwidth is equal to $(0.91 \times 10^{-9})^2$ V^2.

The equation $S_v(f) = 4kTR$ suggests that thermal noise is white. In reality, $S_v(f)$ is flat for up to roughly 100 THz, dropping at higher frequencies. For our purposes, the white spectrum is quite accurate.

Since noise is a random quantity, the polarity used for the voltage source in Fig. 7.12 is unimportant. Nevertheless, once a polarity is chosen, it must be retained throughout the analysis of the circuit so as to obtain consistent results.

Example 7.1

Consider the RC circuit shown in Fig. 7.13. Calculate the noise spectrum and the total noise power in V_{out}.

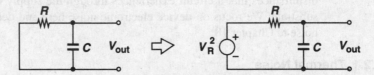

Figure 7.13 Noise generated in a low-pass filter.

Solution

Modeling the noise of R by a series voltage source V_R, we compute the transfer function from V_R to V_{out}:

$$\frac{V_{out}}{V_R}(s) = \frac{1}{RCs + 1}. \tag{7.13}$$

From the theorem in Section 6.1.1, we have

$$S_{out}(f) = S_R(f)\left|\frac{V_{out}}{V_R}(j\omega)\right|^2 \tag{7.14}$$

$$= 4kTR\frac{1}{4\pi^2 R^2 C^2 f^2 + 1}. \tag{7.15}$$

Thus, the white noise spectrum of the resistor is shaped by a low-pass characteristic (Fig. 7.14). To calculate the total noise power at the output, we write

$$P_{n,out} = \int_0^\infty \frac{4kTR}{4\pi^2 R^2 C^2 f^2 + 1} df, \tag{7.16}$$

which, since

$$\int \frac{dx}{x^2 + 1} = \tan^{-1} x, \tag{7.17}$$

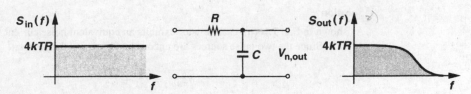

Figure 7.14   Noise spectrum shaping by a low-pass filter.

reduces to

$$P_{n,out} = \frac{2kT}{\pi C} \tan^{-1} u \Big|_{u=0}^{u=\infty} \tag{7.18}$$

$$= \frac{kT}{C}. \tag{7.19}$$

Note that the unit of kT/C is V^2. We may also consider $\sqrt{kT/C}$ as the total rms noise voltage measured at the output. For example, with a 1-pF capacitor, the total noise voltage is equal to 64.3 μV_{rms}.

Equation (7.19) implies that the total noise at the output of the circuit shown in Fig. 7.13 is independent of the value of R. Intuitively, this is because for larger values of R, the associated noise per unit bandwidth increases while the overall bandwidth of the circuit decreases. The fact that kT/C noise can be decreased by only increasing C (if T is fixed) introduces many difficulties in the design of analog circuits (Chapter 12).

The thermal noise of a resistor can be represented by a parallel current source as well (Fig. 7.15). For the representations of Figs. 7.12 and 7.15 to be equivalent, we have $\overline{V_n^2}/R^2 = \overline{I_n^2}$, that is, $\overline{I_n^2} = 4kT/R$. Note that $\overline{I_n^2}$ is expressed in A^2/Hz. Depending on the circuit topology, one model may lead to simpler calculations than the other.

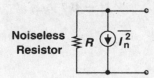

Figure 7.15   Representation of resistor thermal noise by a current source.

Example 7.2

Calculate the equivalent noise voltage of two parallel resistors R_1 and R_2 [Fig. 7.16(a)].

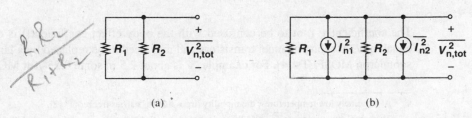

(a)   (b)

Figure 7.16

Solution

As shown in Fig. 7.16(b), each resistor exhibits an equivalent noise current with the spectral density $4kT/R$. Since the two noise sources are uncorrelated, we add the *powers*:

$$\overline{I_{n,tot}^2} = \overline{I_{n1}^2} + \overline{I_{n2}^2} \tag{7.20}$$

$$= 4kT\left(\frac{1}{R_1} + \frac{1}{R_2}\right). \tag{7.21}$$

Thus, the equivalent noise voltage is given by

$$\overline{V_{n,tot}^2} = \overline{I_{n,tot}^2}(R_1 \| R_2)^2 \tag{7.22}$$

$$= 4kT(R_1 \| R_2), \tag{7.23}$$

as intuitively expected. Note that our notation assumes a 1-Hz bandwidth.

The dependence of thermal noise (and some other types of noise) upon T suggests that low-temperature operation can decrease the noise in analog circuits. This approach becomes more attractive with the observation that the mobility of charge carriers in MOS devices increases at low temperatures [2].[6] Nonetheless, the required cooling equipment limits the practicality of low-temperature circuits.

MOSFETs MOS transistors also exhibit thermal noise. The most significant source is the noise generated in the channel. It can be proved [3] that for long-channel MOS devices operating in saturation, the channel noise can be modeled by a current source connected between the drain and source terminals (Fig. 7.17) with a spectral density:[7]

$$\overline{I_n^2} = 4kT\gamma g_m. \tag{7.24}$$

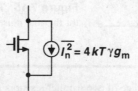

Figure 7.17 Thermal noise of a MOSFET.

The coefficient γ (not to be confused with the body effect coefficient!) is derived to be equal to 2/3 for long-channel transistors and may need to be replaced by a larger value for submicron MOSFETs [4]. For example, γ is about 2.5 in some 0.25-μm MOS devices. It

---

[6]At extremely low temperatures, the mobility drops due to "carrier freezeout" [2].

[7]The actual equation reads $\overline{I_n^2} = 4kT\gamma g_{ds}$, where g_{ds} is the drain-source conductance with $V_{DS} = 0$, i.e., the same as R_{on}^{-1}. For long-channel devices, g_{ds} with $V_{DS} = 0$ is equal to g_m in saturation.

also varies to some extent with the drain-source voltage. The theoretical determination of γ is still under active research.

Example 7.3

Find the maximum noise voltage that a single MOSFET can generate.

Solution

As shown in Fig. 7.18, the maximum output noise occurs if the transistor sees only its own output

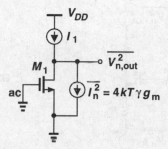

Figure 7.18

impedance as the load, i.e., if the external load is an ideal current source. The output noise voltage is then given by

$$\overline{V_n^2} = \overline{I_n^2} r_O^2 \tag{7.25}$$

$$= 4kT \left(\frac{2}{3} g_m \right) r_O^2. \tag{7.26}$$

Equation (7.26) suggests that the noise *current* of a MOS transistor decreases if the transconductance drops. For example, if the transistor operates as a constant current source, it is desirable to minimize its transconductance.

Another important conclusion is that the noise measured at the output of the circuit does not depend on where the input terminal is because for output noise calculation, the input is set to zero.[8] For example, the circuit of Fig. 7.18 may be a common-source or a common-gate stage, exhibiting the same output noise.

The ohmic sections of a MOSFET also contribute thermal noise. As conceptually illustrated in the top view of Fig. 7.19(a), the gate, source, and drain materials exhibit finite resistivity, thereby introducing noise. For a relatively wide transistor, the source and drain resistance is typically negligible whereas the gate distributed resistance may become noticeable.

In the noise model of Fig. 7.19(b), a lumped resistor R_1 represents the distributed gate resistance. Viewing the overall transistor as the distributed structure shown in Fig. 7.19(c),

---

[8]Of course, if the input voltage or current source has an output impedance that generates noise, this statement must be interpreted carefully.

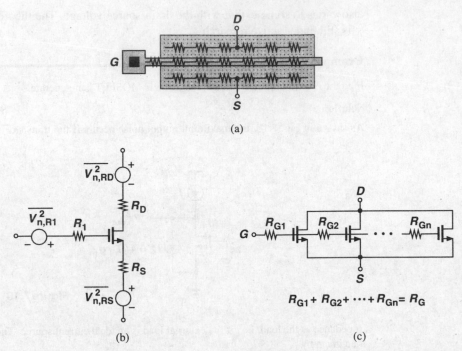

(a)

(b)   (c)

$$R_{G1} + R_{G2} + \cdots + R_{Gn} = R_G$$

Figure 7.19 Layout of a MOSFET indicating the terminal resistances, (b) circuit model, (c) distributed gate resistance.

we observe that the unit transistors near the left end see the noise of only a fraction of R_G whereas those near the right end see the noise of most of R_G. We therefore expect the lumped resistor in the noise model to be *less* than R_G. In fact, it can be proved that $R_1 = R_G/3$ (Problem 7.3).

While the thermal noise generated in the channel is controlled by only the transconductance of the device, the effect of R_G can be reduced by proper layout. Shown in Fig. 7.20 are two examples. In Fig. 7.20(a), the gate is contacted on both ends and in Fig. 7.20(b), the

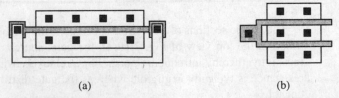

(a)   (b)

Figure 7.20 Reduction of gate resistance by (a) adding contacts to both sides or (b) folding.

device is folded (Chapter 2), each technique reducing R_G by a factor of 4. We will hereafter neglect the thermal noise due to the ohmic sections of MOS devices.

Example 7.4 _____

Find the maximum thermal noise voltage that the gate resistance of a single MOSFET can generate.

Solution

If the total distributed gate resistance is R_G, then from Fig. 7.18, the output noise voltage due to R_G is given by

$$\overline{V_{n,out}^2} = 4kT\frac{R_G}{3}(g_m r_O)^2. \tag{7.27}$$

_____

7.2.2 Flicker Noise *[handwritten: ∝ 1/f noise (since the trap & release phenomenon of charge carriers occurs at low freq)]*

The interface between the gate oxide and the silicon substrate in a MOSFET entails an interesting phenomenon. Since the silicon crystal reaches an end at this interface, many "dangling" bonds appear, giving rise to extra energy states (Fig.7.21). As charge carriers move at the interface, some are randomly trapped and later released by such energy states, introducing "flicker" noise in the drain current. In addition to trapping, several other mechanisms are believed to generate flicker noise [3].

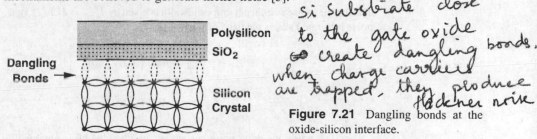

Polysilicon
SiO$_2$

Dangling Bonds →

Silicon Crystal

[handwritten: Si substrate close to the gate oxide create dangling bonds. when charge carriers are trapped, they produce flicker noise]

Figure 7.21 Dangling bonds at the oxide-silicon interface.

Unlike thermal noise, the average power of flicker noise cannot be predicted easily. Depending on the "cleanness" of the oxide-silicon interface, flicker noise may assume considerably different values and as such varies from one CMOS technology to another. The flicker noise is more easily modeled as a voltage source in series with the gate and roughly given by

$$\overline{V_n^2} = \frac{K}{C_{ox}WL}\cdot\frac{1}{f}, \tag{7.28}$$

[handwritten: $\overline{I_n^2} = \overline{V_n^2} \cdot g_m^2$]

where K is a process-dependent constant on the order of 10^{-25} V^{2}F. Note that our notation assumes a bandwidth of 1 Hz. Interestingly, as shown in Fig. 7.22, the noise spectral density is inversely proportional to the frequency. For example, the trap-and-release phenomenon associated with the dangling bonds occurs at low frequencies more often. For this reason, flicker noise is also called $1/f$ noise. Note that (7.28) does not depend on the bias current or the temperature. This is only an approximation and in reality, the flicker noise equation is somewhat more complex [3].

The inverse dependence of (7.28) on WL suggests that to decrease $1/f$ noise, the device *area* must be increased. It is therefore not surprising to see devices having areas of several

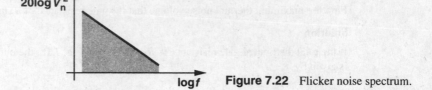

large area → lesser flicker noise.
since $\overline{V_n^2} \propto \frac{1}{WL}$.

Figure 7.22   Flicker noise spectrum.

thousand square microns in low-noise applications. It is also believed that PMOS devices exhibit less $1/f$ noise than NMOS transistors because the former carry the holes in a "buried channel," i.e., at some distance from the oxide-silicon interface. Nonetheless, this difference between PMOS and NMOS transistors is not consistently observed [3].

Example 7.5

For an NMOS current source, calculate the total thermal and $1/f$ noise in the drain current for a band from 1 kHz to 1 MHz.

Solution

The thermal noise current per unit bandwidth is given by $\overline{I_{n,th}^2} = 4kT(2/3)g_m$. Thus, the total thermal noise integrated across the band of interest is

$$\overline{I_{n,th,tot}^2} = 4kT\left(\frac{2}{3}g_m\right)(10^6 - 10^3) \tag{7.29}$$

$$\approx 4kT\left(\frac{2}{3}g_m\right) \times 10^6 \text{ A}^2. \tag{7.30}$$

For $1/f$ noise, the drain noise current per unit bandwidth is obtained by multiplying the noise voltage at the gate by the device transconductance:

$$\overline{I_{n,1/f}^2} = \frac{K}{C_{ox}WL} \cdot \frac{1}{f} \cdot g_m^2. \tag{7.31}$$

The total $1/f$ noise is then equal to

$$\overline{I_{n,1/f,tot}^2} = \frac{Kg_m^2}{C_{ox}WL} \int_{1\text{ kHz}}^{1\text{ MHz}} \frac{df}{f} \tag{7.32}$$

$$= \frac{Kg_m^2}{C_{ox}WL} \ln 10^3 \tag{7.33}$$

$$= \frac{6.91Kg_m^2}{C_{ox}WL}. \tag{7.34}$$

The above example raises an interesting question. What happens to $\overline{I_{n,1/f,tot}^2}$ if the lower end of the band, f_L, is zero rather than 1 kHz? Equation (7.33) then contains the natural logarithm of zero, yielding an infinite value for the total noise. To overcome the fear of infinite noise, we make two observations. First, extending f_L to zero means that we are interested in *arbitrarily* slow noise components. A noise component at 0.01 Hz varies significantly in roughly 10 s and one at 10^{-6} in roughly one week. Second, the infinite flicker noise power simply means that if we observe the circuit for a very long time, the very slow noise components can randomly assume a very large power level. At such slow rates, noise becomes indistinguishable from thermal drift or aging of devices.

The foregoing observations lead to the following conclusions. First, since the *signals* encountered in most applications do not contain significant low-frequency components, our observation window need not be very long. For example, voice signals display negligible energy below 20 Hz and if a noise component varies at a lower rate, it does not corrupt the voice significantly. Second, the logarithmic dependence of the flicker noise power upon f_L allows some margin for error in selecting f_L. For example, if the band of interest is so wide that the total integrated thermal noise power is comparable with the flicker noise contribution, then the choice of f_L is quite relaxed.

In order to quantify the significance of $1/f$ noise with respect to thermal noise for a given device, we plot both spectral densities on the same axes (Fig. 7.23). Called the $1/f$ noise "corner frequency," the intersection point serves as a measure of what part of the band

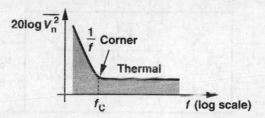

Figure 7.23 Concept of flicker noise corner frequency.

is mostly corrupted by flicker noise. In the above example, the $1/f$ noise corner, f_C, of the output current is determined as

$$4kT\left(\frac{2}{3}g_m\right) = \frac{K}{C_{ox}WL} \cdot \frac{1}{f_C} \cdot g_m^2, \qquad (7.35)$$

that is,

$$f_C = \frac{K}{C_{ox}WL}g_m\frac{3}{8kT}. \qquad (7.36)$$

This result implies that f_C generally depends on device dimensions and bias current. Nonetheless, since for a given L, the dependence is relatively weak, the $1/f$ noise corner is relatively constant, falling in the vicinity of 500 kHz to 1 MHz for submicron transistors.

Example 7.6 ──

For a 100-μm/0.5-μm MOS device with $g_m = 1/(100\ \Omega)$, the $1/f$ noise corner frequency is measured to be 500 kHz. If $t_{ox} = 90$ Å, what is the flicker noise coefficient, K, in this technology?

Solution

For $t_{ox} = 90$ Å, we have $C_{ox} = 3.84$ fF/μm^2. Using Eq. (7.36), we write

$$500 \text{ kHz} = \frac{K}{3.84 \times 100 \times 0.5 \times 10^{-15}} \cdot \frac{1}{100} \cdot \frac{3}{8 \times 1.38 \times 10^{-23} \times 300}. \quad (7.37)$$

That is, $K = 1.06 \times 10^{-25}$ V^{2}F.

7.3 Representation of Noise in Circuits

Consider a general circuit with one input port and one output port (Fig. 7.24). How do we quantify the effect of noise here? The natural approach would be to set the input to zero and calculate the total noise at the output due to various sources of noise in the circuit. This is indeed how the noise is measured in the laboratory or in simulations.

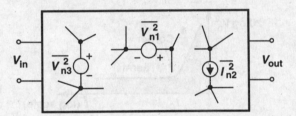

Figure 7.24   Noise sources in a circuit.

Example 7.7

What is the total output noise voltage of the common-source stage shown in Fig. 7.25(a)?

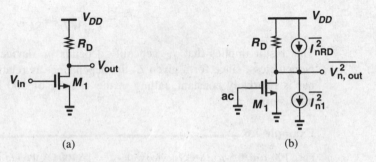

(a)   (b)

Figure 7.25   (a) CS stage, (b) circuit including noise sources.

Solution

We model the thermal and flicker noise of M_1 by two current sources: $\overline{I_{n,th}^2} = 4kT(2/3)g_m$ and $\overline{I_{n,1/f}^2} = Kg_m^2/(C_{ox}WLf)$. We also represent the thermal noise of R_D by a current source $\overline{I_{n,RD}^2} = 4kT/R_D$. The output noise voltage per unit bandwidth is therefore equal to

$$\overline{V_{n,out}^2} = \left(4kT\frac{2}{3}g_m + \frac{K}{C_{ox}WL} \cdot \frac{1}{f} \cdot g_m^2 + \frac{4kT}{R_D} \right) R_D^2. \tag{7.38}$$

Note that the noise mechanisms are added as "power" quantities because they are uncorrelated. The value given by (7.38) represents the noise power in 1 Hz at a frequency f. The total output noise can be obtained by integration over the bandwidth of interest.

While intuitively appealing, the output-referred noise does not allow a fair comparison of the performance of different circuits because it depends on the gain. For example, as depicted in Fig. 7.26, if a common-source stage is followed by a noiseless amplifier having

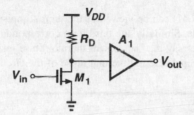

Figure 7.26   Addition of gain stage to a CS stage.

a voltage gain A_1, then the output noise is equal to the expression in (7.38) multiplied by A_1^2. Considering only the output noise, we may conclude that as A_1 increases, the circuit becomes noisier, an incorrect result because a larger A_1 also provides a proportionally higher *signal* level at the output. That is, the output signal-to-noise ratio (SNR) does not depend on A_1.

To overcome the above quandary, we usually specify the "input-referred noise" of circuits. Illustrated conceptually in Fig. 7.27, the idea is to represent the effect of all noise sources in the circuit by a single source, $\overline{V_{n,in}^2}$, at the input such that the output noise

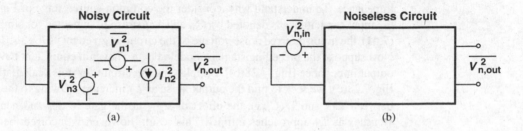

Figure 7.27   Determination of input-referred noise voltage.

in Fig. 7.27(b) equals that in Fig. 7.27(a). If the voltage gain is A_v, then we must have $\overline{V_{n,out}^2} = A_v^2 \overline{V_{n,in}^2}$, that is, the input-referred noise voltage in this simple case is given by the output noise voltage divided by the gain.

Example 7.8

For the circuit of Fig. 7.25, calculate the input-referred noise voltage.

Solution

We have

$$\overline{V_{n,in}^2} = \frac{\overline{V_{n,out}^2}}{A_v^2} \tag{7.39}$$

$$= \left(4kT \frac{2}{3} g_m + \frac{K}{C_{ox} WL} \cdot \frac{1}{f} \cdot g_m^2 + \frac{4kT}{R_D} \right) R_D^2 \frac{1}{g_m^2 R_D^2} \tag{7.40}$$

$$= 4kT \frac{2}{3g_m} + \frac{K}{C_{ox} WL} \cdot \frac{1}{f} + \frac{4kT}{g_m^2 R_D}. \tag{7.41}$$

Note that the first term in (7.41) can be viewed as the thermal noise of a resistor equal to $2/(3g_m)$ placed in series with the gate. Similarly, the third term corresponds to the noise of a resistor equal to $(g_m^2 R_D)^{-1}$. We sometimes say the "equivalent thermal noise resistance" of a circuit is equal to R_T, meaning that the total input-referred thermal noise of the circuit in unit bandwidth is equal to $4kT R_T$.

At this point of our study, we make two observations. First, the input-referred noise and the input signal are both multiplied by the gain as they are processed by the circuit. Thus, the input-referred noise indicates how much the input signal is corrupted by the circuit's noise, i.e., how small an input the circuit can detect with acceptable SNR. For this reason, input-referred noise allows a fair comparison of different circuits. Second, the input-referred noise is a fictitious quantity in that it cannot be *measured* at the input of the circuit. The two circuits of Figs. 7.27(a) and (b) are *mathematically* equivalent but the physical circuit is still that in Fig. 7.27(a).

In the foregoing discussion, we have assumed that the input-referred noise can be modeled by a single voltage source in series with the input. This is generally an incomplete representation if the circuit has a finite input impedance and is driven by a finite source impedance. To understand why, consider the common-source stage of Fig. 7.28(a), where the input capacitance is denoted by C_{in} and $1/f$ noise is neglected for simplicity. From Eq. (7.41), the input-referred noise voltage of the circuit is given by $8kT/(3g_m) + 4kT/(g_m^2 R_D)$. Now suppose the preceding stage is modeled by a Thevenin equivalent having an inductive output impedance [Fig. 7.28(b)]. Simplifying the circuit for noise calculations as shown in Fig. 7.28(c), we seek to find the output noise as L_1 increases. Owing to the voltage division between $L_1 s$ and $1/(C_{in} s)$, the effect of $\overline{V_{n,in}^2}$ at the gate of M_1 and hence at the output vanishes as L_1 approaches infinity. This result, however, is incorrect because the output noise of the circuit is equal to $(8kT/3)g_m R_D^2 + 4kT R_D$ and independent of L_1 and C_{in}.

Noise Factor = $\dfrac{\text{Total o/p noise}}{\text{Noise generated by sig. source}}$

$\overline{V_x^2} \rightarrow$ *noise generated by signal source*

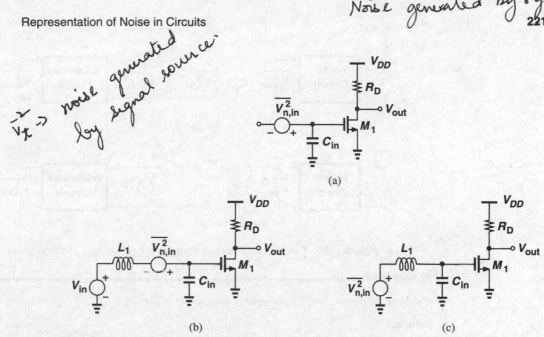

(a)

(b)   (c)

Figure 7.28 CS stage including input capacitance, (b) CS stage stimulated by a finite source impedance, (c) Effect of single noise source.

Let us summarize the problem. If the circuit has a finite input impedance, modeling the input-referred noise by merely a voltage source implies that the output noise vanishes as the source impedance becomes large, an incorrect conclusion. To resolve this issue, we model the input-referred noise by both a series voltage source and a parallel current source (Fig. 7.29) so that if the output impedance of the preceding stage assumes large values—thereby reducing the effect of $\overline{V_{n,in}^2}$—the noise current source still flows through a finite impedance, producing noise at the input. It can be proved that $\overline{V_{n,in}^2}$ and $\overline{I_{n,in}^2}$ are necessary and sufficient to represent the noise of any linear two-port circuit [5].

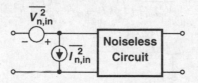

Figure 7.29 Representation of noise by voltage and current sources.

How do we calculate $\overline{V_{n,in}^2}$ and $\overline{I_{n,in}^2}$? Since the model is valid for any source impedance, we consider two extreme cases: zero and infinite source impedances. As shown in Fig. 7.30(a), if the source impedance is zero, $\overline{I_{n,in}^2}$ flows through $\overline{V_{n,in}^2}$ and has no effect on the output. Thus, the output noise measured in this case arises solely from $\overline{V_{n,in}^2}$. Similarly, if the input is open [Fig. 7.30(b)], then $\overline{V_{n,in}^2}$ has no effect and the output noise is due to only $\overline{I_{n,in}^2}$. Let us apply this method to the circuit of Fig. 7.28.

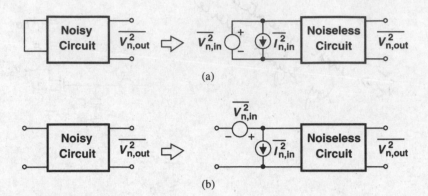

(a)

(b)

Figure 7.30 Calculation of input-referred noise (a) voltage, and (b) current.

Example 7.9

Calculate the input-referred noise voltage and current of Fig. 7.28.

Solution

From (7.41), the input-referred noise voltage (excluding $1/f$ noise) is simply

$$\overline{V_{n,in}^2} = 4kT\frac{2}{3g_m} + \frac{4kT}{g_m^2 R_D}. \tag{7.42}$$

As depicted in Fig. 7.31(a), this voltage generates the same output noise as the actual circuit if the input is shorted.

To obtain the input-referred noise current, we open the input and find the output noise in terms of $\overline{I_{n,in}^2}$ [Fig. 7.31(b)]. The noise current flows through C_{in}, generating at the output

$$\overline{V_{n,out}^2} = \overline{I_{n,in}^2}\left(\frac{1}{C_{in}\omega}\right)^2 g_m^2 R_D^2. \tag{7.43}$$

This value must be equal to the output of the noisy circuit when its input is open:

$$\overline{V_{n,out}^2} = \left(4kT\frac{2}{3}g_m + \frac{4kT}{R_D}\right)R_D^2. \tag{7.44}$$

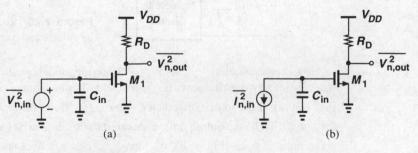

(a)   (b)

Figure 7.31

From (7.43) and (7.44), it follows that

$$\overline{I_{n,in}^2} = (C_{in}\omega)^2 \frac{4kT}{g_m^2} \left(\frac{2}{3} g_m + \frac{1}{R_D} \right).$$ (7.45)

The reader may wonder if the use of both a voltage source and a current source to represent the input-referred noise "counts the noise twice." We utilize the circuit of Fig. 7.28 as an example to demonstrate that this is not so. Considering the environment depicted in Fig. 7.32, we prove that the output noise is correct for any source impedance Z_S. Assuming Z_S is noise-

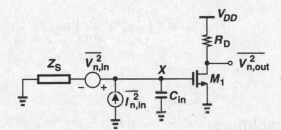

Figure 7.32 CS stage stimulated by a source impedance.

less for simplicity, we first calculate the total noise voltage at the gate of M_1 due to $\overline{V_{n,in}^2}$ and $\overline{I_{n,in}^2}$. How is this voltage obtained: by superposition of voltages or powers? The two sources $\overline{V_{n,in}^2}$ and $\overline{I_{n,in}^2}$ are in general *correlated* simply because they may represent the same noise mechanisms in the circuit. In fact, Eqs. (7.42) and (7.45) can be respectively rewritten as

$$V_{n,in} = V_{n,M1} + \frac{1}{g_m R_D} V_{n,RD}$$ (7.46)

$$I_{n,in} = C_{in}s V_{n,M1} + \frac{C_{in}s}{g_m R_D} V_{n,RD},$$ (7.47)

where $V_{n,M1}$ denotes the gate-referred noise voltage of M_1 and $V_{n,RD}$ the noise voltage of R_D. We recognize that $V_{n,M1}$ and $V_{n,RD}$ appear in both $V_{n,in}$ and $I_{n,in}$, creating a strong correlation between the two. Thus, the calculations must use superposition of voltages—as if $V_{n,in}$ and $I_{n,in}$ were deterministic quantities.

Adding the contributions of $V_{n,in}$ and $I_{n,in}$ at node X in Fig. 7.32, we have

$$V_{n,X} = V_{n,in} \frac{\dfrac{1}{C_{in}s}}{\dfrac{1}{C_{in}s} + Z_S} + I_{n,in} \frac{\dfrac{Z_S}{C_{in}s}}{\dfrac{1}{C_{in}s} + Z_S}$$ (7.48)

$$= \frac{V_{n,in} + I_{n,in} Z_S}{Z_S C_{in}s + 1}.$$ (7.49)

Substituting for $V_{n,in}$ and $I_{n,in}$ from (7.46) and (7.47), respectively, we obtain

$$V_{n,X} = \frac{1}{Z_S C_{in} s + 1} \left[V_{n,M1} + \frac{1}{g_m R_D} V_{n,RD} + C_{in} s Z_S \left(V_{n,M1} + \frac{1}{g_m R_D} V_{n,RD} \right) \right] \quad (7.50)$$

$$= V_{n,M1} + \frac{1}{g_m R_D} V_{n,RD}. \quad (7.51)$$

Note that $V_{n,X}$ is independent of Z_S and C_{in}. It follows that

$$\overline{V_{n,out}^2} = g_m^2 R_D^2 \overline{V_{n,X}^2} \quad (7.52)$$

$$= 4kT \left(\frac{2}{3} g_m + \frac{1}{R_D} \right) R_D^2, \quad (7.53)$$

the same as (7.44).

7.4 Noise in Single-Stage Amplifiers

Having developed basic mathematical tools and models for noise analysis, we now study the noise performance of single-stage amplifiers at low frequencies. Before considering specific topologies, we describe a lemma that simplifies noise calculations.

Lemma   The circuits shown in Fig. 7.33(a) and (b) are equivalent at low frequencies if $\overline{V_n^2} = \overline{I_n^2}/g_m^2$ and the circuits are driven by a finite impedance.

Proof.   Since the circuits have equal output impedances, we simply examine the output short-circuit currents [Figs. 7.33(c) and (d)]. It can be proved (Problem 7.4) that the output noise current of the circuit in Fig. 7.33(c) is given by

$$I_{n,out1} = \frac{I_n}{Z_S(g_m + 1/r_O) + 1} \quad (7.54)$$

and that of Fig. 7.33(d) is

$$I_{n,out2} = \frac{g_m V_n}{Z_S(g_m + 1/r_O) + 1}. \quad (7.55)$$

Equating (7.54) and (7.55), we have $V_n = I_n/g_m$.   ☐

This lemma suggests that the noise source can be transformed from a drain-source current to a gate series voltage for arbitrary Z_S.

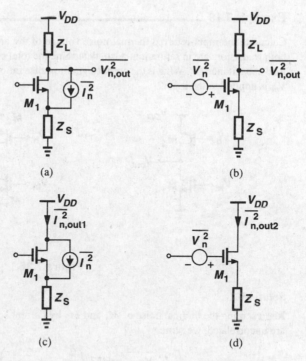

Figure 7.33 Equivalent CS stages

7.4.1 Common-Source Stage

From Example 7.8, the input-referred noise voltage per unit bandwidth of a simple CS stage is equal to

$$\overline{V_{n,in}^2} = 4kT\left(\frac{2}{3g_m} + \frac{1}{g_m^2 R_D}\right) + \frac{K}{C_{ox}WL}\frac{1}{f}. \tag{7.56}$$

From the above lemma, we recognize that the term $4kT[2/(3g_m)]$ is in fact the <u>thermal noise current of M_1 expressed as a voltage in series with the gate.</u>

How can we reduce the input-referred noise voltage? Equation (7.56) implies that the transconductance of M_1 must be maximized. Thus, the transconductance must be maximized if the transistor is to amplify a voltage signal applied to its gate [Fig. 7.34(a)] whereas it must be minimized if the transistor operates as a current source [Fig. 7.34(b)].

Input referred noise can be reduced by increasing gm.

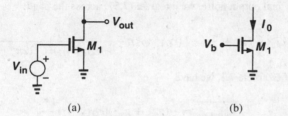

Figure 7.34 Voltage amplification versus current generation.

Example 7.10

Calculate the input-referred thermal noise voltage of the amplifier shown in Fig. 7.35(a), assuming both transistors are in saturation. Also, determine the total output thermal noise if the circuit drives a load capacitance C_L. What is the output signal-to-noise ratio if a low-frequency sinusoid of amplitude V_m is applied to the input?

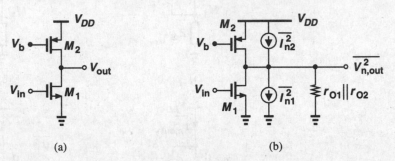

(a)   (b)

Figure 7.35

Solution

Representing the thermal noise of M_1 and M_2 by current sources [Fig. 7.35(b)] and noting that they are uncorrelated, we write

$$\overline{V_{n,out}^2} = 4kT \left(\frac{2}{3}g_{m1} + \frac{2}{3}g_{m2} \right)(r_{O1}\|r_{O2})^2. \tag{7.57}$$

Since the voltage gain is equal to $g_{m1}(r_{O1}\|r_{O2})$, the total noise voltage referred to the gate of M_1 is

$$\overline{V_n^2} = 4kT \left(\frac{2}{3}g_{m1} + \frac{2}{3}g_{m2} \right)\frac{1}{g_{m1}^2} \tag{7.58}$$

$$= 4kT \left(\frac{2}{3g_{m1}} + \frac{2}{3}\frac{g_{m2}}{g_{m1}^2} \right). \tag{7.59}$$

Equation (7.59) reveals the dependence of $\overline{V_{n,in}^2}$ upon g_{m1} and g_{m2}, confirming that g_{m2} must be minimized because M_2 serves as a current source.

The reader may wonder why M_1 and M_2 in Fig. 7.35 exhibit different noise effects. After all, if the noise currents of both transistors flow through $r_{O1}\|r_{O2}$, why should g_{m1} be maximized and g_{m2} minimized? This is simply because, as g_{m1} increases, the output noise *voltage* rises in proportion to $\sqrt{g_{m1}}$ whereas the *voltage gain* of the stage increases in proportion to g_{m1}. As a result, the input-referred noise voltage decreases.

To compute the total output noise, we integrate (7.57) across the band:

$$\overline{V_{n,out,tot}^2} = \int_0^\infty 4kT \left(\frac{2}{3}g_{m1} + \frac{2}{3}g_{m2} \right)(r_{O1}\|r_{O2})^2 \frac{df}{1 + (r_{O1}\|r_{O2})^2 C_L^2 (2\pi f)^2}. \tag{7.60}$$

Using the results of Example 7.1, we have

$$\overline{V_{n,out,tot}^2} = \frac{2}{3}(g_{m1} + g_{m2})(r_{O1}\|r_{O2})\frac{kT}{C_L}. \tag{7.61}$$

An input sinusoid of amplitude V_m yields an output amplitude equal to $g_{m1}(r_{O1}\|r_{O2})V_m$. The output SNR is equal to the ratio of the signal power and the noise power:

$$SNR_{out} = \left[\frac{g_{m1}(r_{O1}\|r_{O2})V_m}{\sqrt{2}}\right]^2 \cdot \frac{1}{(2/3)(g_{m1}+g_{m2})(r_{O1}\|r_{O2})(kT/C_L)} \tag{7.62}$$

$$= \frac{3C_L}{4kT} \cdot \frac{g_{m1}^2(r_{O1}\|r_{O2})}{g_{m1}+g_{m2}}V_m^2. \tag{7.63}$$

We note that to maximize the output SNR, C_L must be maximized, i.e., the bandwidth must be minimized. Of course, the bandwidth is also dictated by the input signal spectrum. This example indicates that it becomes exceedingly difficult to design broadband circuits while maintaining a low noise.

It is also important to observe from (7.56) that the noise contributed by R_D in Fig. 7.25(a) decreases as R_D increases. This is again because the noise voltage due to R_D at the output is proportional to $\sqrt{R_D}$ while the voltage gain of the circuit is proportional to R_D.

Example 7.11

Calculate the input-referred $1/f$ and thermal noise voltage of the circuit depicted in Fig. 7.36(a) assuming M_1 and M_2 are in saturation.

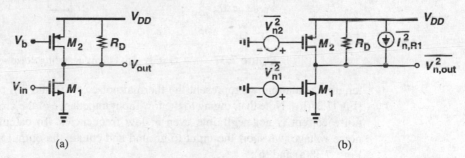

(a)   (b)

Figure 7.36

Solution

We model the $1/f$ and thermal noise of the transistors as voltage sources in series with their gates [Fig. 7.36(b)]. The noise voltage at the gate of M_2 experiences a gain of $g_{m2}(R_D\|r_{O1}\|r_{O2})$ as it appears at the output. The result must then be divided by $g_{m1}(R_D\|r_{O1}\|r_{O2})$ to be referred to the main input. The noise current of R_D is multiplied by $R_D\|r_{O1}\|r_{O2}$ and divided by $g_{m1}(R_D\|r_{O1}\|r_{O2})$. Thus, the overall input-referred noise voltage is given by

$$\overline{V_{n,in}^2} = 4kT\frac{2}{3}\left(\frac{g_{m2}}{g_{m1}^2}+\frac{1}{g_{m1}}\right) + \frac{1}{C_{ox}}\left[\frac{K_P g_{m2}^2}{(WL)_2 g_{m1}^2}+\frac{K_N}{(WL)_1}\right]\frac{1}{f} + \frac{4kT}{g_{m1}^2 R_D}, \tag{7.64}$$

where K_P and K_N denote the flicker noise coefficients of PMOS and NMOS devices, respectively. As expected, the input-referred noise voltage increases if g_{m2} increases.

How do we design a common-source stage for low-noise operation? For thermal noise in the simple topology of Fig. 7.34, we must maximize g_{m1} by increasing the drain current or the device width. A higher I_D translates to greater power dissipation and limited output voltage swings while a wider device leads to larger input and output capacitance. We can also increase R_D, but at the cost of limiting the voltage headroom and lowering the speed.

For $1/f$ noise, the primary approach is to increase the area of the transistor. If WL is increased while W/L remains constant, then the device transconductance and hence its thermal noise do not change but the device capacitances increase. These observations point to the trade-offs between noise, power dissipation, voltage headroom, and speed.

7.4.2 Common-Gate Stage

Consider the common-gate configuration shown in Fig. 7.37(a). Neglecting channel-

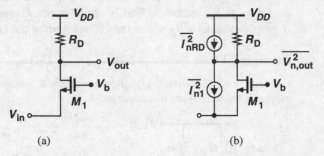

(a)   (b)

Figure 7.37   (a) CG stage, (b) circuit including noise sources.

length modulation, we represent the thermal noise of M_1 and R_D by two current sources [Fig. 7.37(b)]. Note that, owing to the low input impedance of the circuit, the input-referred noise current is not negligible even at low frequencies. To calculate the input-referred noise voltage, we short the input to ground and equate the output noise of the circuits in Figs. 7.38(a) and (b):

$$\left(4kT\frac{2}{3}g_m + \frac{4kT}{R_D} \right) R_D^2 = \overline{V_{n,in}^2}(g_m + g_{mb})^2 R_D^2. \tag{7.65}$$

That is,

$$\overline{V_{n,in}^2} = \frac{4kT(2g_m/3 + 1/R_D)}{(g_m + g_{mb})^2}. \tag{7.66}$$

Similarly, equating the output noise of the circuits in Figs. 7.38(c) and (d) yields the input-referred noise current. What is the effect of $\overline{I_{n1}^2}$ at the output in Fig. 7.38(c)? Since the sum of the currents at the source of M_1 is zero, $I_{n1} + I_{D1} = 0$. Consequently, I_{n1} creates an equal and opposite current in M_1, producing *no* noise at the output. The output noise voltage of Fig. 7.37(a) is therefore equal to $4kTR_D$ and hence $\overline{I_{n,in}^2}R_D^2 = 4kTR_D$. That is,

$$\overline{I_{n,in}^2} = \frac{4kT}{R_D}. \tag{7.67}$$

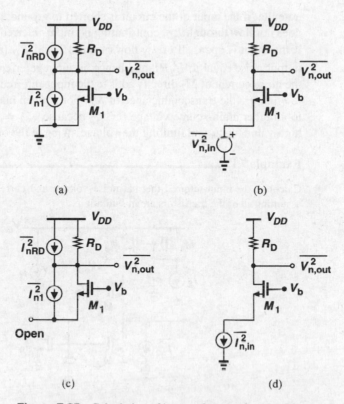

(a)   (b)

(c)   (d)

Figure 7.38   Calculation of input-referred noise of a CG stage.

An important drawback of common-gate topologies is that they directly refer the noise current produced by the load to the input. Exemplified by (7.67), this effect arises because such circuits provide no *current* gain, a point of contrast to common-source amplifiers.

We have thus far neglected the noise contributed by the bias current source of a common-gate stage. Shown in Fig. 7.39 is a simple mirror arrangement establishing the bias current of M_1 as a multiple of I_1. Capacitor C_0 shunts the noise generated by M_0 to ground. We

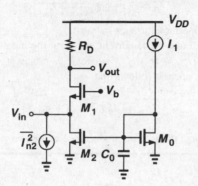

Figure 7.39   Noise contributed by bias current source.

note that if the input of the circuit is shorted to ground, then the drain noise current of M_2 does not flow through R_D, contributing no input-referred noise voltage. On the other hand, if the input is open, all of $\overline{I_{n2}^2}$ flows from M_1 and R_D (at low frequencies), producing an output noise equal to $\overline{I_{n2}^2}R_D^2$ and hence an input-referred noise current of $\overline{I_{n2}^2}$. As a result, the noise current of M_2 directly adds to the input-referred noise current, making it desirable to *minimize* the transconductance of M_2. For a given bias current, however, this translates to a higher drain-source voltage for M_2 because $g_{m2} = 2I_{D2}/(V_{GS2} - V_{TH2})$, requiring a high value for V_b and limiting the voltage swing at the output node.

Example 7.12

Calculate the input-referred thermal noise voltage and current of the circuit shown in Fig. 7.40 assuming all of the transistors are in saturation.

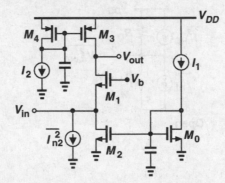

Figure 7.40

Solution

To compute the input-referred noise voltage, we short the input to ground, obtaining

$$\overline{V_{n,out}^2} = 4kT\frac{2}{3}(g_{m1} + g_{m3})(r_{O1}\|r_{O3})^2. \tag{7.68}$$

Thus, the input-referred noise voltage, $V_{n,in}$, must satisfy this relationship:

$$\overline{V_{n,in}^2}(g_{m1} + g_{mb1})^2(r_{O1}\|r_{O3})^2 = 4kT\frac{2}{3}(g_{m1} + g_{m3})(r_{O1}\|r_{O3})^2, \tag{7.69}$$

and hence

$$\overline{V_{n,in}^2} = 4kT\frac{2}{3}\frac{(g_{m1} + g_{m3})}{(g_{m1} + g_{mb1})^2}. \tag{7.70}$$

As expected, the noise is proportional to g_{m3}.

To calculate the input-referred noise current, we open the input and note that the output noise voltage is simply given by $(\overline{I_{n2}^2} + \overline{I_{n3}^2})R_{out}^2$, where $R_{out} \approx r_{O3}\|(g_{m1}r_{O1}r_{O2})$ denotes the output impedance when the input is open. It follows that

$$\overline{I_{n,in}^2} = 4kT\frac{2}{3}(g_{m2} + g_{m3}). \tag{7.71}$$

Again, the noise is proportional to the transconductance of the two current sources.

The effect of $1/f$ noise in a common-gate topology is also of interest. As a typical case, we compute the input-referred $1/f$ noise voltage and current of the circuit shown in Fig. 7.40. Illustrated in Fig. 7.41, each $1/f$ noise generator is modeled by a voltage source in series with the gate of the corresponding transistor. Note that the $1/f$ noise of M_0 and M_4 is neglected. A more realistic case is studied in Problem 7.10.

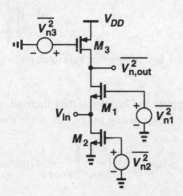

Figure 7.41 Flicker noise in a CG stage.

With the input shorted to ground, we have

$$\overline{V_{n,out}^2} = \frac{1}{C_{ox} f} \left[\frac{g_{m1}^2 K_N}{(WL)_1} + \frac{g_{m3}^2 K_P}{(WL)_3} \right] (r_{O1} \| r_{O3})^2, \tag{7.72}$$

where K_N and K_P denote the flicker noise coefficient of NMOS and PMOS devices, respectively. Thus,

$$\overline{V_{n,in}^2} = \frac{1}{C_{ox} f} \left[\frac{g_{m1}^2 K_N}{(WL)_1} + \frac{g_{m3}^2 K_P}{(WL)_3} \right] \frac{1}{(g_{m1} + g_{mb1})^2}. \tag{7.73}$$

With the input open, the output noise voltage is given by

$$\overline{V_{n,out}^2} = \frac{1}{C_{ox} f} \left[\frac{g_{m2}^2 K_N}{(WL)_2} + \frac{g_{m3}^2 K_P}{(WL)_3} \right] R_{out}^2, \tag{7.74}$$

yielding

$$\overline{I_{n,in}^2} = \frac{1}{C_{ox} f} \left[\frac{g_{m2}^2 K_N}{(WL)_2} + \frac{g_{m3}^2 K_P}{(WL)_3} \right]. \tag{7.75}$$

Equations (7.73) and (7.75) describe the $1/f$ noise behavior of the circuit and must be added to (7.70) and (7.71), respectively, to obtain the overall noise per unit bandwidth.

7.4.3 Source Followers

Consider the source follower depicted in Fig. 7.42(a), where M_2 serves as the bias current source. Since the input impedance of the circuit is quite high, even at relatively high frequencies, the input-referred noise current can usually be neglected for moderate driving

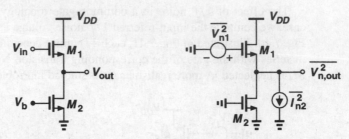

Figure 7.42   (a) Source follower, (b) circuit including noise sources.

source impedances. To compute the input-referred thermal noise voltage, we employ the representation in Fig. 7.42(b), expressing the output noise due to M_2 as

$$\overline{V_{n,out}^2}|_{M2} = \overline{I_{n2}^2} \left(\frac{1}{g_{m1}} \left\| \frac{1}{g_{mb1}} \right\| r_{O1}\|r_{O2} \right)^2. \tag{7.76}$$

From Chapter 3,

$$A_v = \frac{\dfrac{1}{g_{mb1}} \left\| r_{O1}\|r_{O2} \right.}{\dfrac{1}{g_{mb1}} \left\| r_{O1}\|r_{O2} + \dfrac{1}{g_{m1}} \right.}. \tag{7.77}$$

Thus, the total input-referred noise voltage is

$$\overline{V_{n,in}^2} = \overline{V_{n1}^2} + \frac{\overline{V_{n,out}^2}|_{M2}}{A_v^2} \tag{7.78}$$

$$= 4kT \frac{2}{3} \left(\frac{1}{g_{m1}} + \frac{g_{m2}}{g_{m1}^2} \right). \tag{7.79}$$

Note the similarity between (7.59) and (7.79).

Since source followers add noise to the input signal while providing a voltage gain less than unity, they are usually avoided in low-noise amplification. The $1/f$ noise performance of source followers is studied in Problem 7.11.

7.4.4 Cascode Stage

Consider the cascode stage of Fig. 7.43(a). Since at low frequencies the noise currents of M_1 and R_D flow through R_D, the noise contributed by these two devices is quantified as in a common-source stage:

$$\overline{V_{n,in}^2}|_{M1,RD} = 4kT \left(\frac{2}{3g_{m1}} + \frac{1}{g_{m1}^2 R_D} \right), \tag{7.80}$$

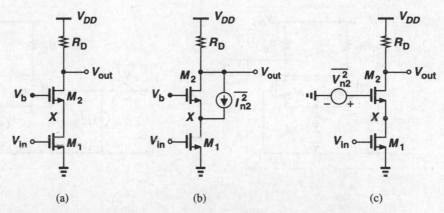

Figure 7.43 (a) Cascode stage, (b) noise of M_2 modeled by a current source, (c) noise of M_2 modeled by a voltage source.

where $1/f$ noise of M_1 is ignored. What is the effect of noise of M_2? Modeled as in Fig. 7.43(b), this noise contributes negligibly to the output, especially at low frequencies. This is because, if channel length modulation in M_1 is neglected, then $I_{n2} + I_{D2} = 0$, and hence M_2 does not affect $V_{n,out}$. From another point of view, using the lemma of Fig. 7.33 to construct the equivalent in Fig. 7.43(c), we note that the voltage gain from V_{n2} to the output is quite small if the impedance at node X is large. At high frequencies, on the other hand, the total capacitance at node X, C_X, gives rise to a gain:

$$\frac{V_{n,out}}{V_{n2}} \approx \frac{-R_D}{1/g_{m2} + 1/(C_X s)},\tag{7.81}$$

increasing the output noise. This capacitance also decreases the gain from the main input to the output by shunting the signal current produced by M_1 to ground. As a result, the input-referred noise of a cascode stage may rise considerably at high frequencies.

7.5 Noise in Differential Pairs

With our understanding of noise in basic amplifiers, we can now study the noise behavior of differential pairs. Shown in Fig. 7.44(a), a differential pair can be viewed as a two-port circuit. It is therefore possible to model the overall noise as depicted in Fig. 7.44(b). For low-frequency operation, the magnitude of $\overline{I_{n,in}^2}$ is typically negligible.

To calculate the thermal component of $\overline{V_{n,in}^2}$, we first obtain the total output noise with the inputs shorted together [Fig. 7.45(a)], noting that superposition of power quantities is possible because the noise sources in the circuit are uncorrelated. Since I_{n1} and I_{n2} are uncorrelated, node P cannot be considered a virtual ground, making it difficult to use the half-circuit concept. Thus, we simply derive the effect of each source individually. Depicted in Fig. 7.45(b), the contribution of I_{n1} is obtained by first reducing the circuit to that in Fig. 7.45(c). With the aid of this figure and neglecting channel-length modulation, the reader can prove that half of I_{n1} flows through R_{D1} and the other half through M_2 and R_{D2}.

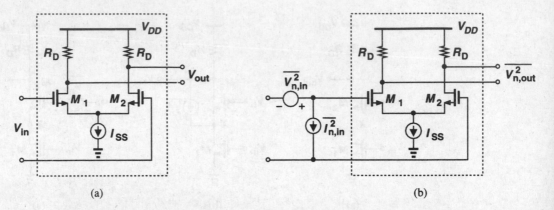

Figure 7.44 (a) Differential pair, (b) circuit including input-referred noise sources.

As shown in Fig. 7.45(d), this can also be proved by decomposing I_{n1} into two (correlated) current sources and calculating their effect at the output. Thus,

$$V_{n,out}|_{M1} = \frac{I_{n1}}{2}R_{D1} + \frac{I_{n1}}{2}R_{D2}. \tag{7.82}$$

Note that the two noise voltages are directly added because they both arise from I_{n1} and are therefore correlated. It follows that, if $R_{D1} = R_{D2} = R_D$,

$$\overline{V_{n,out}^2}|_{M1} = \overline{I_{n1}^2}R_D^2. \tag{7.83}$$

Similarly,

$$\overline{V_{n,out}^2}|_{M2} = \overline{I_{n2}^2}R_D^2, \tag{7.84}$$

yielding

$$\overline{V_{n,out}^2}|_{M1,M2} = \left(\overline{I_{n1}^2} + \overline{I_{n2}^2}\right)R_D^2. \tag{7.85}$$

Taking into account the noise of R_{D1} and R_{D2}, we have for the total output noise:

$$\overline{V_{n,out}^2} = \left(\overline{I_{n1}^2} + \overline{I_{n2}^2}\right)R_D^2 + 2(4kTR_D) \tag{7.86}$$

$$= 8kT\left(\frac{2}{3}g_m R_D^2 + R_D\right). \tag{7.87}$$

Dividing the result by the square of the differential gain, $g_m^2 R_D^2$, we have

$$\overline{V_{n,in}^2} = 8kT\left(\frac{2}{3g_m} + \frac{1}{g_m^2 R_D}\right). \tag{7.88}$$

This is simply twice the input noise voltage squared of a common-source stage.

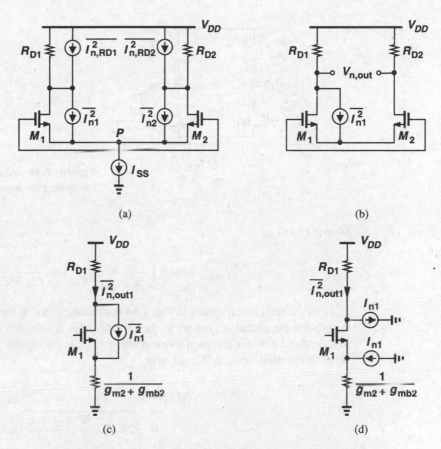

(a)   (b)

(c)   (d)

Figure 7.45   Calculation of input-referred noise of a differential pair.

The input-referred noise voltage can also be calculated by exploiting the lemma illustrated in Fig. 7.33. As shown in Fig. 7.46, the noise of M_1 and M_2 is modeled as a voltage source in series with their gates, and the noise of R_{D1} and R_{D2} is divided by $g_m^2 R_D^2$, thereby resulting in (7.88).

It is instructive to compare the noise performance of a differential pair and a common-source stage, as expressed by (7.56) and (7.88). We conclude that, if each transistor has a transconductance g_m, then the input-referred noise *voltage* of a differential pair is $\sqrt{2}$ times that of a common-source stage. This is simply because the former includes twice as many devices in the signal path, as exemplified by the two series voltage sources in Fig. 7.46. (Since the noise sources are uncorrelated, their powers add.) It is also important to recognize that, with the assumption of equal device transconductances, a differential pair consumes twice as much power as a common-source stage if the transistors have the same dimensions.

The noise modeling of Fig. 7.46 can readily account for $1/f$ noise of the transistors as well. Placing the voltage sources given by $K/(C_{ox}WL)$ in series with each gate, we can

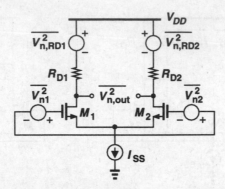

Figure 7.46 Alternative method of calculating the input-referred noise.

rewrite (7.88) as

$$\overline{V_{n,in,tot}^2} = 8kT \left(\frac{2}{3g_m} + \frac{1}{g_m^2 R_D} \right) + \frac{2K}{C_{ox} WL} \frac{1}{f}. \qquad (7.89)$$

Does the tail current source in Fig. 7.44 contribute noise? If the differential input signal is zero and the circuit is symmetric, then the noise in I_{SS} divides equally between M_1 and M_2, producing only a common-mode noise voltage at the output. On the other hand, for a small differential input, ΔV_{in}, we have

$$\Delta I_{D1} - \Delta I_{D2} = g_m \Delta V_{in} \qquad (7.90)$$

$$= \sqrt{2\mu_n C_{ox} \frac{W}{L} (\frac{I_{SS} + I_n}{2})} \Delta V_{in}, \qquad (7.91)$$

where I_n denotes the noise in I_{SS} and $I_n \ll I_{SS}$. In essence, the noise modulates the transconductance of each device. Equation (7.91) can be written as

$$\Delta I_{D1} - \Delta I_{D2} \approx \sqrt{2\mu_n C_{ox} \frac{W}{L} \cdot \frac{I_{SS}}{2}} \left(1 + \frac{I_n}{2I_{SS}} \right) \Delta V_{in} \qquad (7.92)$$

$$= g_{m0} \left(1 + \frac{I_n}{2I_{SS}} \right) \Delta V_{in}, \qquad (7.93)$$

where g_{m0} is the transconductance of the noiseless circuit. Equation (7.93) suggests that as the circuit departs from equilibrium, I_n is more unevenly divided between M_1 and M_2, thereby generating differential noise at the output. This effect is nonetheless usually negligible.

Example 7.13 ─────────────────────────────────

Assuming the devices in Fig. 7.47(a) operate in saturation and the circuit is symmetric, calculate the input-referred noise voltage.

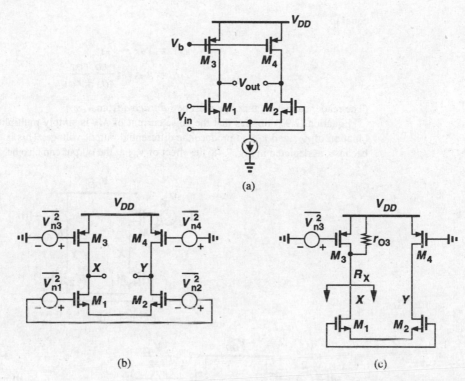

(a)

(b)   (c)

Figure 7.47

Solution

Since the thermal and $1/f$ noise of M_1 and M_2 can be modeled as voltage sources in series with the input, we only need to refer the noise of M_3 and M_4 to the input. Let us calculate the output noise contributed by M_3. The drain noise current of M_3 is divided between r_{O3} and the resistance seen looking into the drain of M_1 [Fig. 7.47(c)]. From Chapter 5, this resistance equals $R_X = r_{O4} + 2r_{O1}$. Denoting the resulting noise currents flowing through r_{O3} and R_X by I_{nA} and I_{nB}, respectively, we have

$$I_{nA} = g_{m3} V_{n3} \frac{r_{O4} + 2r_{O1}}{2r_{O4} + 2r_{O1}} \qquad (7.94)$$

and

$$I_{nB} = g_{m3} V_{n3} \frac{r_{O3}}{2r_{O4} + 2r_{O1}}. \qquad (7.95)$$

The former produces a noise voltage $g_{m3} V_{n3} r_{O3} (r_{O4} + 2r_{O1})/(2r_{O4} + 2r_{O1})$ at node X with respect to ground whereas the latter flows through M_1, M_2, and r_{O4}, generating $g_{m3} V_{n3} r_{O3} r_{O4}/(2r_{O4} + 2r_{O1})$ at node Y with respect to ground. Thus, the total differential output noise due to M_3 is equal to

$$V_{nXY} = V_{nX} - V_{nY} \tag{7.96}$$

$$= g_{m3} V_{n3} \frac{r_{O3} r_{O1}}{r_{O3} + r_{O1}}. \tag{7.97}$$

(The reader can verify that V_{nY} must be *subtracted* from V_{nX}.)

Equation (7.97) implies that the noise current of M_3 is simply multiplied by the parallel combination of r_{O1} and r_{O3} to produce the differential output voltage. This is of course not surprising because, as depicted in Fig. 7.48, the effect of V_{n3} at the output can also be derived by decomposing

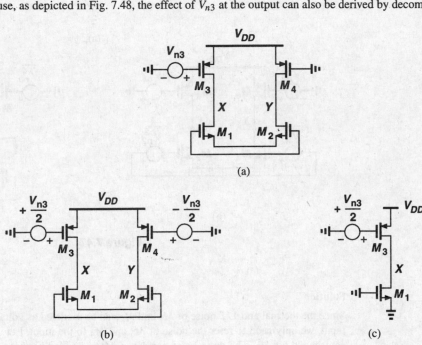

(a)

(b)   (c)

Figure 7.48 Calculation of input-referred noise in a differential pair with current-source loads.

V_{n3} into two differential components applied to the gates of M_3 and M_4 and subsequently using the half-circuit concept. Since this calculation relates to a *single* noise source, we can temporarily ignore the random nature of noise and treat V_{n3} and the circuit as familiar deterministic, linear components.

Applying (7.97) to M_4 as well and adding the resulting powers, we have

$$\overline{V_{n,out}^2}|_{M3,M4} = g_{m3}^2 (r_{O1} \| r_{O3})^2 \overline{V_{n3}^2} + g_{m4}^2 (r_{O2} \| r_{O4})^2 \overline{V_{n4}^2} \tag{7.98}$$

$$= 2 g_{m3}^2 (r_{O1} \| r_{O3})^2 \overline{V_{n3}^2}. \tag{7.99}$$

To refer the noise to the input, we divide (7.99) by $g_{m1}^2 (r_{O1} \| r_{O3})^2$, obtaining the *total* input-referred noise voltage per unit bandwidth as

$$\overline{V_{n,in}^2} = 2\overline{V_{n1}^2} + 2 \frac{g_{m3}^2}{g_{m1}^2} \overline{V_{n3}^2}, \tag{7.100}$$

which, upon substitution for $\overline{V_{n1}^2}$ and $\overline{V_{n3}^2}$, reduces to:

$$\overline{V_{n,in}^2} = 8kT\left(\frac{2}{3g_{m1}} + \frac{2g_{m3}}{3g_{m1}^2}\right) + \frac{2K_N}{C_{ox}(WL)_1 f} + \frac{2K_P}{C_{ox}(WL)_3 f}\frac{g_{m3}^2}{g_{m1}^2}. \tag{7.101}$$

The effect of noise must be studied for many other analog circuits as well. For example, feedback systems, op amps, and bandgap references exhibit interesting and important noise characteristics. We return to these topics in other chapters.

7.6 Noise Bandwidth

The total noise corrupting a signal in a circuit results from all of the frequency components that fall in the bandwidth of the circuit. Consider a multipole circuit having the output noise spectrum shown in Fig. 7.49(a). Since the noise components above ω_{p1} are not negligible,

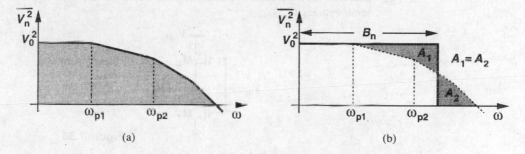

Figure 7.49   (a) Output noise spectrum of a circuit, (b) concept of noise bandwidth.

the total output noise must be evaluated by calculating the total area under the spectral density:

$$\overline{V_{n,out,tot}^2} = \int_0^\infty \overline{V_{n,out}^2}\,df. \tag{7.102}$$

However, as depicted in Fig. 7.49(b), it is sometimes helpful to represent the total noise simply as $V_0^2 \cdot B_n$, where the bandwidth B_n is chosen such that

$$V_0^2 \cdot B_n = \int_0^\infty \overline{V_{n,out}^2}\,df. \tag{7.103}$$

Called the "noise bandwidth," B_n allows a fair comparison of circuits that exhibit the same low-frequency noise, V_0^2, but different high-frequency transfer functions. As an exercise, the reader can prove that the noise bandwidth of a one-pole system is equal to $\pi/2$ times the pole frequency.

Problems

Unless otherwise stated, in the following problems, use the device data shown in Table 2.1 and assume $V_{DD} = 3$ V where necessary. Also, assume all transistors are in saturation.

7.1. A common-source stage incorporates a 50-μm/0.5-μm NMOS device biased at $I_D = 1$ mA along with a load resistor of 2 kΩ. What is the total input-referred thermal noise voltage in a 100-MHz bandwidth?

7.2. Consider the common-source stage of Fig. 7.35. Assume $(W/L)_1 = 50/0.5$, $I_{D1} = I_{D2} = 0.1$ mA, and $V_{DD} = 3$ V. If the contribution of M_2 to the input-referred noise voltage (not voltage squared) must be one-fifth of that of M_1, what is the maximum output voltage swing of the amplifier?

7.3. Using the distributed model of Fig. 7.19(c) and ignoring the channel thermal noise, prove that, for gate noise calculations, a distributed gate resistance of R_G can be replaced by a lumped resistance equal to $R_G/3$. (Hint: model the noise of R_{Gj} by a series voltage source and calculate the total drain noise current. Watch for correlated sources of noise.)

7.4. Prove that the output noise current of Fig. 7.33(c) is given by Eq. (7.54).

7.5. Calculate the input-referred noise voltage of the circuit shown in Fig. 7.50 and compare the result with Eq. (7.59).

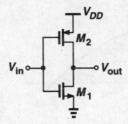

Figure 7.50

7.6. Calculate the input-referred thermal noise voltage of each circuit in Fig. 7.51. Assume $\lambda = \gamma = 0$.

7.7. Calculate the input-referred thermal noise voltage of each circuit in Fig. 7.52. Assume $\lambda = \gamma = 0$.

7.8. Calculate the input-referred thermal noise voltage and current of each circuit in Fig. 7.53. Assume $\lambda = \gamma = 0$.

7.9. Calculate the input-referred thermal noise voltage and current of each circuit in Fig. 7.54. Assume $\lambda = \gamma = 0$.

7.10. Calculate the input-referred $1/f$ noise voltage and current of Fig. 7.40 if the two capacitors are removed.

7.11. Calculate the input-referred $1/f$ noise voltage of the source follower shown in Fig. 7.42.

7.12. Assuming $\lambda = \gamma = 0$, calculate the input-referred thermal noise voltage of each circuit in Fig. 7.55. For part (a), assume $g_{m3,4} = 0.5g_{m5,6}$.

7.13. Consider the degenerated common-source stage shown in Fig. 7.56.
   (a) Calculate the input-referred thermal noise voltage if $\lambda = \gamma = 0$.
   (b) Suppose linearity requirements necessitate that the dc voltage drop across R_S be equal to the overdrive voltage of M_1. How does the thermal noise contributed by R_S compare with that contributed by M_1?

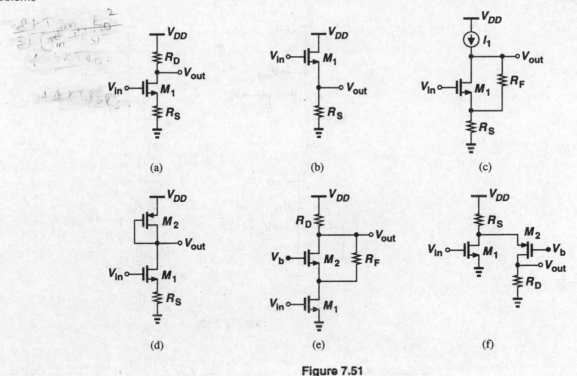

Figure 7.51

(a)   (b)   (c)   (d)   (e)   (f)

(a)   (b)

(c)   (d)

Figure 7.52

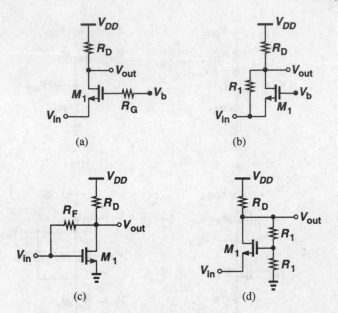

Figure 7.53

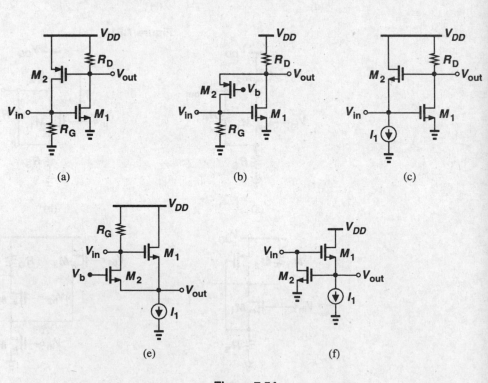

Figure 7.54

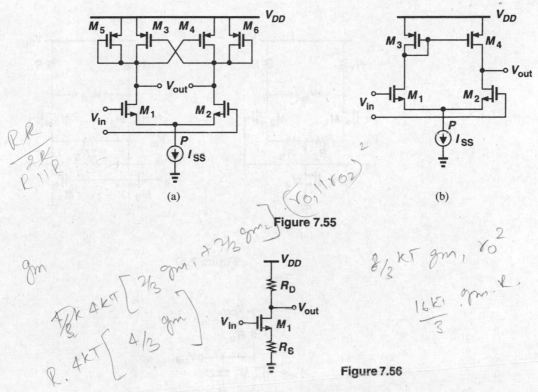

Figure 7.55

Figure 7.56

7.14. Explain why Miller's theorem cannot be applied to calculate the effect of the thermal noise of a floating resistor.

7.15. The circuit of Fig. 7.18 is designed with $(W/L)_1 = 50/0.5$ and $I_{D1} = 0.05$ mA. Calculate the total rms thermal noise voltage at the output in a 50-MHz bandwidth.

7.16. For the circuit shown in Fig. 7.58, calculate the total output thermal and $1/f$ noise in a bandwidth $[f_L, f_H]$. Assume $\lambda \neq 0$ but neglect other capacitances.

7.17. Suppose in the circuit of Fig. 7.35, $(W/L)_{1,2} = 50/0.5$ and $I_{D1} = |I_{D2}| = 0.5$ mA. What is the input-referred thermal noise voltage?

7.18. The circuit of Fig. 7.35 is modified as depicted in Fig. 7.59.
   (a) Calculate the input-referred thermal noise voltage.
   (b) For a given bias current and output voltage swing, what value of R_S minimizes the input-referred thermal noise?

7.19. A common-gate stage incorporates an NMOS device with $W/L = 50/0.5$ biased at $I_D = 1$ mA and a load resistor of 1 kΩ. Calculate the input-referred thermal noise voltage and current.

7.20. The circuit of Fig. 7.39 is designed with $(W/L)_1 = 50/0.5$ and $I_{D1} = I_{D2} = 0.05$ mA and $R_D = 1$ kΩ.
   (a) Determine $(W/L)_2$ such that the contribution of M_2 to the input-referred thermal noise current (not current squared) is one-fifth of that due to R_D.
   (b) Now calculate the minimum value of V_b to place M_2 at the edge of the triode region. What is the maximum allowable output voltage swing?

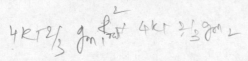

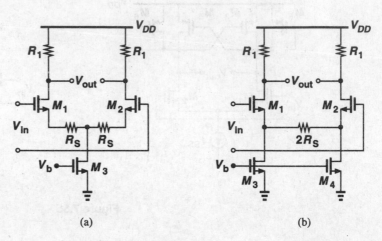

(a)   (b)

Figure 7.57

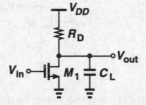

Figure 7.58

7.21. Design the circuit of Fig. 7.39 for an input-referred thermal noise voltage of 3 nV/$\sqrt{\text{Hz}}$ and maximum output swing. Assume $I_{D1} = I_{D2} = 0.5$ mA.

7.22. Consider the circuit of Fig. 7.40. If $(W/L)_{1-3} = 50/0.5$ and $I_{D1-3} = 0.5$ mA, determine the input-referred thermal noise voltage and current.

7.23. The circuit of Fig. 7.40 is designed with $(W/L)_1 = 50/0.5$ and $I_{D1-3} = 0.5$ mA. If an output swing of 2 V is required, estimate by iteration the dimensions of M_2 and M_3 such that the input-referred thermal noise current is minimum.

7.24. The source follower of Fig. 7.42 is to provide an output resistance of 100 Ω with a bias current of 0.1 mA.
(a) Calculate $(W/L)_1$.

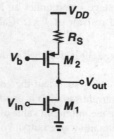

Figure 7.59

(b) Determine $(W/L)_2$ such that the input-referred thermal noise voltage (not voltage squared) contributed by M_2 is one-fifth of that due to M_1. What is the maximum output swing?

7.25. The cascode stage of Fig. 7.43(a) exhibits a capacitance C_X from node X to ground. Neglecting other capacitances, determine the input-referred thermal noise voltage.

7.26. Determine the input-referred thermal and $1/f$ noise voltages of the circuits shown in Fig. 7.57 and compare the results. Assume the circuits draw equal supply currents.

References

1. L. W. Couch, *Digital and Analog Communication Systems,* Fourth Ed., New York: Macmillan Co., 1993.
2. S. M. Sze, *Physics of Semiconductor Devices,* Second Ed., New York: Wiley, 1981.
3. Y. Tsividis, *Operation and Modeling of the MOS Transistor,* Second Ed., Boston: McGraw-Hill, 1999.
4. A. A. Abidi, "High-Frequency Noise Measurements on FETs with Small Dimensions," *IEEE Trans. Electron Devices,* vol. 33, pp. 1801–1805, Nov. 1986.
5. H. A. Haus et al., "Representation of Noise in Linear Twoports," *Proc. IRE,* vol. 48, pp. 69–74, Jan. 1960.

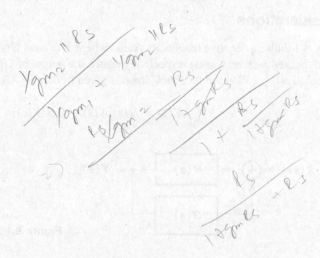

Chapter 8

Feedback

On a mild August morning in 1921, Harold Black was riding the ferry from New York to New Jersey, where he worked at Bell Laboratories. Black and many other researchers had been investigating the problem of nonlinearity in amplifiers used in long-distance telephone networks, seeking a practical solution. While reading the newspaper on the ferry, Black was suddenly struck by an idea and began to draw a diagram on the newspaper, which would later be used as the evidence in his patent application. The idea is known to us as the negative feedback amplifier.

Feedback is a powerful technique that finds wide application in analog circuits. For example, negative feedback allows high-precision signal processing and positive feedback makes it possible to build oscillators. In this chapter, we consider only negative feedback and use the term feedback to mean that.

We begin with a general view of feedback circuits, describing important benefits that result from feedback. Next, we study four feedback topologies and their properties. Finally, we examine the effects of loading in feedback amplifiers.

8.1 General Considerations

Fig. 8.1 shows a negative feedback system, where $H(s)$ and $G(s)$ are called the feedforward and the feedback networks, respectively. Since the output of $G(s)$ is equal to $G(s)Y(s)$, the input to $H(s)$, called the feedback error, is given by $X(s) - G(s)Y(s)$. That is,

$$Y(s) = H(s)[X(s) - G(s)Y(s)]. \tag{8.1}$$

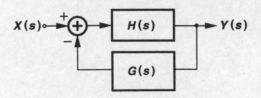

$X(s)$ H(s)   Y(s)

G(s)

Figure 8.1 General feedback system.

Thus,

$$\frac{Y(s)}{X(s)} = \frac{H(s)}{1 + G(s)H(s)}. \tag{8.2}$$

We call $H(s)$ the "open-loop" transfer function and $Y(s)/X(s)$ the "closed-loop" transfer function. In most cases of interest in this book, $H(s)$ represents an amplifier and $G(s)$ is a frequency-independent quantity. In other words, a fraction of the output signal is sensed and compared with the input, generating an error term. In a well-designed negative feedback system, the error term is minimized, thereby making the output of $G(s)$ an accurate "copy" of the input and hence the output of the system a faithful replica of the input (Fig. 8.2). We also say the input of $H(s)$ is a "virtual ground" because the signal amplitude at this point is very small. In subsequent developments, we replace $G(s)$ by a frequency-independent quantity β and call it the "feedback factor."

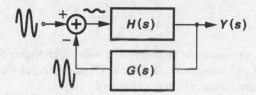

Figure 8.2 Similarity between output of feedback network and input signal.

It is instructive to identify four elements in the feedback system of Fig. 8.1: (1) the feedforward amplifier, (2) a means of sensing the output, (3) the feedback network, (4) a means of generating the feedback error. These elements exist in every feedback system, even though they may not be obvious in cases such as a simple common-source stage with resistive degeneration.

8.1.1 Properties of Feedback Circuits

Before proceeding to the analysis of feedback circuits, we study some simple examples to describe the benefits of negative feedback as well as identify the above four elements in each case.

Gain Desensitization   Consider the common-source stage shown in Fig. 8.3(a), where the voltage gain is equal to $g_{m1}r_{O1}$. A critical drawback of this circuit is the poor definition of the gain: both g_{m1} and r_{O1} vary with process and temperature. Now suppose the circuit is configured as in Fig. 8.3(b), where the gate bias of M_1 is set by means not shown here (Chapter 12). Let us calculate the overall voltage gain of the circuit at relatively low frequencies such that C_2 does not load the output node, i.e., $V_{out}/V_X = -g_{m1}r_{O1}$. Since $(V_{out} - V_X)C_2s = (V_X - V_{in})C_1s$, we have

$$\frac{V_{out}}{V_{in}} = -\frac{1}{\left(1 + \dfrac{1}{g_{m1}r_{O1}}\right)\dfrac{C_2}{C_1} + \dfrac{1}{g_{m1}r_{O1}}}. \tag{8.3}$$

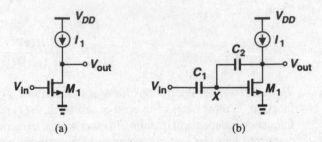

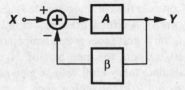

Figure 8.3 (a) Simple common-source stage, (b) circuit of (a) with feedback.

If $g_{m1}r_{O1}$ is sufficiently large, the $1/(g_{m1}r_{O1})$ terms in the denominator can be neglected, yielding

$$\frac{V_{out}}{V_{in}} = -\frac{C_1}{C_2}. \tag{8.4}$$

Compared to $g_{m1}r_{O1}$, this gain can be controlled with much higher accuracy because it is given by the *ratio* of two capacitors. If C_1 and C_2 are made of the same material, then process and temperature variations do not change C_1/C_2.

The above example reveals that negative feedback provides gain "desensitization," i.e., the closed-loop gain is much less sensitive to device parameters than the open-loop gain is. Illustrated for a more general case in Fig. 8.4, this property can be quantified by writing

$$\frac{Y}{X} = \frac{A}{1 + \beta A} \tag{8.5}$$

$$\approx \frac{1}{\beta}\left(1 - \frac{1}{\beta A}\right), \tag{8.6}$$

where we have assumed $\beta A \gg 1$. We note that the closed-loop gain is determined, to the first order, by the feedback factor, β. More importantly, even if the open-loop gain, A, varies by a factor of, say, 2, Y/X varies by a small percentage because $1/(\beta A) \ll 1$.

Figure 8.4 Simple feedback system.

Called the "loop gain," the quantity βA plays an important role in feedback systems.[1] We see from (8.6) that the higher βA is, the less sensitive Y/X will be to variations in A.

---

[1]Loop gain and open-loop gain must not be confused with each other.

From another perspective, the accuracy of the closed-loop gain improves by maximizing A or β. Note that as β increases, the closed-loop gain, $Y/X \approx 1/\beta$, decreases, suggesting a trade-off between precision and the closed-loop gain. In other words, we begin with a high-gain amplifier and apply feedback to obtain a low, but less sensitive closed-loop gain. Another conclusion here is that the output of the feedback network is equal to $X \cdot A/(1 + \beta A)$, approaching A as βA becomes much greater than unity. This result agrees with the illustration in Fig. 8.2.

The calculation of the loop gain usually proceeds as follows. As illustrated in Fig. 8.5, we set the main input to zero, break the loop at some point, inject a test signal in the "right direction," follow the signal around the loop, and obtain the value that returns to the break point. The negative of the transfer function thus derived is the loop gain. Note that the loop gain is a dimensionless quantity. In Fig. 8.5, we have $V_t \beta(-1)A = V_F$ and hence $V_F/V_t = -\beta A$. Similarly, as depicted in Fig. 8.6, for the simple feedback circuit, we can write

$$V_t \frac{C_2}{C_1 + C_2}(-g_{m1}r_{O1}) = V_F, \tag{8.7}$$

that is,

$$\frac{V_F}{V_t} = -\frac{C_2}{C_1 + C_2}g_{m1}r_{O1}. \tag{8.8}$$

Note that the loading of C_2 on the output is neglected here. This issue will be addressed in Section 8.3.

It is also interesting to identify the four elements of feedback in the circuit of Fig. 8.3(b). Transistor M_1 and current source I_1 constitute the feedforward amplifier. Capacitor C_2

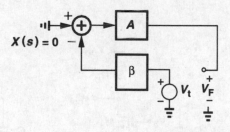

Figure 8.5 Computation of loop gain.

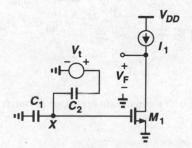

Figure 8.6 Computation of loop gain in a simple feedback circuit.

senses the output voltage and converts it to a current feedback signal, which is then added to the current produced by V_{in} through C_1. Note that the feedback is negative even though the currents through C_1 and C_2 are *added* because the feedforward amplifier itself provides a negative gain.

We should emphasize that the desensitization of gain by feedback leads to many other properties of feedback systems. Our examination of Eq. 8.6 indicates that large variations in A affect Y/X negligibly if βA is large. Such variations can arise from different sources: process, temperature, frequency, and loading. For example, if A drops at high frequencies, Y/X varies to a lesser extent, and the flat bandwidth is increased. Similarly, if A decreases because the amplifier drives a heavy load, Y/X is not affected much. These concepts become clearer below.

Terminal Impedance Modification   As a second example, let us study the circuit shown in Fig. 8.7(a), where a capacitive voltage divider senses the output voltage of a common-gate stage, applying the result to the gate of current source M_2 and hence returning a current feedback signal to the input.[2] Our objective is to compute the input resistance at relatively low frequencies with and without feedback. Neglecting channel-length modulation and breaking the feedback loop [Fig. 8.7(b)], we have

$$R_{in,open} = \frac{1}{g_{m1} + g_{mb1}}. \tag{8.9}$$

For the closed-loop circuit, as depicted in Fig. 8.7(c), we write: $V_{out} = (g_{m1} + g_{mb1})V_X R_D$ and

$$V_P = V_{out} \frac{C_1}{C_1 + C_2} \tag{8.10}$$

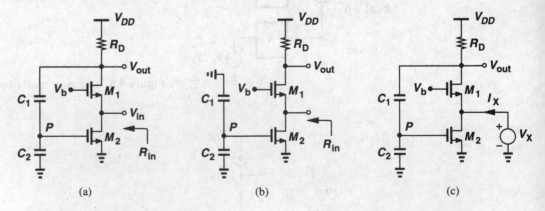

(a)   (b)   (c)

Figure 8.7   (a) Common-gate circuit with feedback, (b) open-loop circuit, (c) calculation of input resistance.

---

[2]The bias network for M_2 is not shown.

$$= (g_{m1} + g_{mb1})V_X R_D \frac{C_1}{C_1 + C_2}. \tag{8.11}$$

Thus, the small-signal drain current of M_2 equals $g_{m2}(g_{m1} + g_{mb1})V_X R_D C_1/(C_1 + C_2)$. Adding this current to the drain current of M_1 with proper polarity yields I_X:

$$I_X = (g_{m1} + g_{mb1})V_X + g_{m2}(g_{m1} + g_{mb1})\frac{C_1}{C_1 + C_2} R_D V_X \tag{8.12}$$

$$= (g_{m1} + g_{mb1})\left(1 + g_{m2}R_D \frac{C_1}{C_1 + C_2}\right)V_X. \tag{8.13}$$

It follows that

$$R_{in,closed} = V_X/I_X \tag{8.14}$$

$$= \frac{1}{g_{m1} + g_{mb1}} \frac{1}{1 + g_{m2}R_D \dfrac{C_1}{C_1 + C_2}}. \tag{8.15}$$

We therefore conclude that this type of feedback reduces the input resistance by a factor of $1 + g_{m2}R_D C_1/(C_1 + C_2)$. The reader can prove that the quantity $g_{m2}R_D C_1/(C_1 + C_2)$ is the loop gain.

We also identify the four elements of feedback in the circuit of Fig. 8.7(a). The feedforward amplifier consists of M_1 and R_D, the output is sensed by C_1 and C_2, the feedback network comprises C_1, C_2, and M_2, and the subtraction occurs in the current domain at the input terminal.

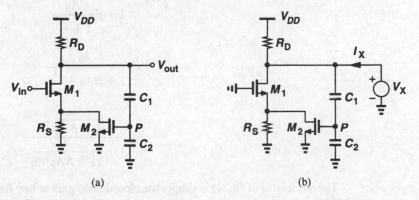

(a)   (b)

Figure 8.8   (a) CS stage with feedback, (b) calculation of output resistance.

Let us now consider the circuit of Fig. 8.8(a) as an example of output impedance modification by feedback. Here M_1, R_S, and R_D constitute a common-source stage and C_1, C_2, and M_2 sense the output voltage,[3] returning a current equal to $[C_1/(C_1 + C_2)]V_{out}g_{m2}$ to the source of M_1. The reader can prove that the feedback is indeed negative. To compute the output resistance at relatively low frequencies, we set the input to zero [Fig. 8.8(b)] and write

$$I_{D1} = V_X \frac{C_1}{C_1 + C_2} g_{m2} \frac{R_S}{R_S + \dfrac{1}{g_{m1} + g_{mb1}}}. \tag{8.16}$$

Since $I_X = V_X/R_D + I_{D1}$, we have

$$\frac{V_X}{I_X} = \frac{R_D}{1 + \dfrac{g_{m2}R_S(g_{m1} + g_{mb1})R_D}{(g_{m1} + g_{mb1})R_S + 1} \dfrac{C_1}{C_1 + C_2}}. \tag{8.17}$$

Equation (8.17) implies that this type of feedback decreases the output resistance. The denominator of (8.17) is indeed equal to one plus the loop gain.

Bandwidth Modification.   The next example illustrates the effect of negative feedback on the bandwidth. Suppose the feedforward amplifier has a one-pole transfer function:

$$A(s) = \frac{A_0}{1 + \dfrac{s}{\omega_0}}, \tag{8.18}$$

where A_0 denotes the low-frequency gain and ω_0 is the 3-dB bandwidth. What is the transfer function of the closed-loop system? From (8.5), we have

$$\frac{Y}{X}(s) = \frac{\dfrac{A_0}{1 + \dfrac{s}{\omega_0}}}{1 + \beta \dfrac{A_0}{1 + \dfrac{s}{\omega_0}}} \tag{8.19}$$

$$= \frac{A_0}{1 + \beta A_0 + \dfrac{s}{\omega_0}} \tag{8.20}$$

$$= \frac{\dfrac{A_0}{1 + \beta A_0}}{1 + \dfrac{s}{(1 + \beta A_0)\omega_0}}. \tag{8.21}$$

The numerator of (8.21) is simply the closed-loop gain at low frequencies—as predicted by (8.5)—and the denominator reveals a pole at $(1+\beta A_0)\omega_0$. Thus, the 3-dB bandwidth has increased by a factor $1+\beta A_0$, albeit at the cost of a proportional reduction in the gain (Fig. 8.9).

---

[3]Biasing of M_2 is not shown.

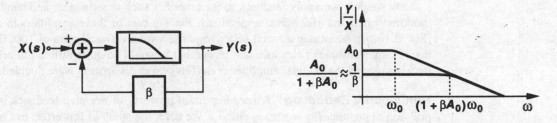

Figure 8.9 Bandwidth modification as a result of feedback.

The increase in the bandwidth fundamentally originates from the gain desensitization property of feedback. Recall from (8.6) that, if A is large enough, the closed-loop gain remains approximately equal to $1/\beta$ even if A experiences substantial variations. In the example of Fig. 8.9, A varies with frequency rather than process or temperature, but negative feedback still suppresses the effect of this variation. Of course, at high frequencies A drops to such low levels that βA becomes comparable with unity and the closed-loop gain falls below $1/\beta$.

Equation (8.21) suggests that the gain-bandwidth product of a one-pole system does not change with feedback, making the reader wonder how feedback improves the speed if a high gain is required. Suppose we need to amplify a 20-MHz square wave by a factor of 100 and maximum bandwidth but we have only a single-pole amplifier with an open-loop gain of 100 and 3-dB bandwidth of 10 MHz. If the input is applied to the open-loop amplifier, the response appears as shown in Fig. 8.10(a), exhibiting long risetime and falltime because the time constant is equal to $1/(2\pi f_{3-\mathrm{dB}}) \approx 16$ ns.

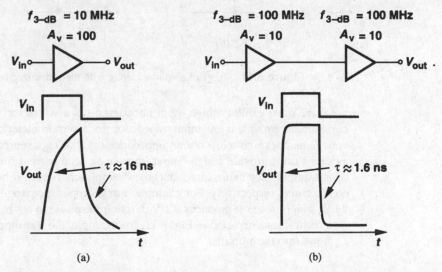

Figure 8.10 Amplification of a 20-MHz squarewave by (a) 20-MHz amplifier and (b) cascade of two 100-MHz feedback amplifiers.

Now suppose we apply feedback to the amplifier such that the gain and bandwidth are modified to 10 and 100 MHz, respectively. Placing two of these amplifiers in a cascade [Fig. 8.10(b)], we obtain a much faster response with an overall gain of 100. Of course, the cascade consumes twice as much power, but it would be quite difficult to achieve this performance by the original amplifier even if its power dissipation were doubled.

Nonlinearity Reduction   A very important property of negative feedback is the suppression of nonlinearity in analog circuits. We defer the study of this effect to Chapter 13.

8.1.2 Types of Amplifiers

Most of the circuits studied thus far can be considered "voltage amplifiers" because they sense a voltage at the input and produce a voltage at the output. However, three other types of amplifiers can also be constructed such that they sense or produce currents. Shown in Fig. 8.11, the four configurations have quite different properties: (1) circuits sensing

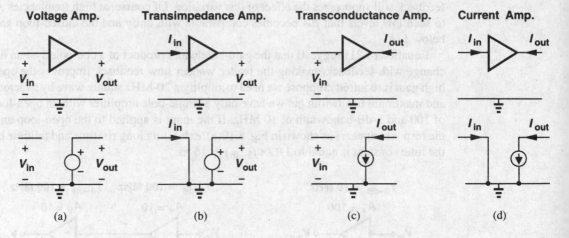

Figure 8.11   Types of amplifiers along with their idealized models.

a voltage must exhibit a high input impedance (as a voltmeter) whereas those sensing a current must provide a low input impedance (as a current meter); (2) circuits generating a voltage must exhibit a low output impedance (as a voltage source) while those generating a current must provide a high output impedance (as a current source). Note that the gains of transimpedance and transconductance[4] amplifiers have a dimension of resistance and conductance, respectively. For example, a transimpedance amplifier may have a gain of 2 kΩ, which means it produces a 2-V output in response to a 1-mA input. Also, we use the sign conventions depicted in Fig. 8.11, for example, the transimpedance $R_0 = V_{out}/I_{in}$ if I_{in} flows *into* the amplifier.

---

[4]This terminology is standard but not consistent. One should use either transimpedance and transadmittance or transresistance and transconductance.

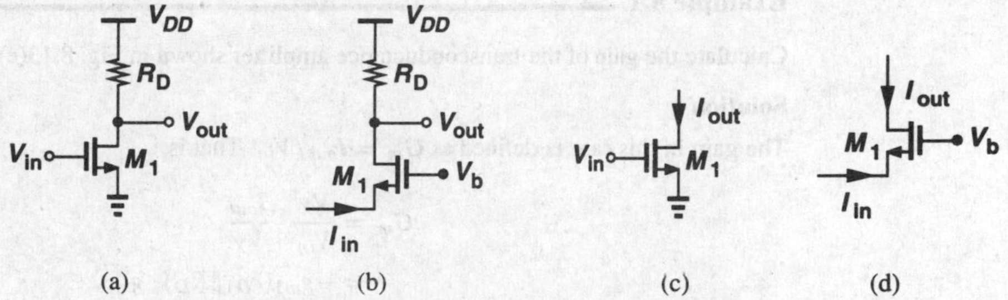

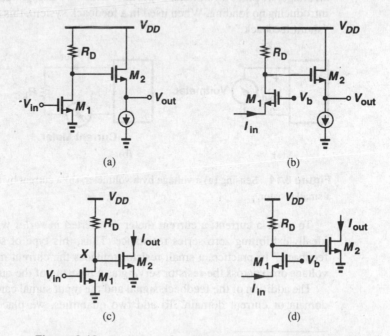

Figure 8.12 Simple implementations of four types of amplifiers.

Figure 8.12 illustrates simple implementations of each amplifier. In Fig. 8.12(a), a common-source stage senses and produces voltages and in Fig. 8.12(b), a common-gate circuit serves as a transimpedance amplifier, converting the source current to a voltage at the drain. In Fig. 8.12(c), a common-source transistor operates as a transconductance amplifier, generating an output current in response to an input voltage, and in Fig. 8.12(d), a common-gate device senses and produces currents.

The circuits of Fig. 8.12 may not provide adequate performance in many applications. For example, the circuits of Figs. 8.12(a) and (b) suffer from a relatively high output impedance. Fig. 8.13 depicts modifications that alter the output impedance or increase the gain.

Figure 8.13 Four types of amplifiers with improved performance.

Example 8.1

Calculate the gain of the transconductance amplifier shown in Fig. 8.13(c).

Solution

The gain in this case is defined as $G_m = I_{out}/V_{in}$. That is,

$$G_m = \frac{V_X}{V_{in}} \cdot \frac{I_{out}}{V_X} \tag{8.22}$$

$$= -g_{m1}(r_{O1}\|R_D) \cdot g_{m2}. \tag{8.23}$$

While most familiar amplifiers are of voltage-voltage type, the other three configurations do find usage. For example, transimpedance amplifiers are an integral part of optical fiber receivers because they must sense the current produced by a photodiode, eventually generating a voltage that can be processed by subsequent circuits.

8.1.3 Sense and Return Mechanisms

Placing a circuit in a feedback loop requires sensing the output signal and returning (a fraction) of the result to the summing node at the input. With voltage or current quantities as input and output signals, we can identify four types of feedback: voltage-voltage, voltage-current, current-current, and current-voltage, where the first entry in each case denotes the quantity sensed at the *output* and the second the type of signal returned to the input.[5]

It is instructive to review methods of sensing and summing voltages or currents. To sense a voltage, we place a voltmeter *in parallel* with the corresponding port [Fig. 8.14(a)], ideally introducing no loading. When used in a feedback system, this type of sensing is also called "shunt feedback."

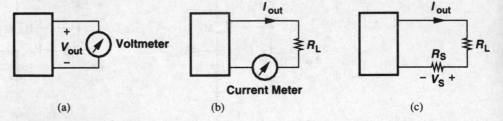

(a)   (b)   (c)

Figure 8.14 Sensing (a) a voltage by a voltmeter, (b) a current by a current meter, (c) a current by a small resistor.

To sense a current, a current meter is inserted *in series* with the signal [Fig. 8.14(b)], ideally exhibiting zero series resistance. Thus, this type of sensing is also called "series feedback." In practice, a small resistor replaces the current meter [Fig. 8.14(c)], with the voltage drop across the resistor serving as a measure of the output current.

The addition of the feedback signal and the input signal can be performed in the voltage domain or current domain. To add two quantities, we place them in series if they are

---

[5]Different authors use different orders or terminologies for the four types of feedback.

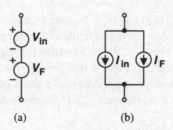

Figure 8.15 Addition of (a) voltages and (b) currents.

voltages and in parallel if they are currents (Fig. 8.15). While ideally having no influence on the operation of the open-loop amplifier itself, the feedback network in reality introduces loading effects that must be taken into account. This issue is discussed in Section 8.3.

To visualize the methods of Figs. 8.14 and 8.15, we consider a number of practical implementations. A voltage can be sensed by a resistive (or capacitive) divider in parallel with the port [Fig. 8.16(a)] and a current by placing a resistor in series with the wire and sensing

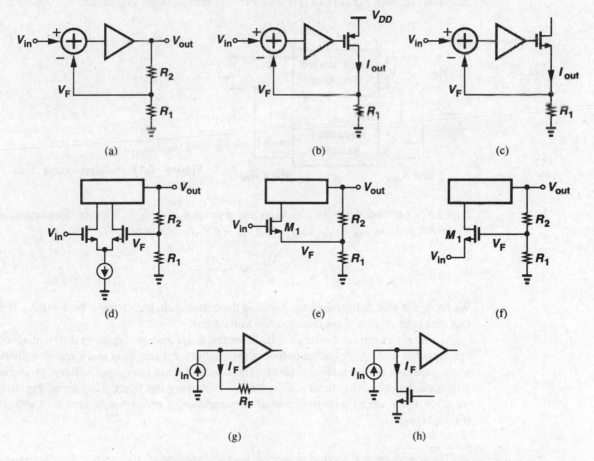

Figure 8.16 Practical means of sensing and adding voltages and currents.

the voltage across it [Figs. 8.16(b) and (c)]. To subtract two voltages, a differential pair can be used [Fig. 8.16(d)]. Alternatively, a single transistor can perform voltage subtraction as shown in Figs. 8.16(e) and (f) because I_{D1} is a function of $V_{in} - V_F$. Subtraction of currents can be accomplished as depicted in Figs. 8.16(g) or (h). Note that for voltage subtraction, the input and feedback signals are applied to *two* distinct nodes whereas for current subtraction they are applied to a single node. This observation proves helpful in identifying the type of feedback.

8.2 Feedback Topologies

8.2.1 Voltage-Voltage Feedback

This topology samples the output voltage and returns the feedback signal as a voltage.[6] Following the conceptual illustrations of Figs. 8.14 and 8.15, we note that the feedback network is connected in *parallel* with the output and in *series* with the input port (Fig. 8.17). An ideal feedback network in this case exhibits infinite input impedance and zero output

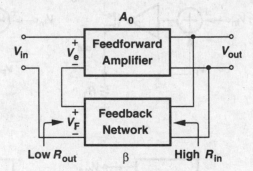

Figure 8.17 Voltage-voltage feedback.

impedance because it senses a voltage and generates a voltage. We can therefore write: $V_F = \beta V_{out}$, $V_e = V_{in} - V_F$, $V_{out} = A_0(V_{in} - \beta V_{out})$, and hence

$$\frac{V_{out}}{V_{in}} = \frac{A_0}{1 + \beta A_0}. \tag{8.24}$$

We recognize that βA_0 is the loop gain and the overall gain has dropped by $1 + \beta A_0$. Note that here both A_0 and β are dimensionless quantities.

As a simple example of voltage-voltage feedback, suppose we employ a differential voltage amplifier with single-ended output as the feedforward amplifier and a resistive divider as the feedback network [Fig. 8.18(a)]. The divider senses the output voltage, producing a fraction thereof as the feedback signal V_F. Following the block diagram of Fig. 8.17, we place V_F in series with the input of the amplifier to perform subtraction of voltages [Fig. 8.18(b)].

---

[6]This configuration is also called "series-shunt" feedback, where the first term refers to the *input* connection and the second to the *output* connection.

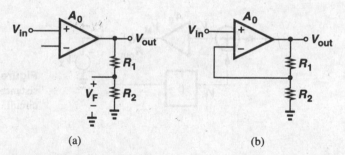

(a)   (b)

Figure 8.18 (a) Amplifier with output sensed by a resistive divider, (b) voltage-voltage feedback amplifier.

How does voltage-voltage feedback modify the input and output impedances? Let us first consider the output impedance. Recall that a negative feedback system attempts to make the output an accurate replica of the input. Now suppose, as shown in Fig. 8.19, we load the output by a resistor, gradually decreasing its value. While in the open-loop configuration the

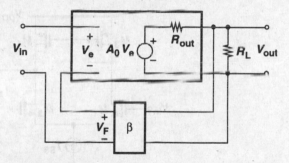

Figure 8.19 Effect of voltage-voltage feedback on output resistance.

output would simply drop in proportion to $R_L/(R_L + R_{out})$, in the feedback system, V_{out} is maintained as a reasonable replica of V_{in} even though R_L decreases. That is, so long as the loop gain remains much greater than unity, $V_{out}/V_{in} \approx 1/\beta$, regardless of the value of R_L. From another point of view, since the circuit stabilizes the output voltage amplitude despite load variations, it behaves as a *voltage* source, thus exhibiting a low output impedance. This property fundamentally originates from the gain desensitization provided by feedback.

In order to formally prove that voltage feedback lowers the output impedance, we consider the simple model in Fig. 8.20, where R_{out} represents the output impedance of the feedforward amplifier. Setting the input to zero and applying a voltage at the output, we write $V_F = \beta V_X$, $V_e = -\beta V_X$, $V_M = -\beta A_0 V_X$, and hence $I_X = [V_X - (-\beta A_0 V_X)]/R_{out}$ (if the current drawn by the feedback network is neglected). It follows that

$$\frac{V_X}{I_X} = \frac{R_{out}}{1 + \beta A_0}. \tag{8.25}$$

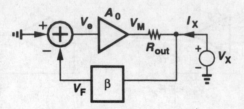

Figure 8.20 Calculation of output resistance of a voltage-voltage feedback circuit.

Thus, the output impedance and the gain are lowered by the same factor. In the circuit of Fig. 8.18(b), for example, the output impedance is lowered by $1 + A_0 R_2/(R_1 + R_2)$.

Example 8.2

The circuit shown in Fig. 8.21(a) is an implementation of the feedback configuration depicted in Fig. 8.18(b), but with the resistors replaced by capacitors. (The bias network of M_2 is not shown.) Calculate the closed-loop gain and output resistance of the amplifier at relatively low frequencies.

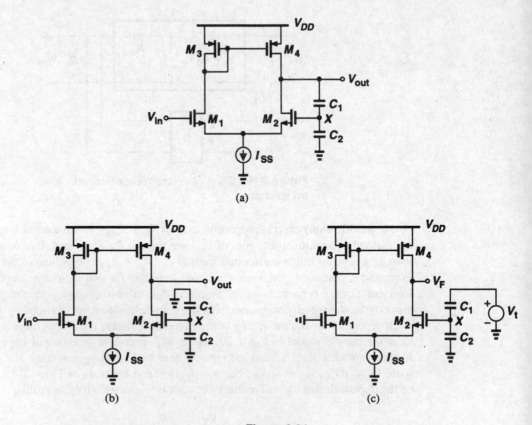

Figure 8.21

Solution

At low frequencies, C_1 and C_2 load the amplifier negligibly. To find the open-loop voltage gain, we break the feedback loop as shown in Fig. 8.21(b), grounding the top plate of C_1 to ensure zero voltage feedback. The open-loop gain is thus equal to $g_{m1}(r_{O2}\|r_{O4})$.

We must also compute the loop gain. With the aid of Fig. 8.21(c), we have

$$V_F = -V_t \frac{C_1}{C_1 + C_2} g_{m1}(r_{O2}\|r_{O4}). \tag{8.26}$$

That is,

$$\beta A_0 = \frac{C_1}{C_1 + C_2} g_{m1}(r_{O2}\|r_{O4}) \tag{8.27}$$

and hence

$$A_{closed} = \frac{g_{m1}(r_{O2}\|r_{O4})}{1 + \dfrac{C_1}{C_1 + C_2} g_{m1}(r_{O2}\|r_{O4})}. \tag{8.28}$$

As expected, if $\beta A_0 \gg 1$, then $A_{closed} \approx 1 + C_2/C_1$.

The open-loop output resistance of the circuit is equal to $r_{O2}\|r_{O4}$ (Chapter 5). It follows that

$$R_{out,closed} = \frac{r_{O2}\|r_{O4}}{1 + \dfrac{C_1}{C_1 + C_2} g_{m1}(r_{O2}\|r_{O4})}. \tag{8.29}$$

It is interesting to note that, if $\beta A_0 \gg 1$, then

$$R_{out,closed} \approx \left(1 + \frac{C_2}{C_1}\right) \frac{1}{g_{m1}}. \tag{8.30}$$

In other words, even if the open-loop amplifier suffers from a *high* output resistance, the closed-loop output resistance is independent of $r_{O2}\|r_{O4}$, simply because the open-loop *gain* scales with $r_{O2}\|r_{O4}$ as well.

Voltage-voltage feedback also modifies the input impedance. Comparing the configurations in Fig. 8.22, we note that the input impedance of the feedforward amplifier sustains the entire input voltage in Fig. 8.22(a), but only a fraction of V_{in} in Fig. 8.22(b). As a result, the current drawn by R_{in} in the feedback topology is *less* than that in the open-loop system, suggesting that returning a voltage quantity to the input *increases* the input impedance.

The foregoing observation can be confirmed analytically with the aid of Fig. 8.23. Since $V_e = I_X R_{in}$ and $V_F = \beta A_0 I_X R_{in}$, we have $V_e = V_X - V_F = V_X - \beta A_0 I_X R_{in}$. Thus, $I_X R_{in} = V_X - \beta A_0 I_X R_{in}$, and

$$\frac{V_X}{I_X} = R_{in}(1 + \beta A_0). \tag{8.31}$$

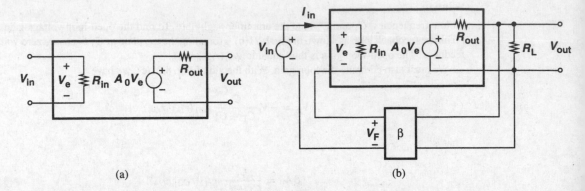

(a)   (b)

Figure 8.22 Effect of voltage-voltage feedback on input resistance.

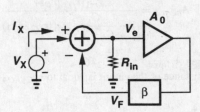

Figure 8.23 Calculation of input impedance of a voltage-voltage feedback circuit.

The input impedance therefore increases by the ubiquitous factor $1 + \beta A_0$, bringing the circuit closer to an ideal voltage amplifier.

Example 8.3

Fig. 8.24(a) shows a common-gate topology placed in a voltage-voltage feedback configuration. Note that the summation of the feedback voltage and the input voltage is accomplished by applying the former to the gate and the latter to the source.[7] Calculate the input resistance at low frequencies if channel-length modulation is negligible.

Solution

Breaking the loop as depicted in Fig. 8.24(b), we recognize that the open-loop input resistance is equal to $(g_{m1} + g_{mb1})^{-1}$. To find the loop gain, we set the input to zero and inject a test signal in the loop [Fig. 8.24(c)], obtaining $V_F/V_t = -g_{m1}R_D C_1/(C_1 + C_2)$. The closed-loop input impedance is then equal to

$$R_{in,closed} = \frac{1}{g_{m1} + g_{mb1}} \left(1 + \frac{C_1}{C_1 + C_2} g_{m1} R_D\right). \tag{8.32}$$

The increase in the input impedance can be explained as follows. Suppose the input voltage decreases by ΔV, causing the output voltage to (momentarily) fall. As a result, the gate voltage of M_1 *decreases*,

---

[7]This circuit is similar to the right half of the topology shown in Fig. 8.21(a).

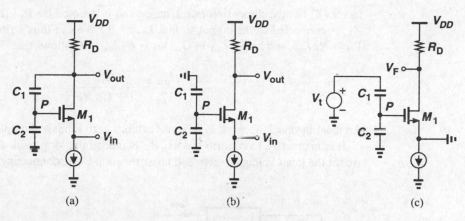

(a)   (b)   (c)

Figure 8.24

thereby lowering the gate-source voltage of M_1 and yielding a change in V_{GS1} that is *less* than ΔV. By contrast, if the gate of M_1 were connected to a constant potential, the gate-source voltage would change by ΔV, resulting in a larger current change.

In summary, voltage-voltage feedback decreases the output impedance and increases the input impedance, thereby proving useful as a "buffer" stage that can be interposed between a high-impedance source and a low-impedance load.

8.2.2 Current-Voltage Feedback

In some circuits, it is desirable or simpler to sense the output current to perform feedback. The current is actually sensed by placing a small resistor in series with the output and using the voltage drop across the resistor as the feedback information. This voltage may even serve as the return signal that is directly subtracted from the input.

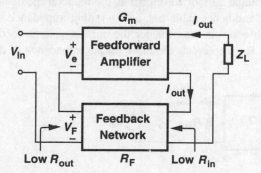

Figure 8.25 Current-voltage feed-back.

Let us consider the general current-voltage feedback system illustrated in Fig. 8.25.[8] Since the feedback network senses the output current and returns a voltage, its feedback

---

[8]This topology is also called "series-series" feedback.

factor (β) has the dimension of resistance and is denoted by R_F. (Note that a finite load, Z_L, is connected to the output so that $I_{out} \neq 0$.) We can thus write $V_F = R_F I_{out}$, $V_e = V_{in} - R_F I_{out}$, and hence $I_{out} = G_m(V_{in} - R_F I_{out})$. It follows that

$$\frac{I_{out}}{V_{in}} = \frac{G_m}{1 + G_m R_F}. \tag{8.33}$$

An ideal feedback network in this case exhibits zero input and output impedances.

It is instructive to confirm that $G_m R_F$ is indeed the loop gain. As shown in Fig. 8.26, we set the input voltage to zero and break the loop by disconnecting the feedback network

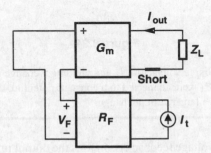

Figure 8.26 Calculation of loop gain for current-voltage feedback.

from the output and replacing it with a *short* at the output (if the feedback network is ideal). We then inject the test signal I_t, producing $V_F = R_F I_t$ and hence $I_{out} = -G_m R_F I_t$. Thus, the loop gain is equal to $G_m R_F$ and the transconductance of the amplifier is reduced by $1 + G_m R_F$ when feedback is applied.

Sensing the current at the output of a feedback system *increases* the output impedance. This is because the system attempts to make the output *current* a faithful replica of the input signal (with a proportionality factor if the input is a voltage quantity). Consequently, the system delivers the same current waveform as the load varies, in essence approaching an ideal current source and hence exhibiting a high output impedance.

To prove the above result, we consider the current-voltage feedback topology shown in Fig. 8.27, where R_{out} represents the finite output impedance of the feedforward ampli-

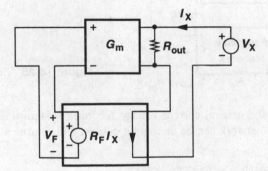

Figure 8.27 Calculation of output resistance of a current-voltage feedback amplifier.

fier.[9] The feedback network produces a voltage V_F proportional to $I_X : V_F = R_F I_X$, and the current generated by G_m equals $-R_F I_X G_m$. As a result, $-R_F I_X G_m = I_X - V_X/R_{out}$, yielding

$$\frac{V_X}{I_X} = R_{out}(1 + G_m R_F). \tag{8.34}$$

The output impedance therefore increases by a factor of $1 + G_m R_F$.

Example 8.4

Suppose we need to increase the output impedance of a common-source stage by current feedback. As shown in Fig. 8.28(a), we insert a small resistor r in the output current path, apply the voltage

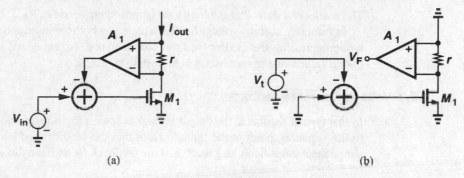

(a)   (b)

Figure 8.28

across r to an amplifier A_1, and subtract the output of A_1 from the input voltage. Calculate the output impedance of this circuit.

Solution

Using the circuit of Fig. 8.28(b) to determine the loop gain, we have

$$\frac{V_F}{V_t} = -g_m r A_1. \tag{8.35}$$

Thus, the overall output impedance is given by

$$R_{out,closed} = (1 + g_m r A_1) r_{O1}. \tag{8.36}$$

As with voltage-voltage feedback, current-voltage feedback increases the input impedance by a factor equal to one plus the loop gain. As illustrated in Fig. 8.29, we have $I_X R_{in} G_m = I_{out}$. Thus, $V_e = V_X - G_m R_F I_X R_{in}$ and

$$\frac{V_X}{I_X} = R_{in}(1 + G_m R_F). \tag{8.37}$$

---

[9]Note that R_{out} is placed in *parallel* with the output because the ideal transimpedance amplifier is modeled by a voltage-dependent current source.

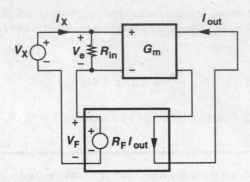

Figure 8.29 Calculation of input resistance of a current-voltage feedback amplifier.

The reader can show that the loop gain is indeed equal to $G_m R_F$.

In summary, current-voltage feedback increases both the input and the output impedances while decreasing the feedforward transconductance. As explained in Chapter 9, the high output impedance proves useful in high-gain op amps.

8.2.3 Voltage-Current Feedback

In this type of feedback, the output voltage is sensed and a proportional current is returned to the summing point at the input.[10] Note that the feedforward path incorporates a transimpedance amplifier with gain R_0 and the feedback factor has a dimension of conductance.

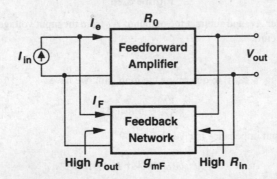

Figure 8.30 Voltage-current feedback.

A voltage-current feedback topology is shown in Fig. 8.30. Sensing a voltage and producing a current, the feedback network is characterized by a transconductance g_{mF}, ideally exhibiting infinite input and output impedances. Since $I_F = g_{mF}V_{out}$ and $I_e = I_{in} - I_F$, we have $V_{out} = R_0 I_e = R_0(I_{in} - g_{mF}V_{out})$. It follows that

$$\frac{V_{out}}{I_{in}} = \frac{R_0}{1 + g_{mF}R_0}. \tag{8.38}$$

---

[10]This topology is also called "shunt-shunt" feedback.

The reader can prove that $g_{mF}R_0$ is indeed the loop gain, concluding that this type of feedback lowers the transimpedance by a factor equal to one plus the loop gain.

Example 8.5

Calculate the transimpedance, V_{out}/I_{in}, of the circuit shown in Fig. 8.31(a) at relatively low frequencies.

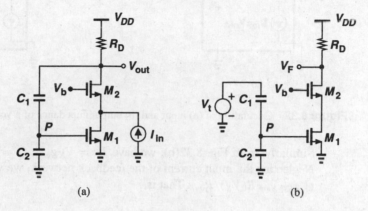

(a)   (b)

Figure 8.31

Solution

In this circuit, the capacitive divider C_1-C_2 senses the output voltage, applying the result to the gate of M_1 and producing a current that is subtracted from I_{in}. The open-loop transimpedance equals that of the core common-gate stage, R_D. The loop gain is obtained by setting I_{in} to zero and breaking the loop at the output [Fig. 8.31(b)]:

$$-V_t \frac{C_1}{C_1 + C_2} g_{m1} R_D = V_F.$$

(8.39)

Thus, the overall transimpedance is equal to

$$R_{tot} = \frac{R_D}{1 + \dfrac{C_1}{C_1 + C_2} g_{m1} R_D}.$$

(8.40)

Following our reasoning for the other two types of feedback studied above, we surmise that voltage-current feedback decreases both the input and the output impedances. As shown in Fig. 8.32(a), the input resistance of R_0 is placed in *series* because an ideal transimpedance amplifier exhibits a zero input impedance. We write $I_F = I_X - V_X/R_{in}$ and $(V_X/R_{in})R_0 g_{mF} = I_F$. Thus,

$$\frac{V_X}{I_X} = \frac{R_{in}}{1 + g_{mF} R_0}.$$

(8.41)

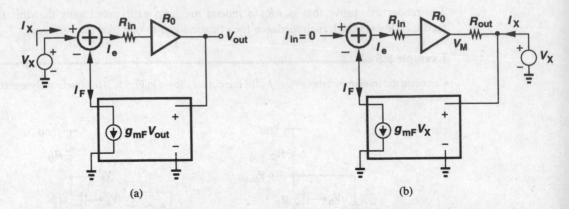

Figure 8.32 Calculation of (a) input and (b) output impedance of a voltage-current feedback amplifier.

Similarly, from Fig. 8.32(b), we have $I_F = V_X g_{mF}$, $I_e = -I_F$, and $V_M = -R_0 g_{mF} V_X$. Neglecting the input current of the feedback network, we write $I_X = (V_X - V_M)/R_{out} = (V_X + g_{mF} R_0 V_X)/R_{out}$. That is,

$$\frac{V_X}{I_X} = \frac{R_{out}}{1 + g_{mF} R_0}. \tag{8.42}$$

Example 8.6

Calculate the input and output impedances of the circuit shown in Fig. 8.33(a). For simplicity, assume $R_F \gg R_D$.

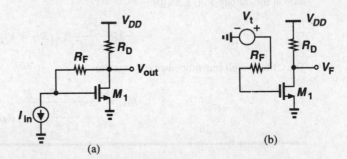

Figure 8.33

Solution

In this circuit, R_F senses the output voltage and returns a current to the input. Breaking the loop as depicted in Fig. 8.33(b), we calculate the loop gain as $g_m R_D$. Thus, the open-loop input impedance, R_F, is divided by $1 + g_m R_D$:

$$R_{in,closed} = \frac{R_F}{1 + g_m R_D}. \tag{8.43}$$

Similarly,

$$R_{out,closed} = \frac{R_D}{1 + g_m R_D}. \tag{8.44}$$

Note $R_{out,closed}$ is in fact the parallel combination of a diode-connected transistor and R_D.

An important application of amplifiers with *low* input impedance is in fiber optic receivers, where light received through a fiber is converted to a *current* by a reverse-biased photodiode. This current is typically converted to a voltage for further amplification and processing. Shown in Fig. 8.34(a), such conversion can be accomplished by a simple resis-

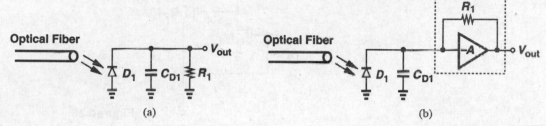

Figure 8.34 Detection of current produced by a photodiode by (a) resistor R_1 and (b) a transimpedance amplifier.

tor but at the cost of bandwidth because the diode suffers from a relatively large junction capacitance. For this reason, the feedback topology of Fig. 8.34(b) is usually employed, where R_1 is placed around the voltage amplifier A to form a transimpedance circuit. The input impedance is $R_1/(1 + A)$ and the output voltage is approximately $R_1 I_{D1}$.

8.2.4 Current-Current Feedback

Fig. 8.35 illustrates this type of feedback.[11] Here, the feedforward amplifier is characterized by a current gain, A_I, and the feedback network by a current ratio, β. In a fashion similar to the previous derivations, the reader can easily prove that the closed-loop current gain is equal to $A_I/(1+\beta A_I)$, the input impedance is divided by $1+\beta A_I$ and the output impedance is multiplied by $1 + \beta A_I$.

Fig. 8.36 illustrates an example of current-current feedback. Here, since the source and drain currents of M_2 are equal (at low frequencies), resistor R_S is inserted in the source network to monitor the output current. Resistor R_F plays the same role as in Fig. 8.33.

---

[11]This topology is also called "shunt-series" feedback, where the first term refers to the input connection and the second to the output connection.

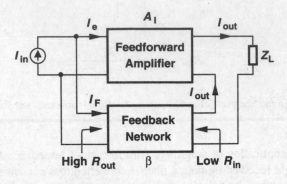

Figure 8.35 Current-current feedback.

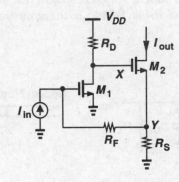

Figure 8.36

8.3 Effect of Loading

In our analysis of feedback systems thus far, we have tacitly assumed that the feedback network does not "load" the feedforward amplifier at the input or output. For example, in the voltage-voltage feedback topology of Fig. 8.21, we assumed C_1 and C_2 do not load the amplifier so that the *open-loop* gain could still be written as $g_{m1}(r_{O2}\|r_{O4})$. In reality, however, the loading due to the feedback network may not be negligible, complicating the analysis.

The problem of loading manifests itself when we need to break the feedback loop so as to identify the open-loop system, e.g., calculate the open-loop gain and the input and output impedances. To arrive at the proper procedure for including the feedback network terminal impedances, we first review models of two-port networks.

8.3.1 Two-Port Network Models

The feedback network placed around the feedforward amplifier can be considered a two-port circuit sensing and producing voltages or currents. Recall from basic circuit theory that a two-port linear (and time-invariant) network can be represented by any of the four models shown in Fig. 8.37. The "Z model" in Fig. 8.37(a) consists of input and output impedances in series with current-dependent voltage sources whereas the "Y model" in Fig. 8.37(b) comprises input and output admittances in parallel with voltage-dependent current sources. The "hybrid models" of Figs. 8.37(c) and (d) incorporate a combination of impedances

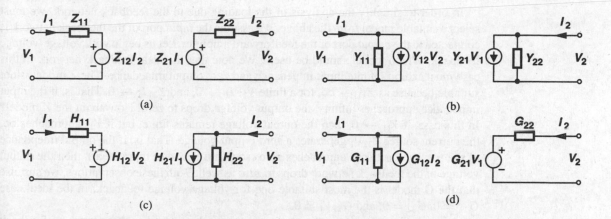

Figure 8.37 Linear two-port network models.

and admittances and voltage sources and current sources. Each model is described by two equations. For the Z model, we have

$$V_1 = Z_{11}I_1 + Z_{12}I_2 \tag{8.45}$$

$$V_2 = Z_{21}I_1 + Z_{22}I_2. \tag{8.46}$$

Each Z parameter has a dimension of impedance and is obtained by leaving one port open, e.g., $Z_{11} = V_1/I_1$ when $I_2 = 0$. Similarly, for the Y model,

$$I_1 = Y_{11}V_1 + Y_{12}V_2 \tag{8.47}$$

$$I_2 = Y_{21}V_1 + Y_{22}V_2, \tag{8.48}$$

where each Y parameter is calculated by shorting one port, e.g., $Y_{11} = I_1/V_1$ when $V_2 = 0$. For the H model,

$$V_1 = H_{11}I_1 + H_{12}V_2 \tag{8.49}$$

$$I_2 = H_{21}I_1 + H_{22}V_2, \tag{8.50}$$

and for the G model,

$$I_1 = G_{11}V_1 + G_{12}I_2 \tag{8.51}$$

$$V_2 = G_{21}V_1 + G_{22}I_2. \tag{8.52}$$

Note that, for example, Y_{11} may not be equal to the inverse of Z_{11} because the two are obtained under different conditions: the output is shorted for the former but left open for the latter.

In order to simplify the analysis of the loading due to the feedback network, we must select a suitable model from the above. We assume the *input* port of the feedback network is connected to the *output* port of the feedforward amplifier. Let us begin with voltage-voltage feedback. Which model should be used? We note that the ideal feedback network in this case must exhibit infinite input impedance and zero output impedance. The Z model is not suitable because as $Z_{11} \to \infty$, for a finite V_1, $I_1 \to 0$, and $Z_{21}I_1 \to 0$. That is, if the input impedance approaches infinity, the output voltage drops to zero. How about the Y model? In this case, if $Y_{11} \to 0$, then the output voltage remains finite, but if Y_{22} approaches ∞, the current source $Y_{21}V_1$ generates a zero output voltage. That is, if the output impedance of the feedback network approaches zero (so that it becomes more ideal), then the output voltage of the feedback network drops to zero as well. With these observations, we surmise that the G model is the most suitable one for voltage-voltage feedback; in the ideal case $G_{11} = 0$, $G_{22} = 0$, and $G_{21}V_1 \neq 0$.

Using similar arguments, the reader can show that the other three types of feedback require the following network models: voltage-current: Y model; current-voltage: Z model; current-current: H model.

8.3.2 Loading in Voltage-Voltage Feedback

Replacing the feedback network by a G model, we arrive at the representation in Fig. 8.38(a). Unlike the simple models used in previous sections, this circuit incorporates *two* dependent sources in the feedback path: $G_{12}I_2$ and $G_{21}V_{out}$. What is the effect of $G_{12}I_2$? This current flows through the parallel combination of Z_{out} and G_{11}, contributing to the output voltage. However, if A_0 is large, the signal amplified by A_0 is much greater than the contribution of $G_{12}I_2$. In other words, the *forward* gain of the main amplifier overwhelms the *reverse* gain of the feedback network. Since this condition holds in most circuits of interest, we can neglect $G_{12}I_2$, obtaining the circuit in Fig. 8.38(b). A rigorous analysis of Fig. 8.38(a) (Problem 8.8) reveals that if $G_{12} \ll A_0 Z_{in}/Z_{out}$, then the "reverse transmission" through the feedback circuit is negligible. It is indeed expected that Z_{in} and Z_{out} play a role here. If Z_{in} is small, the voltage division between Z_{in} and G_{22} reduces the signal through the feedforward path. Similarly, if Z_{out} is large, then the voltage division between Z_{out} and G_{11} lowers the contribution of A_0V_e to the output.

Let us now compute the closed-loop gain of the circuit shown in Fig. 8.38(b). We have

$$V_e = (V_{in} - G_{21}V_{out})\frac{Z_{in}}{Z_{in} + G_{22}}, \tag{8.53}$$

and hence

$$(V_{in} - G_{21}V_{out})\frac{Z_{in}}{Z_{in} + G_{22}}A_0\frac{G_{11}^{-1}}{G_{11}^{-1} + Z_{out}} = V_{out}. \tag{8.54}$$

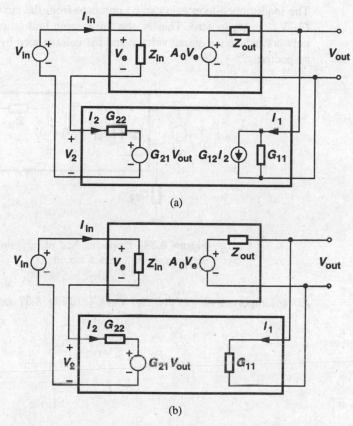

Figure 8.38 Voltage-voltage feedback circuit with (a) feedback network represented by a G model and (b) simplified G model.

It follows that

$$\frac{V_{out}}{V_{in}} = \frac{A_0 \dfrac{Z_{in}}{Z_{in} + G_{22}} \dfrac{G_{11}^{-1}}{G_{11}^{-1} + Z_{out}}}{1 + \dfrac{Z_{in}}{Z_{in} + G_{22}} \dfrac{G_{11}^{-1}}{G_{11}^{-1} + Z_{out}} G_{21} A_0}. \tag{8.55}$$

Note that if the feedback network is ideal, i.e., if $G_{11}^{-1} = \infty$ and $G_{22} = 0$, then $V_{out}/V_{in} = A_0/(1 + G_{21}A_0)$, as expected.

Equation 8.55 assumes the standard form of a feedback transfer function if we define the open-loop gain in the presence of loading as

$$A_{v,open} = \frac{Z_{in}}{Z_{in} + G_{22}} \frac{G_{11}^{-1}}{G_{11}^{-1} + Z_{out}} A_0. \tag{8.56}$$

The loaded open-loop gain can be obtained from the circuit depicted in Fig. 8.39, where $G_{21}V_{out}$ is set to zero. That is, the finite input and output impedances of the feedback network reduce the output voltage and the voltage seen by the input of the main amplifier, respectively.

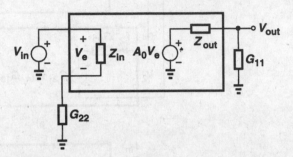

Figure 8.39 Proper method of including loading in a voltage-voltage feedback circuit.

It is important to note that G_{11} and G_{22} in Fig. 8.37 are computed as follows:

$$G_{11} = \left.\frac{I_1}{V_1}\right|_{I2=0}$$ (8.57)

$$G_{22} = \left.\frac{V_2}{I_2}\right|_{V1=0}$$ (8.58)

Thus, as illustrated in Fig. 8.40, G_{11} is obtained by leaving the output of the feedback

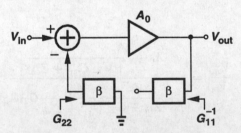

Figure 8.40 Conceptual view of opening a voltage-voltage feedback loop with proper loading.

network open whereas G_{22} is calculated by *shorting* the input of the feedback network.

Another important result of the foregoing analysis is that the loop gain, i.e., the second term in the denominator of (8.55) is simply equal to the loaded open-loop gain multiplied by G_{21}. Thus, a separate calculation of the loop gain is not necessary. Also, the open-loop input and output impedances obtained from Fig. 8.39 are scaled by $1 + G_{21}A_{v,open}$ to yield the closed-loop values.

Example 8.7

For the circuit shown in Fig. 8.41(a), calculate the open-loop and closed-loop gains.

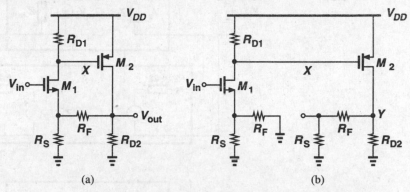

Figure 8.41

Solution

The circuit consists of two common-source stages, with R_F and R_S sensing the output voltage and returning a fraction thereof to the source of M_1. The reader can prove that the feedback is indeed negative. Following the procedure illustrated in Fig. 8.40, we identify R_F and R_S as the feedback network and construct the open-loop circuit as shown in Fig. 8.41(b). Note that the loading effect in the input network is obtained by shorting the right terminal of R_F to ground and that in the output by leaving the left terminal of R_F open. Neglecting channel-length modulation and body effect for simplicity, we have

$$A_{v,open} = \frac{V_Y}{V_{in}} = \frac{-R_{D1}}{R_F \| R_S + 1/g_{m1}} \{-g_{m2}[R_{D2} \| (R_F + R_S)]\}. \tag{8.59}$$

To compute the closed-loop gain, we first find the loop gain as $G_{21}A_{v,open}$. Recall from (8.52) that $G_{21} = V_2/V_1$ with $I_2 = 0$. For the voltage divider consisting of R_F and R_D, $G_{21} = R_S/(R_F + R_S)$. The closed-loop gain is simply equal to $A_{v,closed} = A_{v,open}/(1 + G_{21}A_{v,open})$.

8.3.3 Loading in Current-Voltage Feedback

Replacing the feedback network by a Z model, we obtain the circuit shown in Fig. 8.42(a). Using an argument similar to that for voltage-voltage feedback, we neglect the source $Z_{12}I_2$, thereby arriving at the circuit in Fig. 8.42(b). We thus have

$$(V_{in} - Z_{21}I_{out})\frac{Z_{in}}{Z_{in} + Z_{22}}G_m\frac{Z_{out}}{Z_{out} + Z_{11}} = I_{out}. \tag{8.60}$$

That is,

$$\frac{I_{out}}{V_{in}} = \frac{\dfrac{Z_{in}}{Z_{in} + Z_{22}}\dfrac{Z_{out}}{Z_{out} + Z_{11}}G_m}{1 + \dfrac{Z_{in}}{Z_{in} + Z_{22}}\dfrac{Z_{out}}{Z_{out} + Z_{11}}G_m Z_{21}}. \tag{8.61}$$

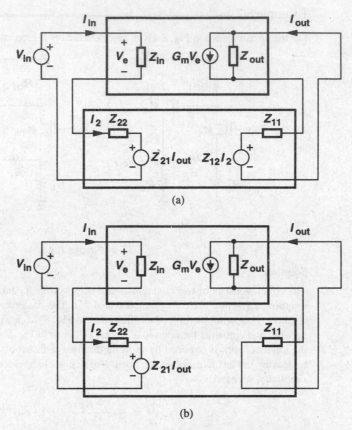

Figure 8.42 Current-voltage feedback circuit with (a) feedback network represented by a Z model and (b) simplified Z model.

Equation (8.61) suggests that the loaded open-loop gain is equal to

$$G_{m,open} = \frac{Z_{in}}{Z_{in} + Z_{22}} \frac{Z_{out}}{Z_{out} + Z_{11}} G_m, \tag{8.62}$$

revealing voltage division at the input and current division at the output (Fig. 8.43). Since Z_{22} and Z_{11} are obtained by opening the input and output ports of the feedback network, respectively, the open-loop circuit can be visualized as in Fig. 8.44. Note that the loop gain is equal to $Z_{21} G_{m,open}$.

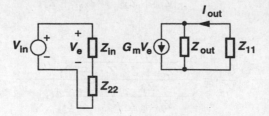

Figure 8.43 Current-voltage feedback circuit with proper loading of feedback network.

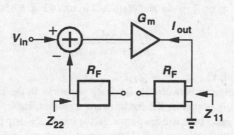

Figure 8.44 Conceptual view of opening the loop in current-voltage feedback.

Example 8.8

Calculate the open-loop and closed-loop gain of the circuit shown in Fig. 8.45(a).

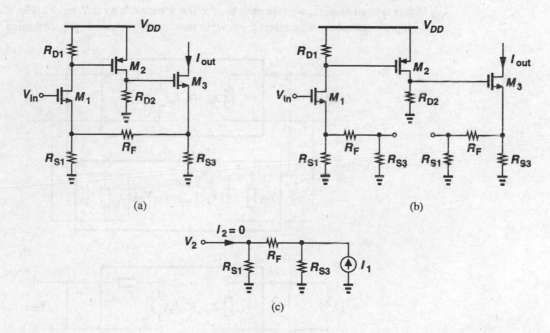

(a)

(b)

(c)

Figure 8.45

Solution

This circuit consists of two voltage gain stages, M_1 and M_2, and a voltage-to-current converter, M_3. Since the drain and source currents of M_3 are equal, the output current is monitored by R_{S3}. Thus, R_{S3}, R_F, and R_{S1} sense the output current and return a proportional voltage to the input.

If $\lambda = \gamma = 0$, the open-loop gain is equal to

$$G_{m,open} = \frac{-R_{D1}}{R_{S1}\|(R_F + R_{S3}) + 1/g_{m1}} \cdot \frac{-g_{m2}R_{D2}}{R_{S3}\|(R_F + R_{S1}) + 1/g_{m3}}. \tag{8.63}$$

The loop gain is given by $Z_{21}G_{m,open}$, where, from (8.46), $Z_{21} = V_2/I_1$ with $I_2 = 0$. For the feedback network consisting of R_{S2}, R_F and R_{S1}, the circuit of Fig. 8.45(c) gives

$$Z_{21} = \frac{R_{S3}}{R_{S3} + R_{S1} + R_F} R_{S1}. \qquad (8.64)$$

The closed-loop gain equals $G_{m,open}/(1 + Z_{21}G_{m,open})$.

It is important to distinguish between the feedback networks in the circuits of Fig. 8.41(a) and 8.45(a). In the former, R_{D2} is part of the feedforward amplifier, rather than part of the feedback network, because it must generate a *voltage* output. In the latter, R_{S3} is part of the feedback network because it is used to *sense* the output current. If the output of interest in Fig. 8.45(a) is the voltage at the source of M_3, then R_{S3} is part of the feedforward amplifier rather than the feedback network.

8.3.4 Loading in Voltage-Current Feedback

In this type of system, we represent the feedback network by a Y model [Fig. 8.46(a)]. As with previous cases, we neglect the reverse transmission term, $Y_{12}V_2$, obtaining the circuit

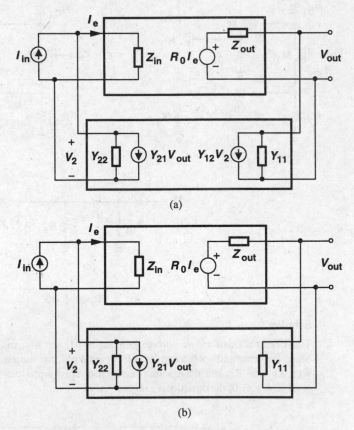

(a)

(b)

Figure 8.46 Voltage-current feedback circuit with (a) feedback network represented by a Y model and (b) simplified Y model.

in Fig. 8.46(b). Writing

$$(I_{in} - Y_{21} V_{out}) \frac{Y_{22}^{-1}}{Y_{22}^{-1} + Z_{in}} R_0 \frac{Y_{11}^{-1}}{Y_{11}^{-1} + Z_{out}} = V_{out}, \tag{8.65}$$

we have

$$\frac{V_{out}}{I_{in}} = \frac{\dfrac{Y_{22}^{-1}}{Y_{22}^{-1} + Z_{in}} R_0 \dfrac{Y_{11}^{-1}}{Y_{11}^{-1} + Z_{out}}}{1 + \dfrac{Y_{22}^{-1}}{Y_{22}^{-1} + Z_{in}} R_0 \dfrac{Y_{11}^{-1}}{Y_{11}^{-1} + Z_{out}} Y_{21}}. \tag{8.66}$$

It is therefore possible to define the loaded open-loop gain as

$$R_{0,open} = \frac{Y_{22}^{-1}}{Y_{22}^{-1} + Z_{in}} \frac{Y_{11}^{-1}}{Y_{11}^{-1} + Z_{out}} R_0. \tag{8.67}$$

The loading manifests itself as current division between Y_{22}^{-1} and Z_{in} and voltage division

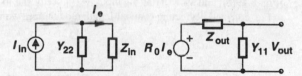

Figure 8.47 Voltage-current feedback circuit with proper loading of feedback network.

between Z_{out} and Y_{11}^{-1} (Fig. 8.47). Since Y_{22} and Y_{11} are obtained by shorting the input and output ports of the feedback network, respectively, the procedure for including the loading can be illustrated as in Fig. 8.48. The loop gain is given by $Y_{21} R_{0,open}$.

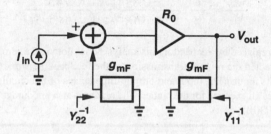

Figure 8.48 Conceptual view of opening the loop in voltage-current feedback.

Example 8.9 _____

Calculate the voltage gain of the circuit shown in Fig. 8.49(a).

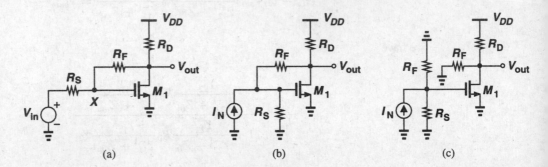

Figure 8.49

Solution

What type of feedback is used in this circuit? Resistor R_F senses the output voltage and returns a proportional current to node X. Thus, the feedback can be considered as the voltage-current type. However, in the general representation of Fig. 8.46(a), the input signal is a current quantity, whereas in this example, it is a voltage quantity. For this reason, we replace V_{in} and R_S by a Norton equivalent [Fig. 8.49(b)] and view R_S as the input resistance of the main amplifier. Opening the loop according to Fig. 8.48 and neglecting channel-length modulation, we write the open-loop gain from Fig. 8.49(c) as

$$R_{0,open} = \frac{V_{out}}{I_N}\bigg|_{open} \tag{8.68}$$

$$= -(R_S \| R_F)g_m(R_F \| R_D), \tag{8.69}$$

where $I_N = V_{in}/R_S$. We also calculate the loop gain as $Y_{21}R_{0,open}$. From (8.48), $Y_{21} = I_2/V_1$ with $V_2 = 0$, and since the feedback network consists of only R_F, we have $Y_{21} = -1/R_F$. Thus, the circuit of Fig. 8.49(a) exhibits a voltage gain of

$$\frac{V_{out}}{V_{in}} = \frac{1}{R_S} \cdot \frac{-(R_S \| R_F)g_m(R_F \| R_D)}{1 + g_m(R_F \| R_D)R_S/(R_S + R_F)}. \tag{8.70}$$

Interestingly, if R_F is replaced by a capacitor, this analysis does not yield a zero in the transfer function because we have neglected the reverse transmission of the feedback network (from the output of the feedback network to its input.) The input and output impedances of the circuit are also interesting to calculate. This is left as an exercise for the reader. The reader is also encouraged to apply this solution to the circuit of Fig. 8.3(b).

8.3.5 Loading in Current-Current Feedback

Fig. 8.50(a) depicts a current-current feedback system with the feedback network represented by an H model. Neglecting the effect of $H_{12}V_2$ compared to the forward gain of the amplifier and drawing the circuit as in Fig. 8.50(b), we write

$$(I_{in} - H_{21}I_{out})\frac{H_{22}^{-1}}{H_{22}^{-1} + Z_{in}}A_I\frac{Z_{out}}{H_{11} + Z_{out}} = I_{out}. \tag{8.71}$$

It follows that

$$\frac{I_{out}}{I_{in}} = \frac{\dfrac{H_{22}^{-1}}{H_{22}^{-1} + Z_{in}}A_I\dfrac{Z_{out}}{H_{11} + Z_{out}}}{1 + \dfrac{H_{22}^{-1}}{H_{22}^{-1} + Z_{in}}A_I\dfrac{Z_{out}}{H_{11} + Z_{out}}H_{21}}. \tag{8.72}$$

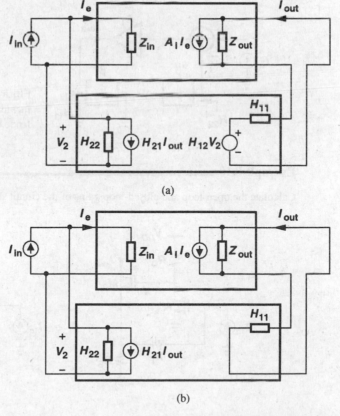

(a)

(b)

Figure 8.50 Current-current feedback circuit with (a) feedback network represented by an H model and (b) simplified H model.

We can thus define the loaded open-loop gain as

$$A_{I,open} = \frac{H_{22}^{-1}}{H_{22}^{-1} + Z_{in}} \frac{Z_{out}}{H_{11} + Z_{out}} A_I, \tag{8.73}$$

concluding that the feedback network introduces current division at both the input and the output of the system (Fig. 8.51). Note that H_{22} and H_{11} are measured with the input and the

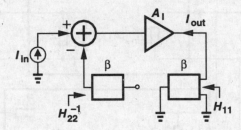

Figure 8.51 Current-current feed-back circuit with proper loading of feedback network.

output ports of the feedback network open and shorted, respectively (Fig. 8.52). The loop gain is obtained as $H_{21}A_{I,open}$.

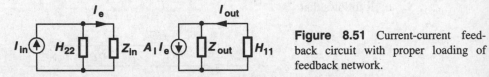

Figure 8.52 Conceptual view of including loading in current-current feedback.

Example 8.10

Calculate the open-loop and closed-loop gains of the circuit shown in Fig. 8.53(a).

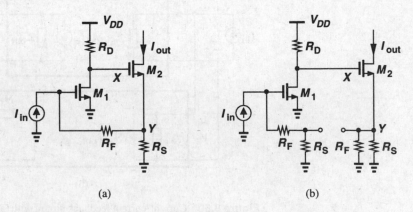

(a)   (b)

Figure 8.53

Solution

In this circuit, R_S and R_F sense the output current and return a fraction thereof to the input. Breaking the loop according to Fig. 8.52, we arrive at the circuit in Fig. 8.53(b), where we have

$$A_{I,open} = -(R_F + R_S)g_{m1}R_D \frac{1}{R_S \| R_F + 1/g_{m2}}. \tag{8.74}$$

The loop gain is given by $H_{21}A_{I,open}$, where, from (8.50), $H_{21} = I_2/I_1$ with $V_2 = 0$. For the feedback network consisting of R_S and R_F, we have $H_{21} = -R_S/(R_S + R_F)$. The closed-loop gain equals $A_{I,open}/(1 + H_{21}A_{I,open})$.

8.3.6 Summary of Loading Effects

The results of our study of loading are summarized in Fig. 8.54. The analysis is carried out in three steps: (1) open the loop with proper loading and calculate the open-loop gain, A_{OL}, and the open-loop input and output impedances; (2) determine the feedback ratio, β, and hence the loop gain, βA_{OL}; (3) calculate the closed-loop gain and input and output impedances by scaling the open-loop values by a factor of $1 + \beta A_{OL}$. Note that in the equations defining β, the subscripts 1 and 2 refer to the input and output ports of the feedback network, respectively.

In this chapter, we have described two methods of obtaining the loop gain: (1) by breaking the loop at an arbitrary point as shown in Fig. 8.5 and (2) by calculating A_{OL} and β as illustrated in Fig. 8.54. The two methods may yield slightly different results because the

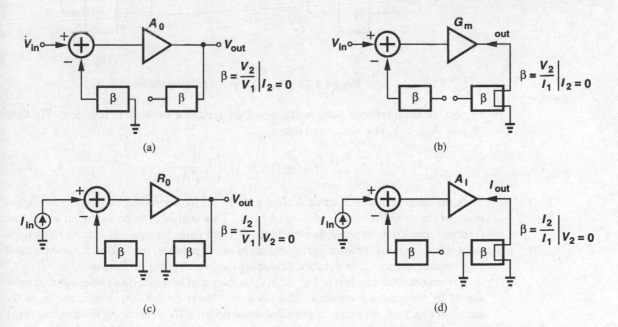

Figure 8.54 Summary of loading effects.

latter neglects the reverse transmission through the feedback network. However, the first method may be difficult to apply if loading effects must be taken into account because, if the loop can be broken at an arbitrary point, then the actual input and output ports of the overall system are unknown and the type of feedback unimportant. For example, the loop gain of the circuit of Fig. 8.53(a) does not depend on whether the output of interest is I_{out} or V_Y. In other words, since the first method does not distinguish between different types of feedback, it generally cannot utilize the loading calculations depicted in Fig. 8.54. For this reason, the second method is preferable.

We should also mention that some feedback circuits do not fall under any of the four types studied in this chapter because we have restricted our attention to cases where the output of interest is directly sensed by the feedback network. For example, if I_{out} in Fig. 8.53(a) flows through a resistor tied from the drain of M_2 to V_{DD}, then the resulting voltage is not inside the feedback loop. These cases are usually analyzed individually.

8.4 Effect of Feedback on Noise

Feedback does not improve the noise performance of circuits. Let us first consider the simple case illustrated in Fig. 8.55(a), where the open-loop voltage amplifier A_1 is characterized

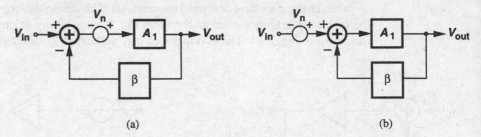

(a)   (b)

Figure 8.55   Feedback around a noisy circuit.

by only an input-referred noise voltage and the feedback network is noiseless. We have $(V_{in} - \beta V_{out} + V_n)A_1 = V_{out}$ and hence

$$V_{out} = (V_{in} + V_n)\frac{A_1}{1 + \beta A_1}. \tag{8.75}$$

Thus, the circuit can be simplified as shown in Fig. 8.55(b), revealing that the input-referred noise of the overall circuit is still equal to V_n. This analysis can be extended to all four feedback topologies to prove that the input-referred noise voltage and current remain the same if the feedback network introduces no noise. In practice, the feedback network itself may contain resistors or transistors, degrading the overall noise performance.

It is important to note that in Fig. 8.55(a) the output of interest is the same as the quantity sensed by the feedback network. This need not always be the case. For example, in the circuit of Fig. 8.56, the output is provided at the drain of M_1 whereas the feedback network senses the voltage at the source of M_1. In such cases, the input-referred noise of the closed-loop circuit may not be equal to that of the open-loop circuit even if the feedback network

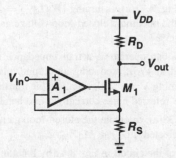

Figure 8.56 Noisy circuit with feedback sensing the source voltage.

is noiseless. As an example, let us consider the topology of Fig. 8.56 and, for simplicity, take only the noise of R_D, $V_{n,RD}$, into account. The reader can prove that the closed-loop voltage gain is equal to $-A_1 g_m R_D / [1 + (1 + A_1) g_m R_S]$ and hence the input-referred noise voltage due to R_D is

$$\left| V_{n,in,closed} \right| = \frac{|V_{n,RD}|}{A_1 R_D} \left[\frac{1}{g_m} + (1 + A_1) R_S \right]. \tag{8.76}$$

For the open-loop circuit, on the other hand, the input-referred noise is

$$\left| V_{n,in,open} \right| = \frac{|V_{n,RD}|}{A_1 R_D} \left[\frac{1}{g_m} + R_S \right]. \tag{8.77}$$

Interestingly, as $A_1 \to \infty$, $|V_{n,in,closed}| \to |V_{n,RD}| R_S / R_D$ whereas $|V_{n,in,open}| \to 0$.

Problems

Unless otherwise stated, in the following problems, use the device data shown in Table 2.1 and assume $V_{DD} = 3$ V where necessary. Also, assume all transistors are in saturation.

8.1. Consider the circuit of Fig. 8.3(b), assuming I_1 is ideal and $g_{m1} r_{O1}$ cannot exceed 50. If a gain error of less than 5% is required, what is the maximum closed-loop voltage gain that can be achieved by this topology? What is the low-frequency closed-loop output impedance under this condition?

8.2. In the circuit of Fig. 8.7(a), assume $(W/L)_1 = 50/0.5$, $(W/L)_2 = 100/0.5$, $R_D = 2$ kΩ, and $C_2 = C_1$. Neglecting channel-length modulation and body effect, determine the bias current of M_1 and M_2 such that the input resistance at low frequencies is equal to 50 Ω.

8.3. Calculate the output impedance of the circuit shown in Fig. 8.8(a) at relatively low frequencies if R_D is replaced by an ideal current source.

8.4. Consider the example illustrated in Fig. 8.10. Suppose an overall voltage gain of 500 is required with maximum bandwidth. How many stages with what gain per stage must be placed in a cascade? (Hint: first find the 3-dB bandwidth of a cascade of n identical stages in terms of that of each stage.)

8.5. If in Fig. 8.18(b), amplifier A_0 exhibits an output impedance of R_0, calculate the closed-loop voltage gain and output impedance, taking into account loading effects.

8.6. Consider the circuit of Fig. 8.21(a), assuming $(W/L)_{1,2} = 50/0.5$ and $(W/L)_{3,4} = 100/0.5$. If $I_{SS} = 1$ mA, what is the maximum closed-loop voltage gain that can be achieved if the gain error is to remain below 5%?

8.7. The circuit of Fig. 8.36 can operate as a transimpedance amplifier if I_{out} flows through a resistor, R_{D2}, connected to V_{DD}, producing an output voltage. Replacing R_S with an ideal current source and assuming $\lambda = \gamma = 0$, calculate the transimpedance of the resulting circuit. Also, calculate the input-referred noise current per unit bandwidth.

8.8. For the circuit of Fig. 8.38(a), calculate the closed-loop gain without neglecting $G_{12}I_2$. Prove that this term can be neglected if $G_{12} \ll A_0 Z_{in}/Z_{out}$.

8.9. Calculate the loop gain of the circuit in Fig. 8.41 by breaking the loop at node X. Why is this result somewhat different from $G_{21}A_{v,open}$?

8.10. Using feedback techniques, calculate the input and output impedance and voltage gain of each circuit in Fig. 8.57.

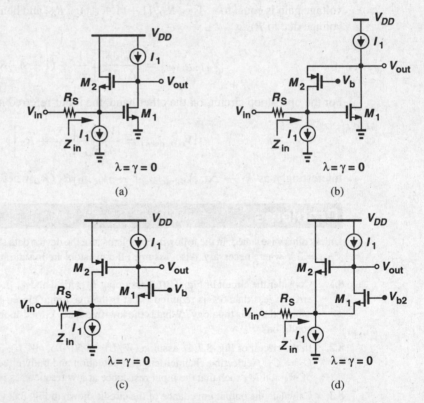

(a)   (b)

(c)   (d)

Figure 8.57

8.11. Using feedback techniques, calculate the input and output impedances of each circuit in Fig. 8.58.

8.12. Consider the circuit of Fig. 8.41(a), assuming $(W/L)_1 = (W/L)_2 = 50/0.5$, $\lambda = \gamma = 0$, and each resistor is equal to 2 kΩ. If $I_{D2} = 1$ mA, what is the bias current of M_1? What value of V_{in} gives such a current? Calculate the overall voltage gain.

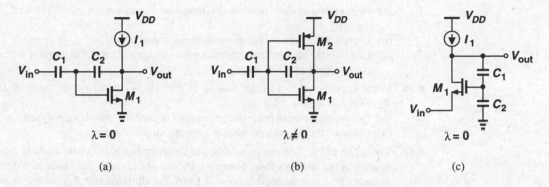

Figure 8.58

8.13. Suppose the amplifier of the circuit shown in Fig. 8.18 has an open-loop transfer function $A_0/(1 + s/\omega_0)$ and an output resistance R_0. Calculate the output impedance of the closed-loop circuit and plot the magnitude as a function of frequency. Explain the behavior.

8.14. Calculate the input-referred noise voltage of the circuit shown in Fig. 8.21(a) at relatively low frequencies.

8.15. A differential pair with current-source loads can be represented as in Fig. 8.59(a), where $R_0 = r_{ON} \| r_{OP}$ and r_{ON} and r_{OP} denote the output resistance of NMOS and PMOS devices, respectively. Consider the circuit shown in Fig. 8.59(b), where G_{m1} and G_{m2} are placed in a negative feedback loop.

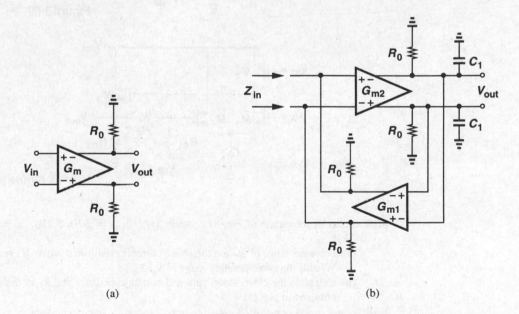

Figure 8.59

(a) Neglecting all other capacitances, derive an expression for Z_{in}. Sketch $|Z_{in}|$ versus frequency.

(b) Explain intuitively the behavior observed in part (a).

(c) Calculate the input-referred thermal noise voltage and current in terms of the input-referred noise voltage each G_m stage.

8.16. In the circuit of Fig. 8.60, $(W/L)_{1-3} = 50/0.5, I_{D1} = |I_{D2}| = |I_{D3}| = 0.5$ mA, and $R_{S1} = R_F = R_{D2} = 3$ kΩ.

(a) Determine the input bias voltage required to establish the above currents.

(b) Calculate the closed-loop voltage gain and output resistance.

8.17. The circuit of Fig. 8.60 can be modified as shown in Fig. 8.61, where a source follower, M_4, is inserted in the feedback loop. Note that M_1 and M_4 can also be viewed as a differential pair. Assume $(W/L)_{1-4} = 50/0.5$, $I_D = 0.5$ mA, for all transistors $R_{S1} = R_F = R_{D2} = 3$ kΩ, and $V_{b2} = 1.5$ V. Calculate the closed-loop voltage gain and output resistance and compare the results with those obtained in Problem 8.16(b).

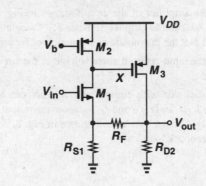

Figure 8.60

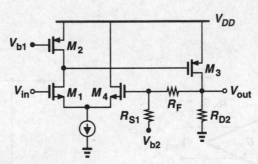

Figure 8.61

8.18. Consider the circuit of Fig. 8.62, where $(W/L)_{1-4} = 50/0.5, |I_{D1-4}| = 0.5$ mA and $R_2 = 3$ kΩ.

(a) For what range of R_1 are the above currents established while M_2 remains in saturation? What is the corresponding range of V_{in}?

(b) Calculate the closed-loop gain and output impedance for R_1 in the middle of the range obtained in part (a).

8.19. In the circuit of Fig. 8.63, suppose all resistors are equal to 2 kΩ and $g_{m1} = g_{m2} = 1/(200\ \Omega)$. Assuming $\lambda = \gamma = 0$, calculate the closed-loop gain and output impedance.

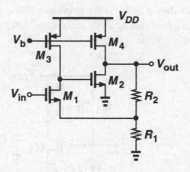

Figure 8.62

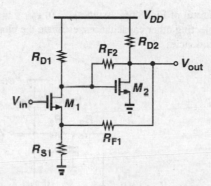

Figure 8.63

8.20. A CMOS inverter can be used as an amplifier with or without feedback (Fig. 8.64). Assume $(W/L)_{1,2} = 50/0.5$, $R_1 = 1$ kΩ, $R_2 = 10$ kΩ, and the dc levels of V_{in} and V_{out} are equal.
   (a) Calculate the voltage gain and the output impedance of each circuit.
   (b) Calculate the sensitivity of each circuit's output with respect to the supply voltage. That is, calculate the small-signal "gain" from V_{DD} to V_{out}. Which circuit exhibits less sensitivity?

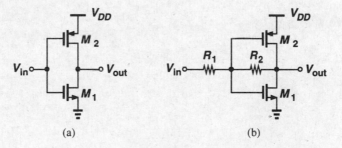

Figure 8.64

8.21. Calculate the input-referred thermal noise voltage of the circuits shown in Fig. 8.64.

8.22. The circuit shown in Fig. 8.65 employs positive feedback to produce a negative input capacitance. Using feedback analysis techniques, determine Z_{in} and identify the negative capacitance component. Assume $\lambda = \gamma = 0$.

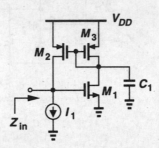

Figure 8.65

8.23. In the circuit of Fig. 8.66, assume $\lambda = 0$, $g_{m1,2} = 1/(200\ \Omega)$, $R_{1-3} = 2\ \text{k}\Omega$, and $C_1 = 100$ pF. Neglecting other capacitances, estimate the closed-loop voltage gain at very low and very high frequencies.

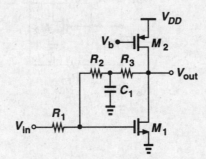

Figure 8.66

Chapter 9

Operational Amplifiers

Operational amplifiers (op amps) are an integral part of many analog and mixed-signal systems. Op amps with vastly different levels of complexity are used to realize functions ranging from dc bias generation to high-speed amplification or filtering. The design of op amps continues to pose a challenge as the supply voltage and transistor channel lengths scale down with each generation of CMOS technologies.

This chapter deals with the analysis and design of CMOS op amps. Following a review of performance parameters, we describe simple op amps such as telescopic and folded cascode topologies. Next, we study two-stage and gain-boosting configurations and the problem of common-mode feedback. Finally, we introduce the concept of slew rate and analyze the effect of supply rejection and noise in op amps.

9.1 General Considerations

We loosely define an op amp as a "high-gain differential amplifier." By "high," we mean a value that is adequate for the application, typically in the range of 10^1 to 10^5. Since op amps are usually employed to implement a feedback system, their open-loop gain is chosen according to the precision required of the closed-loop circuit.

Up to two decades ago, most op amps were designed to serve as "general-purpose" building blocks, satisfying the requirements of many different applications. Such efforts sought to create an "ideal" op amp, e.g., with very high voltage gain (several hundred thousand), high input impedance, and low output impedance, but at the cost of many other aspects of the performance, e.g., speed, output voltage swings, and power dissipation.

By contrast, today's op amp design proceeds with the recognition that the trade-offs between the parameters eventually require a multi-dimensional compromise in the overall implementation, making it necessary to know the *adequate* value that must be achieved for each parameter. For example, if the speed is critical while the gain error is not, a topology is chosen that favors the former, possibly sacrificing the latter.

9.1.1 Performance Parameters

In this section, we describe a number of op amp design parameters, providing an understanding of why and where each may become important. For this discussion, we

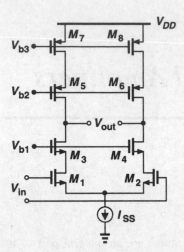

Figure 9.1   Cascode op amp.

consider the differential cascode circuit shown in Fig. 9.1 as a representative op amp design.[1] The voltages V_{b1}-V_{b3} are generated by the current mirror techniques described in Chapter 5.

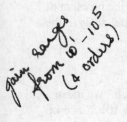

Gain   The open-loop gain of an op amp determines the precision of the feedback system employing the op amp. As mentioned before, the required gain may vary by four orders of magnitude according to the application. Trading with such parameters as speed and output voltage swings, the minimum required gain must therefore be known. As explained in Chapter 13, a high open-loop gain may also be necessary to suppress nonlinearity.

Example 9.1

The circuit of Fig. 9.2 is designed for a nominal gain of 10, i.e., $1 + R_1/R_2 = 10$. Determine the minimum value of A_1 for a gain error of 1%.

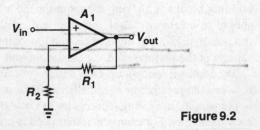

Figure 9.2

[1]Since op amps of this type have a high output resistance, they are sometimes called "operational transconductance amplifiers" (OTAs). In the limit, the circuit can be represented by a single voltage-dependent current source and called a "G_m stage."

Solution

The closed-loop gain is obtained from Chapter 8 as:

$$\frac{V_{out}}{V_{in}} = \frac{A_1}{1 + \dfrac{R_2}{R_1 + R_2}A_1} \qquad \left| \; \frac{V_{out}}{V_{in}} = \frac{A}{1 + \beta A} \; . \right. \tag{9.1}$$

$$= \frac{R_1 + R_2}{R_2} \frac{A_1}{\dfrac{R_1 + R_2}{R_2} + A_1}. \tag{9.2}$$

Predicting that $A_1 \gg 10$, we approximate (9.2) as:

$$\frac{V_{out}}{V_{in}} \approx \left(1 + \frac{R_1}{R_2}\right)\left(1 - \frac{R_1 + R_2}{R_2}\frac{1}{A_1}\right). \tag{9.3}$$

The term $(R_1 + R_2)/(R_2 A_1) = (1 + R_1/R_2)/A_1$ represents the relative gain error. To achieve a gain error less than 1%, we must have $A_1 > 1000$.

It is instructive to compare the circuit of Fig. 9.2 with an open-loop implementation such as that in Fig. 9.3. While it is possible to obtain a nominal gain of $g_m R_D = 10$ by a common-source stage, it is extremely difficult to guarantee an error less than 1%. The variations in the mobility and gate oxide thickness of the transistor and the value of the resistor typically yield an error greater than 20%. // $\frac{\circ}{\circ} \! \circ$

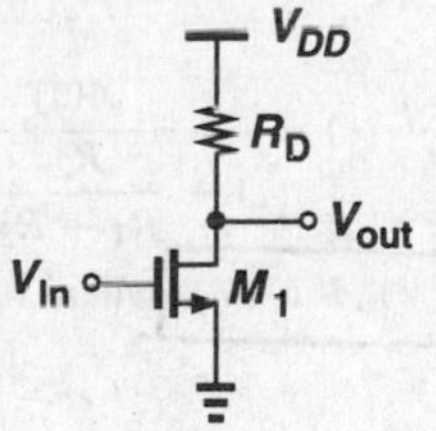

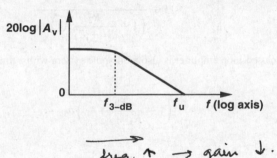

Figure 9.3 Simple common-source stage.

Small-Signal Bandwidth   The high-frequency behavior of op amps plays a critical role in many applications. For example, as the frequency of operation increases, the open-loop gain begins to drop (Fig. 9.4), creating larger errors in the feedback system. The small-signal bandwidth is usually defined as the "unity-gain" frequency, f_u, which exceeds 1 GHz in

Figure 9.4 Gain roll-off with frequency.

freq ↑ → gain ↓.

small signal
BW – unity gain freq – fu

*CMOS op amps
fu exceeds 1GHz.*

today's CMOS op amps. The 3-dB frequency, f_{3-dB}, may also be specified to allow easier prediction of the closed-loop frequency response.

Example 9.2

In the circuit of Fig. 9.5, assume the op amp is a single-pole voltage amplifier. If V_{in} is a small step,

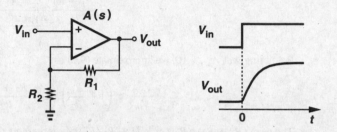

Figure 9.5

calculate the time required for the output voltage to reach within 1% of its final value. What unity-gain bandwidth must the op amp provide if $1 + R_1/R_2 \approx 10$ and the settling time is to be less than 5 ns? For simplicity, assume the low-frequency gain is much greater than unity.

Solution

Since

$$\left(V_{in} - V_{out} \frac{R_2}{R_1 + R_2} \right) A(s) = V_{out}, \tag{9.4}$$

we have

$$\frac{V_{out}}{V_{in}}(s) = \frac{A(s)}{1 + \frac{R_2}{R_1 + R_2}A(s)}. \tag{9.5}$$

For a one-pole system, $A(s) = A_0/(1 + s/\omega_0)$, where ω_0 is the 3-dB bandwidth and $A_0\omega_0$ the unity-gain bandwidth. Thus,

$$\frac{V_{out}}{V_{in}}(s) = \frac{A_0}{1 + \frac{R_2}{R_1 + R_2}A_0 + \frac{s}{\omega_0}} \tag{9.6}$$

$$= \frac{\dfrac{A_0}{1 + \dfrac{R_2}{R_1 + R_2}A_0}}{1 + \dfrac{s}{\left(1 + \dfrac{R_2}{R_1 + R_2}A_0\right)\omega_0}}, \tag{9.7}$$

indicating that the closed-loop amplifier is also a one-pole system with a time constant equal to

$$\tau = \frac{1}{\left(1 + \dfrac{R_2}{R_1 + R_2}A_0\right)\omega_0}. \tag{9.8}$$

Recognizing that the quantity $R_2 A_0/(R_1 + R_2)$ is the low-frequency loop gain and usually much greater than unity, we have

$$\tau \approx \left(1 + \frac{R_1}{R_2}\right) \frac{1}{A_0 \omega_0}. \tag{9.9}$$

The output step response for $V_{in} = au(t)$ can now be expressed as

step response·

$$V_{out}(t) \approx a \left(1 + \frac{R_1}{R_2}\right) \left(1 - \exp \frac{-t}{\tau}\right) u(t), \tag{9.10}$$

with the final value $V_F \approx a(1 + R_1/R_2)$. For 1% settling, $V_{out} = 0.99 V_F$ and hence

$$1 - \exp \frac{-t_{1\%}}{\tau} = 0.99, \tag{9.11}$$

yielding $t_{1\%} = \tau \ln 100 \approx 4.6\tau$. For a 1% settling of 5 ns, $\tau \approx 1.09$ ns, and from (9.9), $A_0 \omega_0 = (1 + R_1/R_2)/\tau = 9.21$ Grad/s (1.47 GHz).

The key point in the above example is that the required bandwidth depends on both the settling accuracy and the closed-loop gain that must be provided.

Large-Signal Bandwidth   In many of today's applications, op amps must operate with large transient signals. Under these conditions, nonlinear phenomena make it difficult to characterize the speed by merely small-signal properties such as the open-loop response shown in Fig. 9.4. As an example, suppose the feedback circuit of Fig. 9.5 incorporates a realistic op amp (i.e., with finite output impedance) while driving a large load capacitance. How does the circuit behave if we apply a 1-V step at the input? Since the output voltage cannot change instantaneously, the voltage difference sensed by the op amp itself at $t \geq 0$ is equal to 1 V. Such a large difference momentarily drives the op amp into a nonlinear region of operation. (Otherwise, with an open-loop gain of, say, 1000, the op amp would produce 1000 V at the output.)

As explained in Section 9.8, the large-signal behavior is usually quite complex, mandating careful simulations.

Output Swing   Most systems employing op amps require large voltage swings to accommodate a wide range of signal amplitudes. For example, a high-quality microphone that senses the music produced by an orchestra may generate instantaneous voltages that vary by more than four orders of magnitude, demanding that subsequent amplifiers and filters handle large swings (and/or achieve a low noise).

The need for large output swings has made fully differential op amps quite popular. Similar to the circuits described in Chapter 4, such op amps generate "complementary" outputs, roughly doubling the available swing. Nonetheless, as mentioned in Chapters 3 and 4 and explained later in this chapter, the maximum voltage swing trades with device size and bias currents and hence speed. Achieving large swings is the principal challenge in today's op amp design.

⊕ . max voltage swings trades with device size, bias currents and hence speed.

Linearity   Open-loop op amps suffer from substantial nonlinearity. In the circuit of Fig. 9.1, for example, the input pair M_1-M_2 exhibits a nonlinear relationship between its differential drain current and input voltage. As explained in Chapter 13, the issue of nonlinearity is tackled by two approaches: using fully differential implementations to suppress even-order harmonics and allowing sufficient open-loop gain such that the closed-loop feedback system achieves adequate linearity. It is interesting to note that in many feedback circuits, the linearity requirement, rather than the gain error requirement, governs the choice of the open-loop gain.

[handwritten in margin: linearity requirement governs the o/o. the o/R open loop gain]

Noise and Offset   The input noise and offset of op amps determine the minimum signal level that can be processed with reasonable quality. In a typical op amp topology, several devices contribute noise and offset, necessitating large dimensions or bias currents. For example, in the circuit of Fig. 9.1, M_1-M_2 and M_7-M_8 contribute the most.

We should also recognize a trade-off between noise and *output swing*. For a given bias current, as the overdrive voltage of M_7 and M_8 in Fig. 9.1 is lowered to allow larger swings at the output, their transconductance increases and so does their drain noise current.

Supply Rejection   Op amps are often employed in mixed-signal systems and sometimes connected to noisy digital supply lines. Thus, the performance of op amps in the presence of supply noise, especially as the noise frequency increases, is quite important. For this reason, fully differential topologies are preferred.

9.2 One-Stage Op Amps

All of the differential amplifiers studied in Chapters 4 and 5 can be considered as op amps. Fig. 9.6 shows two such topologies with single-ended and differential outputs. The small-signal, low-frequency gain of both circuits is equal to $g_{mN}(r_{ON}\|r_{OP})$, where the subscripts N and P denote NMOS and PMOS, respectively. This value hardly exceeds 20 in submicron devices with typical current levels. The bandwidth is usually determined by the load capacitance, C_L. Note that the circuit of Fig. 9.6(a) exhibits a mirror pole (Chapter 5)

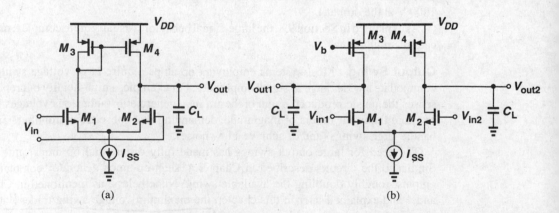

Figure 9.6   Simple op amp topologies.

whereas that of Fig. 9.6(b) does not, a critical difference in terms of the stability of feedback systems using these topologies (Chapter 10).

The circuits of Fig. 9.6 suffer from noise contributions of M_1-M_4, as calculated in Chapter 7. Interestingly, in all op amp topologies, at least four devices contribute to the input noise: two input transistors and two "load" transistors.

Example 9.3

Calculate the input common-mode voltage range and the closed-loop output impedance of the unity-gain buffer depicted in Fig. 9.7.

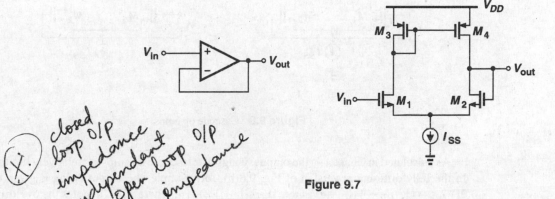

Figure 9.7

Solution

The minimum allowable input voltage is equal to $V_{CSS} + V_{GS1}$, where V_{CSS} is the voltage required across the current source. The maximum voltage is given by the level that places M_1 at the edge of the triode region: $V_{in,max} = V_{DD} - |V_{GS3}| + V_{TH1}$. For example, if each device (including the current source) has a threshold voltage of 0.7 V and an overdrive of 0.3 V, then $V_{in,min} = 0.3 + 0.3 + 0.7 = 1.3$ V and $V_{in,max} = 3 - (0.3 + 0.7) + 0.7 = 2.7$ V. Thus, the input CM range equals 1.4 V with a 3-V supply.

Since the circuit employs voltage feedback at the output, the output impedance is equal to the open-loop value, $r_{OP}\|r_{ON}$, divided by one plus the loop gain, $1 + g_{mN}(r_{OP}\|r_{ON})$. In other words, for large open-loop gain, the closed-loop output impedance is approximately equal to $(r_{OP}\|r_{ON})/[g_{mN}(r_{OP}\|r_{ON})] = 1/g_{mN}$.

It is interesting to note that the closed-loop output impedance is relatively *independent* of the open-loop output impedance. This is an important observation, allowing us to design high-gain op amps by *increasing* the open-loop output impedance while still achieving a relatively low closed-loop output impedance.

In order to achieve a high gain, the differential cascode topologies of Chapters 4 and 5 can be used. Shown in Figs. 9.8(a) and (b) for single-ended and differential output generation, respectively, such circuits display a gain on the order of $g_{mN}[(g_{mN}r_{ON}^2)\|(g_{mP}r_{OP}^2)]$, but at the cost of output swing and additional poles. These configurations are also called "telescopic" cascode op amps to distinguish them from another cascode op amp described below. The circuit providing a single-ended output suffers from a mirror pole at node X, creating stability issues (Chapter 10).

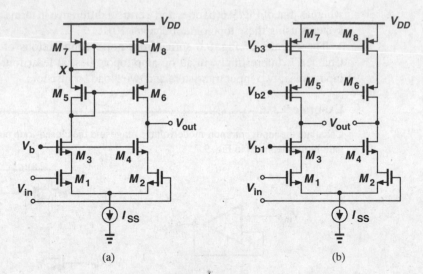

Figure 9.8 Cascode op amps.

As calculated in Chapter 4 , the output swings of telescopic op amps are relatively limited. In the fully differential version of Fig. 9.8(b), for example, the output swing is given by $2[V_{DD} - (V_{OD1} + V_{OD3} + V_{CSS} + |V_{OD5}| + |V_{OD7}|)]$, where V_{ODj} denotes the overdrive voltage of M_j.

Another drawback of telescopic cascodes is the difficulty in shorting their inputs and outputs, e.g., to implement a unity-gain buffer similar to the circuit of Fig. 9.7. To understand the issue, let us consider the unity-gain feedback topology shown in Fig. 9.9. Under what conditions are both M_2 and M_4 in saturation? We must have $V_{out} \leq V_X + V_{TH2}$ and

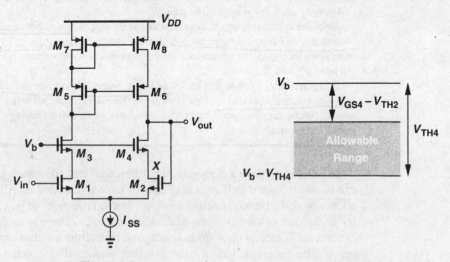

Figure 9.9 Cascode op amp with input and output shorted.

$V_{out} \geq V_b - V_{TH4}$. Since $V_X = V_b - V_{GS4}$, $V_b - V_{TH4} \leq V_{out} \leq V_b - V_{GS4} + V_{TH2}$. Depicted in Fig. 9.9, this voltage range is simply equal to $V_{max} - V_{min} = V_{TH4} - (V_{GS4} - V_{TH2})$, maximized by minimizing the overdrive of M_4 but always less than V_{TH2}.

Example 9.4

For the circuit of Fig. 9.9, explain in which region each transistor operates as V_{in} varies from below $V_b - V_{TH4}$ to above $V_b - V_{GS4} + V_{TH2}$.

Solution

Since the op amp attempts to force V_{out} to be equal to V_{in}, for $V_{in} < V_b - V_{TH4}$, we have $V_{out} \approx V_{in}$ and M_4 is in the triode region while other transistors are saturated. Under this condition, the open-loop gain of the op amp is reduced. As V_{in} and hence V_{out} exceed $V_b - V_{TH4}$, M_4 enters saturation and the open-loop gain reaches a maximum. For $V_b - V_{TH4} < V_{in} < V_b - (V_{GS4} - V_{TH2})$, both M_2 and M_4 are saturated and for $V_{in} > V_b - (V_{GS4} - V_{TH2})$, M_2 and M_1 enter the triode region, degrading the gain.

While a cascode op amp is rarely used as a unity-gain buffer, some other topologies such as the switched-capacitor circuits of Chapter 12 require that the input and output of the op amp be shorted for part of the operation period.

At this point, the reader may wonder how exactly we design an op amp. With so many devices and performance parameters, it may not be clear where the starting point is and how the numbers are chosen. Indeed, the actual design methodology of an op amp somewhat depends on the specifications that the circuit must meet. For example, a high-gain op amp may be designed quite differently from a low-noise op amp. Nevertheless, in most cases, some aspects of the performance, e.g., output voltage swings and open-loop gain, are of primary concern, pointing to a specific design procedure. The following example illustrates these ideas.

Example 9.5

Design a fully differential telescopic op amp with the following specifications: $V_{DD} = 3$ V, differential output swing = 3 V, power dissipation = 10 mW, voltage gain = 2000. Assume $\mu_n C_{ox} = 60$ μA/V^2, $\mu_p C_{ox} = 30$ μA/V^2, $\lambda_n = 0.1$ V^{-1}, $\lambda_p = 0.2$ V^{-1} (for an effective channel length of 0.5 μm), $\gamma = 0$, $V_{THN} = |V_{THP}| = 0.7$ V.

Solution

Fig. 9.10 shows the op amp topology along with two current mirrors defining the drain currents of M_7-M_9. We begin with the power budget, allocating 3 mA to M_9 and the remaining 330 μA to M_{b1} and M_{b2}. Thus, each cascode branch of the op amp carries a current of 1.5 mA. Next, we consider the required output swings. Each of nodes X and Y must be able to swing by 1.5 V without driving M_3-M_6 into the triode region. With a 3-V supply, therefore, the total voltage available for M_9 and each cascode branch is equal to 1.5 V, i.e., $|V_{OD7}| + |V_{OD5}| + V_{OD3} + V_{OD1} + V_{OD9} = 1.5$ V. Since M_9 carries the largest current, we choose $V_{OD9} \approx 0.5$ V, leaving 1 V for the four transistors in the cascode. Moreover, since M_5-M_8 suffer from low mobility, we allocate an overdrive of approximately 300 mV to each, obtaining 400 mV for $V_{OD1} + V_{OD3}$. As an initial guess, $V_{OD1} = V_{OD3} = 200$ mV.

here we assume (or) allot more voltage for the foot transistor M_9, since that takes maximum current in the circuit.

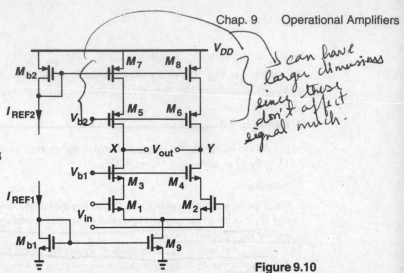

↳ can have larger dimensions since these don't affect signal much.

Figure 9.10

To minimize the device capacitance we ~~reduce the~~ choose minimum length for each transistor.

✳ To increase gain we can increase W(or)length (or) decrease I_D. for det.

With the bias current and overdrive voltage of each transistor known, we can easily determine the aspect ratios from $I_D = (1/2)\mu C_{ox}(W/L)(V_{GS} - V_{TH})^2$. To minimize the device capacitances, we choose the minimum length for each transistor, obtaining a corresponding width. We then have $(W/L)_{1-4} = 1250$, $(W/L)_{5-8} = 1111$, $(W/L)_9 = 400$.

The design has thus far satisfied the swing, power dissipation, and supply voltage specifications. But, how about the gain? Using $A_v \approx g_{m1}[(g_{m3}r_{O3}r_{O1})\|(g_{m5}r_{O5}r_{O7})]$ and assuming minimum channel length for all of the transistors, we have $A_v = 1416$, quite lower than the required value.

In order to increase the gain, we recognize that $g_m r_O = \sqrt{2\mu C_{ox}(W/L)I_D}/(\lambda I_D)$. Now, recall that $\lambda \propto 1/L$, and hence $g_m r_O \propto \sqrt{WL/I_D}$. We can therefore increase the width or length or *decrease* the bias current of the transistors. In practice, speed or noise requirements may dictate the bias current, leaving only the dimensions as the variables. Of course, the width of each transistor must at least scale with its length so as to maintain a constant overdrive voltage.

Which transistors in the circuit of Fig. 9.10 should be made longer? Since M_1-M_4 appear in the signal path, it is desirable to keep their capacitances to a minimum. The PMOS devices, M_5-M_8, on the other hand, affect the signal to a much lesser extent and can therefore have larger dimensions.[2] Doubling the (effective) length and width of each of these transistors in fact *doubles* their $g_m r_O$ because g_m remains constant while r_O increases by a factor of 2. Choosing $(W/L)_{5-8} = 1111\,\mu m/1.0\,\mu m$ and hence $\lambda_p = 0.1\,V^{-1}$, we obtain $A_v \approx 4000$. Thus, the PMOS dimensions can be somewhat smaller. Note that with such large dimensions for PMOS transistors, we may revisit our earlier distribution of the overdrive voltages, possibly reducing that of M_9 by 100 to 200 mV and allocating more to the PMOS devices.

In the op amp of Fig. 9.10, the input CM level and the bias voltages V_{b1} and V_{b2} must be chosen so as to allow maximum output swings. The minimum allowable input CM level equals $V_{GS1} + V_{OD9} = V_{TH1} + V_{OD1} + V_{OD9} = 1.4$ V. The minimum value of V_{b1} is given by $V_{GS3} + V_{OD1} + V_{OD9} = 1.6$ V, placing M_1-M_2 at the edge of the triode region. Similarly, $V_{b2,max} = V_{DD} - (|V_{GS5}| + |V_{OD7}|) = 1.7$ V. In practice, some margin must be included in the value of V_{b1} and V_{b2} to allow for process variations. Also, the increase in the threshold voltages due to body effect must be taken into account.

---

[2]This point is studied in Chapter 10.

Transistors in the signal path should have minimum capacitance, so ~~don't change~~ keep their lengths minimum.

In order to alleviate the drawbacks of telescopic cascode op amps, namely, limited output swings and difficulty in shorting the input and output, a "folded cascode" op amp can be used. As described in Chapter 3 and illustrated in Fig. 9.11, in an NMOS or PMOS cascode amplifier, the input device is replaced by the opposite type while still converting the

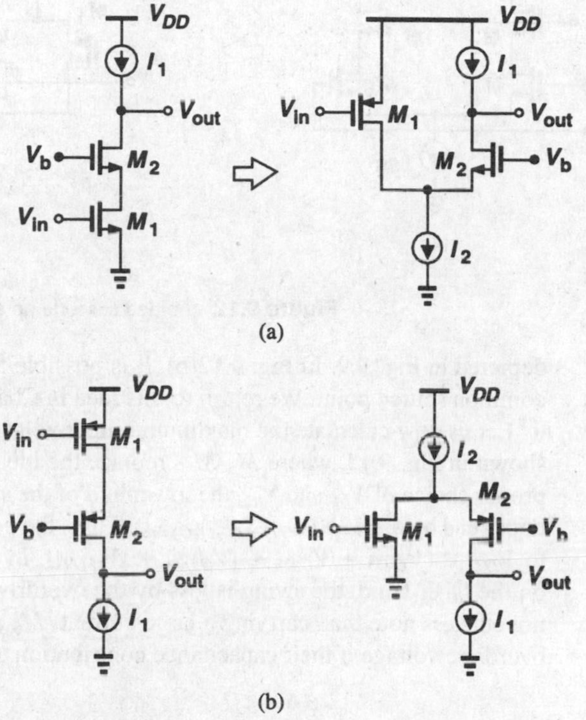

(a)

(b)

Figure 9.11   Folded cascode circuits.

input voltage to a current. In the four circuits shown in Fig. 9.11, the small-signal current generated by M_1 flows through M_2 and subsequently the load, producing an output voltage approximately equal to $g_{m1} R_{out} V_{in}$. The primary advantage of the folded structure lies in the choice of the voltage levels because it does not "stack" the cascode transistor on top of the input device. We will return to this point later.

The folding idea depicted in Fig. 9.11 can easily be applied to differential pairs and hence operational amplifiers as well. Shown in Fig. 9.12, the resulting circuit replaces the input NMOS pair with a PMOS counterpart. Note two important differences between the two circuits. (1) In Fig. 9.12(a), one bias current, I_{SS}, provides the drain current of both the input transistors and the cascode devices, whereas in Fig. 9.12(b) the input pair requires an additional bias current. In other words, $I_{SS1} = I_{SS}/2 + I_{D3}$. Thus, the folded-cascode configuration generally consumes higher power. (2) In Fig. 9.12(a), the input CM level cannot exceed $V_{b1} - V_{GS3} + V_{TH1}$, whereas in Fig. 9.12(b), it cannot be *less* than $V_{b1} - V_{GS3} + |V_{THP}|$. It is therefore possible to design the latter to allow shorting its input and output terminals with negligible swing limitation. This is in contrast to the behavior

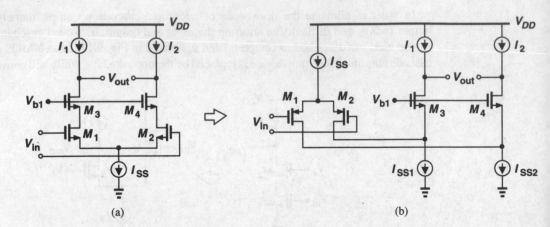

Figure 9.12 Folded cascode op amp topology.

depicted in Fig. 9.9. In Fig. 9.12(b), it is possible to tie the n-well of M_1 and M_2 to their common source point. We return to this idea in Chapters 13 and 18.

Let us now calculate the maximum output voltage swing of the folded-cascode op amp shown in Fig. 9.13, where M_5-M_{10} replace the ideal current sources of Fig. 9.12(b). With proper choice of V_{b1} and V_{b2}, the lower end of the swing is given by $V_{OD3} + V_{OD5}$ and the upper end by $V_{DD} - (|V_{OD7}| + |V_{OD9}|)$. Thus, the peak-to-peak swing on each side is equal to $V_{DD} - (V_{OD3} + V_{OD5} + |V_{OD7}| + |V_{OD9}|)$. In the telescopic cascode of Fig. 9.12(a), on the other hand, the swing is less by the overdrive of the tail current source. We should nonetheless note that, carrying a large current, M_5 and M_6 in Fig. 9.13 may require a high overdrive voltage if their capacitance contribution to nodes X and Y is to be minimized.

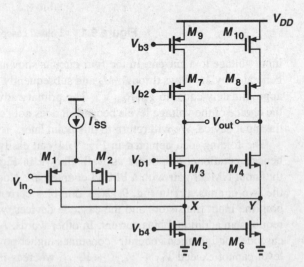

Figure 9.13 Folded cascode op amp with cascode PMOS loads.

We now determine the small-signal voltage gain of the folded-cascode op amp of Fig. 9.13. Using the half circuit depicted in Fig. 9.14(a) and writing $|A_v| = G_m R_{out}$, we must calculate G_m and R_{out}. As shown in Fig. 9.14(b), the output short-circuit current is approximately equal to the drain current of M_1 because the impedance seen looking into the source of M_3, that is, $(g_{m3} + g_{mb3})^{-1} \| r_{O3}$, is typically much lower than $r_{O1} \| r_{O5}$. Thus, $G_m \approx g_{m1}$. To calculate R_{out}, we use Fig. 9.14(c), with $R_{OP} \approx (g_{m7} + g_{mb7}) r_{O7} r_{O9}$, to write $R_{out} \approx R_{OP} \| [(g_{m3} + g_{mb3}) r_{O3} (r_{O1} \| r_{O5})]$. It follows that

$$|A_v| \approx g_{m1} \{ [(g_{m3} + g_{mb3}) r_{O3} (r_{O1} \| r_{O5})] \| [(g_{m7} + g_{mb7}) r_{O7} r_{O9}] \}. \tag{9.12}$$

How does this value compare with the gain of a telescopic op amp? For comparable device dimensions and bias currents, the PMOS input differential pair exhibits a lower transconductance than does an NMOS pair. Furthermore, r_{O1} and r_{O5} appear in parallel,

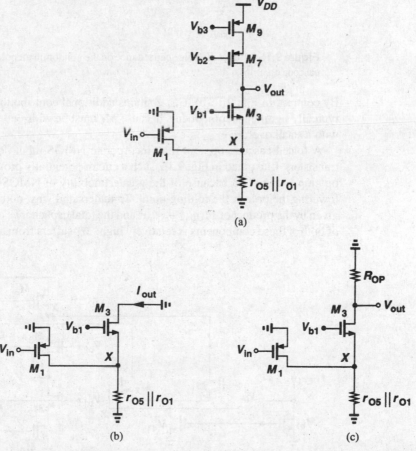

Figure 9.14   (a) Half circuit of folded cascode op amp, (b) equivalent circuit with output shorted to ground, (c) equivalent circuit with output open.

reducing the output impedance, especially because M_5 carries the currents of both the input device and the cascode branch. As a consequence, the gain in (9.12) is usually two to three times lower than that of a comparable telescopic cascode.

It is also worth noting that the pole at the "folding point," i.e., the sources of M_3 and M_4, is quite closer to the origin than that associated with the source of cascode devices in a telescopic topology. In Fig. 9.15(a), C_{tot} arises from C_{GS3}, C_{SB3}, C_{DB1}, and C_{GD1}.

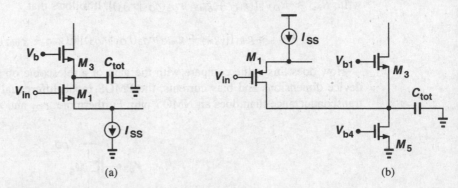

Figure 9.15 Effect of device capacitance on the nondominant pole in telescopic and folded-cascode op amps.

By contrast, in Fig. 9.15(b), C_{tot} contains additional contributions due to C_{GD5} and C_{DB5}, typically significant components because M_5 must be wide enough to carry a large current with a small overdrive.

A folded-cascode op amp may incorporate NMOS input devices and PMOS cascode transistors. Illustrated in Fig. 9.16, such a circuit potentially provides a higher gain than the op amp of Fig. 9.13 because of the greater mobility of NMOS devices, but at the cost of lowering the pole at the folding point. To understand why, note that the pole at node X is given by the product of $1/(g_{m3} + g_{mb3})$ and the total capacitance at this node. The magnitude of both of these components is relatively high: M_3 suffers from a low transconductance and

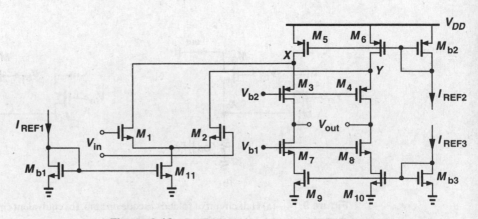

Figure 9.16 Realization of a folded-cascode op amp.

M_5 contributes substantial capacitance because it must be wide enough to carry the drain currents of both M_1 and M_3. In fact, for comparable bias currents, M_5-M_6 in Fig. 9.16 may be several times wider than M_5-M_6 in Fig. 9.13.

Our study thus far suggests that the overall voltage swing of a folded-cascode op amp is only slightly higher than that of a telescopic configuration. This advantage comes at the cost of higher power dissipation, lower voltage gain, lower pole frequencies, and, as explained in Section 9.10, higher noise. Nonetheless, folded-cascode op amps are used quite widely, even more than telescopic topologies, because the inputs and outputs can be shorted together and the choice of the input common-mode level is easier. In a telescopic op amp, *three* voltages must be defined carefully: the input CM level and the gate bias voltages of the PMOS and NMOS cascode transistors, whereas in folded-cascode configurations only the latter two are critical.

We now carry out the design of a folded-cascode op amp to reinforce the foregoing concepts.

Example 9.6

Design a folded-cascode op amp with an NMOS input pair (Fig. 9.16) to satisfy the following specifications: $V_{DD} = 3$ V, differential output swing $= 3$ V, power dissipation $= 10$ mW, voltage gain $= 2000$. Use the same device parameters as in Example 9.5.

Solution

As with the telescopic cascode of the previous example, we begin with the power and swing specifications. Allocating 1.5 mA to the input pair, 1.5 mA to the two cascode branches, and the remaining 330 μA to the three current mirrors, we first consider the devices in each cascode branch. Since M_5 and M_6 must each carry 1.5 mA, we allow an overdrive of 500 mV for these transistors so as to keep their width to a reasonable value. To M_3-M_4, we allocate 400 mV and to M_7-M_{10}, 300 mV. Thus, $(W/L)_{5,6} = 400, (W/L)_{3,4} = 313, (W/L)_{7-10} = 278$. Since the minimum and maximum output levels are equal to 0.6 V and 2.1 V, respectively, the optimum output common-mode level is 1.35 V.

The minimum dimensions of M_1-M_2 are dictated by the minimum input common-mode level, $V_{GS1} + V_{OD11}$. For example, if the input and the output are shorted for part of the operation period (Fig. 9.17), then $V_{GS2} + V_{OD11} = 1.35$ V. With $V_{OD11} = 0.4$ V as an initial guess, we have $V_{GS1} =$

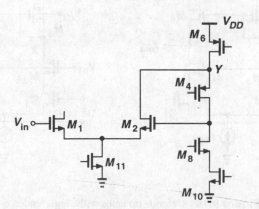

Figure 9.17 Folded-cascode op amp with input and output shorted.

0.95 V, obtaining $V_{OD1,2} = 0.95 - 0.7 = 0.25$ V and hence $(W/L)_{1,2} = 400$. The maximum dimensions of M_1 and M_2 are determined by the tolerable input capacitance and the capacitance at nodes X and Y in Fig. 9.16.

We now calculate the small-signal gain. Using $g_m = 2I_D/(V_{GS} - V_{TH})$, we have $g_{m1,2} = 0.006$ A/V, $g_{m3,4} = 0.0038$ A/V, and $g_{m7,8} = 0.05$ A/V. For $L = 0.5$ μm, $r_{O1,2} = r_{O7-10} = 13.3$ kΩ, and $r_{O3,4} = 2r_{O5,6} = 6.67$ kΩ. It follows that the impedance seen looking into the drain of M_7 (or M_8) is equal to 8.8 MΩ whereas, owing to the limited intrinsic gain of M_3 (or M_4), that seen looking into the drain of M_3 is equal to 66.5 kΩ. The overall gain is therefore limited to about 400.

In order to increase the gain, we first observe that $r_{O5,6}$ is quite lower than $r_{O1,2}$. Thus, the length of M_5-M_6 must be increased. Also, the transconductance of M_1-M_2 is relatively low and can be increased by widening these transistors. Finally, we may decide to double the intrinsic gain of M_3 and M_4 by doubling both their length and width, but at the cost of increasing the capacitance at nodes X and Y. We leave the exact choice of the device dimensions as an exercise for the reader.

An important property of folded-cascode op amps is the capability of handling input common-mode levels close to one of the supply rails. In Fig. 9.16, for example, the CM voltage at the gates of M_1 and M_2 can be equal to V_{DD} because $V_X = V_Y = V_{DD} - 500$ mV. By the same token, a similar topology using a PMOS input pair can accommodate input CM levels as low as zero.

Telescopic and folded-cascode op amps can also be designed to provide a single-ended output. Shown in Fig. 9.18(a) is an example, where a PMOS cascode current mirror converts the differential currents of M_3 and M_4 to a single-ended output voltage. In this implementation, however, $V_X = V_{DD} - |V_{GS5}| - |V_{GS7}|$, limiting the maximum value of V_{out} to $V_{DD} - |V_{GS5}| - |V_{GS7}| + |V_{TH6}|$ and "wasting" one PMOS threshold voltage in the swing (Chapter 5). To resolve this issue, the PMOS load can be modified as shown in Fig. 9.18(b).

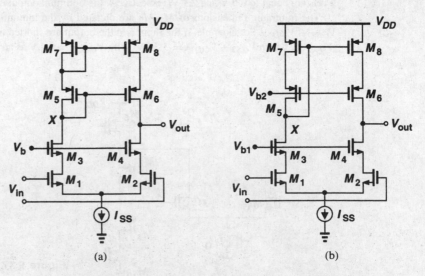

Figure 9.18   Cascode op amps with single-ended output.

so that M_7 and M_8 are biased at the edge of the triode region. Similar ideas apply to folded-cascode op amps as well.

The circuit of Fig. 9.18(a) suffers from two disadvantages with respect to its differential counterpart in Fig. 9.8(b). First, it provides only half the output voltage swing. Second, it contains a mirror pole at node X (Chapter 5), thus limiting the speed of feedback systems employing such an amplifier. It is therefore preferable to use the differential topology, although it requires a feedback loop to define the output common-mode level (Section 9.6).

As a final note, we recognize that to achieve a higher gain, additional cascode devices can be inserted in each branch. Shown in Fig. 9.19 is a "triple cascode," providing a gain on

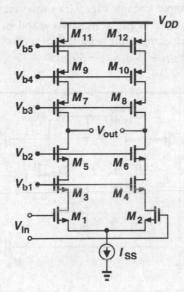

Figure 9.19 Triple-cascode op amp.

the order of $(g_m r_O)^3/2$ but further limiting the output swings. With six overdrive voltages subtracted from V_{DD} in this circuit, it is difficult to operate the amplifier from a supply voltage of 3 V or lower while obtaining reasonable output swings.

9.3 Two-Stage Op Amps

The op amps studied thus far exhibit a "one-stage" nature in that they allow the small-signal current produced by the input pair to flow directly through the output impedance. The gain of these topologies is therefore limited to the product of the input pair transconductance and the output impedance. We have also observed that cascoding in such circuits increases the gain while limiting the output swings.

In some applications, the gain and/or the output swings provided by cascode op amps are not adequate. For example, an op amp used in a hearing aid must operate with supply voltages as low as 0.9 V while delivering single-ended output swings as large as 0.5 V.

In such cases, we resort to "two-stage" op amps, with the first stage providing a high gain and the second, large swings (Fig. 9.20). In contrast to cascode op amps, a two-stage configuration isolates the gain and swing requirements.

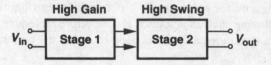

Figure 9.20   Two-stage op amp.

Each stage in Fig. 9.20 can incorporate various amplifier topologies studied in previous sections, but the second stage is typically configured as a simple common-source stage so as to allow maximum output swings. Fig. 9.21 shows an example, where the first and second stages exhibit gains equal to $g_{m1,2}(r_{O1,2} \| r_{O3,4})$ and $g_{m5,6}(r_{O5,6} \| r_{O7,8})$, respectively. The overall gain is therefore comparable with that of a cascode op amp, but the swing at V_{out1} and V_{out2} is equal to $V_{DD} - |V_{OD5,6}| - V_{OD7,8}$.

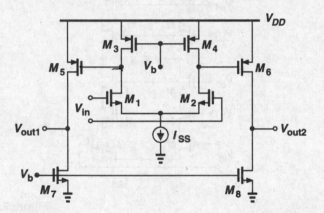

Figure 9.21   Simple implementation of a two-stage op amp.

To obtain a higher gain, the first stage can incorporate cascode devices, as depicted in Fig. 9.22. With a gain of, say, 10 in the output stage, the voltage swings at X and Y are quite small, allowing optimization of M_1-M_8 for higher gain. The overall voltage gain can be expressed as

$$A_v \approx \{g_{m1,2}[(g_{m3,4} + g_{mb3,4})r_{O3,4}r_{O1,2}] \| [(g_{m5,6} + g_{mb5,6})r_{O5,6}r_{O7,8}]\}$$
$$\times [g_{m9,10}(r_{O9,10} \| r_{O11,12})]. \tag{9.13}$$

A two-stage op amp may provide a single-ended output. One method is to convert the differential currents of the two output stages to a single-ended voltage. Illustrated in Fig. 9.23, this approach maintains the differential nature of the first stage, using only the current mirror M_7-M_8 to generate a single-ended output. Note, however, that if the gate of M_1 is shorted to V_{out2} to form a unity-gain buffer, then the minimum allowable output level is equal to $V_{GS1} + V_{ISS}$, severely limiting the output swing.

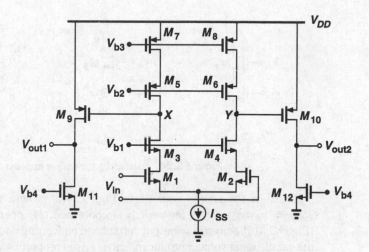

Figure 9.22 Two-stage op amp employing cascoding.

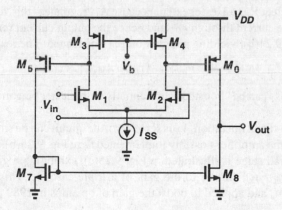

Figure 9.23 Two-stage op amp with single-ended output.

Can we cascade more than two stages to achieve a higher gain? As explained in Chapter 10, each gain stage introduces at least one pole in the open-loop transfer function, making it difficult to guarantee stability in a feedback system using such an op amp. For this reason, op amps having more than two stages are rarely used. Exceptions are described in [1, 2, 3].

9.4 Gain Boosting

The limited gain of one-stage op amps studied in Section 9.2 and the difficulties in using two-stage op amps at high speeds have motivated extensive work on new topologies. Recall that in one-stage op amps such as telescopic and folded-cascode topologies the objective is to maximize the output impedance so as to attain a high voltage gain. The idea behind gain boosting is to further increase the output impedance without adding more cascode devices.

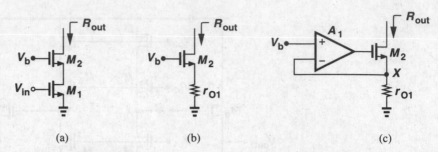

Figure 9.24 Increasing the output impedance by feedback.

Consider the simple cascode in Fig. 9.24(a), whose output impedance is given by $R_{out} = g_{m2}r_{O2}r_{O1}$. As far as R_{out} is concerned, M_1 operates as a degeneration resistor [Fig. 9.24(b)], sensing the output current and converting it to a voltage. The observation that the small-signal voltage produced across r_{O1} is proportional to the output current suggests that this voltage can be subtracted from V_b so as to place M_2 in current-voltage feedback, thereby increasing the output impedance. Illustrated in Fig. 9.24(c), the idea is to drive the gate of M_2 by an amplifier that forces V_X to be equal to V_b. Thus, voltage variations at the drain of M_2 now affect V_X to a lesser extent because A_1 "regulates" this voltage. With smaller variations at X, the current through r_{O1} and hence the output current remain more constant than those in Fig. 9.24(b), yielding a higher output impedance. The reader can prove that

$$R_{out} \approx A_1 g_{m2}r_{O2}r_{O1}, \tag{9.14}$$

concluding that R_{out} can be "boosted" substantially without stacking more cascode devices on top of M_2.

Since for small-signal operation, V_b is set to zero, the circuit can be simplified as shown in Fig. 9.25(a), with the amplifier possibly implemented as in Fig. 9.25(b). Called a "regulated cascode," the overall stage is illustrated in Fig. 9.25(c), exhibiting a gain equal to $|A_v| \approx g_{m1}(g_{m2}r_{O2}r_{O1})(g_{m3}r_{O3})$, similar to the gain of a *triple* cascode. This topology was first invented in 1976 [4] and applied to boost the gain of op amps in 1989 [5, 6].

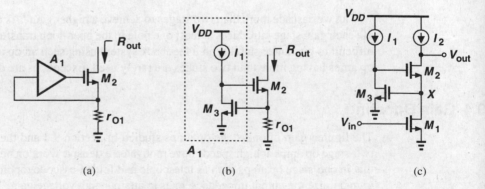

Figure 9.25 Gain boosting in cascode stage.

Before incorporating the technique of Fig. 9.25(c) in an op amp, let us examine the output voltage swings, in particular, the minimum allowable level. Since $V_X = V_{GS3}$, the minimum value of V_{out} is $V_{OD2} + V_{GS3}$, whereas, in a simple cascode with proper choice of V_{G2}, it would be $V_{OD2} + V_{OD1}$. Thus, the auxiliary amplifier in this case limits the output swing.

We now apply gain boosting to a differential cascode stage, as shown in Fig. 9.26(a). Since the signals at nodes X and Y are differential, we surmise that the two single-ended gain boosting amplifiers A_1 and A_2 can be replaced by one differential amplifier [Fig. 9.26(b)]. Following the topology of Fig. 9.25(c), we implement the differential auxiliary amplifier as shown in Fig. 9.26(c), but noting that the minimum level at the drain of M_3 is equal to $V_{OD3} + V_{GS5} + V_{ISS2}$, where V_{ISS2} denotes the voltage required across I_{SS2}. In a simple differential cascode, on the other hand, the minimum would be approximately one threshold voltage lower.

The voltage swing limitation in Fig. 9.26(c) results from the fact that the gain-boosting amplifier incorporates an NMOS differential pair. If nodes X and Y are sensed by a PMOS pair, the minimum value of V_X and V_Y is not dictated by the gain-boosting amplifier. Now

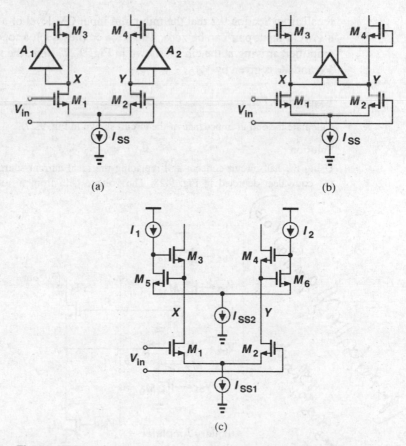

Figure 9.26   Boosting the output impedance of a differential cascode stage.

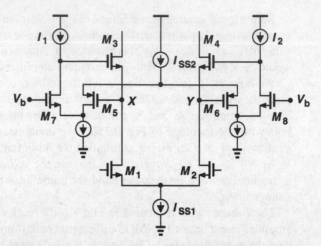

Figure 9.27 Folded-cascode circuit used as auxiliary amplifier.

recall from Section 9.2 that the minimum input CM level of a folded-cascode stage using a PMOS input pair can be zero. Thus, we employ such a topology for the gain-boosting amplifier, arriving at the circuit shown in Fig. 9.27. Here, the minimum allowable level of V_X and V_Y is given by $V_{OD1,2} + V_{ISS1}$.

Example 9.7

Calculate the output impedance of the circuit shown in Fig. 9.27.

Solution

Using the half-circuit concept and replacing the ideal current sources with transistors, we obtain the equivalent depicted in Fig. 9.28. The voltage gain from X to P is approximately equal to

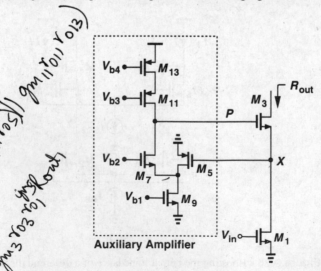

Auxiliary Amplifier

Figure 9.28

$R_{out_1} \approx (g_{m7} r_{07} (g_9 || r_{05}) || (r_{07} || r_{09}) g_{m11} || r_{011}) r_{013})$

$R_{out} = g_{m3} r_{03} r_{01} g_{m} R_{out}$

$g_{m5}R_{out1}$, where $R_{out1} \approx [g_{m7}r_{O7}(r_{O9}\|r_{O5})]\|(g_{m11}r_{O11}r_{O13})$. Thus, $R_{out} \approx g_{m3}r_{O3}r_{O1}g_{m5}R_{out1}$. In essence, since the output impedance of a cascode is boosted by a folded-cascode stage, the overall output impedance is similar to that of a "quadruple" cascode.

Regulated cascodes can also be utilized in the load current sources of a cascode op amp. Shown in Fig. 9.29(a), such a topology boosts the output impedance of the PMOS current sources as well, thereby achieving a very high voltage gain. To allow maximum swings at the output, amplifier A_2 must employ an NMOS input differential pair. Similar ideas apply to folded-cascode op amps [Fig. 9.29(b)].

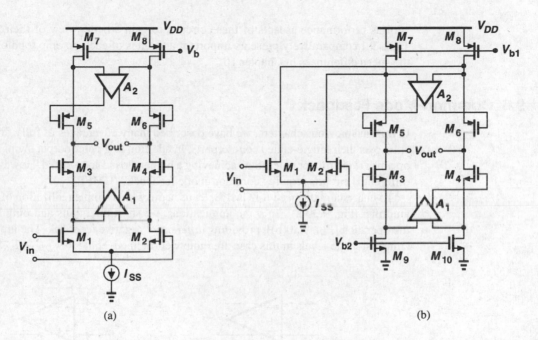

Figure 9.29   Gain boosting applied to both signal path and load devices.

Now recall that the premise behind gain boosting is to increase the gain without adding a second stage or more cascode devices. Does this mean that the op amps of Fig. 9.29 have a one-stage nature? After all, the gain-boosting amplifier introduces its own poles. In contrast to two-stage op amps, where the entire signal experiences the poles associated with each stage, in a gain-boosted op amp, most of the signal directly flows through the cascode devices to the output. Only a small "error" component is processed by the gain-boosting amplifier and "slowed down."

9.5   Comparison

Our study of op amps in this chapter has introduced four principal topologies: telescopic cascode, folded cascode, two-stage op amp, and gain boosting. It is instructive to compare

Table 9.1 Comparison of performance of various op amp topologies.

	Gain	Output Swing	Speed	Power Dissipation	Noise
Telescopic	Medium	Medium	Highest	Low	Low
Folded-Cascode	Medium	Medium	High	Medium	Medium
Two-Stage	High	Highest	Low	Medium	Low
Gain-Boosted	High	Medium	Medium	High	Medium

various performance aspects of these circuits to gain a better view of their applicability. Table 9.1 comparatively presents important attributes of each op amp topology. We study the speed differences in Chapter 10.

9.6 Common-Mode Feedback

In this and previous chapters, we have described many advantages of fully differential circuits over their single-ended counterparts. In addition to greater output swings, differential op amps avoid mirror poles, thus achieving a higher closed-loop speed. However, high-gain differential circuits require "common-mode feedback" (CMFB).

To understand the need for CMFB, let us begin with a simple realization of a differential amplifier [Fig. 9.30(a)]. In some applications, we short the inputs and outputs for part of the operation [Fig. 9.30(b)], providing *differential* negative feedback. The input and output common-mode levels in this case are quite well-defined, equal to $V_{DD} - I_{SS}R_D/2$.

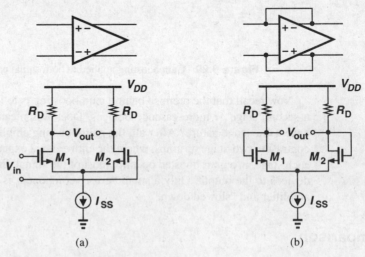

Figure 9.30   (a) Simple differential pair, (b) circuit with inputs shorted to outputs.

Now suppose the load resistors are replaced by PMOS current sources so as to increase the differential voltage gain [Fig. 9.31(a)]. What is the common-mode level at nodes X

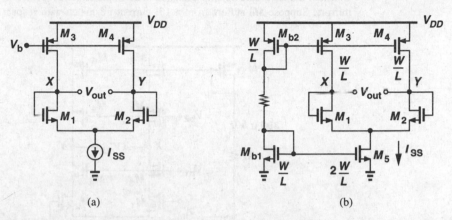

Figure 9.31 (a) High-gain differential pair with inputs shorted to outputs, (b) effect of current mismatches.

and Y? Since each of the input transistors carries a current of $I_{SS}/2$, the CM level depends on how close I_{D3} and I_{D4} are to this value. In practice, as exemplified by Fig. 9.31(b), mismatches in the PMOS and NMOS current mirrors defining I_{SS} and $I_{D3,4}$ create a finite error between $I_{D3,4}$ and $I_{SS}/2$. Suppose, for example, that the drain currents of M_3 and M_4 in the saturation region are slightly greater than $I_{SS}/2$. As a result, to satisfy Kirchoff current law at nodes X and Y, both M_3 and M_4 must enter the triode region so that their drain currents fall to $I_{SS}/2$. Conversely, if $I_{D3,4} < I_{SS}/2$, then both V_X and V_Y must drop so that M_5 enters the triode region, thereby producing only $2I_{D3,4}$.

The above difficulties fundamentally arise because in high-gain amplifiers, we wish a p-type current source to balance an n-type current source. As illustrated in Fig. 9.32, the difference between I_P and I_N must flow through the intrinsic output impedance of the

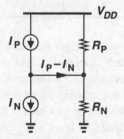

Figure 9.32 Simplified model of high-gain amplifier.

→ high value.

amplifier, creating an output voltage change of $(I_P - I_N)(R_P \| R_N)$. Since the current error depends on mismatches and $R_P \| R_N$ is quite high, the voltage error may be large, thus driving the p-type or n-type current source into the triode region. As a general rule, if the output CM level cannot be determined by "visual inspection" and requires calculations based on device properties, then it is poorly defined. This is the case in Fig. 9.31 but not in Fig. 9.30. We emphasize that differential feedback cannot define the CM level.

Example 9.8

Consider the telescopic op amp designed in Example 9.5 and repeated in Fig. 9.33 with bias current mirrors. Suppose M_9 suffers from a 1% current mismatch with respect to M_{10}, producing $I_{SS} =$

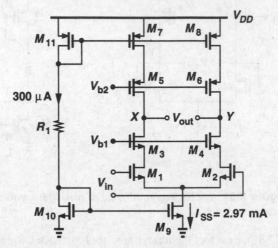

Figure 9.33

2.97 mA rather than 3 mA. Assuming perfect matching for other transistors, explain what happens in the circuit.

Solution

From Example 9.5, the single-ended output impedance of the circuit equals 266 kΩ. Since the difference between the drain currents of M_3 and M_5 (and M_4 and M_6) is 30 μA/2 = 15 μA, the output voltage error would be 266 kΩ × 15 μA= 3.99 V. Since this large error cannot be produced, V_X and V_Y must rise so much that M_5-M_6 and M_7-M_8 enter the triode region, yielding $I_{D7,8} = 1.485$ mA. We should also mention that another important source of CM error in the simple biasing scheme of Fig. 9.33 is the *deterministic* error between $I_{D7,8}$ and I_{11} (and also between I_{D9} and I_{D10}) due to their different drain-source voltages. This error can nonetheless be reduced by means of the current mirror techniques of Chapter 5.

The foregoing study implies that in high-gain amplifiers, the output CM level is quite sensitive to device properties and mismatches and it cannot be stabilized by means of *differential* feedback. Thus, a common-mode feedback network must be added to sense the CM level of the two outputs and accordingly adjust one of the bias currents in the amplifier. Following our view of feedback systems in Chapter 8, we divide the task of CMFB into three operations: sensing the output CM level, comparison with a reference, and returning the error to the amplifier's bias network. Fig. 9.34 conceptually illustrates the idea.

In order to sense the output CM level, we recall that $V_{out,CM} = (V_{out1} + V_{out2})/2$ where V_{out1} and V_{out2} are the single-ended outputs. It therefore seems plausible to employ a resistive divider as shown in Fig. 9.35, generating $V_{out,CM} = (R_1 V_{out2} + R_2 V_{out1})/(R_1 + R_2)$, which reduces to $(V_{out1} + V_{out2})/2$ if $R_1 = R_2$. The difficulty, however, is that R_1 and R_2 must be much greater than the output impedance of the op amp so as to avoid lowering the open-loop gain. For example, in the design of Fig. 9.33, the output impedance equals 266 kΩ,

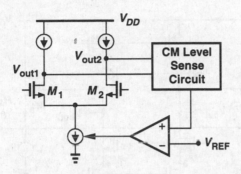

Figure 9.34 Conceptual topology for common-mode feedback.

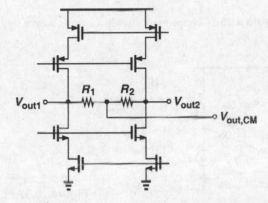

Figure 9.35 Common-mode feedback with resistive sensing.

necessitating a value of several megaohms for R_1 and R_2. As explained in Chapter 17, such large resistors occupy a very large area and, more importantly, suffer from substantial parasitic capacitance to the substrate.

To eliminate the resistive loading, we can interpose source followers between each output and its corresponding resistor. Illustrated in Fig. 9.36, this technique produces a CM level that is in fact lower than the output CM level by $V_{GS7,8}$, but this shift can be taken into account in the comparison operation. Note that R_1 and R_2 or I_1 and I_2 must be large enough to ensure that M_7 or M_8 is not "starved" when a large differential swing appears at the output. As conceptually depicted in Fig. 9.37, if, say, V_{out2} is quite higher than V_{out1}, then I_1 must sink both $I_X \approx (V_{out2} - V_{out1})/(R_1 + R_2)$ and I_{D7}. Consequently, if $R_1 + R_2$ or I_1 is not sufficiently large, I_{D7} drops to zero and $V_{out,CM}$ no longer represents the true output CM level.

The sensing method of Fig. 9.36 nevertheless suffers from an important drawback: it limits the differential output swings (even if $R_{1,2}$ and $I_{1,2}$ are large enough.) To understand why, let us determine the minimum allowable level of V_{out1} (and V_{out2}), noting that without CMFB it would be equal to $V_{OD3} + V_{OD5}$. With the source followers in place, $V_{out1,min} = V_{GS7} + V_{I1}$, where V_{I1} denotes the minimum voltage required across I_1. This is roughly equal to two overdrive voltages plus one threshold voltage. Thus, the swing at each output is reduced by approximately V_{TH}, a significant value in low-voltage design.

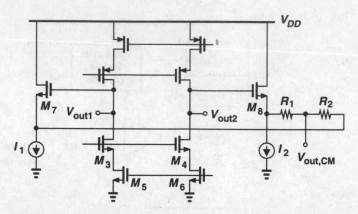

Figure 9.36 Common-mode feedback using source followers.

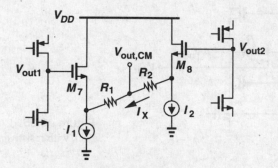

Figure 9.37 Current starvation of source followers for large swings.

Looking at Fig. 9.35, the reader may wonder if the output CM level can be sensed by means of *capacitors*, rather than resistors, so as to avoid degrading the low-frequency open-loop gain of the op amp. This is indeed possible in some cases and will be studied in Chapter 12.

Another type of CM sensing is depicted in Fig. 9.38. Here, identical transistors M_7 and M_8 operate in deep triode region, introducing a total resistance between P and ground equal to

$$R_{tot} = R_{on7} \| R_{on8} \tag{9.15}$$

$$= \frac{1}{\mu_n C_{ox} \dfrac{W}{L} (V_{out1} - V_{TH})} \left\| \frac{1}{\mu_n C_{ox} \dfrac{W}{L} (V_{out2} - V_{TH})} \right. \tag{9.16}$$

$$= \frac{1}{\mu_n C_{ox} \dfrac{W}{L} (V_{out2} + V_{out1} - 2V_{TH})}, \tag{9.17}$$

where W/L denotes the aspect ratio of M_7 and M_8. Equation (9.17) indicates that R_{tot} is a function of $V_{out2} + V_{out1}$ but independent of $V_{out2} - V_{out1}$. From Fig. 9.38, we observe that if the outputs rise together, then R_{tot} drops, whereas if they change differentially, one R_{on} increases and the other decreases.

AAIC.

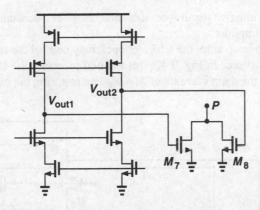

Figure 9.38 Common-mode sensing using MOSFETs operating in deep triode region.

In the circuit of Fig. 9.38, the use of M_7 and M_8 limits the output voltage swings. Here, it may seem that $V_{out,min} = V_{TH7,8}$, which is relatively close to two overdrive voltages, but the difficulty arises from the assumption above that both M_7 and M_8 operate in deep triode region. In fact, if, say, V_{out1} drops from the equilibrium CM level to one threshold voltage above ground and V_{out2} rises by the same amount, then M_7 enters the saturation region, thus exhibiting a variation in its on-resistance that is not counterbalanced by that of M_8.

We now study techniques of comparing the measured CM level with a reference and returning the error to the op amp's bias network. In the circuit of Fig. 9.39, we employ a simple amplifier to detect the difference between $V_{out,CM}$ and a reference voltage, V_{REF}, applying the result to the NMOS current sources with negative feedback. If both V_{out1} and V_{out2} rise, so does V_E, thereby increasing the drain currents of M_3-M_4 and lowering the output CM level. In other words, if the loop gain is large, the feedback network forces the CM level of V_{out1} and V_{out2} to approach V_{REF}. Note that the feedback can be applied to the PMOS current sources as well. Also, the feedback may control only a fraction of the current to allow optimization of the settling behavior. For example, each of M_3 and M_4

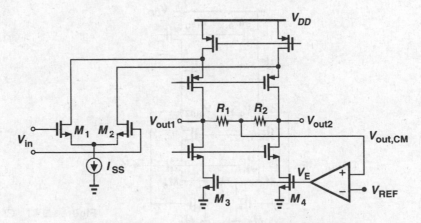

Figure 9.39 Sensing and controlling output CM level.

can be decomposed into two parallel devices, one biased at a constant current and the other driven by the error amplifier.

In a folded-cascode op amp, the CM feedback may control the tail current of the input differential pair. Illustrated in Fig. 9.40, this method increases the tail current if V_{out1} and V_{out2} rise, lowering the drain currents of M_5-M_6 and restoring the output CM level.

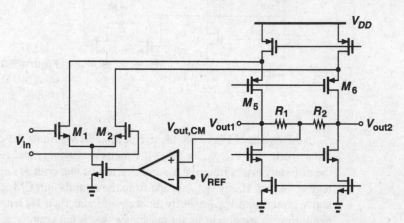

Figure 9.40   Alternative method of controlling output CM level.

How do we perform comparison and feedback with the sensing scheme of Fig. 9.38? Here, the output CM voltage is directly converted to a resistance or a current, prohibiting comparison with a reference voltage. A simple feedback topology utilizing this technique is depicted in Fig. 9.41, where $R_{on7} \| R_{on8}$ adjusts the bias current of M_5 and M_6. The output CM level sets $R_{on7} \| R_{on8}$ such that I_{D5} and I_{D6} exactly balance I_{D9} and I_{D10}, respectively. Assuming $I_{D9} = I_{D10} = I_D$, we must have $V_b - V_{GS5} = 2I_D(R_{on7} \| R_{on8})$ and hence

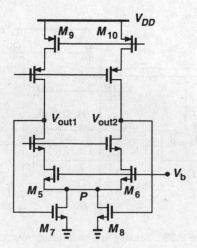

Figure 9.41 CMFB using triode devices.

$R_{on7} \| R_{on8} = (V_b - V_{GS5})/(2I_D)$. From (9.17),

$$\frac{1}{\mu_n C_{ox} \left(\dfrac{W}{L}\right)_{7,8} (V_{out2} + V_{out1} - 2V_{TH})} = \frac{V_b - V_{GS5}}{2I_D}, \qquad (9.18)$$

that is,

$$V_{out1} + V_{out2} = \frac{2I_D}{\mu_n C_{ox} \left(\dfrac{W}{L}\right)_{7,8}} \frac{1}{V_b - V_{GS5}} + 2V_{TH}. \qquad (9.19)$$

The CM level can thus be obtained by noting that $V_{GS5} = \sqrt{2I_D/[\mu_n C_{ox}(W/L)_5]} + V_{TH5}$.

The CMFB network of Fig. 9.41 suffers from several drawbacks. First, the value of the output CM level is a function of device parameters. Second, the voltage drop across $R_{on7} \| R_{on8}$ limits the output voltage swings. Third, to minimize this drop, M_7 and M_8 are usually quite wide devices, introducing substantial capacitance at the output. The second issue can be alleviated by applying the feedback to the tail current of the input differential pair (Fig. 9.42), but the other two remain.

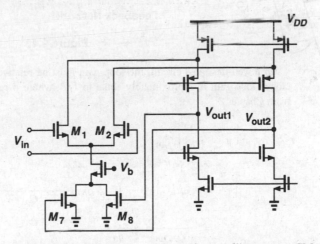

Figure 9.42 Alternative method of controlling output CM level.

How is V_b generated in Fig. 9.42? We note that $V_{out,CM}$ is somewhat sensitive to the value of V_b: if V_b is higher than expected, the tail current of M_1 and M_2 increases and the output CM level falls. Since the feedback through M_7 and M_8 attempts to correct this error, the overall change in $V_{out,CM}$ depends on the loop gain in the CMFB network. This is studied in the following example.

Example 9.9

For the circuit of Fig. 9.42, determine the sensitivity of $V_{out,CM}$ to V_b, i.e., $dV_{out,CM}/dV_b$.

Solution

Setting V_{in} to zero and following the procedure depicted in Fig. 4.25, we simplify the circuit as shown in Fig. 9.43. Note that g_{m7} and g_{m8} must be calculated in the triode region: $g_{m7} = g_{m8} = \mu_n C_{ox}(W/L)_{7,8}V_{DS7,8}$, where $V_{DS7,8}$ denotes the bias value of the drain-source voltage of M_7 and M_8. Since M_7 and M_8 operate in deep triode region, $V_{DS7,8}$ typically does not exceed a few hundred millivolts.

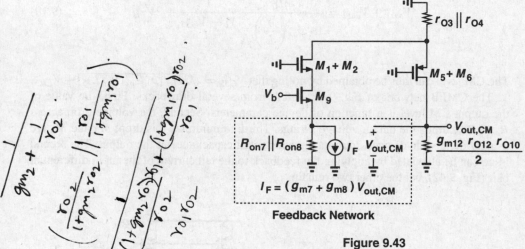

Feedback Network

Figure 9.43

In a well-designed circuit, the loop gain must be relatively high. We therefore surmise that the closed-loop gain is approximately equal to $1/\beta$, where β represents the feedback factor. We write from Chapter 8:

$$\beta = \left.\frac{V_2}{V_1}\right|_{I2=0} \tag{9.20}$$

$$= -(g_{m7} + g_{m8})(R_{on7}\|R_{on8}) \tag{9.21}$$

$$= -2\mu_n C_{ox}\left(\frac{W}{L}\right)_{7,8} V_{DS7,8} \cdot \frac{1}{2\mu_n C_{ox}(W/L)_{7,8}(V_{GS7,8} - V_{TH7,8})} \tag{9.22}$$

$$= -\frac{V_{DS7,8}}{V_{GS7,8} - V_{TH7,8}}, \tag{9.23}$$

where $V_{GS7,8} - V_{TH7,8}$ denotes the overdrive voltage of M_7 and M_8. Thus,

$$\left|\frac{dV_{out,CM}}{dV_b}\right|_{closed} \approx \frac{V_{GS7,8} - V_{TH7,8}}{V_{DS7,8}}. \tag{9.24}$$

This is an important result. Since $V_{GS7,8}$ (i.e., the output CM level) is typically in the vicinity of $V_{DD}/2$, the above equation suggests that $V_{DS7,8}$ must be maximized.

We now introduce a modification to the circuit of 9.42 that both makes the output level relatively independent of device parameters and lowers the sensitivity to the value of V_b. Illustrated in Fig. 9.44(a), the idea is to define V_b by a current mirror arrangement such that I_{D9} "tracks" I_1 and V_{REF}. For simplicity, suppose $(W/L)_{15} = (W/L)_9$ and

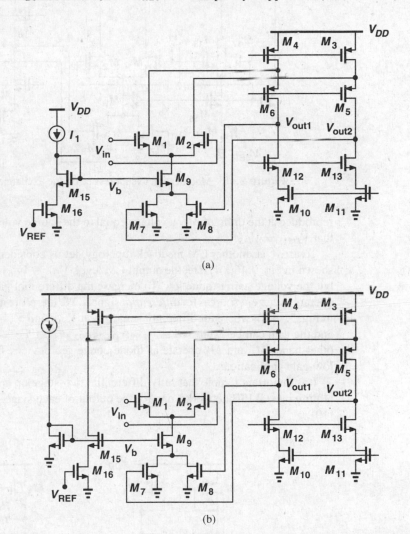

(a)

(b)

Figure 9.44 Modification of CMFB for more accurate definition of output MC level.

$(W/L)_{16} = (W/L)_7 + (W/L)_8$. Thus, $I_{D9} = I_1$ only if $V_{out,CM} = V_{REF}$. In other words, as with Fig. 9.40, the circuit produces an output CM level equal to a reference but it requires no resistors in sensing $V_{out,CM}$. The overall design can be simplified as shown in Fig. 9.44(b).

In practice, since $V_{DS15} \neq V_{DS9}$, channel-length modulation results in a finite error. Figure 9.45 depicts a modification that suppresses this error. Here, transistors M_{17} and M_{18}

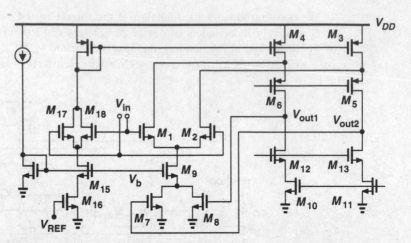

Figure 9.45 Modification to suppress error due to channel-length modulation.

reproduce at the drain of M_{15} a voltage equal to the source voltage of M_1 and M_2, ensuring that $V_{DS15} = V_{DS9}$.

To arrive at another CM feedback topology, let us consider the simple differential pair shown in Fig. 9.46(a). Here, the output CM level, $V_{DD} - V_{GS3,4}$, is relatively well-defined, but the voltage gain is quite low. To increase the differential gain, the PMOS devices must operate as current sources for *differential* signals. We therefore modify the circuit as depicted in Fig. 9.46(b), where for differential changes at V_{out1} and V_{out2}, node P is a virtual ground and the gain can be expressed as $g_{m1,2}(r_{O1,2}\|r_{O3,4}\|R_F)$. For common-mode levels, on the other hand, M_3 and M_4 operate as diode-connected devices. The circuit proves useful in low-gain applications.

It is important to note that fully-differential two-stage op amps such as that in Fig. 9.22 require *two* CMFB networks, one for the output of each stage. An example is described in [10].

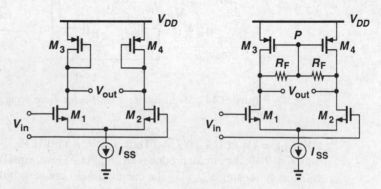

Figure 9.46 (a) Differential pair using diode-connected loads, (b) resistive CMFB.

9.7 Input Range Limitations

The op amp circuits studied thus far have evolved to achieve large differential output swings. While the differential input swings are usually much smaller (by a factor equal to the open-loop gain), the input *common-mode* level may need to vary over a wide range in some applications. For example, consider the simple unity-gain buffer shown in Fig. 9.47, where the input swing is nearly equal to the output swing. Interestingly, in this case the voltage swings are limited by the input differential pair rather than the output cascode branch. Specifically, $V_{in,min} \approx V_{out,min} = V_{GS1,2} + V_{ISS}$, approximately one threshold voltage higher than the allowable minimum provided by M_5-M_8.

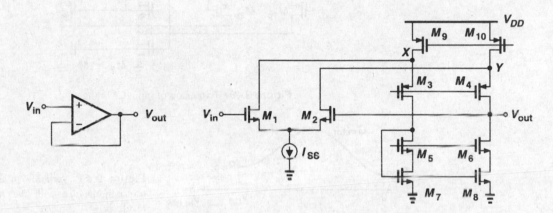

Figure 9.47   Unity-gain buffer.

What happens if V_{in} falls below the minimum given above? The MOS transistor operating as I_{SS} enters the triode region, decreasing the bias current of the differential pair and hence lowering the transconductance. We then postulate that the limitation is overcome if the transconductance can somehow be restored.

A simple approach to extending the input CM range is to incorporate both NMOS and PMOS differential pairs such that when one is "dead," the other is "alive." Illustrated in Fig. 9.48, the idea is to combine two folded-cascode op amps with NMOS and PMOS input differential pairs. Here, as the input CM level approaches the ground potential, the NMOS pair's transconductance drops, eventually falling to zero. Nonetheless, the PMOS pair remains active, allowing normal operation. Conversely, if the input CM level approaches V_{DD}, M_{1P} and M_{2P} begin to turn off but M_1 and M_2 function properly.

An important concern in the circuit of Fig. 9.48 is the *variation* of the overall transconductance of the two pairs as the input CM level changes. Considering the operation of each pair, we anticipate the behavior depicted in Fig. 9.49. Thus, many properties of the circuit, including gain, speed, and noise, vary. More sophisticated techniques of minimizing this variation are described in [7].

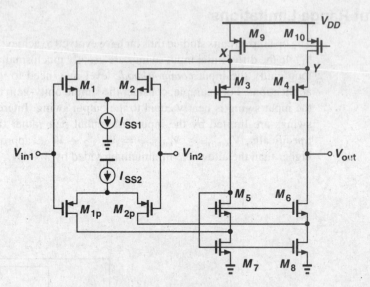

Figure 9.48 Extension of input CM range.

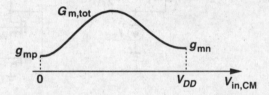

Figure 9.49 Variation of equivalent transconductance with the input CM level.

9.8 Slew Rate

Op amps used in feedback circuits exhibit a large-signal behavior called "slewing." We first describe an interesting property of *linear* systems that vanishes during slewing. Consider the simple RC network shown in Fig. 9.50, where the input is an ideal voltage step of height V_0. Since $V_{out} = V_0[1 - \exp(-t/\tau)]$, where $\tau = RC$, we have

$$\frac{dV_{out}}{dt} = \frac{V_0}{\tau} \exp \frac{-t}{\tau}. \tag{9.25}$$

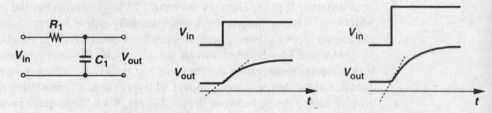

Figure 9.50 Response of a linear circuit to input step.

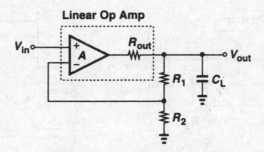

Figure 9.51 Response of linear op amp to step response.

That is, the slope of the step response is proportional to the final value of the output; if we apply a larger input step, the output rises more rapidly. This is a fundamental property of linear systems: if the input amplitude is, say, doubled while other parameters remain constant, the output signal level must double at *every* point, leading to a twofold increase in the slope.

The foregoing observation applies to linear feedback systems as well. Shown in Fig. 9.51 is an example, where the op amp is assumed linear. Here, we can write

$$\left[\left(V_{in} - V_{out}\frac{R_2}{R_1 + R_2}\right)A - V_{out}\right]\frac{1}{R_{out}} = \frac{V_{out}}{R_1 + R_2} + V_{out}C_L s. \tag{9.26}$$

Assuming $R_1 + R_2 \gg R_{out}$, we have

$$\frac{V_{out}}{V_{in}}(s) \approx \frac{A}{\left(1 + A\frac{R_2}{R_1 + R_2}\right)\left[1 + \frac{R_{out}C_L}{1 + AR_2/(R_1 + R_2)}s\right]}. \tag{9.27}$$

As expected, both the low-frequency gain and the time constant are divided by $1 + AR_2/(R_1 + R_2)$. The step response is therefore given by

$$V_{out} = V_0\frac{A}{1 + A\frac{R_2}{R_1 + R_2}}\left(1 - \exp\frac{-t}{\frac{C_L R_{out}}{1 + AR_2/(R_1 + R_2)}}\right)u(t), \tag{9.28}$$

indicating that the slope is proportional to the final value. This type of response is called "linear settling."

With a realistic op amp, on the other hand, the step response of the circuit begins to deviate from (9.28) as the input amplitude increases. Illustrated in Fig. 9.52, the response to sufficiently small inputs follows the exponential of Eq. (9.28), but with large input steps, the output displays a linear *ramp* having a *constant slope*. Under this condition, we say the op amp experiences slewing and call the slope of the ramp the "slew rate."

To understand the origin of slewing, let us replace the op amp of Fig. 9.52 by a simple CMOS implementation (Fig. 9.53), assuming for simplicity that $R_1 + R_2$ is quite large. We first examine the circuit with a small input step. If V_{in} experiences a change of ΔV, I_{D1} increases by $g_m\Delta V/2$ and I_{D2} decreases by $g_m\Delta V/2$. Since the mirror action of M_3

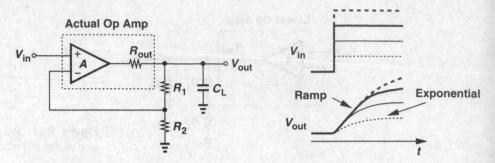

Figure 9.52 Slewing in an op amp circuit.

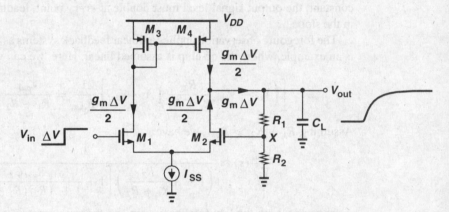

Figure 9.53 Small-signal operation of a simple op amp.

and M_4 raises $|I_{D4}|$ by $g_m \Delta V/2$, the total small-signal current provided by the op amp equals $g_m \Delta V$. This current begins to charge C_L, but as V_{out} rises, so does V_X, reducing the difference between V_{G1} and V_{G2} and hence the output current of the op amp. As a result, V_{out} varies according to (9.28).

Now suppose ΔV is so large that M_1 absorbs all of I_{SS}, turning off M_2. The circuit then reduces to that shown in Fig. 9.54, generating a ramp output with a slope equal to I_{SS}/C_L (if the channel-length modulation of M_4 and the current drawn by $R_1 + R_2$ are neglected). Note that so long as M_2 remains off, the feedback loop is broken and the current charging C_L is constant and independent of the input level. As V_{out} rises, V_X eventually approaches V_{in}, M_2 turns on, and the circuit returns to linear operation.

In Fig. 9.53, slewing occurs for falling edges at the input as well. If the input drops so much that M_1 turns off, then the circuit is simplified as in Fig. 9.55, discharging C_L by a current approximately equal to I_{SS}. After V_{out} decreases sufficiently, the difference between V_X and V_{in} is small enough to allow M_1 to turn on, leading to linear behavior thereafter.

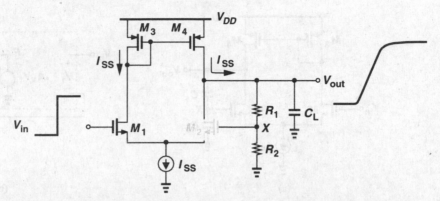

Figure 9.54 Slewing during low-to-high transition.

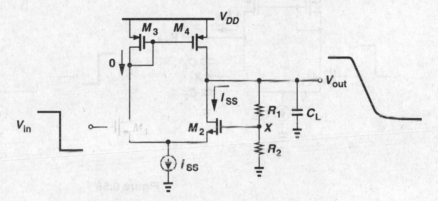

Figure 9.55 Slewing during high-to-low transition.

The foregoing observations explain why slewing is a nonlinear phenomenon. If the input amplitude, say, doubles, the output level does not double at *all* points because the ramp exhibits a slope independent of the input.

Slewing is an undesirable effect in high-speed circuits that process large signals. While the small-signal bandwidth of a circuit may suggest a fast time-domain response, the large-signal speed may be limited by the slew rate simply because the current available to charge and discharge the dominant capacitor in the circuit is small. Moreover, since the input-output relationship during slewing is nonlinear, the output of a slewing amplifier exhibits substantial distortion. For example, if a circuit is to amplify a sinusoid $V_0 \sin \omega_0 t$ (in the steady state), then its slew rate must exceed $V_0 \omega_0$.

Example 9.10

Consider the feedback amplifier depicted in Fig. 9.56(a), where C_1 and C_2 set the closed-loop gain. (The bias network for the gate of M_2 is not shown.) (a) Determine the small-signal step response of the circuit. (b) Calculate the positive and negative slew rates.

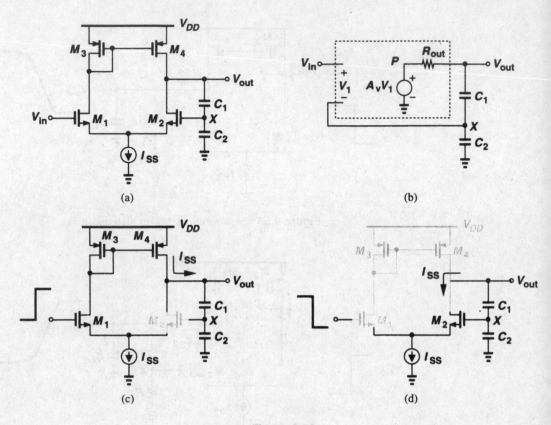

Figure 9.56

Solution

(a) Modeling the op amp as in Fig. 9.56(b), where $A_v = g_{m1,2}(r_{O2} \| r_{O4})$ and $R_{out} = r_{O2} \| r_{O4}$, we have $V_X = C_1 V_{out}/(C_1 + C_2)$ and hence

$$V_P = \left(V_{in} - \frac{C_1}{C_1 + C_2} V_{out} \right) A_v, \tag{9.29}$$

obtaining

$$\left[\left(V_{in} - \frac{C_1}{C_1 + C_2} V_{out} \right) A_v - V_{out} \right] \frac{1}{R_{out}} = V_{out} \frac{C_1 C_2}{C_1 + C_2} s. \tag{9.30}$$

It follows that

$$\frac{V_{out}}{V_{in}}(s) = \frac{A_v}{1 + A_v \dfrac{C_1}{C_1 + C_2} + \dfrac{C_1 C_2}{C_1 + C_2} R_{out} s} \tag{9.31}$$

$$= \frac{A_v/\left(1 + A_v \dfrac{C_1}{C_1 + C_2}\right)}{1 + \dfrac{C_1 C_2}{C_1 + C_2} R_{out} s / \left(1 + A_v \dfrac{C_1}{C_1 + C_2}\right)}, \tag{9.32}$$

revealing that both the low-frequency gain and the time constant of the circuit have decreased by a factor of $1 + A_v C_1/(C_1 + C_2)$. The response to a unity step is thus given by

$$V_{out}(t) = \frac{A_v}{1 + A_v \dfrac{C_1}{C_1 + C_2}} V_0 \left(1 - \exp\frac{-t}{\tau}\right) u(t), \tag{9.33}$$

where

$$\tau = \frac{C_1 C_2}{C_1 + C_2} R_{out} / \left(1 + A_v \frac{C_1}{C_1 + C_2}\right). \tag{9.34}$$

(b) Suppose a large positive step is applied to the gate of M_1 in Fig. 9.56(a) while the initial voltage across C_1 is zero. Then, M_2 turns off and, as shown in Fig. 9.56(c), V_{out} rises according to $V_{out}(t) = I_{SS}/[C_1 C_2/(C_1 + C_2)]t$. Similarly, for a large negative step at the input, Fig. 9.56(d) yields $V_{out} = -I_{SS}/[C_1 C_2/(C_1 + C_2)]t$.

As another example, let us find the slew rate of the telescopic op amp shown in Fig. 9.57(a). When a large differential input is applied, M_1 or M_2 turns off, reducing the circuit to that shown in Fig. 9.57(b). Thus, V_{out1} and V_{out2} appear as ramps with slopes equal to $\pm I_{SS}/(2C_L)$, and consequently $V_{out1} - V_{out2}$ exhibits a slew rate equal to I_{SS}/C_L. (Of course, the circuit is usually used in closed-loop form.)

It is also instructive to study the slewing behavior of a folded-cascode op amp with single-ended output [Fig. 9.58(a)]. Figs. 9.58(a) and (b) depict the equivalent circuit for

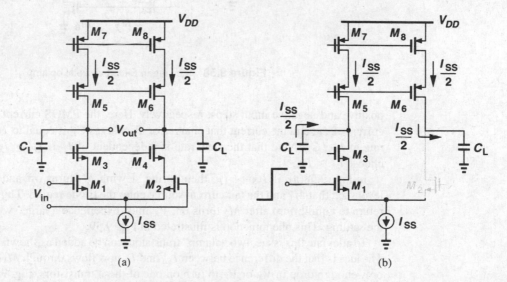

Figure 9.57 Slewing in telescopic op amp.

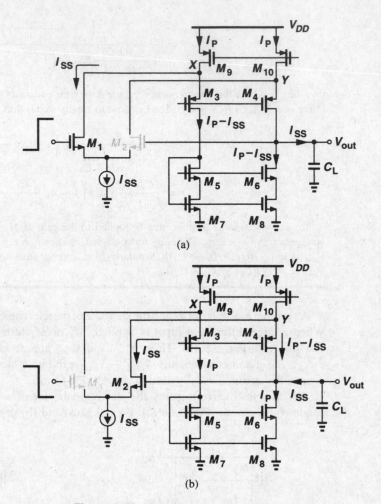

Figure 9.58 Slewing in folded-cascode op amp.

positive and negative input steps, respectively. Here, the PMOS current sources provide a
current of I_P, and the current that charges or discharges C_L is equal to I_{SS}, yielding a slew
rate of I_{SS}/C_L. Note that the slew rate is independent of I_P if $I_P \geq I_{SS}$. In practice, we
choose $I_P \approx I_{SS}$.

In Fig. 9.58(a), if $I_{SS} > I_P$, then during slewing M_3 turns off and V_X falls to a low
level such that M_1 and the tail current source enter the triode region. Thus, for the circuit to
return to equilibrium after M_2 turns on, V_X must experience a large swing, slowing down
the settling. This phenomenon is illustrated in Fig. 9.59.

To alleviate this issue, two "clamp" transistors can be added as shown in Fig. 9.60(a) [8].
The idea is that the difference between I_{SS} and I_P now flows through M_{11} or M_{12}, requiring
only enough drop in V_X or V_Y to turn on one of these transistors. Fig. 9.60(b) illustrates a
more aggressive approach, where M_{11} and M_{12} clamp the two nodes directly to V_{DD}. Since

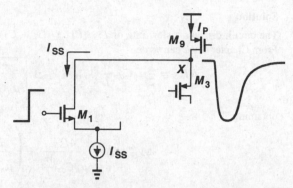

Figure 9.59 Long settling due to overdrive recovery after slewing.

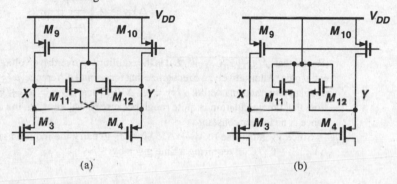

Figure 9.60 Clamp circuit to limit swings at X and Y.

the equilibrium value of V_X and V_Y is usually higher than $V_{DD} - V_{THN}$, M_{11} and M_{12} are off during small-signal operation.

What trade-offs are encountered in increasing the slew rate? In the examples of Figs. 9.57 and 9.58, for a given load capacitance, I_{SS} must be increased and to maintain the same maximum output swing, all of the transistors must be made proportionally wider. As a result, the power dissipation and the input capacitance are increased. Note that if the device currents and widths scale together, $g_m r_O$ of each transistor and hence the open-loop gain of the op amp remain constant.

How does an op amp leave the slewing regime and enter the linear-settling regime? Since the point at which one of the input transistors "turns on" is ambiguous, the distinction between slewing and linear settling is somewhat arbitrary. The following example illustrates the point.

Example 9.11

Consider the circuit of Fig. 9.56(a) in the slewing regime [Fig. 9.56(c)]. As V_{out} rises, so does V_X, eventually turning M_2 on. As I_{D2} increases from zero, the differential pair becomes more linear. Considering M_1 and M_2 to operate linearly if the difference between their drain currents is less than αI_{SS} (e.g., $\alpha = 0.1$), determine how long the circuit takes to enter linear settling. Assume the input step has an amplitude of V_0.

Solution

The circuit displays a slew rate of $I_{SS}/[C_1 C_2/(C_1 + C_2)]$ until $|V_{in1} - V_{in2}|$ is sufficiently small. From Chapter 4, we can write

$$\alpha I_{SS} = \frac{1}{2}\mu_n C_{ox}\frac{W}{L}(V_{in1} - V_{in2})\sqrt{\frac{4I_{SS}}{\mu_n C_{ox}\dfrac{W}{L}} - (V_{in1} - V_{in2})^2}, \tag{9.35}$$

obtaining

$$\Delta V_G^4 - \Delta V_G^2 \frac{4I_{SS}}{\mu_n C_{ox}\dfrac{W}{L}} + \left(\frac{2\alpha I_{SS}}{\mu_n C_{ox}\dfrac{W}{L}}\right)^2 = 0, \tag{9.36}$$

where $\Delta V_G = V_{in1} - V_{in2}$. Thus,

$$\Delta V_G \approx \alpha \sqrt{\frac{I_{SS}}{\mu_n C_{ox}\dfrac{W}{L}}}. \tag{9.37}$$

(Recall that $\sqrt{I_{SS}/[\mu_n C_{ox}(W/L)]}$ is the equilibrium overdrive voltage of each transistor in the differential pair.) Alternatively, we recognize that for a small difference, αI_{SS}, between I_{D1} and I_{D2}, a small-signal approximation is valid: $\alpha I_{SS} = g_m \Delta V_G$. Thus, $\Delta V_G = \alpha I_{SS}/g_m \approx \alpha I_{SS}/\sqrt{\mu_n C_{ox}(W/L)I_{SS}}$. Note that this calculation is quite rough because as M_2 turns on, the current charging the load capacitance is no longer constant.

Since V_X must rise to $V_0 - \Delta V_G$ for M_2 to carry the required current, V_{out} increases by $(V_0 - \Delta V_G)(1 + C_2/C_1)$, requiring a time given by

$$t = \frac{C_2}{I_{SS}}\left(V_0 - \alpha\sqrt{\frac{I_{SS}}{\mu_n C_{ox}\dfrac{W}{L}}}\right). \tag{9.38}$$

In the above example, the value of α that determines the onset of linear settling depends, among other things, on the actual required linearity. In other words, for a nonlinearity of 1%, α can be quite larger than for a nonlinearity of 0.1%.

The slewing behavior of two-stage op amps is somewhat different from that of the circuits studied above. This case is studied in Chapter 10.

9.9 Power Supply Rejection

As other analog circuits, op amps are often supplied from noisy lines and must therefore "reject" the noise adequately. For this reason, it is important to understand how noise on the supply manifests itself at the output of an op amp.

Let us consider the simple op amp shown in Fig. 9.61, assuming the supply voltage varies slowly. If the circuit is perfectly symmetric, $V_{out} = V_X$. Since the diode-connected device "clamps" node X to V_{DD}, V_X and hence V_{out} experience approximately the same change as does V_{DD}. In other words, the gain from V_{DD} to V_{out} is close to unity. The power supply

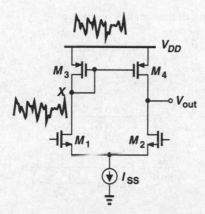

Figure 9.61 Supply rejection of differential pair with active current mirror.

rejection ratio (PSRR) is defined as the gain from the input to the output divided by the gain from the supply to the output. At low frequencies:

$$PSRR \approx g_{mN}(r_{OP} \| r_{ON}). \tag{9.39}$$

Example 9.12

Calculate the low-frequency PSRR of the feedback circuit shown in Fig. 9.62(a).

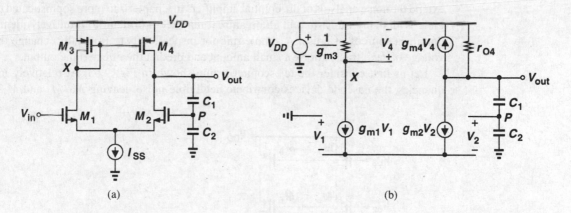

(a)   (b)

Figure 9.62

Solution

From the foregoing analysis, we may surmise that a change ΔV in V_{DD} appears unattenuated at the output. But, we should note that if V_{out} changes, so do V_P and I_{D2}, thereby opposing the change. Using Fig. 9.62(b) and neglecting channel-length modulation in M_1-M_3 for simplicity, we can write:

$$V_{out}\frac{C_1}{C_1 + C_2} - V_2 = -V_1, \tag{9.40}$$

and $g_{m1}V_1 + g_{m2}V_2 = 0$. Thus, if the circuit is symmetric,

$$V_2 = \frac{V_{out}}{2}\frac{C_1}{C_1 + C_2}. \tag{9.41}$$

We also have

$$-\frac{g_{m1}V_1}{g_{m3}}g_{m4} - \frac{V_{DD} - V_{out}}{r_{O4}} + g_{m2}V_2 = 0. \tag{9.42}$$

It follows that

$$\frac{V_{out}}{V_{DD}} = \frac{1}{g_{m2}r_{O4}\dfrac{C_1}{C_1 + C_2} + 1}. \tag{9.43}$$

Thus,

$$PSRR \approx \frac{1 + \dfrac{C_2}{C_1}}{g_{m2}r_{O4}\dfrac{C_1}{C_1 + C_2} + 1}. \tag{9.44}$$

9.10 Noise in Op Amps

In low-noise applications, the input-referred noise of op amps becomes critical. We now extend the noise analysis of differential amplifiers in Chapter 7 to more sophisticated topologies. With many transistors in an op amp, it may seem difficult to intuitively identify the dominant sources of noise. A simple rule for inspection is to (mentally) change the gate voltage of each transistor by a small amount and predict the effect at the output.

Let us first consider the telescopic op amp shown in Fig. 9.63. At relatively low frequencies, the cascode devices contribute negligible noise, leaving M_1-M_2 and M_7-M_8 as

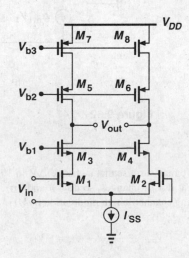

Figure 9.63 Noise in a telescopic op amp.

the primary noise sources. The input-referred noise voltage per unit bandwidth is therefore similar to that in Fig. 7.47(a) and given by:

$$\overline{V_n^2} = 4kT \left(2\frac{2}{3g_{m1,2}} + 2\frac{2g_{m7,8}}{3g_{m1,2}^2} \right) + 2\frac{K_N}{(WL)_{1,2}C_{ox}f} + 2\frac{K_P}{(WL)_{7,8}C_{ox}f} \frac{g_{m7,8}^2}{g_{m1,2}^2}, \qquad (9.45)$$

where K_N and K_P denote the $1/f$ noise coefficients of NMOS and PMOS devices, respectively.

Next, we study the noise behavior of the folded-cascode op amp of Fig. 9.64(a), considering only thermal noise at this point. Again, the noise of the cascode devices is negligible at low frequencies, leaving M_1-M_2, M_7-M_8, and M_9-M_{10} as potentially significant sources. Do both pairs M_7-M_8 and M_9-M_{10} contribute noise? Using our simple rule, we change the gate voltage of M_7 by a small amount [Fig. 9.64(b)], noting that the output indeed

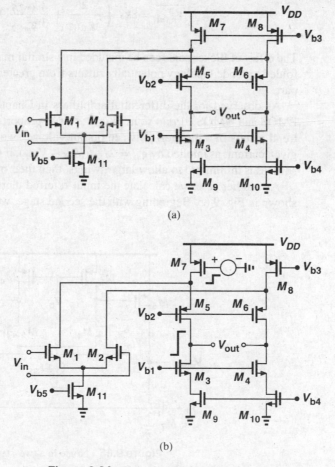

(a)

(b)

Figure 9.64   Noise in a folded-cascode op amp.

changes considerably. The same observation applies to M_8-M_{10} as well. To determine the input-referred thermal noise, we first refer the noise of M_7-M_8 and M_9-M_{10} to the output:

$$\overline{V_{n,out}^2}\big|_{M7,8} = 2\left(4kT\frac{2}{3g_{m7,8}}g_{m7,8}^2 R_{out}^2\right), \tag{9.46}$$

where the factor 2 accounts for (uncorrelated) noise of M_7 and M_8 and R_{out} denotes the open-loop output resistance of the op amp. Similarly,

$$\overline{V_{n,out}^2}\big|_{M9,10} = 2\left(4kT\frac{2}{3g_{m9,10}}g_{m9,10}^2 R_{out}^2\right). \tag{9.47}$$

Dividing these quantities by $g_{m1,2}^2 R_{out}^2$ and adding the contribution of M_1-M_2, we obtain the overall noise:

$$\overline{V_{n,int}^2} = 8kT\left(\frac{2}{3g_{m1,2}} + \frac{2}{3}\frac{g_{m7,8}}{g_{m1,2}^2} + \frac{2}{3}\frac{g_{m9,10}}{g_{m1,2}^2}\right). \tag{9.48}$$

The effect of flicker noise can be included in a similar manner (Problem 9.15). Note that the folded-cascode topology potentially suffers from greater noise than the telescopic counterpart.

As observed for the differential amplifiers in Chapter 7, the noise contribution of the PMOS and NMOS current sources *increases* in proportion to their transconductance. This trend results in a trade-off between output voltage swings and input-referred noise: for a given current, as implied by $g_m = 2I_D/(V_{GS} - V_{TH})$, if the overdrive voltage of the current sources is minimized to allow large swings, then their transconductance is maximized.

As another case, we calculate the input-referred thermal noise of the two-stage op amp shown in Fig. 9.65. Beginning with the second stage, we note that the noise current of M_5

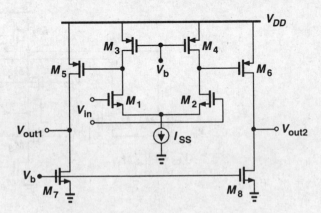

Figure 9.65   Noise in a two-stage op amp.

and M_7 flows through $r_{O5} \| r_{O7}$. Dividing the resulting output noise voltage by the total gain, $g_{m1}(r_{O1} \| r_{O3}) \times g_{m5}(r_{O5} \| r_{O7})$, and doubling the power, we obtain the input-referred contribution of M_5-M_8:

$$\overline{V_n^2}\Big|_{M5-8} = 2 \times 4kT\frac{2}{3}(g_{m5} + g_{m7})(r_{O5} \| r_{O7})^2 \frac{1}{g_{m1}^2(r_{O1} \| r_{O3})^2 g_{m5}^2(r_{O5} \| r_{O7})^2} \qquad (9.49)$$

$$= \frac{16kT}{3}\frac{g_{m5} + g_{m7}}{g_{m1}^2 g_{m5}^2 (r_{O1} \| r_{O3})^2}. \qquad (9.50)$$

The noise due to M_1-M_4 is simply equal to

$$\overline{V_n^2}\Big|_{M1-4} = 2 \times 4kT\frac{2}{3}\frac{g_{m1} + g_{m3}}{g_{m1}^2}. \qquad (9.51)$$

It follows that

$$\overline{V_{n,tot}^2} = \frac{16kT}{3}\frac{1}{g_{m1}^2}\left[g_{m1} + g_{m3} + \frac{g_{m5} + g_{m7}}{g_{m5}^2(r_{O1} \| r_{O3})^2}\right]. \qquad (9.52)$$

Note the noise resulting from the second stage is usually negligible because it is divided by the gain of the first stage when referred to the main input.

Example 9.13 ───

A simple amplifier is constructed as shown in Fig. 9.66. Note that the first stage incorporates diode-connected—rather than current-source—loads. Assuming all of the transistors are in saturation and $(W/L)_{1,2} = 50/0.6$, $(W/L)_{3,4} = 10/0.6$, $(W/L)_{5,6} = 20/0.6$, and $(W/L)_{7,8} = 56/0.6$, calculate the input-referred noise voltage if $\mu_n C_{ox} = 75 \ \mu A/V^2$ and $\mu_p C_{ox} = 30 \ \mu A/V^2$.

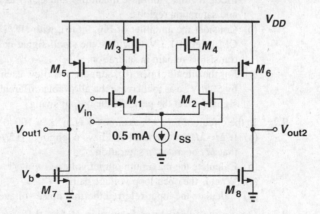

Figure 9.66

Solution

We first calculate the small-signal gain of the first stage.:

$$A_{v1} \approx \frac{g_{m1}}{g_{m3}} \tag{9.53}$$

$$= \sqrt{\frac{50 \times 75}{10 \times 30}} \tag{9.54}$$

$$\approx 3.54. \tag{9.55}$$

The noise of M_5 and M_7 referred to the gate of M_5 is equal to $4kT(2/3)(g_{m5} + g_{m7})/g_{m5}^2 = 2.87 \times 10^{-17}$ V^2/Hz, which is divided by A_{v1}^2 when referred to the main input: $\overline{V_n^2}|_{M5,7} = 2.29 \times 10^{-18}$ V^2/Hz. Transistors M_1 and M_3 produce an input-referred noise of $\overline{V_n^2}|_{M1,3} = (8kT/3)$ $(g_{m3} + g_{m1})/g_{m1}^2 = 1.10 \times 10^{-17}$ V^2/Hz. Thus, the total input-referred noise equals

$$\overline{V_{n,in}^2} = 2(2.29 \times 10^{-18} + 1.10 \times 10^{-17}) \tag{9.56}$$

$$= 2.66 \times 10^{-17} \text{ V}^2/\text{Hz}, \tag{9.57}$$

where the factor 2 accounts for the noise produced by both odd-numbered and even-numbered transistors in the circuit. This value corresponds to an input noise voltage of 5.16 nV/$\sqrt{\text{Hz}}$.

Problems

Unless otherwise stated, in the following problems, use the device data shown in Table 2.1 and assume $V_{DD} = 3$ V where necessary. Also, assume all transistors are in saturation.

9.1. (a) Derive expressions for the transconductance and output resistance of a MOSFET in the triode region. Plot these quantities and $g_m r_O$ as a function of V_{DS}, covering both triode and saturation regions.

  (b) Consider the amplifier of Fig. 9.6(b), with $(W/L)_{1-4} = 50/0.5$, $I_{SS} = 1$ mA, and input CM level of 1.3 V. Calculate the small-signal gain and the maximum output swing if all transistors remain in saturation.

  (c) For the circuit of part (b), suppose we allow each PMOS device to enter the triode region by 50 mV so as to increase the allowable differential swing by 100 mV. What is the small-signal gain at the peaks of the output swing?

9.2. In the circuit of Fig. 9.9, assume $(W/L)_{1-4} = 100/0.5$, $I_{SS} = 1$ mA, $V_b = 1.4$ V, and $\gamma = 0$.
  (a) If M_5-M_8 are identical and have a length of 0.5 μm, calculate their minimum width such that M_3 operates in saturation.

  (b) Calculate the maximum output voltage swing.

  (c) What is the open-loop voltage gain?

  (d) Calculate the input-referred thermal noise voltage.

9.3. Design the folded-cascode op amp of Fig. 9.13 for the following requirements: maximum differential swing = 2.4 V, total power dissipation = 6 mW. If all of the transistors have a channel length of 0.5 μm, what is the overall voltage gain? Can the input common-mode level be as low as zero?

9.4. In the op amp of Fig. 9.18(b), $(W/L)_{1-8} = 100/0.5$, $I_{SS} = 1$ mA, and $V_{b1} = 1.7$ V. Assume $\gamma = 0$.

  (a) What is the maximum allowable input CM level?

  (b) What is V_X?

  (c) What is the maximum allowable output swing if the gate of M_2 is connected to the output?

  (d) What is the acceptable range of V_{b2}?

  (e) What is the input-referred thermal noise voltage?

9.5. Design the op amp of Fig. 9.18(b) for the following requirements: maximum differential swing $= 2.4$ V, total power dissipation $= 6$ mW. (Assume the gate of M_2 is never shorted to the output.)

9.6. If in Fig. 9.21, $(W/L)_{1-8} = 100/0.5$ and $I_{SS} = 1$ mA,

  (a) What CM level must be established at the drains of M_3 and M_4 so that $I_{D5} = I_{D6} = 1$ mA? How does this constrain the maximum input CM level?

  (b) With the choice made in part (a), calculate the overall voltage gain and the maximum output swing.

9.7. Design the op amp of Fig. 9.21 for the following requirements: maximum differential swing $= 4$ V, total power dissipation $= 6$ mW, $I_{SS} = 0.5$ mA.

9.8. Suppose the circuit of Fig. 9.22 is designed with I_{SS} equal to 1 mA, I_{D9}-I_{D12} equal to 0.5 mA, and $(W/L)_{9-12} = 100/0.5$.

  (a) What CM level is required at X and Y?

  (b) If I_{SS} requires a minimum voltage of 400 mV, choose the minimum dimensions of M_1-M_8 to allow a peak-to-peak swing of 200 mV at X and at Y.

  (c) Calculate the overall voltage gain.

9.9. In Fig. 9.25(c), calculate the input-referred thermal noise if I_1 and I_2 are implemented by PMOS devices.

9.10. Suppose in Fig. 9.25(c), $I_1 = 100 \ \mu A$, $I_2 = 0.5$ mA, and $(W/L)_{1-3} = 100/0.5$. Assuming I_1 and I_2 are implemented with PMOS devices having $(W/L)_P = 50/0.5$,

  (a) Calculate the gate bias voltages of M_2 and M_3.

  (b) Determine the maximum allowable output voltage swing.

  (c) Calculate the overall voltage gain and the input-referred thermal noise voltage.

9.11. In the circuit of Fig. 9.41, each branch is biased at a current of 0.5 mA. Choose the dimensions of M_7 and M_8 such that the output CM level is equal to 1.5 V and $V_P = 100$ mV.

9.12. Consider the CMFB network in Fig. 9.39. The amplifier sensing $V_{out,CM}$ is to be implemented as a different pair with active current mirror load.

  (a) Should the input pair of the amplifier use PMOS devices or NMOS devices?

  (b) Calculate the loop gain for the CMFB network.

9.13. Repeat Problem 9.12(b) for the circuit of Fig. 9.40.

9.14. In the circuit of Fig. 9.56(a), assume $(W/L)_{1-4} = 100/0.5$, $C_1 = C_2 = 0.5$ pF, and $I_{SS} = 1$ mA.

  (a) Calculate the small-signal time constant of the circuit.

  (b) With a 1-V step at the input [Fig. 9.56(c)], how long does it take for I_{D2} to reach $0.1 I_{SS}$?

9.15. It is possible to argue that the auxiliary amplifier in the circuit of Fig. 9.24(c) *reduces* the output impedance. Consider the circuit as drawn in Fig. 9.67, where the drain voltage of M_2 is changed by ΔV to measure the output impedance. It seems that, since the feedback provided by A_1 attempts to hold V_X constant, the change in the current through r_{O2} is much *greater* than in the circuit of Fig. 9.24(b), suggesting that $R_{out} \approx r_{O2}$. Explain the flaw in this argument.

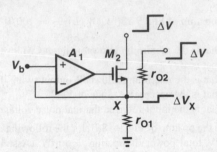

Figure 9.67

9.16. Calculate the CMRR of the circuit shown in Fig. 9.56(a).

9.17. Calculate the input-referred flicker noise of the op amp shown in Fig. 9.64(a).

9.18. In this problem, we design a two-stage op amp based on the topology shown in Fig. 9.68. Assume a power budget of 6 mW, a required output swing of 2.5 V, and $L_{eff} = 0.5$ μm for all devices.

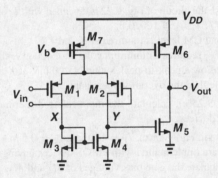

Figure 9.68

(a) Allocating a current of 1 mA to the output stage and roughly equal overdrive voltages to M_5 and M_6, determine $(W/L)_5$ and $(W/L)_6$. Note that the gate-source capacitance of M_5 is in the signal path whereas that of M_6 is not. Thus, M_6 can be quite larger than M_5.

(b) Calculate the small-signal gain of the output stage.

(c) With the remaining 1 mA flowing through M_7, determine the aspect ratio of M_3 (and M_4) such that $V_{GS3} = V_{GS5}$. This is to guarantee that if $V_{in} = 0$ and hence $V_X = V_Y$, then M_5 carries the expected current.

(d) Calculate the aspect ratios of M_1 and M_2 such that the overall voltage gain of the op amp is equal to 500.

9.19. Consider the op amp of Fig. 9.68, assuming that the second stage is to provide a voltage gain of 20 with a bias current of 1 mA.

(a) Determine $(W/L)_5$ and $(W/L)_6$ such that M_5 and M_6 have equal overdrive voltages.

(b) What is the small-signal gain of this stage if M_6 is driven into the triode region by 50 mV?

9.20. The op amp designed in Problem 9.18(d) is placed in unity-gain feedback. Assume $|V_{GS7} - V_{TH7}| = 0.4$ V.

(a) What is the allowable input voltage range?

(b) At what input voltage are the input and output voltages *exactly* equal?

9.21. Calculate the input-referred noise of the op amp designed in Problem 9.18(d).

9.22. It is possible to use the bulk terminal of PMOS devices as an input [9]. Consider the amplifier shown in Fig. 9.69 as an example.

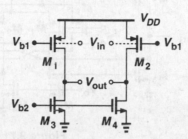

Figure 9.69

(a) Calculate the voltage gain.

(b) What is the acceptable input common-mode range?

(c) How does the small-signal gain vary with the input common-mode level?

(d) Calculate the input-referred thermal noise voltage and compare the result with that of a regular PMOS differential pair having NMOS current-source loads.

9.23. The idea of the active current mirror can be applied to the output stage of a two-stage op amp as well. That is, the load current source can become a function of the signal. Figure 9.70 shows an example [10]. Here, the first stage consists of M_1-M_4 and the output is produced by M_5-M_8. Transistors M_7 and M_8 operate as active current sources because their current varies with the signal voltage at nodes Y and X, respectively.

(a) Calculate the differential voltage gain of the op amp.

(b) Estimate the magnitude of the three major poles of the circuit.

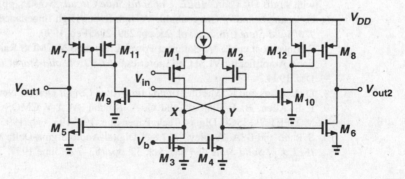

Figure 9.70

9.24. The circuit of Fig. 9.71 employs a fast path (M_1' and M_2') in parallel with the slow path. Calculate the differential voltage gain of the circuit. Which transistors typically limit the output swing?

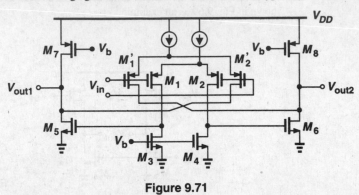

Figure 9.71

9.25. Calculate the input-referred thermal noise of the op amp in Fig. 9.71.

References

1. R. G. Eschauzier, L. P. T. Kerklaan, and J. H. Huising, "A 100-MHz 100-dB Operational Amplifier with Multipath Nested Miller Compensation Structure," *IEEE J. of Solid-State Circuits*, vol. 27, pp. 1709–1717, Dec. 1992.

2. R. M. Ziazadeh, H.-T. Ng, and D. J. Allstot, "A Multistage Amplifier Topology with Embeded Tracking Compensation," *CICC Proc.*, pp. 361–364, May 1998.

3. F. You, S. H. Embabi, and E. Sanchez-Sincencio, "A Multistage Amplifier Topology with Nested G_m-C Compensation for Low-Voltage Application," *ISSCC Dig. of Tech. Papers*, pp. 348–349, Feb. 1997.

4. B. J. Hosticka, "Improvement of the Gain of CMOS Amplifiers," *IEEE J. of Solid-State Circuits*, vol. 14, pp. 1111–1114, Dec. 1979.

5. K. Bult and G. J. G. H. Geelen, "A Fast-Settling CMOS Operational Amplifier for SC Circuits with 90-dB DC Gain," *IEEE J. of Solid-State Circuits*, vol. 25, pp. 1379–1384, Dec. 1990.

6. E. Sackinger and W. Guggenbuhl, "A High-Swing High-Impedance MOS Cascode Circuit," *IEEE J. of Solid-State Circuits*, vol. 25, pp. 289–298, Feb. 1990.

7. R. Hogervost et al., "A Compact Power-Efficient 3-V CMOS Rail-to-Rail Input/Output Operational Amplifier for VLSI Cell Libraries," *IEEE J. of Solid-State Circuits*, vol. 29, pp. 1505–1513, Dec. 1994.

8. D. A. Johns and K. Martin, *Analog Integrated Circuit Design*, New York: Wiley, 1997.

9. P. E. Allen, B. J. Blalock, and G. A. Rincon, "A 1-V CMOS Op Amp Using Bulk-Driven MOSFETs," ISSCC Dig. of Tech. Papers, pp. 192–193, Feb. 1995.

10. S. Rabii and B. A. Wooley, "A 1.8-V Digital-Audio Sigma-Delta Modulator in 0.8-μm CMOS," *IEEE J. of Solid-State Circuits*, vol. 32, pp. 783–796, June 1997.

Chapter 10

Stability and Frequency Compensation

Negative feedback finds wide application in the processing of analog signals. The properties of feedback described in Chapter 8 allow precise operations by suppressing variations of the open-loop characteristics. Feedback systems, however, suffer from potential instability, that is, they may oscillate.

In this chapter, we deal with the stability and frequency compensation of linear feedback systems to the extent necessary to understand design issues of analog feedback circuits. Beginning with a review of stability criteria and the concept of phase margin, we study frequency compensation, introducing various techniques suited to different op amp topologies. We also analyze the impact of frequency compensation on the slew rate of two-stage op amps.

10.1 General Considerations

Let us consider the negative feedback system shown in Fig. 10.1, where β is assumed constant. Writing the closed-loop transfer function as

$$\frac{Y}{X}(s) = \frac{H(s)}{1 + \beta H(s)}, \tag{10.1}$$

we note that if $\beta H(s = j\omega_1) = -1$, the "gain" goes to infinity, and the circuit can amplify its own noise until it eventually begins to oscillate. In other words, if $\beta H(j\omega_1) = -1$, then

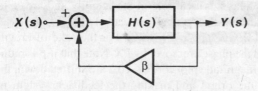

Figure 10.1 Basic negative-feedback system.

345

the circuit may oscillate at frequency ω_1. This condition can be expressed as

$$|\beta H(j\omega_1)| = 1 \tag{10.2}$$

$$\angle \beta H(j\omega_1) = -180°, \tag{10.3}$$

which are called "Barkhausen's Criteria." Note that the total phase shift around the loop at ω_1 is 360° because *negative* feedback itself introduces 180° of phase shift. The 360° phase shift is necessary for oscillation since the feedback signal must add *in phase* to the original noise to allow oscillation buildup. By the same token, a loop gain of unity (or greater) is also required to enable growth of the oscillation amplitude.

In summary, a negative feedback system may oscillate at ω_1 if (1) the phase shift around the loop at this frequency is so much that the feedback becomes *positive* and (2) the loop gain is still enough to allow signal buildup. Illustrated in Fig. 10.2, the situation can be viewed as excessive loop gain at the frequency for which the phase shift reaches −180° or, equivalently, excessive phase at the frequency for which the loop gain drops to unity. Thus, to avoid instability, we must minimize the total phase shift so that for $|\beta H| = 1$, $\angle \beta H$ is still more positive than −180°. In this chapter, we assume β is less than or equal to unity and does not depend on the frequency.

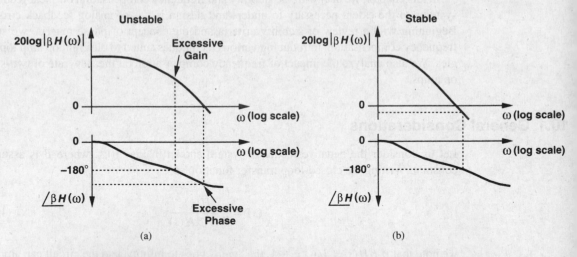

Figure 10.2 Bode plots of loop gain for unstable and stable systems.

The frequencies at which the magnitude and phase of the loop gain are equal to unity and −180°, respectively, play a critical role in the stability and are called the "gain crossover point" and the "phase crossover point," respectively. In a stable system, the gain crossover must occur well before the phase crossover. For the sake of brevity, we denote the gain crossover by GX and the phase crossover by PX. Note that if β is reduced (i.e., less feedback is applied), then the magnitude plots of Fig. 10.2 are shifted down, thereby moving the gain crossover closer to the origin and making the feedback system more stable. Thus, the worst-case stability corresponds to $\beta = 1$, i.e, unity-gain feedback. For this reason, we often analyze the magnitude and phase plots for $\beta H = H$.

Before studying more specific cases, let us review a few basic rules of constructing Bode plots. A Bode plot illustrates the asymptotic behavior of the magnitude and phase of a complex function according to the magnitude of the poles and zeros. The following two rules are used. (1) The slope of the magnitude plot changes by +20 dB/dec at every zero frequency and by −20 dB/dec at every pole frequency. (2) For a pole (zero) frequency of ω_m, the phase begins to fall (rise) at approximately $0.1\omega_m$, experiences a change of −45° (+45°) at ω_m, and approaches a change of −90° (+90°) at approximately $10\omega_m$. The key point here is that the phase may be much more significantly affected by high-frequency poles and zeros than the magnitude is.

It is also instructive to plot the location of the poles of a closed-loop system on a complex plane. Expressing each pole frequency as $s_p = j\omega_p + \sigma_p$ and noting that the impulse response of the system includes a term $\exp(j\omega_p + \sigma_p)t$, we observe that if s_p falls in the right half plane, i.e., if $\sigma_p > 0$, then the system is likely to oscillate because its time-domain response exhibits a growing exponential [Fig. 10.3(a)]. Even if $\sigma_p = 0$, the system may sustain oscillations [Fig. 10.3(b)]. Conversely, if the poles lie in the left half plane, all time-domain exponential terms decay to zero [Fig. 10.3(c)].[1] In practice, we plot

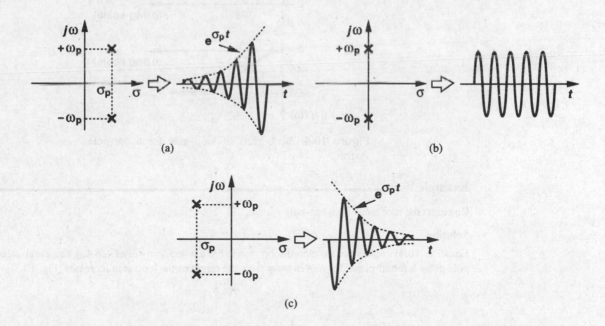

(a)

(b)

(c)

Figure 10.3   Time-domain response of a system versus the position of poles, (a) unstable with growing amplitude, (b) unstable with constant-amplitude oscillation, (c) stable.

---

[1]We ignore the effect of zeros for now.

the location of the poles as the loop gain varies, thereby revealing how close to oscillation the system may come. Such a plot is called a "root locus."

We now study a feedback system incorporating a one-pole feedforward amplifier. Assuming $H(s) = A_0/(1 + s/\omega_0)$, we have from (10.1),

$$\frac{Y}{X}(s) = \frac{\dfrac{A_0}{1 + \beta A_0}}{1 + \dfrac{s}{\omega_0(1 + \beta A_0)}}. \tag{10.4}$$

In order to analyze the stability behavior, we plot $|\beta H(s = j\omega)|$ and $\angle \beta H(s = j\omega)$ (Fig. 10.4), observing that a single pole cannot contribute a phase shift greater than $90°$ and the system is unconditionally stable for all non-negative values of β. Note that $\angle \beta H$ is independent of β.

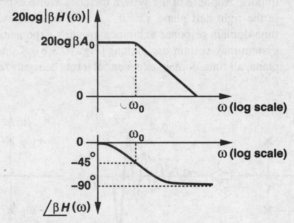

Figure 10.4 Bode plots of loop gain for a one-pole system.

Example 10.1

Construct the root locus for a one-pole system.

Solution

Equation (10.4) implies that the closed-loop system has a pole $s_p = -\omega_0(1 + \beta A_0)$, i.e., a real-valued pole in the left half plane that moves away from the origin as the loop gain increases (Fig. 10.5).

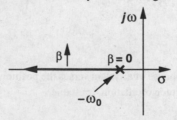

Figure 10.5

10.2 Multipole Systems

Our study of op amps in Chapter 9 indicates that such circuits generally contain multiple poles. In two-stage op amps, for example, each gain stage introduces a "dominant" pole. It is therefore important to study a feedback system whose core amplifier exhibits more than one pole.

Let us consider a two-pole system first. For stability considerations, we plot $|\beta H|$ and $\angle \beta H$ as a function of the frequency. Shown in Fig. 10.6, the magnitude begins to drop at 20 dB/dec at $\omega = \omega_{p1}$ and at 40 dB/dec at $\omega = \omega_{p2}$. Also, the phase begins to change at $\omega = 0.1\omega_{p1}$, reaches $-45°$ at $\omega = \omega_{p1}$ and $-90°$ at $\omega = 10\omega_{p1}$, begins to change again at $\omega = 0.1\omega_{p2}$ (if $0.1\omega_{p2} > 10\omega_{p1}$), reaches $-135°$ at $\omega = \omega_{p2}$, and asymptotically approaches $-180°$. The system is therefore stable because $|\beta H|$ drops to below unity at a frequency for which $\angle \beta H < -180°$.

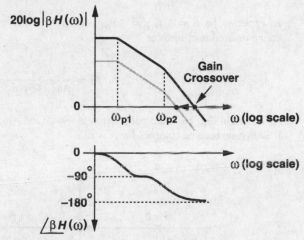

Figure 10.6   Bode plots of loop gain for a two-pole system.

What happens if the feedback is made "weaker?" To reduce the amount of feedback, we decrease β, obtaining the gray magnitude plot in Fig. 10.6. For a logarithmic vertical axis, a change in β translates the magnitude plot vertically. Note that the phase plot does not change. The key point is that as the feedback becomes weaker, the gain crossover point moves toward the origin while the phase crossover point remains constant, resulting in a more stable system. The stability is obtained at the cost of weaker feedback.

Example 10.2

Construct the root locus for a two-pole system.

Solution

Writing the open-loop transfer function as:

$$H(s) = \frac{A_0}{\left(1 + \dfrac{s}{\omega_{p1}}\right)\left(1 + \dfrac{s}{\omega_{p2}}\right)}, \tag{10.5}$$

we have

$$\frac{Y}{X}(s) = \frac{A_0}{\left(1 + \dfrac{s}{\omega_{p1}}\right)\left(1 + \dfrac{s}{\omega_{p2}}\right) + \beta A_0} \tag{10.6}$$

$$= \frac{A_0 \omega_{p1} \omega_{p2}}{s^2 + (\omega_{p1} + \omega_{p2})s + (1 + \beta A_0)\omega_{p1}\omega_{p2}}. \tag{10.7}$$

Thus, the closed-loop poles are given by

$$s_{1,2} = \frac{-(\omega_{p1} + \omega_{p2}) \pm \sqrt{(\omega_{p1} + \omega_{p2})^2 - 4(1 + \beta A_0)\omega_{p1}\omega_{p2}}}{2}. \tag{10.8}$$

As expected, for $\beta = 0$, $s_{1,2} = -\omega_{p1}, -\omega_{p2}$. As β increases, the term under the square root drops, taking on a value of zero for

$$\beta_1 = \frac{1}{A_0} \frac{(\omega_{p1} - \omega_{p2})^2}{4\omega_{p1}\omega_{p2}}. \tag{10.9}$$

As shown in Fig. 10.7, the poles begin at $-\omega_{p1}$ and $-\omega_{p2}$, move toward each other, coincide for $\beta = \beta_1$, and become complex for $\beta > \beta_1$.

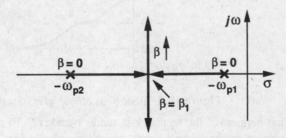

Figure 10.7

The foregoing calculations point to the complexity of the algebra required to construct a root locus for higher-order systems. For this reason, many root locus techniques have been devised so as to minimize such computations [1].

We now study a three-pole system. Shown in Fig. 10.8 are the Bode plots of the magnitude and phase of the loop gain. The third pole gives rise to additional phase shift, possibly moving the phase crossover to frequencies lower than the gain crossover and leading to oscillation. Since the third pole also decreases the *magnitude* of the loop gain at a greater rate, the reader may wonder why the gain crossover does not move as much as the phase crossover does. As mentioned before, the phase begins to change at approximately one-tenth of the pole frequency whereas the magnitude begins to drop only near the pole frequency. For this reason, additional poles (and zeros) impact the phase to a much greater extent than they do the magnitude.

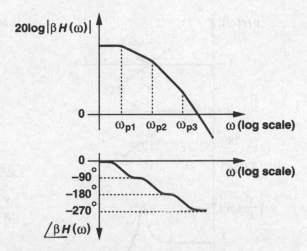

Figure 10.8 Bode plots of loop gain for a three-pole system.

As with a two-pole system, if the feedback factor in Fig. 10.8 decreases, the circuit becomes more stable because the gain crossover moves toward the origin while the phase crossover remains constant.

10.3 Phase Margin

Our foregoing study indicates that to ensure stability, $|\beta H|$ must drop to unity before $\angle \beta H$ crosses $-180°$. We may naturally ask: how far should PX be from GX? Let us first consider a "marginal" case where, as depicted in Fig. 10.9(a), GX is only slightly below PX; sharp peak for example, at GX the phase equals $-175°$. How does the closed-loop system respond in this case? Noting that at GX, $\beta H(j\omega_1) = 1 \times \exp(-j175°)$, we have

$$\frac{Y}{X}(j\omega_1) = \frac{H(j\omega_1)}{1 + \beta H(j\omega_1)} \tag{10.10}$$

$$= \frac{\dfrac{1}{\beta}\exp(-j175°)}{1 + \exp(-j175°)} \tag{10.11}$$

$$= \frac{1}{\beta} \cdot \frac{-0.9962 - j0.0872}{0.0038 - j0.0872}, \tag{10.12}$$

and hence

$$\left| \frac{Y}{X}(j\omega_1) \right| = \frac{1}{\beta} \cdot \frac{1}{0.0872} \tag{10.13}$$

$$\approx \frac{11.5}{\beta}. \tag{10.14}$$

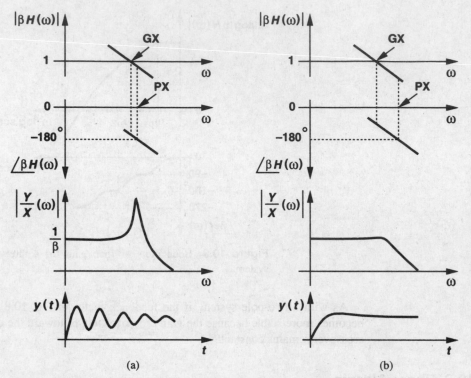

Figure 10.9 Closed-loop frequency and time response for (a) small and (b) large margin between gain and phase crossover points.

Since at low frequencies, $|Y/X| \approx 1/\beta$, the closed-loop frequency response exhibits a sharp peak in the vicinity of $\omega = \omega_1$. In other words, the closed-loop system is near oscillation and its step response exhibits a very underdamped behavior. This point also reveals that a second-order system may suffer from ringing although it is stable.

Now suppose, as shown in Fig. 10.9(b), GX precedes PX by a greater margin. Then, we expect a relatively "well-behaved" closed-loop response in both the frequency domain and the time domain. It is therefore plausible to conclude that the greater the spacing between GX and PX (while GX remains below PX), the more stable the feedback system. Alternatively, the phase of βH at the gain crossover frequency can serve as a measure of stability: the smaller $|\angle \beta H|$ at this point, the more stable the system.

This observation leads us to the concept of "phase margin" (PM), defined as $PM = 180° + \angle \beta H(\omega = \omega_1)$, where ω_1 is the gain crossover frequency.

Example 10.3

A two-pole feedback system is designed such that $|\beta H(\omega_{p2})| = 1$ and $|\omega_{p1}| \ll |\omega_{p2}|$ (Fig. 10.10). What is the phase margin?

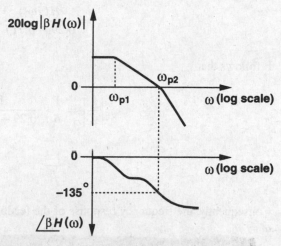

Figure 10.10

Solution

Since $\angle\beta H$ reaches $-135°$ at $\omega = \omega_{p2}$, the phase margin is equal to $45°$.

How much phase margin is adequate? It is instructive to examine the closed-loop frequency response for different phase margins [1]. For $PM = 45°$, at the gain crossover frequency $\angle\beta H(\omega_1) = -135°$ and $|\beta H(\omega_1)| = 1$ (Fig. 10.11), yielding

$$\frac{Y}{X} = \frac{H(j\omega_1)}{1 + 1 \times \exp(-j135°)} \tag{10.15}$$

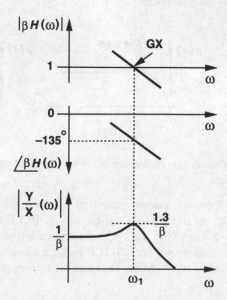

Figure 10.11 Closed-loop frequency response for 45° phase margin.

$$= \frac{H(j\omega_1)}{0.29 - 0.71j}. \tag{10.16}$$

It follows that

$$\left| \frac{Y}{X} \right| = \frac{1}{\beta} \cdot \frac{1}{|0.29 - 0.71j|} \tag{10.17}$$

$$\approx \frac{1.3}{\beta}. \tag{10.18}$$

Consequently, the frequency response of the feedback system suffers from a 30% peak at $\omega = \omega_1$.

It can be shown that for $PM = 60°$, $Y(j\omega_1)/X(j\omega_1) = 1/\beta$, suggesting a negligible frequency peaking. This typically means that the step response of the feedback system exhibits little ringing, providing a fast settling. For greater phase margins, the system is more stable but the time response slows down (Fig. 10.12). Thus, $PM = 60°$ is typically considered the optimum value.

The concept of phase margin is well-suited to the design of circuits that process *small* signals. In practice, the large-signal step response of feedback amplifiers does not follow the illustration of Fig. 10.12. This is not only due to slewing but also because of the nonlinear behavior resulting from large excursions in the bias voltages and currents of the amplifier. Such excursions in fact cause the pole and zero frequencies to *vary* during the transient, leading to a complicated time response. Thus, for large-signal applications, time-domain simulations of the closed-loop system prove more relevant and useful than small-signal ac computations of the open-loop amplifier.

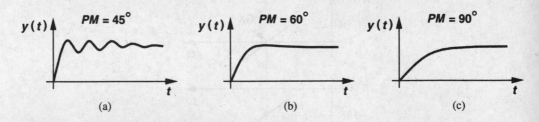

Figure 10.12 Closed-loop time response for 45°, 60°, and 90° phase margins.

As an example of a feedback circuit exhibiting a reasonable phase margin but poor settling behavior, consider the unity-gain amplifier of Fig. 10.13, where the aspect ratio of all transistors is equal to 50 μm / 0.6 μm. With the choice of the device dimensions, bias currents, and capacitor values shown here, SPICE yields a phase margin of approximately 65° and a unity-gain frequency of 150 MHz. The large-signal step response, however, suffers from significant ringing.

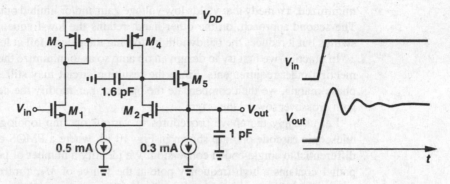

Figure 10.13 Unity-gain buffer.

10.4 Frequency Compensation

Typical op amp circuits contain many poles. In a folded-cascode topology, for example, both the folding node and the output node contribute poles. For this reason, op amps must usually be "compensated," that is, their open-loop transfer function must be modified such that the closed-loop circuit is stable and the time response is well-behaved.

The need for compensation arises because $|\beta H|$ does not drop to unity well before $\angle \beta H$ reaches $-180°$. We then postulate that stability can be achieved by (1) minimizing the overall phase shift, thus pushing the phase crossover *out* [Fig. 10.14(a)]; or (2) dropping the gain, thereby pushing the gain crossover *in* [Fig. 10.14(b)]. The first approach requires that we attempt to minimize the number of poles in the signal path by proper design. Since each additional stage contributes at least one pole, this means the number of stages must be

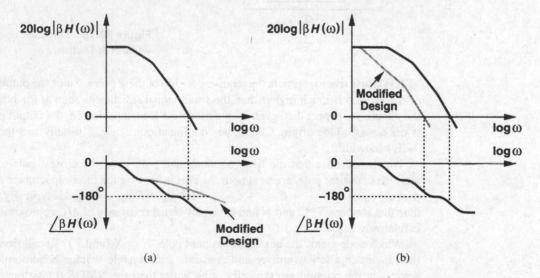

Figure 10.14 Frequency compensation by (a) moving PX out, (b) pushing GX in.

minimized, a remedy that yields low voltage gain and/or limited output swings (Chapter 9). The second approach, on the other hand, retains the low-frequency gain and the output swings but it reduces the bandwidth by forcing the gain to fall at lower frequencies.

In practice, we first try to design an op amp so as to minimize the number of poles while meeting other requirements. Since the resulting circuit may still suffer from insufficient phase margin, we then compensate the op amp, i.e., modify the design so as to move the gain crossover toward the origin.

Let us apply the above procedures to various op amp topologies. We begin with the telescopic cascode op amp shown in Fig. 10.15, where a PMOS current mirror performs differential to single-ended conversion. We identify a number of poles in the signal paths: path 1 contains a high-frequency pole at the source of M_3, a mirror pole at node A, and another high-frequency pole at the source of M_7, whereas path 2 contains a high-frequency pole at the source of M_4. The two paths share a pole at the output.

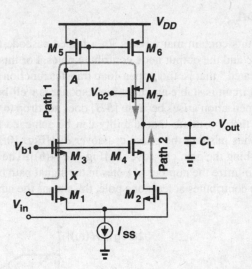

Figure 10.15 Telescopic op amp with single-ended output.

It is instructive to estimate the relative position of these poles. Since the output resistance of the op amp is much higher than the small-signal resistances seen at the other nodes in the circuit, we expect that, even with a moderate load capacitance, the output pole, $\omega_{p,out}$, is the closest to the origin. Called the "dominant pole," $\omega_{p,out}$ usually sets the open-loop 3-dB bandwidth.

We also surmise that the first "nondominant pole," i.e., the closest pole to the origin after the dominant pole, arises at node A. This is because the total capacitance at this node, roughly equal to $C_{GS5} + C_{GS6} + C_{DB5} + 2C_{GD6} + C_{DB3} + C_{GD3}$, is typically quite larger than that at nodes X, Y, and N and the small-signal resistance of M_5, approximately $1/g_{m5}$, is relatively large.

Which node yields the next nondominant pole: N or X (and Y)? Recall from Chapter 9 that, to obtain a low overdrive and consume a reasonable voltage headroom, the PMOS devices in the op amp are typically quite wider than the NMOS transistors. Comparing M_4 and M_7 and neglecting body effect, we note that since $g_m = 2I_D/|V_{GS} - V_{TH}|$, if the two transistors are designed to have the same overdrive, they also exhibit the same transconductance. However, from square-law characteristics, we have $W_4/W_7 = \mu_p/\mu_n$, which is about $1/3$ in today's technologies. Thus, nodes N and X (or Y) see roughly equal small-signal resistances to ground but node N suffers from much more capacitance. It is therefore plausible to assume that node N contributes the next nondominant pole. Figure 10.16 illustrates the results, denoting the capacitance at nodes A, N, and X by C_A, C_N, and C_X, respectively. The poles at nodes X and Y are nearly equal and their

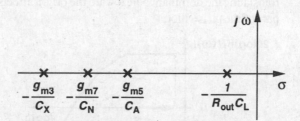

Figure 10.16   Pole locations for the op amp of Fig. 10.15.

corresponding terms in the transfer functions of path 1 and path 2 can be factored out. Thus, they count as one pole rather than two.

With the position of the poles roughly determined, we can construct the magnitude and phase plots for βH, using $\beta = 1$ for the worst case. Shown in Fig. 10.17, such characteristics indicate that the mirror pole typically limits the phase margin because its phase contribution occurs at lower frequencies than that of other nondominant poles.

Recall from Chapter 6 that differential pairs using active current mirrors exhibit a zero located at twice the mirror pole frequency. The circuit of Fig. 10.15 contains such a zero as well. Located at $2\omega_{p,A}$, the zero has some effect on the magnitude and phase characteristics. The analysis is left to the reader.

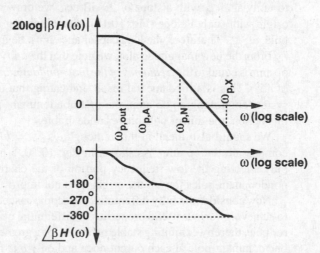

Figure 10.17   Bode plots of loop gain for op amp of Fig. 10.15.

How should we compensate the op amp? Let us assume that the number and location of the nondominant poles and hence the phase plot at frequencies higher than roughly $10\omega_{p,out}$ remain constant. Thus, we must force the loop gain to drop such that the gain crossover point moves toward the origin. To accomplish this, we simply lower the frequency of the dominant pole by increasing the load capacitance. The key point is that the phase contribution of the dominant pole in the vicinity of the gain or phase crossover points is close to 90° and relatively independent of the location of the pole. That is, as illustrated in Fig. 10.18, translating the dominant pole toward the origin affects the magnitude plot but not the critical part of the phase plot.

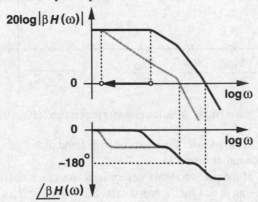

Figure 10.18 Translating the dominant pole toward origin.

In order to understand how much the dominant pole must be shifted down as well as arrive at an important conclusion, let us assume (1) the second nondominant pole ($\omega_{p,N}$) in Fig. 10.15 is quite higher than the mirror pole so that the phase shift at $\omega = \omega_{p,A}$ is equal to −135° and (2) a phase margin of 45° (which is usually inadequate) is necessary. To compensate the circuit, we first identify the frequency at which the phase plot yields the required phase margin, in this case, $\omega_{p,A}$. Since the dominant pole must drop the gain to unity at $\omega_{p,A}$ with a slope of 20 dB/dec, we draw a straight line from $\omega_{p,A}$ toward the origin with such a slope (Fig. 10.19), thus obtaining the new magnitude of the dominant pole, $\omega'_{p,out}$. Therefore, the load capacitance must be increased by a factor of $\omega_{p,out}/\omega'_{p,out}$.

From the new magnitude plot, we note that the unity-gain bandwidth of the compensated op amp is equal to the *frequency of the first nondominant pole* (of course with a phase margin of 45°). This is a fundamental result, indicating that to achieve a wideband in a feedback system employing an op amp, the first nondominant pole must be as far as possible. For this reason, the mirror pole proves undesirable.

We should also mention that although $\omega_{p,out} = (R_{out}C_L)^{-1}$, increasing R_{out} does *not* compensate the op amp. As shown in Fig. 10.20, a higher R_{out} results in a greater gain, only affecting the low-frequency portion of the characteristics. Also, moving one of the nondominant poles toward the origin does not improve the phase margin. (Why?)

Now consider the fully differential telescopic cascode depicted in Fig. 10.21. In addition to achieving various useful properties of differential operation, this topology avoids the mirror pole, thereby exhibiting stable behavior for a greater bandwidth. In fact, we can identify one dominant pole at each output node and only *one* nondominant pole arising from node X (or Y). This suggests that fully differential telescopic cascode circuits are quite stable.

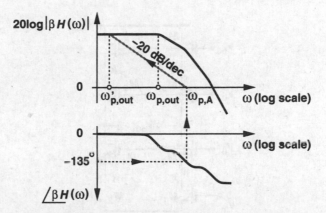

Figure 10.19 Translating the dominant pole toward the origin for 45° phase margin.

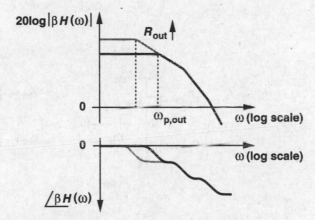

Figure 10.20 Bode plots of loop gain for higher output resistance.

But how about the pole at node N (or K) in Fig. 10.21? Considering one of the PMOS cascodes as shown in Fig. 10.22(a), we may think that the capacitance at node N, $C_N = C_{GS5} + C_{SB5} + C_{GD7} + C_{DB7}$, shunts the output resistance of M_7 at high frequencies, thereby dropping the output impedance of the cascode. To quantify this effect, we first determine Z_{out} in Fig. 10.22(a):

$$Z_{out} = (1 + g_{m5}r_{O5})Z_N + r_{O5}, \tag{10.19}$$

where body effect is neglected and $Z_N = r_{O7}||(C_N s)^{-1}$. Assuming the first term is much greater than the second, we have

$$Z_{out} \approx (1 + g_{m5}r_{O5})\frac{r_{O7}}{r_{O7}C_N s + 1}. \tag{10.20}$$

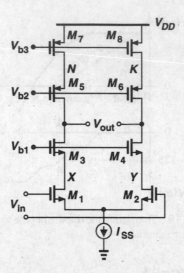

Figure 10.21 Fully differential telescopic op amp.

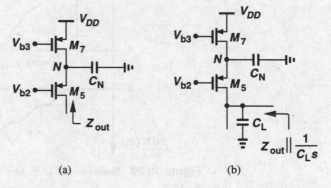

(a)   (b)

Figure 10.22 Effect of device capacitance at internal node of a cascode current source.

Now, as illustrated in Fig. 10.22(b), we take the output load capacitance into account:

$$Z_{out} \| \frac{1}{C_L s} = \frac{(1 + g_{m5} r_{O5}) \dfrac{r_{O7}}{r_{O7} C_N s + 1} \cdot \dfrac{1}{C_L s}}{(1 + g_{m5} r_{O5}) \dfrac{r_{O7}}{r_{O7} C_N s + 1} + \dfrac{1}{C_L s}} \tag{10.21}$$

$$= \frac{(1 + g_{m5} r_{O5}) r_{O7}}{[(1 + g_{m5} r_{O5}) r_{O7} C_L + r_{O7} C_N] s + 1}. \tag{10.22}$$

Thus, the parallel combination of Z_{out} and the load capacitance still contains a single pole corresponding to a time constant $(1 + g_{m5} r_{O5}) r_{O7} C_L + r_{O7} C_N$. Note that $(1 + g_{m5} r_{O5}) r_{O7} C_L$

is simply due to the low-frequency output resistance of the cascode. In other words, the overall time constant equals the "output" time constant plus $r_{O7} C_N$. The key point in this calculation is that the pole in the PMOS cascode is *merged* with the output pole, thus creating no *additional* pole. It merely lowers the dominant pole by a slight amount. For this reason, we loosely say that the signal does not "see" the pole in the cascode current sources.[2]

Comparison of the circuits shown in Figs. 10.15 and 10.21 now reveals that the fully differential configuration avoids both the mirror pole *and* the pole at node N. With the approximation made in (10.22), the circuit of Fig. 10.21 contains only one nondominant pole located at relatively high frequencies owing to the high transconductance of the NMOS transistors. This is a remarkable advantage of fully differential cascode op amps.

We have thus far observed that nondominant poles give rise to instability, requiring frequency compensation. It is possible to cancel one or more of these poles by introducing *zeros* in the transfer function? For example, following the analysis of Fig. 6.31, we surmise that if a low-gain but fast path is placed in parallel with the main amplifier, a zero is created that can be positioned atop the first nondominant pole. However, cancellation of a pole by a zero in the presence of mismatches leads to long settling components in the step response of the closed-loop circuit. This effect is studied in Problem 10.19.

10.5 Compensation of Two-Stage Op Amps

Our study of op amps in Chapter 9 indicates that two-stage topologies may prove inevitable if the output voltage swing must be maximized. Thus, the stability and compensation of such op amps is of interest.

Consider the circuit shown in Fig. 10.23. We identify three poles: a pole at X (or Y), another at E (or F), and a third at A (or B). From our foregoing discussions, we know that the pole at X lies at relatively high frequencies. But how about the other two? Since the small-signal resistance seen at E is quite high, even the capacitances of M_3, M_5, and M_9 can create a pole relatively close to the origin. At node A, the small-signal resistance is lower but the value of C_L may be quite high. Consequently, we say the circuit exhibits *two* dominant poles.

From these observations, we can construct the magnitude and phase plots shown in Fig. 10.24. Here, $\omega_{p,E}$ is assumed more dominant but the relative position of $\omega_{p,E}$ and $\omega_{p,A}$ depends on the design and the load capacitance. Note that, since the poles at E and A are relatively close to the origin, the phase approaches $-180°$ well below the third pole. In other words, the phase margin may be quite close to zero even before the third pole contributes significant phase shift.

Let us now investigate the frequency compensation of two-stage op amps. In Fig. 10.24, one of the dominant poles must be moved toward the origin so as to place the gain crossover well below the phase crossover. However, recall from Section 10.4 that the unity-gain bandwidth after compensation cannot exceed the frequency of the second pole of the open-loop system. Thus, if in Fig. 10.24 the magnitude of $\omega_{p,E}$ is to be reduced, the available

---

[2] If the second term in Eq. (10.19) is included in subsequent derivations, a pole and a zero that are nearly equal appear in the overall output impedance. Nonetheless, for $g_m r_O \gg 1$ and $C_L > C_N$, their effect is negligible.

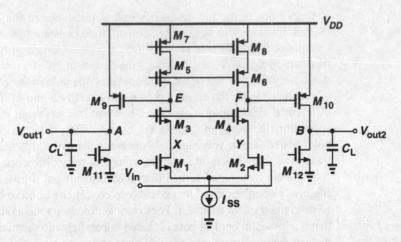

Figure 10.23 Two-stage op amp.

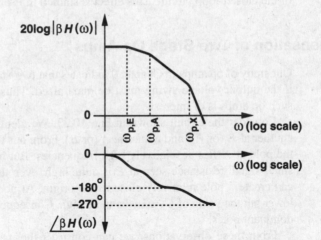

Figure 10.24 Bode plots of loop gain of two-stage op amp.

bandwidth is limited to approximately $\omega_{p,A}$, a low value. Furthermore, the very small magnitude of the required dominant pole translates to a very large compensation capacitor.

Fortunately, a more efficient method of compensation can be applied to the circuit of Fig. 10.23. To arrive at this method, we note that, as illustrated in Fig. 10.25(a), the first stage exhibits a high output impedance and the second stage provides a moderate gain, thereby providing a suitable environment for Miller multiplication of capacitors. Shown in Fig. 10.25(b), the idea is to create a large capacitance at node E, equal to $(1+A_{v2})C_C$, moving the corresponding pole to $R_{out1}^{-1}[C_E + (1 + A_{v2})C_C]^{-1}$, where C_E denotes the capacitance at node E before C_C is added. As a result, a low-frequency pole can be established with a moderate capacitor value, saving considerable chip area. This technique is called "Miller compensation."

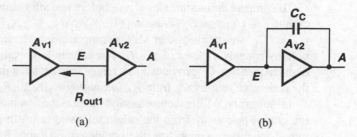

Figure 10.25 Miller compensation of a two-stage op amp.

In addition to lowering the required capacitor value, Miller compensation entails a very important property: it moves the *output* pole *away* from the origin. Illustrated in Fig. 10.26, this effect is called "pole splitting." To understand the underlying principle, we simplify the output stage of Fig. 10.23 as in Fig. 10.27, where R_S denotes the output resistance of the first stage and $R_L = r_{O9} \| r_{O11}$. From our analysis in Chapter 6, we note that this circuit contains two poles:

$$\omega_{p1} \approx \frac{1}{R_S[(1 + g_{m9}R_L)(C_C + C_{GD9}) + C_E] + R_L(C_C + C_{GD9} + C_L)} \quad (10.23)$$

$$\omega_{p2} \approx \frac{R_S[(1 + g_{m9}R_L)(C_C + C_{GD9}) + C_E] + R_L(C_C + C_{GD9} + C_L)}{R_S R_L[(C_C + C_{GD9})C_E + (C_C + C_{GD9})C_L + C_E C_L)]}. \quad (10.24)$$

These expressions are based on the assumption $|\omega_{p1}| \ll |\omega_{p2}|$. Before compensation, however, ω_{p1} and ω_{p2} are of the same order of magnitude. For $C_C = 0$ and relatively large C_L, we may approximate the magnitude of the output pole as $\omega_{p2} \approx 1/(R_L C_L)$.

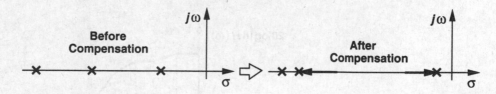

Figure 10.26 Pole splitting as a result of Miller compensation.

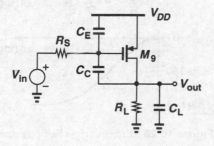

Figure 10.27 Simplified circuit of a two-stage op amp.

To compare the magnitudes of ω_{p2} before and after compensation, we consider a typical case: $C_C + C_{GD9} \gg C_E$, reducing (10.24) to $\omega_{p2} \approx g_{m9}/(C_E + C_L)$. Noting that typically $C_E \ll C_L$, we conclude that Miller compensation increases the magnitude of the output pole by roughly a factor of $g_{m9}R_L$, a relatively large value. Intuitively, this is because at high frequencies, C_C provides a low impedance between the gate and drain of M_9, reducing the resistance seen by C_L from R_L to roughly $R_S \| g_{m9}^{-1} \| R_L \approx g_{m9}^{-1}$.

In summary, Miller compensation moves the interstage pole toward the origin and the output pole away from the origin, allowing a much greater bandwidth than that obtained by merely connecting the compensation capacitor from one node to ground. In practice, the choice of the compensation capacitor for proper phase margin requires some iteration.

Our study of stability and compensation has thus far neglected the effect of *zeros* of the transfer function. While in cascode topologies, the zeros are quite far from the origin, in two-stage op amps incorporating Miller compensation, a nearby zero appears in the circuit. Recall from Chapter 6 that the circuit of Fig. 10.27 contains a right-half-plane zero at $\omega_z = g_{m9}/(C_C + C_{GD9})$. This is because $C_C + C_{GD9}$ forms a "parasitic" signal path from the input to the output. What is the effect of such a zero? The numerator of the transfer function reads $(1 - s/\omega_z)$, yielding a phase of $-\tan^{-1}(\omega/\omega_z)$, a negative value because ω_z is positive. In other words, as with poles in the left half plane, a zero in the right half plane contributes more phase shift, thus moving the phase crossover toward the origin. Furthermore, from Bode approximations, the zero slows down the drop of the magnitude, thereby pushing the gain crossover away from the origin. As a result, the stability degrades considerably.

To better understand the foregoing discussion, let us construct the Bode plots for a third-order system containing a dominant pole ω_{p1}, two nondominant poles ω_{p2} and ω_{p3}, and a zero in the right half plane ω_z. For two-stage op amps, typically $|\omega_{p1}| < |\omega_z| < |\omega_{p2}|$. As shown in Fig. 10.28, the zero introduces significant phase shift while preventing the gain from falling sufficiently.

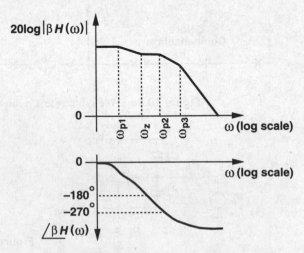

Figure 10.28   Effect of right half plane zero.

The right half plane zero in two-stage CMOS op amps, given by $g_m/(C_C + C_{GD})$, is a serious issue because g_m is relatively small and C_C is chosen large enough to position the dominant pole properly. Various techniques of eliminating or moving the zero have been invented. Illustrated in Fig. 10.29, one approach places a resistor in series with the com-

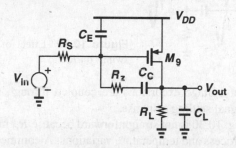

Figure 10.29 Addition of R_z to move the right half plane zero.

pensation capacitor, thereby modifying the zero frequency. The output stage now exhibits *three* poles, but for moderate values of R_z the third pole is located at high frequencies and the first two poles are close to the values calculated with $R_z = 0$. Moreover, it can be shown (Problem 10.8) that the zero frequency is given by

$$\omega_z \approx \frac{1}{C_C \left(g_{m9}^{-1} - R_z\right)}. \tag{10.25}$$

Thus, if $R_z \geq g_{m9}^{-1}$, then $\omega_z \leq 0$. While $R_z = g_{m9}^{-1}$ seems a natural choice, in practice we may even move the zero well into the left half plane so as to cancel the first nondominant pole. This occurs if

$$\frac{1}{C_C \left(g_{m9}^{-1} - R_z\right)} = \frac{-g_{m9}}{C_L + C_E}, \tag{10.26}$$

that is,

$$R_z = \frac{C_L + C_E + C_C}{g_{m9} C_C} \tag{10.27}$$

$$\approx \frac{C_L + C_C}{g_{m9} C_C}, \tag{10.28}$$

because C_E is typically much less than $C_L + C_C$.

The possibility of canceling the nondominant pole makes this technique quite attractive, but in reality two important drawbacks must be considered. First, it is difficult to guarantee the relationship given by (10.28), especially if C_L is unknown or variable. For example, as explained in Chapter 12, the load capacitance seen by an op amp may vary from one part of the period to another in switched-capacitor circuits, necessitating a corresponding change in R_z and complicating the design. The second drawback relates to the actual implementation of R_z. Typically realized by a MOS transistor in the triode region (Fig. 10.30), R_z changes

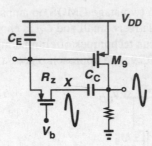

Figure 10.30 Effect of large output swings on R_Z.

substantially as output voltage excursions are coupled through C_C to node X, thereby degrading the large-signal settling response.

Generating V_b in Fig. 10.30 is not straightforward because R_Z must remain equal to $(1 + C_L/C_C)g_{m9}^{-1}$ despite process and temperature variations. A common approach is illustrated in Fig. 10.31 [2], where diode-connected devices M_{13} and M_{14} are placed in series. I_1 is chosen with respect to I_{D9} such that $V_{GS13} = V_{GS9}$, then $V_{GS15} = V_{GS14}$. Since $g_{m14} = \mu_p C_{ox}(W/L)_{14}(V_{GS14} - V_{TH14})$ and $R_{on15} = [\mu_p C_{ox}(W/L)_{15}(V_{GS15} - V_{TH15})]^{-1}$, we have $R_{on15} = g_{m14}^{-1}(W/L)_{14}/(W/L)_{15}$. For pole-zero cancellation to occur,

$$g_{m14}^{-1}\frac{(W/L)_{14}}{(W/L)_{15}} = g_{m9}^{-1}\left(1 + \frac{C_L}{C_C}\right),\tag{10.29}$$

and hence

$$(W/L)_{15} = \sqrt{(W/L)_{14}(W/L)_9}\sqrt{\frac{I_{D9}}{I_{D14}}}\frac{C_C}{C_C + C_L}.\tag{10.30}$$

If C_L is constant, (10.30) can be established with reasonable accuracy because it contains only the *ratio* of quantities.

Another method of guaranteeing Eq. (10.28) is to use a simple resistor for R_Z and define g_{m9} with respect to a resistor that closely matches R_Z [3]. Depicted in Fig. 10.32, this technique incorporates M_{b1}-M_{b4} along with R_S to generate $I_b \propto R_S^{-2}$. (This circuit is studied in detail in Chapter 11.) Thus, $g_{m9} \propto \sqrt{I_{D9}} \propto \sqrt{I_{D11}} \propto R_S^{-1}$. Proper ratioing of R_Z and R_S therefore ensures (10.28) is valid even with temperature and process variations.

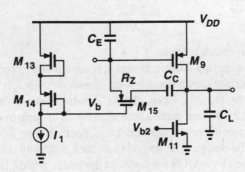

Figure 10.31 Generation of V_b for proper temperature and process tracking.

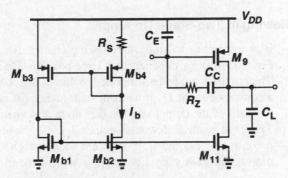

Figure 10.32 Method of defining g_{m9} with respect to R_S.

The principal drawback of the two methods described above is that they assume square-law characteristics for all of the transistors. As described in Chapter 16, short-channel MOSFETs may substantially deviate from the square-law regime, creating errors in the foregoing calculations. In particular, transistor M_9 is typically a short-channel device because it appears in the signal path and its raw speed is critical.

An attribute of two-stage op amps that makes them inferior to "one-stage" op amps is the susceptibility to the load capacitance. Since Miller compensation establishes the dominant pole at the output of the first stage, a higher load capacitance presented to the second stage moves the second pole toward the origin, degrading the phase margin. By contrast, in one-stage op amps, a higher load capacitance brings the *dominant* pole closer to the origin, *improving* the phase margin (albeit making the feedback system more overdamped). Illustrated in Fig. 10.33 is the step response of a unity-gain feedback amplifier employing a one-stage or a two-stage op amp, suggesting that the response approaches an oscillatory behavior if the load capacitance seen by the two-stage op amp increases.

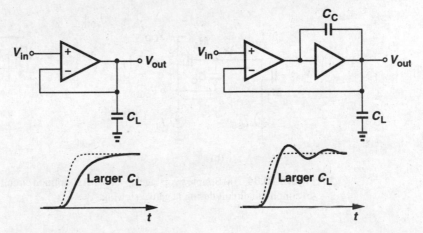

Figure 10.33 Effect of increased load capacitance on step response of one- and two-stage op amps.

10.5.1 Slewing in Two-Stage Op Amps

It is instructive to study the slewing characteristics of two-stage op amps. Suppose in Fig. 10.34(a) V_{in} experiences a large positive step at $t = 0$, turning off M_2, M_4, and M_3. The circuit can then be simplified to that in Fig. 10.34(b), revealing that C_C is charged by a constant current I_{SS} if parasitic capacitances at node X are negligible. Recognizing that the gain of the output stage makes node X a virtual ground, we write: $V_{out} \approx I_{SS}t/C_C$. Thus, the positive slew rate[3] equals I_{SS}/C_C. Note that during slewing, M_5 must provide *two* currents: I_{SS} and I_1. If M_5 is not wide enough to sustain $I_{SS} + I_1$ in saturation, then V_X drops significantly, possibly driving M_1 into the triode region.

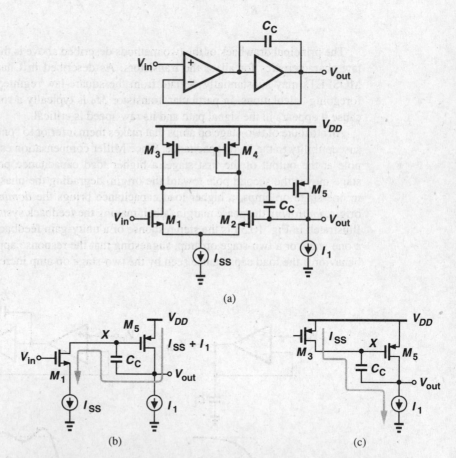

(a)

(b)

(c)

Figure 10.34 (a) Simple two-stage op amp, (b) simplified circuit during positive slewing, (c) simplified circuit during negative slewing.

---

[3]The term positive refers to the slope of the waveform at the output of the op amp.

For the negative slew rate, we simplify the circuit as shown in Fig. 10.34(c). Here I_1 must support both I_{SS} and I_{D5}. For example, if $I_1 = I_{SS}$, then V_X rises so as to turn off M_5. If $I_1 < I_{SS}$, then M_3 enters the triode region and the slew rate is given by I_{D3}/C_C.

10.6 Other Compensation Techniques

The difficulty in compensating two-stage CMOS op amps arises from the feedforward path formed by the compensation capacitor [Fig. 10.35(a)]. If C_C could conduct current from the output node to node X but not vice versa, then the zero would move to a very high

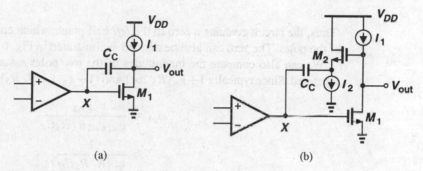

(a)   (b)

Figure 10.35 (a) Two-stage op amp with right half plane zero due to C_C, (b) addition of a source follower to remove the zero.

frequency. As shown in Fig. 10.35(b), this can be accomplished by inserting a source follower in series with the capacitor. Since the gate-source capacitance of M_2 is typically much less than C_C, we expect the right half plane zero to occur at high frequencies. Assuming $\gamma = \lambda = 0$ for the source follower, neglecting some of the device capacitances, and simplifying the circuit as shown in Fig. 10.36, we can write $-g_{m1}V_1 = V_{out}(R_L^{-1} + C_L s)$ and hence

$$V_1 = \frac{-V_{out}}{g_{m1}R_L}(1 + R_L C_L s). \tag{10.31}$$

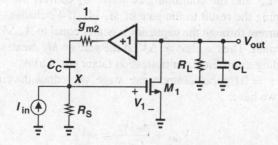

Figure 10.36 Simplified equivalent circuit of Fig. 10.35(b).

We also have

$$\frac{V_{out} - V_1}{\dfrac{1}{g_{m2}} + \dfrac{1}{C_C s}} + I_{in} = \frac{V_1}{R_S}. \tag{10.32}$$

Substituting for V_1 from (10.31) yields:

$$\frac{V_{out}}{I_{in}} = \frac{-g_{m1}R_L R_S (g_{m2} + C_C s)}{R_L C_L C_C (1 + g_{m2}R_S)s^2 + [(1 + g_{m1}g_{m2}R_L R_S)C_C + g_{m2}R_L C_L]s + g_{m2}}. \tag{10.33}$$

Thus, the circuit contains a zero in the *left* half plane, which can be chosen to cancel one of the poles. The zero can also be derived as illustrated in Fig. 6.15.

We can also compute the magnitudes of the two poles assuming that they are widely separated. Since typically $1 + g_{m2}R_S \gg 1$ and $(1 + g_{m1}g_{m2}R_L R_S)C_C \gg g_{m2}R_L C_L$, we have

$$\omega_{p1} \approx \frac{g_{m2}}{g_{m1}g_{m2}R_L R_S C_C} \tag{10.34}$$

$$\approx \frac{1}{g_{m1}R_L R_S C_C}, \tag{10.35}$$

and

$$\omega_{p2} \approx \frac{g_{m1}g_{m2}R_L R_S C_C}{R_L C_L C_C g_{m2}R_S} \tag{10.36}$$

$$\approx \frac{g_{m1}}{C_L}. \tag{10.37}$$

Thus, the new values of ω_{p1} and ω_{p2} are similar to those obtained by simple Miller approximation. For example, the output pole has moved from $(R_L C_L)^{-1}$ to g_{m1}/C_L.

The primary issue in the circuit of Fig. 10.35(b) is that the source follower limits the lower end of the output voltage to $V_{GS2} + V_{I2}$, where V_{I2} is the voltage required across I_2. For this reason, it is desirable to utilize the compensation capacitor to isolate the dc levels in the active feedback stage from that at the output. Such a topology is depicted in Fig. 10.37, where C_C and the common-gate stage M_2 convert the output voltage swing to a current, returning the result to the gate of M_1 [4]. If V_1 changes by ΔV and V_{out} by $A_v \Delta V$, then the current through the capacitor is nearly equal to $A_v \Delta V C_C s$ because $1/g_{m2}$ can be relatively small. Thus, a change ΔV at the gate of M_1 creates a current change of $A_v \Delta V C_C s$, providing a capacitor multiplication factor equal to A_v.

Assuming $\lambda = \gamma = 0$ for the common-gate stage, we redraw the circuit of Fig. 10.37 in Fig. 10.38, where we have:

$$V_{out} + \frac{g_{m2}V_2}{C_C s} = -V_2 \tag{10.38}$$

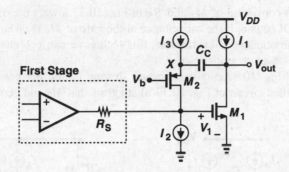

Figure 10.37 Compensation technique using a common-gate stage.

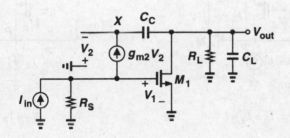

Figure 10.38 Simplified equivalent circuit of Fig. 10.37.

and hence

$$V_2 = -V_{out} \frac{C_C s}{C_C s + g_{m2}}. \tag{10.39}$$

Also,

$$g_{m1} V_1 + V_{out} \left(\frac{1}{R_L} + C_L s \right) = g_{m2} V_2 \tag{10.40}$$

and $I_{in} = V_1/R_S + g_{m2} V_2$. Solving these equations, we obtain

$$\frac{V_{out}}{I_{in}} = \frac{-g_{m1} R_S R_L (g_{m2} + C_C s)}{R_L C_L C_C s^2 + [(1 + g_{m1} R_S) g_{m2} R_L C_C + C_C + g_{m2} R_L C_L] s + g_{m2}}. \tag{10.41}$$

As with the circuit of Fig. 10.35(b), this topology contains a zero in the left half plane. Using similar approximations, we compute the poles as

$$\omega_{p1} \approx \frac{1}{g_{m1} R_L R_S C_C} \tag{10.42}$$

$$\omega_{p2} \approx \frac{g_{m2} R_S g_{m1}}{C_L}. \tag{10.43}$$

Interestingly, the second pole has considerably risen in magnitude — by a factor of $g_{m2} R_S$ with respect to that of the circuit of Fig. 10.35. This is because at very high frequencies, the feedback loop consisting of M_2 and R_S in Fig. 10.37 lowers the output resistance b the same factor. Of course, if the capacitance at the gate of M_1 is taken into account, po splitting is less pronounced. Nevertheless, this technique can potentially provide a hig bandwidth in two-stage op amps.

The op amp of Fig. 10.37 entails important slewing issues. For positive slewing at th output, the simplified circuit of Fig. 10.39(a) suggests that M_2 and hence I_1 must suppo

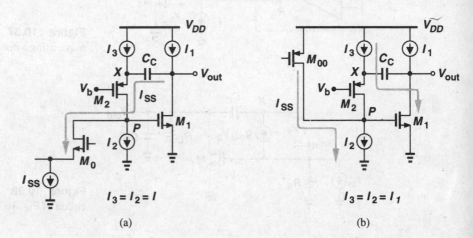

Figure 10.39 Circuit of Fig. 10.37 during (a) positive and (b) negative slewing.

I_{SS}, requiring that $I_1 \geq I_{SS} + I_{D1}$. If I_1 is less, then V_P drops, turning M_1 off, and $I_1 < I_{SS}$, M_0 and its tail current source must enter the triode region, yielding a slew ra equal to I_1/C_C.

For negative slewing, I_2 must support both I_{SS} and I_{D2} [Fig. 10.39(b)]. As I_{SS} flov into node P, V_P tends to rise, increasing I_{D1}. Thus, M_1 absorbs the current produced by through C_C, turning off M_2 and opposing the increase in V_P. We can therefore consider a virtual ground node. This means that, for equal positive and negative slew rates, I_3 (ar hence I_2) must be as large as I_{SS}, raising the power dissipation.

Op amps using a cascode topology as their first stage can incorporate a variant of th technique illustrated in Fig. 10.37. Shown in Fig. 10.40(a), this approach places the con pensation capacitor between the *source* of the cascode devices and the output nodes. Usir the simplified model of Fig. 10.40(b) and the method of Fig. 6.15, the reader can prove th the zero appears at $(g_{m4}R_{eq})(g_{m9}/C_C)$, a much greater magnitude than g_{m9}/C_C. If oth capacitances are neglected, it can also be proved that the dominant pole is located at appro imately $(R_{eq}g_{m9}R_LC_C)^{-1}$, as if C_C were connected to the gate of M_9 rather than the sour of M_4. Also, the first nondominant pole is given by $g_{m4}g_{m9}R_{eq}/C_L$, an effect similar to th described by Eq. (10.43). In reality, the capacitance at X may not be negligible because th resistance seen at this node is quite large. The analysis of the slew rate is left as an exerci for the reader.

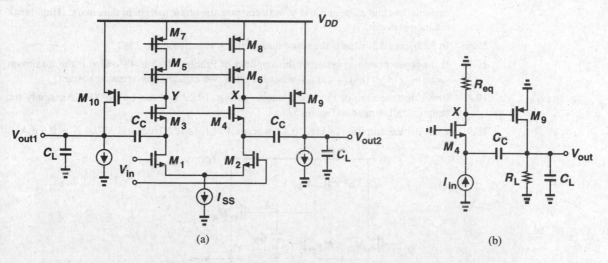

(a)   (b)

Figure 10.40 (a) Alternative method of compensating two-stage op amps, (b) simplified equivalent circuit of (a).

Problems

Unless otherwise stated, in the following problems, use the device data shown in Table 2.1 and assume $V_{DD} = 3$ V where necessary. Also, assume all transistors are in saturation.

10.1. An amplifier with a forward gain of A_0 and two poles at 10 MHz and 500 MHz is placed in a unity-gain feedback loop. Calculate A_0 for a phase margin of $60°$.

10.2. An amplifier with a forward gain of A_0 has two coincident poles at ω_p. Calculate the maximum value of A_0 for a $60°$ phase margin with a closed-loop gain of **(a)** unity, **(b)** 4.

10.3. An amplifier has a forward gain of $A_0 = 1000$ and two poles at ω_{p1} and ω_{p2}. For $\omega_{p1} = 1$ MHz, calculate the phase margin of a unity-gain feedback loop if **(a)** $\omega_{p2} = 2\omega_{p1}$, **(b)** $\omega_{p2} = 4\omega_{p1}$.

10.4. A unity-gain closed-loop amplifier exhibits a frequency peaking of 50% in the vicinity of the gain crossover. What is the phase margin?

10.5. Consider the transimpedance amplifier shown in Fig. 10.41, where $R_D = 1$ kΩ, $R_F = 10$ kΩ, $g_{m1} = g_{m2} = 1/(100\ \Omega)$, and $C_A = C_X = C_Y = 100$ fF. Neglecting all other

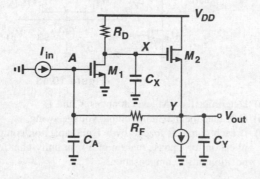

Figure 10.41

capacitances and assuming $\lambda = \gamma = 0$, compute the phase margin of the circuit. (Hint: break the loop at node X.)

10.6. In Problem 10.5, what is the phase margin if R_D is increased to 2 kΩ?

10.7. If the phase margin required of the amplifier of Problem 10.5 is 45°, what is the maximum value of (a) C_Y, (b) C_A, (c) C_X while the other two capacitances remain constant?

10.8. Prove that the zero of the circuit shown in Fig. 10.29 is given by Eq. (10.25). Apply the technique illustrated in Fig. 6.15.

10.9. Consider the amplifier of Fig. 10.42, where $(W/L)_{1-4} = 50/0.5$ and $I_{SS} = I_1 = 0.5$ mA.

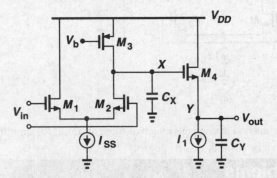

Figure 10.42

(a) Estimate the poles at nodes X and Y by multiplying the small-signal resistance and capacitance to ground. Assume $C_X = C_Y = 0.5$ pF. What is the phase margin for unity-gain feedback?

(b) If $C_X = 0.5$ pF, what is the maximum tolerable value of C_Y that yields a phase margin of 60° for unity-gain feedback?

10.10. Estimate the slew rate of the op amp of Problem 10.9(b) for both parts (a) and (b).

10.11. In the two-stage op amp of Fig. 10.43, $W/L = 50/0.5$ for all transistors except for $M_{5,6}$, for which $W/L = 60/0.5$. Also, $I_{SS} = 0.25$ mA and each output branch is biased at 1 mA.

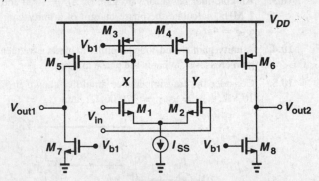

Figure 10.43

(a) Determine the CM level at nodes X and Y.

(b) Calculate the maximum output voltage swing.

(c) If each output is loaded by a 1-pF capacitor, compensate the op amp by Miller multiplication for a phase margin of 60° in unity-gain feedback. Calculate the pole and zero positions after compensation.

(d) Calculate the resistance that must be placed in series with the compensation capacitors to position the zero atop the nondominant pole.

(e) Determine the slew rate.

10.12. In Problem 10.11(e), the pole-zero cancellation resistor is implemented with a PMOS device as in Fig. 10.31. Calculate the dimensions of M_{13}-M_{15} if $I_1 = 100\ \mu A$.

10.13. Calculate the input-referred thermal noise voltage of the op amp shown in Fig. 10.43.

10.14. Figure 10.44 depicts a transimpedance amplifier employing voltage-current feedback. Note that the feedback factor may exceed unity because of M_3. Assume I_1-I_3 are ideal, $I_1 = I_2 = 1$ mA, $I_3 = 10\ \mu A$, $(W/L)_{1,2} = 50/0.5$, and $(W/L)_3 - 5/0.5$.

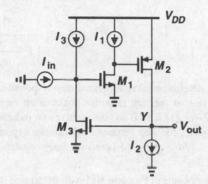

Figure 10.44

(a) Breaking the loop at the gate of M_3, estimate the poles of the open-loop transfer function.

(b) If the circuit is compensated by adding a capacitor C_C between the gate and the drain of M_1, what value of C_C achieves a phase margin of 60°? Determine the poles after compensation.

(c) What resistance must be placed in series with C_C to position the zero of the output stage atop the first nondominant pole?

10.15. Repeat Problem 10.14(c) if the output node is loaded by a 0.5-pF capacitor.

10.16. Suppose in the circuit of Fig. 10.44 a large negative input current is applied such that M_1 turns off momentarily. What is the slew rate at the output?

10.17. Explain why in the circuit of Fig. 10.44, the compensation capacitor should not be placed between the gate and the drain of M_2 or M_3.

10.18. Determine the input-referred noise current of the circuit shown in Fig. 10.44 and described in Problem 10.14(c).

10.19. The cancellation of a pole by a zero, e.g., in a two-stage op amp, entails an issue called the "doublet" problem [5, 6]. If the pole and the zero do not exactly coincide, we say they constitute a doublet. The step response of feedback circuits in the presence of doublets is of great interest. Suppose the open-loop transfer function of a two-stage op amp is expressed as

$$H_{open}(s) = \frac{A_0\left(1 + \dfrac{s}{\omega_z}\right)}{\left(1 + \dfrac{s}{\omega_{p1}}\right)\left(1 + \dfrac{s}{\omega_{p2}}\right)}. \tag{10.44}$$

Ideally, $\omega_z = \omega_{p2}$ and the feedback circuit exhibits a first-order behavior, i.e., its step response contains a single time constant and no overshoot.

(a) Prove that the transfer function of the amplifier in a unity-gain feedback loop is given by

$$H_{closed}(s) = \frac{A_0\left(1 + \dfrac{s}{\omega_z}\right)}{\dfrac{s^2}{\omega_{p1}\omega_{p2}} + \left(\dfrac{1}{\omega_{p1}} + \dfrac{1}{\omega_{p2}} + \dfrac{A_0}{\omega_z}\right)s + A_0 + 1}. \tag{10.45}$$

(b) Determine the two poles of $H_{closed}(s)$, assuming they are widely spaced.

(c) Assuming $\omega_z \approx \omega_{p2}$ and $\omega_{p2} \ll (1 + A_0)\omega_{p1}$, write $H_{closed}(s)$ in the form

$$H_{closed}(s) = \frac{A\left(1 + \dfrac{s}{\omega_z}\right)}{\left(1 + \dfrac{s}{\omega_{pA}}\right)\left(1 + \dfrac{s}{\omega_{pB}}\right)}, \tag{10.46}$$

and determine the small-signal step response of the closed-loop amplifier.

(d) Prove that the step response contains an exponential term of the form $(1 - \omega_z/\omega_{p2})$ $\exp(-\omega_{p2}t)$. This is an important result, indicating that if the zero does not exactly cancel the pole, the step response exhibits an exponential with an amplitude proportional to $1 - \omega_z/\omega_{p2}$ (which depends on the mismatch between ω_z and ω_{p2}) and a time constant of $1/\omega_z$.

10.20. Using the results of Problem 10.19(d), determine the step response of the amplifier described in Problem 10.11(e) with (a) perfect pole-zero cancellation, (b) 10% mismatch between the pole and the zero magnitudes.

References

1. P. R. Gray and R. G. Meyer, *Analysis and Design of Analog Integrated Circuits*, Third Ed., New York: Wiley, 1993.

2. W. C. Black, D. J. Allstot, and R. A. Reed, "A High Performance Low Power CMOS Channel Filter," *IEEE J. of Solid-State Circuits*, vol. 15, pp. 929–938, Dec. 1983.

3. R. M. Ziazadeh, H.-T. Ng, and D. J. Allstot, "A Multistage Amplifier Topology with Embeded Tracking Compensation," *CICC Proc.*, pp. 361–364, May 1998.

4. B. K. Ahuja, "An Improved Frequency Compensation Technique for CMOS Operational Amplifiers," *IEEE J. of Solid-State Circuits*, vol. 18, pp. 629–633, Dec. 1983.

5. P. R. Gray and R. G. Meyer, "MOS Operational Amplifier Design—A Tutorial Overview," *IEEE J. of Solid-State Circuits*, vol. 17, pp. 969–982, Dec. 1982.

6. B. Y. Kamath, R. G. Meyer, and P. R. Gray, "Relationship between Frequency Response and Settling Time of Operational Amplifiers," *IEEE J. of Solid-State Circuits*, vol. 9, pp. 347–352, Dec. 1974.

Chapter 11

Bandgap References

Analog circuits incorporate voltage and current references extensively. Such references are dc quantities that exhibit little dependence on supply and process parameters and a *well-defined* dependence on the temperature. For example, the bias current of a differential pair must be generated according to a reference, for it affects the voltage gain and noise of the circuit. Also, in systems such as A/D and D/A converters, a reference is required to define the input or output full-scale range.

In this chapter, we deal with the design of reference generators in CMOS technology, focusing on well-established "bandgap" techniques. First, we study supply-independent biasing and the problem of start-up. Next, we describe temperature-independent references and examine issues such as the effect of offset voltages. Finally, we present constant-G_m biasing and study an example of state-of-the-art bandgap references.

11.1 General Considerations

As mentioned above, the objective of reference generation is to establish a dc voltage or current that is independent of the supply and process and has a well-defined behavior with temperature. In most applications, the required temperature dependence assumes one of three forms: (a) proportional to absolute temperature (PTAT); (2) constant-G_m behavior, i.e., such that the transconductance of certain transistors remains constant; (3) temperature independent. We can therefore divide the task into two design problems: supply-independent biasing and definition of the temperature variation.

In addition to supply, process, and temperature variability, several other parameters of reference generators may be critical as well. These include output impedance, output noise, and power dissipation. We return to these issues later in this chapter.

11.2 Supply-Independent Biasing

Our use of bias currents and current mirrors in previous chapters has implicitly assumed that a "golden" reference current is available. As shown in Fig. 11.1(a), if I_{REF} does not

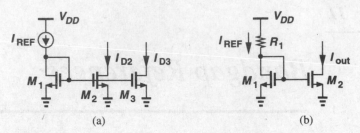

Figure 11.1　Current-mirror biasing using (a) an ideal current source, (b) a resistor.

vary with V_{DD} and channel-length modulation of M_2 and M_3 is neglected, then I_{D2} and I_{D3} remain independent of the supply voltage. The question then is: how do we generate I_{REF}?

As an approximation of a current source, we tie a resistor from V_{DD} to the gate of M_1 [Fig. 11.1(b)]. However, the output current of this circuit is quite sensitive to V_{DD}:

$$\Delta I_{out} = \frac{\Delta V_{DD}}{R_1 + 1/g_{m1}} \cdot \frac{(W/L)_2}{(W/L)_1}. \tag{11.1}$$

In order to arrive at a less sensitive solution, we postulate that the circuit must bias *itself*, i.e., I_{REF} must be somehow derived from I_{out}. The idea is that if I_{out} is to be ultimately independent of V_{DD}, then I_{REF} can be a replica of I_{out}. Fig. 11.2 illustrates an implementation where M_3 and M_4 copy I_{out}, thereby defining I_{REF}. In essence, I_{REF} is "bootstrapped" to I_{out}. With the sizes chosen here, we have $I_{out} = K I_{REF}$ if channel-length modulation is neglected. Note that, since each diode-connected device feeds from a current source, I_{out} and I_{REF} are relatively independent of V_{DD}.

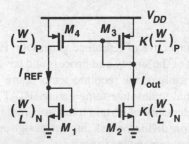

Figure 11.2　Simple circuit to establish supply-independent currents.

Since I_{out} and I_{REF} in Fig. 11.2 display little dependence on V_{DD}, their magnitude is set by other parameters. How do we calculate these currents? Interestingly, if M_1-M_4 operate in saturation and $\lambda \approx 0$, then the circuit is governed by only one equation, $I_{out} = K I_{REF}$, and hence can support *any* current level! For example, if we initially force I_{REF} to be 10 μA, the resulting I_{out} of $K \times 10$ μA "circulates" around the loop, sustaining these current levels in the left and right branches indefinitely.

To uniquely define the currents, we add another constraint to the circuit, e.g., as shown in Fig. 11.3(a). Here, resistor R_S decreases the current of M_2 while the PMOS devices

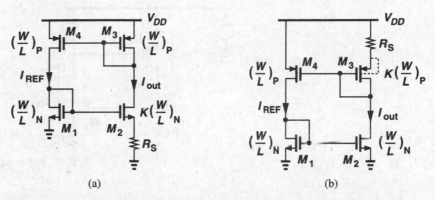

Figure 11.3 (a) Addition of R_S to define the currents, (b) alternative implementation eliminating body effect.

require that $I_{out} = I_{REF}$ because they have identical dimensions. We can write $V_{GS1} = V_{GS2} + I_{D2}R_S$, or

$$\sqrt{\frac{2I_{out}}{\mu_n C_{ox}(W/L)_N}} + V_{TH1} = \sqrt{\frac{2I_{out}}{\mu_n C_{ox}K(W/L)_N}} + V_{TH2} + I_{out}R_S. \tag{11.2}$$

Neglecting body effect, we have

$$\sqrt{\frac{2I_{out}}{\mu_n C_{ox}(W/L)_N}}\left(1 - \frac{1}{\sqrt{K}}\right) = I_{out}R_S, \tag{11.3}$$

and hence

$$I_{out} = \frac{2}{\mu_n C_{ox}(W/L)_N} \cdot \frac{1}{R_S^2}\left(1 - \frac{1}{\sqrt{K}}\right)^2. \tag{11.4}$$

As expected, the current is independent of the supply voltage (but still a function of process and temperature).

The assumption $V_{TH1} = V_{TH2}$ introduces some error in the foregoing calculations because the sources of M_1 and M_2 are at different voltages. Shown in Fig. 11.3(b), a simple remedy is to place the resistor in the source of M_3 while eliminating body effect by tying the source and bulk of each PMOS transistor. Another solution is described in Problem 11.1.

The circuits of Figs. 11.3(a) and (b) exhibit little supply dependence if channel-length modulation is negligible. For this reason, relatively long channels are used for all of the transistors in the circuit.

Example 11.1 _____

Assuming $\lambda \neq 0$ in Fig. 11.3(a), estimate the change in I_{out} for a small change ΔV_{DD} in the supply voltage.

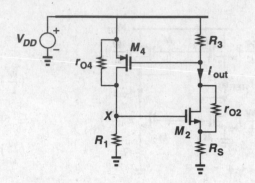

Figure 11.4

Solution

Simplifying the circuit as depicted in Fig. 11.4, where $R_1 = r_{O1} \| (1/g_{m1})$ and $R_3 = r_{O3} \| (1/g_{m3})$, we calculate the "gain" from V_{DD} to I_{out}. The small-signal gate-source voltage of M_4 equals $-I_{out} R_3$ and the current through r_{O4} is $(V_{DD} - V_X)/r_{O4}$. Thus,

$$\frac{V_{DD} - V_X}{r_{O4}} + I_{out} R_3 g_{m4} = \frac{V_X}{R_1}. \tag{11.5}$$

If we denote the equivalent transconductance of M_2 and R_S by $G_{m2} = I_{out}/V_X$, then

$$\frac{I_{out}}{V_{DD}} = \frac{1}{r_{O4}} \left[\frac{1}{G_{m2}(r_{O4} \| R_1)} - g_{m4} R_3 \right]^{-1}. \tag{11.6}$$

Note from Chapter 3 that

$$G_{m2} = \frac{g_{m2} r_{O2}}{R_S + r_{O2} + (g_{m2} + g_{mb2}) R_S r_{O2}}. \tag{11.7}$$

Interestingly, the sensitivity vanishes if $r_{O4} = \infty$.

In some applications, the sensitivity given by (11.6) is prohibitively large. Also, owing to various capacitive paths, the supply sensitivity of the circuit typically rises at high frequencies. For these reasons, the supply voltage of the core is often derived from a locally-generated, less sensitive voltage. We return to this point in Section 11.7.

An important issue in supply-independent biasing is the existence of "degenerate" bias points. In the circuit of Fig. 11.3(a), for example, if all of the transistors carry zero current when the supply is turned on, they may remain off indefinitely because the loop can support a zero current in both branches. This condition is not predicted by (11.4) because in manipulating (11.3) we divided both sides by $\sqrt{I_{out}}$, tacitly assuming $I_{out} \neq 0$. In other words, the circuit can settle in one of *two* different operating conditions.

Called the "start-up" problem, the above issue is resolved by adding a mechanism that drives the circuit out of the degenerate bias point when the supply is turned on. Shown in Fig. 11.5 is a simple example, where the diode-connected device M_5 provides a current path from V_{DD} through M_3 and M_1 to ground upon start-up. Thus, M_3 and M_1, and hence M_2

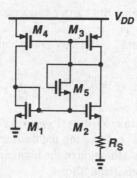

Figure 11.5 Addition of start-up device to the circuit of Fig. 11.3(a).

and M_4, cannot remain off. Of course, this technique is practical only if $V_{TH1} + V_{TH5} + |V_{TH3}| < V_{DD}$ and $V_{GS1} + V_{TH5} + |V_{GS3}| > V_{DD}$, the latter to ensure M_5 remains off after start-up. Another start-up circuit is analyzed in Problem 11.2.

The problem of start-up generally requires careful analysis and simulation. The supply voltage must be ramped from zero in a dc sweep simulation (such that parasitic capacitances do not cause false start-up) as well as in a transient simulation and the behavior of the circuit examined for each supply voltage. In complex implementations, more than one degenerate point may exist.

11.3 Temperature-Independent References

Reference voltages or currents that exhibit little dependence on temperature prove essential in many analog circuits. It is interesting to note that, since most process parameters vary with temperature, if a reference is temperature-independent, then it is usually process-independent as well.

How do we generate a quantity that remains constant with temperature? We postulate that if two quantities having opposite temperature coefficients (TCs) are added with proper weighting, the result displays a zero TC. For example, for two voltages V_1 and V_2 that vary in opposite directions with temperature, we choose α_1 and α_2 such that $\alpha_1 \partial V_1/\partial T + \alpha_2 \partial V_2/\partial T = 0$, obtaining a reference voltage, $V_{REF} = \alpha_1 V_1 + \alpha_2 V_2$, with zero TC.

We must now identify two voltages that have positive and negative TCs. Among various device parameters in semiconductor technologies, the characteristics of bipolar transistors have proven the most reproducible and well-defined quantities that can provide positive and negative TCs. Even though many parameters of MOS devices have been considered for the task of reference generation [1, 2], bipolar operation still forms the core of such circuits.

11.3.1 Negative-TC Voltage

The base-emitter voltage of bipolar transistors or, more generally, the forward voltage of a pn-junction diode exhibits a negative TC. We first obtain the expression for the TC in terms of readily-available quantities.

For a bipolar device we can write $I_C = I_S \exp(V_{BE}/V_T)$, where $V_T = kT/q$. The saturation current I_S is proportional to $\mu k T n_i^2$, where μ denotes the mobility of minority carriers and n_i is the intrinsic minority carrier concentration of silicon. The temperature dependence of these quantities is represented as $\mu \propto \mu_0 T^m$, where $m \approx -3/2$, and $n_i^2 \propto T^3 \exp[-E_g/(kT)]$, where $E_g \approx 1.12$ eV is the bandgap energy of silicon. Thus,

$$I_S = bT^{4+m} \exp \frac{-E_g}{kT}, \tag{11.8}$$

where b is a proportionality factor. Writing $V_{BE} = V_T \ln(I_C/I_S)$, we can now compute the TC of the base-emitter voltage. In taking the derivative of V_{BE} with respect to T, we must know the behavior of I_C as a function of the temperature. To simplify the analysis, we assume for now that I_C is held constant. Thus,

$$\frac{\partial V_{BE}}{\partial T} = \frac{\partial V_T}{\partial T} \ln \frac{I_C}{I_S} - \frac{V_T}{I_S} \frac{\partial I_S}{\partial T}. \tag{11.9}$$

From (11.8), we have

$$\frac{\partial I_S}{\partial T} = b(4+m)T^{3+m} \exp \frac{-E_g}{kT} + bT^{4+m} \left(\exp \frac{-E_g}{kT} \right) \left(\frac{E_g}{kT^2} \right). \tag{11.10}$$

Therefore,

$$\frac{V_T}{I_S} \frac{\partial I_S}{\partial T} = (4+m)\frac{V_T}{T} + \frac{E_g}{kT^2} V_T. \tag{11.11}$$

With the aid of (11.9) and (11.11), we can write

$$\frac{\partial V_{BE}}{\partial T} = \frac{V_T}{T} \ln \frac{I_C}{I_S} - (4+m)\frac{V_T}{T} - \frac{E_g}{kT^2} V_T \tag{11.12}$$

$$= \frac{V_{BE} - (4+m)V_T - E_g/q}{T}. \tag{11.13}$$

Equation (11.13) gives the temperature coefficient of the base-emitter voltage at a given temperature T, revealing dependence on the magnitude of V_{BE} itself. With $V_{BE} \approx 750$ mV and $T = 300°$K, $\partial V_{BE}/\partial T \approx -1.5$ mV/°K.

From (11.13), we note that the temperature coefficient of V_{BE} itself depends on the temperature, creating error in constant reference generation if the positive-TC quantity exhibits a *constant* temperature coefficient.

11.3.2 Positive-TC Voltage

It was recognized in 1964 [3] that if two bipolar transistors operate at unequal current densities, then the *difference* between their base-emitter voltages is directly proportional to the absolute temperature. For example, as shown in Fig. 11.6, if two identical transistors ($I_{S1} = I_{S2}$) are biased at collector currents of nI_0 and I_0 and their base currents are neg-

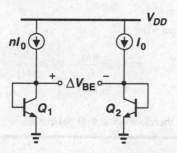

Figure 11.6 Generation of PTAT voltage.

ligible, then

$$\Delta V_{BE} = V_{BE1} - V_{BE2} \tag{11.14}$$

$$= V_T \ln \frac{nI_0}{I_{S1}} - V_T \ln \frac{I_0}{I_{S2}} \tag{11.15}$$

$$= V_T \ln n. \tag{11.16}$$

Thus, the V_{BE} difference exhibits a positive temperature coefficient:

$$\frac{\partial \Delta V_{BE}}{\partial T} = \frac{k}{q} \ln n. \tag{11.17}$$

Interestingly, this TC is independent of the temperature or behavior of the collector currents.[1]

Example 11.2 ———

Calculate ΔV_{BE} in the circuit of Fig. 11.7.

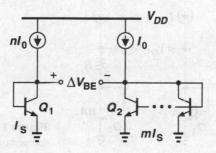

Figure 11.7

---

[1]Nonidealities in the characteristics of bipolar transistors introduce a small temperature dependence in this TC.

Solution

Neglecting base currents, we can write

$$= V_T \ln \frac{nI_0}{I_S} - V_T \ln \frac{I_0}{mI_S} \tag{11.18}$$

$$= V_T \ln(nm). \tag{11.19}$$

The temperature coefficient is therefore equal to $(k/q)\ln(nm)$.

11.3.3 Bandgap Reference

With the negative- and positive-TC voltages obtained above, we can now develop a reference having a nominally zero temperature coefficient. We write $V_{REF} = \alpha_1 V_{BE} + \alpha_2(V_T \ln n)$, where $V_T \ln n$ is the difference between the base-emitter voltages of the two bipolar transistors operating at different current densities. How do we choose α_1 and α_2? Since at room temperature $\partial V_{BE}/\partial T \approx -1.5$ mV/°K whereas $\partial V_T/\partial T \approx +0.087$ mV/°K, we may set $\alpha_1 = 1$ and choose $\alpha_2 \ln n$ such that $(\alpha_2 \ln n)(0.087$ mV/°K$) = 1.5$ mV/°K. That is, $\alpha_2 \ln n \approx 17.2$, indicating that for zero TC:

$$V_{REF} \approx V_{BE} + 17.2V_T \tag{11.20}$$

$$\approx 1.25 \text{ V}. \tag{11.21}$$

Let us now devise a circuit that adds V_{BE} to $17.2V_T$. First, consider the circuit shown in Fig. 11.8, where base currents are assumed negligible, transistor Q_2 consists of n unit transistors in parallel, and Q_1 is a unit transistor. Suppose we somehow force V_{O1} and V_{O2} to be equal. Then, $V_{BE1} = RI + V_{BE2}$ and $RI = V_{BE1} - V_{BE2} = V_T \ln n$. Thus, $V_{O2} = V_{BE2} + V_T \ln n$, suggesting that V_{O2} can serve as a temperature-independent reference if $\ln n \approx 17.2$ (while V_{O1} and V_{O2} remain equal).

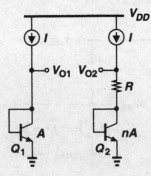

Figure 11.8 Conceptual generation of temperature-independent voltage.

The circuit of Fig. 11.8 requires two modifications to become practical. First, a mechanism must be added to guarantee $V_{O1} = V_{O2}$. Second, since $\ln n = 17.2$ translates to a prohibitively large n, the term $RI = V_T \ln n$ must be scaled up by a reasonable factor. Shown in Fig. 11.9 is an implementation accomplishing both tasks [4]. Here, amplifier

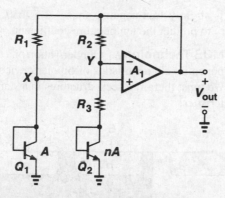

Figure 11.9 Actual implementation of the concept shown in Fig. 11.8.

A_1 senses V_X and V_Y, driving the top terminals of R_1 and R_2 ($R_1 = R_2$) such that X and Y settle to approximately equal voltages. The reference voltage is obtained at the output of the amplifier (rather than at node Y). Following the analysis of Fig. 11.8, we have $V_{BE1} - V_{BE2} = V_T \ln n$, arriving at a current equal to $V_T \ln n / R_3$ through the right branch and hence an output voltage of

$$V_{out} = V_{BE2} + \frac{V_T \ln n}{R_3}(R_3 + R_2) \tag{11.22}$$

$$= V_{BE2} + (V_T \ln n)\left(1 + \frac{R_2}{R_3}\right). \tag{11.23}$$

For a zero TC, we must have $(1 + R_2/R_3) \ln n \approx 17.2$. For example, we may choose $n = 31$ and $R_2/R_3 = 4$. Note these results do not depend on the TC of the resistors.

The circuit of Fig. 11.9 entails a number of design issues. We consider each one below.

Collector Current Variation   The circuit of Fig. 11.9 violates one of our earlier assumptions: the collector currents of Q_1 and Q_2, given by $(V_T \ln n)/R_3$, are proportional to T, whereas $\partial V_{BE}/\partial T \approx -1.5$ mV/°K was derived for a constant current. What happens to the temperature coefficient of V_{BE} if the collector currents are PTAT? As a first-order iterative solution, let us assume $I_{C1} = I_{C2} \approx (V_T \ln n)/R_3$. Returning to Eq. (11.9) and including $\partial I_C/\partial T$, we have

$$\frac{\partial V_{BE}}{\partial T} = \frac{\partial V_T}{\partial T}\ln \frac{I_C}{I_S} + V_T\left(\frac{1}{I_C}\frac{\partial I_C}{\partial T} - \frac{1}{I_S}\frac{\partial I_S}{\partial T}\right). \tag{11.24}$$

Since $\partial I_C/\partial T \approx (V_T \ln n)/(R_3 T) = I_C/T$, we can write

$$\frac{\partial V_{BE}}{\partial T} = \frac{\partial V_T}{\partial T}\ln \frac{I_C}{I_S} + \frac{V_T}{T} - \frac{V_T}{I_S}\frac{\partial I_S}{\partial T}. \tag{11.25}$$

Equation (11.13) is therefore modified as

$$\frac{\partial V_{BE}}{\partial T} = \frac{V_{BE} - (3+m)V_T - E_g/q}{T}, \tag{11.26}$$

indicating that the TC is slightly less negative than -1.5 mV/°K. In practice, accurate simulations are necessary to predict the temperature coefficient.

Compatibility with CMOS Technology   Our derivation of a temperature-independent voltage relies on the exponential characteristics of bipolar devices for both negative- and positive-TC quantities. We must therefore seek structures in a standard CMOS technology that exhibit such characteristics.

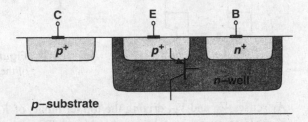

Figure 11.10 Realization of a *pnp* bipolar transistor in CMOS technology.

In *n*-well processes, a *pnp* transistor can be formed as depicted in Fig. 11.10. A p^+ region (the same as the S/D region of PFETs) inside an *n*-well serves as the emitter and the *n*-well itself as the base. The *p*-type substrate acts as the collector and it is inevitably connected to the most negative supply (usually ground). The circuit of Fig. 11.9 can therefore be redrawn as shown in Fig. 11.11.

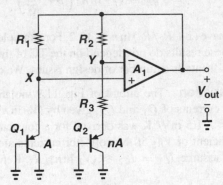

Figure 11.11 Circuit of Fig. 11.9 implemented with *pnp* transistors.

Op Amp Offset and Output Impedance   As explained in Chapter 13, owing to asymmetries, op amps suffer from input "offsets," i.e., the output voltage of the op amp is not zero if the input is set to zero. The input offset voltage of the op amp in Fig. 11.9 introduces error in the output voltage. Included in Fig. 11.12, the effect is quantified as $V_{BE1} - V_{OS} \approx V_{BE2} + R_3 I_{C2}$ (if A_1 is large) and $V_{out} = V_{BE2} + (R_3 + R_2) I_{C2}$. Thus,

$$V_{out} = V_{BE2} + (R_3 + R_2) \frac{V_{BE1} - V_{BE2} - V_{OS}}{R_3} \tag{11.27}$$

$$= V_{BE2} + \left(1 + \frac{R_2}{R_3}\right)(V_T \ln n - V_{OS}), \tag{11.28}$$

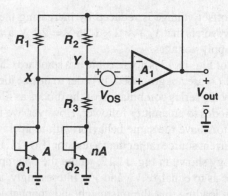

Figure 11.12 Effect of op amp offset on the reference voltage.

where we have assumed $I_{C2} \approx I_{C1}$ despite the offset voltage. The key point here is that V_{OS} is amplified by $1 + R_2/R_3$, introducing error in V_{out}. More importantly, as explained in Chapter 13, V_{OS} itself varies with temperature, raising the temperature coefficient of the output voltage.

Several methods are employed to lower the effect of V_{OS}. First, the op amp incorporates large devices in a carefully chosen topology so as to minimize the offset (Chapter 18). Second, as illustrated in Fig. 11.7, the collector currents of Q_1 and Q_2 can be ratioed by a factor of m such that $\Delta V_{BE} = V_T \ln(mn)$. Third, each branch may use two pn junctions in series to double ΔV_{BE}. Fig. 11.13 depicts a realization using the last two techniques. Here,

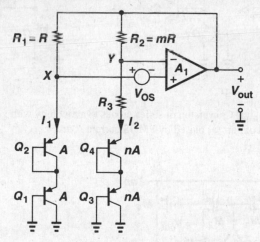

Figure 11.13 Reduction of the effect of op amp offset.

R_1 and R_2 are ratioed by a factor of m, producing $I_1 \approx m I_2$. Neglecting base currents and assuming A_1 is large, we can now write $V_{BE1} + V_{BE2} - V_{OS} = V_{BE3} + V_{BE4} + R_3 I_2$ and $V_{out} = V_{BE3} + V_{BE4} + (R_3 + R_2)I_2$. It follows that

$$V_{out} = V_{BE3} + V_{BE4} + (R_3 + R_2)\frac{2V_T \ln(mn) - V_{OS}}{R_3} \tag{11.29}$$

$$= 2V_{BE} + \left(1 + \frac{R_2}{R_3}\right)[2V_T \ln(mn) - V_{OS}]. \tag{11.30}$$

Thus, the effect of the offset voltage is reduced by increasing the first term in the square brackets. The issue, however, is that $V_{out} \approx 2 \times 1.25 \text{ V} = 2.5 \text{ V}$, a value difficult to generate by the op amp at low supply voltages.

The implementation of Fig. 11.13 is not feasible in a standard CMOS technology because the collectors of Q_2 and Q_4 are not grounded. In order to utilize the bipolar structure shown in Fig. 11.10, we modify the series combination of the diodes as illustrated in Fig. 11.14(a), converting one of the diodes to an emitter follower. However, we must ensure that the bias currents of both transistors have the same behavior with temperature. Thus, we bias each transistor by a PMOS current source rather than a resistor [Fig. 11.14(b)]. The overall circuit then assumes the topology shown in Fig. 11.15, where the op amp adjusts the gate voltage of the PMOS devices so as to equalize V_X and V_Y. Interestingly, in this circuit the op amp experiences no resistive loading, but the mismatch and channel-length modulation of the PMOS devices introduce error at the output [Problem 11.3(d)].

An important concern in the circuit of Fig. 11.15 is the relatively low current gain of the "native" pnp transistors. Since the base currents of Q_2 and Q_4 generate an error in

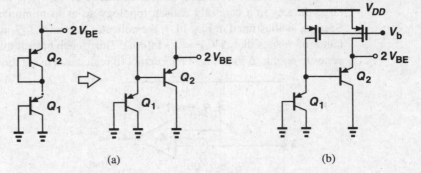

Figure 11.14 (a) Conversion of series diodes to a topology with grounded collectors, (b) circuit of part (a) biased by PMOS current sources.

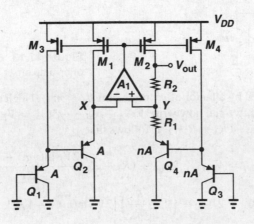

Figure 11.15 Reference generator incorporating two series base-emitter voltages.

the emitter currents of Q_1 and Q_3, a means of base current cancellation may be necessary (Problem 11.5).

Feedback Polarity   In the circuit of Fig. 11.9, the feedback signal produced by the op amp returns to both of its inputs. The negative feedback factor is given by

$$\beta_N = \frac{1/g_{m2} + R_3}{1/g_{m2} + R_3 + R_2},$$ (11.31)

and the positive feedback factor by

$$\beta_P = \frac{1/g_{m1}}{1/g_{m1} + R_1}.$$ (11.32)

To ensure an overall negative feedback, β_P must be less than β_N, preferably by roughly a factor of two so that the circuit's transient response remains well-behaved with large capacitive loads.

Bandgap Reference   The voltage generated according to (11.20) is called a "bandgap reference." To understand the origin of this terminology, let us write the output voltage as

$$V_{REF} = V_{BE} + V_T \ln n$$ (11.33)

and hence:

$$\frac{\partial V_{REF}}{\partial T} = \frac{\partial V_{BE}}{\partial T} + \frac{V_T}{T} \ln n.$$ (11.34)

Setting this to zero and substituting for $\partial V_{BE}/\partial T$ from (11.13), we have

$$\frac{V_{BE} - (4+m)V_T - E_g/q}{T} = -\frac{V_T}{T} \ln n.$$ (11.35)

If $V_T \ln n$ is found from this equation and inserted in (11.33), we obtain:

$$V_{REF} = \frac{E_g}{q} + (4+m)V_T.$$ (11.36)

Thus, the reference voltage exhibiting a nominally-zero TC is given by a few *fundamental* numbers: the bandgap voltage of silicon, E_g/q, the temperature exponent of mobility, m, and the thermal voltage, V_T. The term "bandgap" is used here because as $T \to 0$, $V_{REF} \to E_g/q$.

Supply Dependence and Start-Up   In the circuit of Fig. 11.9, the output voltage is relatively independent of the supply voltage so long as the open-loop gain of the op amp is sufficiently high. The circuit may require a start-up mechanism because if V_X and V_Y are equal to zero, the input differential pair of the op amp may turn off. Start-up techniques similar to those of Fig. 11.5 can be added to ensure the op amp turns on when the supply is applied.

The supply rejection of the circuit typically degrades at high frequencies owing to the op amp's rejection properties, often mandating "supply regulation." An example is described in Section 11.7.

Curvature Correction   If plotted as a function of temperature, bandgap voltages exhibit a finite "curvature," i.e., their TC is typically zero at one temperature and positive or negative at other temperatures (Fig. 11.16). The curvature arises from temperature variation of base-emitter voltages, collector currents, and offset voltages.

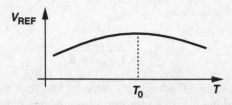

Figure 11.16   Curvature in temperature dependence of a bandgap voltage.

Many curvature correction techniques have been devised to suppress the variation of V_{REF} [5, 6] in bipolar bandgap circuits but they are seldom used in CMOS counterparts. This is because, due to large offsets and process variations, samples of a bandgap reference display substantially different zero-TC temperatures (Fig. 11.17), making it difficult to correct the curvature reliably.

Figure 11.17   Variation of the zero-TC temperature for difference samples.

11.4 PTAT Current Generation

In the analysis of bandgap circuits, we noted that the bias currents of the bipolar transistors are in fact proportional to absolute temperature. Useful in many applications, PTAT currents can be generated by a topology such as that shown in Fig. 11.18. Alternatively, we can combine the supply-independent biasing scheme of Fig. 11.2 with a bipolar core, arriving at Fig. 11.19.[2] Assuming for simplicity that M_1-M_2 and M_3-M_4 are identical pairs, we note that for $I_{D1} = I_{D2}$, the circuit must ensure that $V_X = V_Y$. Thus, $I_{D1} = I_{D2} = (V_T \ln n)/R_1$, yielding the same behavior for I_{D5}. In practice, due to mismatches between the transistors and, more importantly, the temperature coefficient of R_1 the variation of I_{D5} deviates from the ideal equation.

The circuit of Fig. 11.19 can be readily modified to provide a bandgap reference voltage as well. Illustrated in Fig. 11.20, the idea is to add a PTAT voltage $I_{D5}R_2$ to a base-emitter

---

[2]The the two circuits in Figs. 11.18 and 11.19 exhibit difference supply rejections. With a carefully designed op amp, the former achieves a higher rejection.

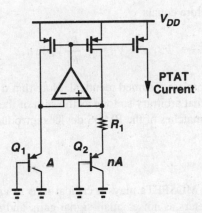

Figure 11.18 Generation of a PTAT current.

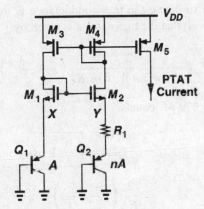

Figure 11.19 Generation of a PTAT current using a simple amplifier.

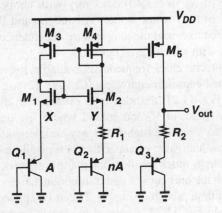

Figure 11.20 Generation of a temperature-independent voltage.

voltage. The output therefore equals

$$V_{REF} = V_{BE3} + \frac{R_2}{R_1} V_T \ln n, \tag{11.37}$$

where all PMOS transistors are assumed identical. Note that the value of V_{BE3} and hence the size of Q_3 are somewhat arbitrary so long as the sum of the two terms in (11.37) gives a zero TC. In reality, mismatches of the PMOS devices introduce error in V_{out}.

11.5 Constant-G_m Biasing

The transconductance of MOSFETs plays a critical role in analog circuits, determining such performance parameters as noise, small-signal gain, and speed. For this reason, it is often desirable to bias the transistors such that their transconductance does not depend on the temperature, process, or supply voltage.

A simple circuit used to define the transconductance is the supply-independent bias topology of Fig. 11.3. Recall that the bias current is given by

$$I_{out} = \frac{2}{\mu_n C_{ox}(W/L)_N} \frac{1}{R_S^2} \left(1 - \frac{1}{\sqrt{K}}\right)^2. \tag{11.38}$$

Thus, the transconductance of M_1 equals

$$g_{m1} = \sqrt{2\mu_n C_{ox} \left(\frac{W}{L}\right)_N I_{D1}} \tag{11.39}$$

$$= \frac{2}{R_S} \left(1 - \frac{1}{\sqrt{K}}\right), \tag{11.40}$$

a value independent of the supply voltage and MOS device parameters.

In reality, the value of R_S in (11.40) does vary with temperature and process. If the temperature coefficient of the resistor is known, bandgap and PTAT reference generation techniques can be utilized to cancel the temperature dependence. Process variations, however, limit the accuracy with which g_{m1} is defined.

In systems where a precise clock frequency is available, the resistor R_S in Fig. 11.3 can be replaced by a switched-capacitor equivalent (Chapter 12) to achieve a somewhat higher accuracy. Depicted in Fig. 11.21, the idea is to establish an average resistance equal to $(C_S f_{CK})^{-1}$ between the source of M_2 and ground, where f_{CK} denotes the clock frequency. Capacitor C_B is added to shunt the high-frequency components resulting from switching to ground. Since the absolute value of capacitors is typically more tightly controlled and since the TC of capacitors is much smaller than that of resistors, this technique provides a higher reproducibility in the bias current and transconductance.

The switched-capacitor approach of Fig. 11.21 can be applied to other circuits as well. For example, as shown in Fig. 11.22, a voltage-to-current converter with a relatively high accuracy can be constructed.

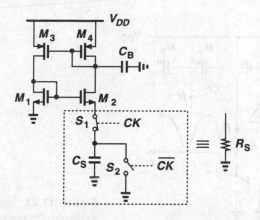

Figure 11.21   Constant-G_m biasing by means of a switched-capacitor "resistor."

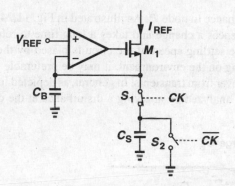

Figure 11.22   Voltage-to-current conversion by means of a switched-capacitor resistor.

11.6 Speed and Noise Issues

Even though reference generators are low-frequency circuits, they may impact the speed of the circuits that they feed. Furthermore, various building blocks may experience "crosstalk" through reference lines. These difficulties arise because of the finite output impedance of reference voltage generators, especially if they incorporate op amps. As an example, let us consider the configuration shown in Fig. 11.23, assuming the voltage at node N is heavily disturbed by the circuit fed by M_5. For fast changes in V_N, the op amp cannot maintain V_P constant and the bias currents of M_5 and M_6 experience large transient changes. Also, the duration of the transient at node P may be quite long if the op amp suffers from a slow response. For this reason, many applications may require a high-speed op amp in the reference generator.

In systems where the power consumed by the reference circuit must be small, the use of a high-speed op amp may not be feasible. Alternatively, the critical node, e.g., node P in Fig. 11.23, can be bypassed to ground by means of a large capacitor (C_B) so as to suppress the effect of external disturbances. This approach involves two issues. First, the stability of the op amp must not degrade with the addition of the capacitor, requiring the op amp to be of one-stage nature (Chapter 10). Second, since C_B generally slows down the transient response of the op amp, its value must be much greater than the capacitance

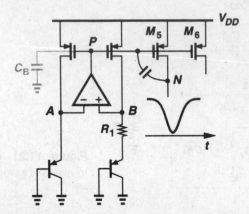

Figure 11.23 Effect of circuit transients on reference voltages and currents.

that couples the disturbance to node P. As illustrated in Fig. 11.24, if C_B is not sufficiently large, then V_P experiences a change and takes a long time to return to its original value, possibly degrading the settling speed of the circuits biased by the reference generator. In other words, depending on the environment, it may be preferable to leave node P agile so that it can quickly recover from transients. In general, as depicted in Fig. 11.25, the response of the circuit must be analyzed by applying a disturbance at the output and observing the settling behavior.

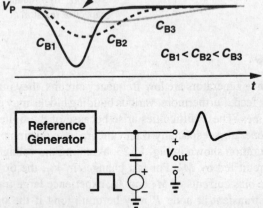

Figure 11.24 Effect of increasing bypass capacitor on the response of reference generator.

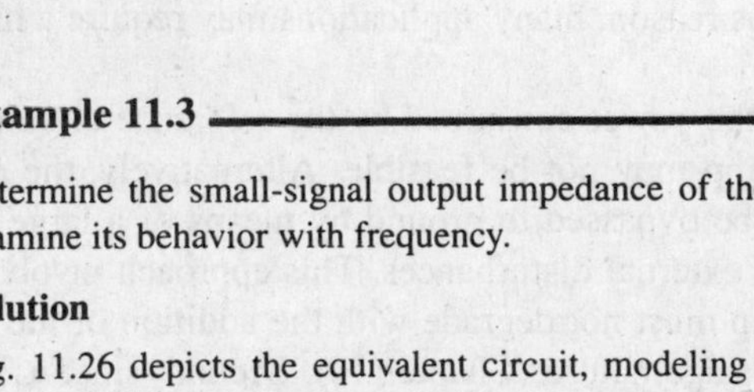

Figure 11.25 Setup for testing the transient response of a reference generator.

Example 11.3

Determine the small-signal output impedance of the bandgap reference shown in Fig. 11.23 and examine its behavior with frequency.

Solution

Fig. 11.26 depicts the equivalent circuit, modeling the open-loop op amp by a one-pole transfer function $A(s) = A_0/(1 + s/\omega_0)$ and an output resistance R_{out} and each bipolar transistor by a

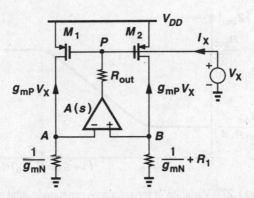

Figure 11.26 Circuit for calculation of the output impedance of a reference generator.

resistance $1/g_{mN}$. If M_1 and M_2 are identical, each having a transconductance of g_{mP}, then their drain currents are equal to $g_{mP}V_X$, producing a differential voltage at the input of the op amp equal to

$$V_{AB} = -g_{mP}V_X \frac{1}{g_{mN}} + g_{mP}V_X \left(\frac{1}{g_{mN}} + R_1 \right) \tag{11.41}$$

$$= g_{mP}V_X R_1. \tag{11.42}$$

The current flowing through R_{out} is therefore given by

$$I_X = \frac{V_X + g_{mP}V_X R_1 A(s)}{R_{out}}, \tag{11.43}$$

yielding

$$\frac{V_X}{I_X} = \frac{R_{out}}{1 + g_{mP}R_1 A(s)} \tag{11.44}$$

$$= \frac{R_{out}}{1 + g_{mP}R_1 \dfrac{A_0}{1 + s/\omega_0}} \tag{11.45}$$

$$= \frac{R_{out}}{1 + g_{mP}R_1 A_0} \frac{1 + \dfrac{s}{\omega_0}}{1 + \dfrac{s}{(1 + g_{mP}R_1 A_0)\omega_0}}. \tag{11.46}$$

Thus, the output impedance exhibits a zero at ω_0 and a pole at $(1 + g_{mP}R_1 A_0)\omega_0$, with the magnitude behavior plotted in Fig. 11.27. Note that $|Z_{out}|$ is quite low for $\omega < \omega_0$, but it rises to a high value as the frequency approaches the pole. In fact, setting $\omega = (1 + g_{mP}R_1 A_0)\omega_0$ and assuming $g_{mP}R_1 A_0 \gg 1$, we have

$$|Z_{out}| = \frac{R_{out}}{1 + g_{mP}R_1 A_0} \left| \frac{1 + j(1 + g_{mP}R_1 A_0)}{1 + j} \right| \tag{11.47}$$

$$= \frac{R_{out}}{\sqrt{2}}, \tag{11.48}$$

which is only 30% lower than the open-loop value.

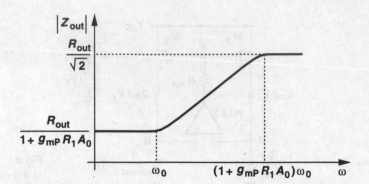

Figure 11.27 Variation of the reference generator output impedance with frequency.

The output noise of reference generators may impact the performance of low-noise circuits considerably. For example, if a high-precision A/D converter employs a bandgap voltage as the reference with which the analog input signal is compared (Fig. 11.28), then the noise in the reference is directly added to the input.

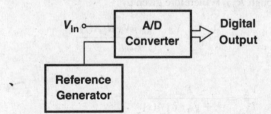

Figure 11.28 A/D converter using a reference generator.

As a simple example, let us calculate the output noise voltage of the circuit shown in Fig. 11.29, taking into account only the input-referred noise voltage of the op amp, $V_{n,op}$. Since the small-signal drain currents of M_1 and M_2 are equal to $V_{n,out}/(R_1 + g_{mN}^{-1})$, we have $V_P = -g_{mP}^{-1}V_{n,out}/(R_1 + g_{mN}^{-1})$, obtaining the differential voltage at the input of the op amp as $-g_{mP}^{-1}A_0^{-1}V_{n,out}/(R_1 + g_{mN}^{-1})$. Beginning from node A, we can then write

$$\frac{V_{n,out}}{R_1 + g_{mN}^{-1}} \cdot \frac{1}{g_{mN}} - \frac{V_{n,out}}{g_{mP}A_0(R_1 + g_{mN}^{-1})} = V_{n,op} + V_{n,out} \qquad (11.49)$$

and hence

$$V_{n,out}\left[\frac{1}{R_1 + g_{mN}^{-1}}\left(\frac{1}{g_{mN}} - \frac{1}{g_{mP}A_0}\right) - 1\right] = V_{n,op}. \qquad (11.50)$$

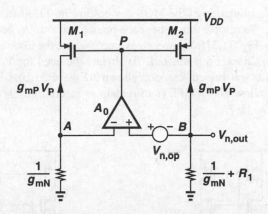

Figure 11.29 Circuit for calculation of noise in a reference generator.

Since typically $g_{mP}A_0 \gg g_{mN} \gg R_1^{-1}$,

$$|V_{n,out}| \approx V_{n,op}, \qquad (11.51)$$

suggesting that the noise of the op amp directly appears at the output. Note that even the addition of a large capacitor from the output to ground may not suppress low-frequency $1/f$ noise components, a serious difficulty in low-noise applications. The noise contributed by other devices in the circuit is studied in Problem 11.6.

11.7 Case Study

In this section, we study a bandgap reference circuit designed for high-precision analog systems [7]. The reference generator incorporates the topology of Fig. 11.19 but with two series base-emitter voltages in each branch so as to reduce the effect of MOSFET mismatches. A simplified version of the core is depicted in Fig. 11.30, where the PMOS current mirror arrangement ensures equal collector currents for Q_1-Q_4.

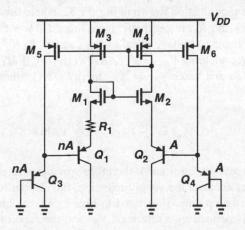

Figure 11.30 Simplified core of the bandgap circuit reported in [7].

Channel-length modulation of the MOS devices in Fig. 11.30 still results in significant supply dependence. To resolve this issue, each branch can employ both NMOS and PMOS cascode topologies. Fig. 11.31(a) shows an example where the low-voltage cascode current mirror described in Chapter 5 is utilized. To obviate the need for V_{b1} and V_{b2}, this design actually introduces a "self-biased" cascode, shown in Fig. 11.31(b), where R_2 and R_3 sustain proper voltages to allow all MOSFETs to remain in saturation. This cascode topology is analyzed in Problem 11.7.

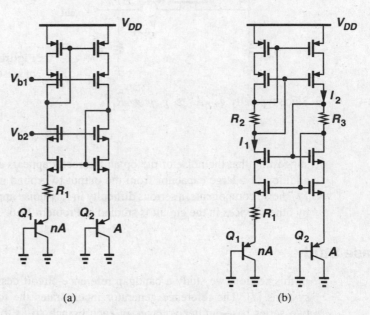

Figure 11.31   (a) Addition of cascode devices to improve supply rejection, (b) use of self-biased cascode to eliminate V_{b1} and V_{b2}.

The bandgap circuit reported in [7] is designed to generate a *floating* reference. This is accomplished by the modification shown in Fig. 11.32, where the drain currents of M_9 and M_{10} flow through R_4 and R_5, respectively. Note that M_{11} sets the gate voltage of M_9 at $V_{BE4} + V_{GS11}$, establishing a voltage equal to V_{BE4} across R_6 if M_9 and M_{11} are identical. Thus, $I_{D9} = V_{BE4}/R_6$, yielding $V_{R4} = V_{BE4}(R_4/R_6)$. Also, if M_{10} is identical to M_2, then $|I_{D10}| = 2(V_T \ln n)/R_1$ and hence $V_{R5} = 2(V_T \ln n)(R_5/R_1)$. Since the op amp ensures that $V_E \approx V_F$, we have

$$V_{out} = \frac{R_4}{R_6} V_{BE4} + 2\frac{R_5}{R_1} V_T \ln n. \tag{11.52}$$

Proper choice of the resistor ratios and n therefore provides a zero temperature coefficient.

In order to further enhance the supply rejection, this design regulates the supply voltage of the core and the op amp. Illustrated in Fig. 11.33, the idea is to generate a local supply, V_{DDL}, that is defined by a reference V_{R1} and the ratio of R_{r1} and R_{r2} and hence

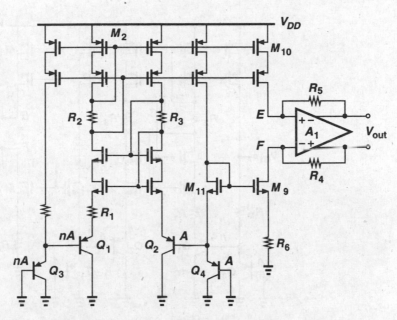

Figure 11.32 Generation of a floating reference voltage.

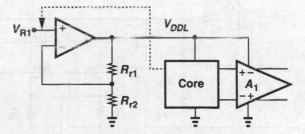

Figure 11.33 Regulation of the supply voltage of the core
and op amp to improve supply rejection.

remains relatively independent of the global supply voltage. But how is V_{R1} itself generated? To minimize the dependence of V_{R1} upon the supply, this voltage is established *inside* the core, as depicted in Fig. 11.34. In fact, R_M is chosen such that V_{R1} is a bandgap reference.

Fig. 11.35 shows the overall implementation, omitting a few details for simplicity. A start-up circuit is also used. Operating from a 5-V supply, the reference generator produces a 2.00-V output while consuming 2.2 mW. The supply rejection is 94 dB at low frequencies, dropping to 58 dB at 100 kHz [7].

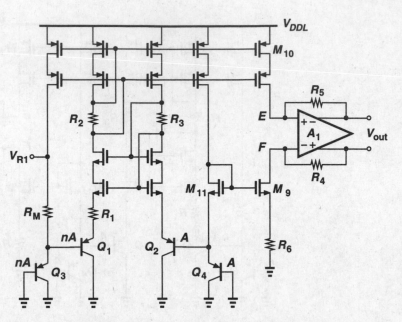

Figure 11.34 Generation of V_{R1}, used in Fig. 11.33.

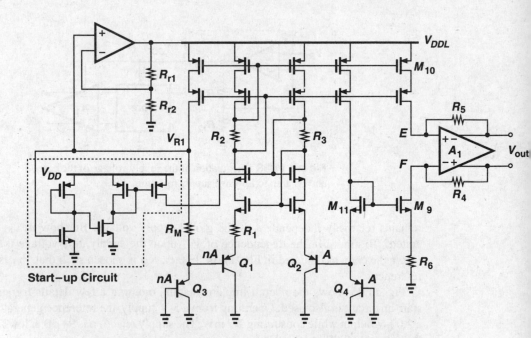

Figure 11.35 Overall circuit of the bandgap generator reported in [7].

Problems

Unless otherwise stated, in the following problems, use the device data shown in Table 2.1 and assume $V_{DD} = 3$ V where necessary.

11.1. Derive an expression for I_{out} in Fig. 11.36.

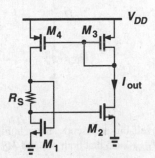

Figure 11.36

11.2. Explain how the start-up circuit shown in Fig. 11.37 operates. Derive a relationship that guarantees $V_X < V_{TH}$ after the circuit turns on.

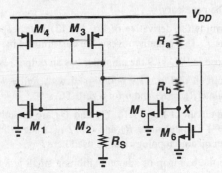

Figure 11.37

11.3. Consider the circuit of Fig. 11.15.
   (a) If M_1 and M_2 suffer from channel-length modulation, what is the error in the output voltage?
   (b) Repeat part (a) for M_3 and M_4.
   (c) If M_1 and M_2 have a threshold mismatch of ΔV, i.e., $V_{TH1} = V_{TH}$ and $V_{TH2} = V_{TH} + \Delta V$, what is the error in the output voltage?
   (d) Repeat part (c) for M_3 and M_4.

11.4. In Fig. 11.15, if the open-loop gain of the op amp A_1 is not sufficiently large, then $|V_X - V_Y|$ exceeds V_e, where V_e is the maximum tolerable error. Calculate the minimum value of A_1 in terms of V_e such that the condition $|V_X - V_Y| < V_e$ is satisfied.

11.5. In the circuit of Fig. 11.15, assume Q_2 and Q_4 have a finite current gain β. Calculate the error in the output voltage.

11.6. Calculate the output noise voltage of the circuit shown in Fig. 11.29 due to the thermal and flicker noise of M_1 and M_2.

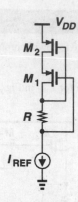

Figure 11.38

11.7. Consider the self-biased cascode shown in Fig. 11.38. Determine the minimum and maximum values of RI_{REF} such that both M_1 and M_2 remain in saturation.

11.8. The circuit of Fig. 11.3(a) sometimes turns on even with no explicit start-up mechanism. Identify the capacitive path(s) that couple the transition on V_{DD} to the internal nodes and hence provide the start-up current.

11.9. Sketch the temperature coefficient of V_{BE} [Eq. (11.13)] versus temperature. Some iteration may be necessary.

11.10. Determine the derivative of Eq. (11.13) with respect to temperature and sketch the result versus T. This quantity reveals the curvature of the voltage.

11.11. Suppose in Fig. 11.9 the amplifier has an output resistance R_{out}. Calculate the error in V_{out}.

11.12. The circuit of Fig. 11.9 is designed with $R_3 = 1$ kΩ and a current of 50 μA through it. Calculate $R_1 = R_2$ and n for a zero TC.

11.13. In the circuit of Fig. 11.15, Q_1 and Q_2 are biased at 100 μA and Q_3 and Q_4 at 50 μA. If $R_1 = 1$ kΩ, calculate R_2 and $(W/L)_{1-4}$ such that the circuit operates with $V_{DD} = 3$ V. Which op amp topology can be used here?

11.14. Since the bandgap of silicon exhibits a small temperature coefficient, Eq. (11.36) suggests that $\partial V_{REF}/\partial T \propto (4 + m)k/q$, a relatively large value, whereas we derived V_{REF} such that it has a zero TC. Explain the flaw in this argument.

11.15. A differential pair with resistive loads is designed such that its voltage gain, $g_m R_D$, has a zero TC at room temperature. If only the temperature dependence of the mobility is considered, determine the required temperature behavior of the tail current. Design a circuit that roughly approximates this behavior.

11.16. In Problem 11.15, assume the tail current is constant but the load resistors exhibit a finite TC. What resistor temperature coefficient cancels the variation of the mobility at room temperature?

11.17. Equation (11.36) suggests that a zero-TC voltage cannot be generated if the supply voltage is as low as, say, 1 V. Figure 11.39 shows a bandgap reference that can operate with low supply voltages [8]. If $R_2 = R_3$, derive an expression for V_{out}.

11.18. Repeat Problem 11.17, if the op amp has an offset voltage V_{OS}.

11.19. Figure 11.40 illustrates a "single-junction" bandgap design [9]. Here, switches S_1 and S_2 are driven by complementary clocks.

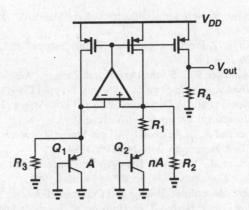

Figure 11.39

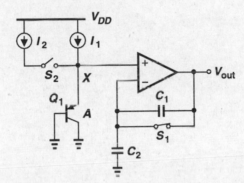

Figure 11.40

(a) What is V_{out} when S_1 is on and S_2 is off?

(b) What is the change in V_{out} when S_1 turns off and S_2 turns on?

(c) How are I_1, I_2, C_1, and C_2 chosen to produce a zero-TC output when S_1 is off?

11.20. Suppose in Fig. 11.40, I_2/I_1 deviates from its nominal value by a small error ϵ. Calculate V_{out} when S_1 is off.

11.21. The circuit of Fig. 11.20 is designed with $(W/L)_{1-4} = 50/0.5$, $I_{D1} = I_{D2} = 50\ \mu A$, $R_1 = 1\ k\Omega$, and $R_2 = 2\ k\Omega$. Assume $\lambda = \gamma = 0$ and Q_3 is identical to Q_1.

(a) Determine n and $(W/L)_5$ such that V_{out} has a zero TC at room temperature.

(b) Neglecting the noise contribution of Q_1-Q_3, calculate the output thermal noise.

11.22. Consider the circuit of Fig. 11.21. Assume $K = 4$, $f_{CK} = 50$ MHz, and a power budget of 1 mW. Determine the aspect ratio of M_1-M_4 and the value of C_S such that $g_{m1} = 1/(500\ \Omega)$.

References

1. R. A. Blauschild et al., "A New NMOS Temperature-Stable Voltage Reference," *IEEE J. of Solid-State Circuits,* vol. 13, pp. 767–774, Dec. 1978.

2. Y. P. Tsividis and R. W. Ulmer, "A CMOS Voltage Reference," *IEEE J. of Solid-State Circuits,* vol. 13, pp. 774–778, Dec. 1978.

3. D. Hilbiber, "A New Semiconductor Voltage Standard," *ISSCC Dig. of Tech. Papers,* pp. 32–33, Feb. 1964.

4. K. E. Kujik, "A Precision Reference Voltage Source," *IEEE J. of Solid-State Circuits,* vol. 8, pp. 222–226, June 1973.

5. G. C. M. Meijer, P. C. Schmall and K. van Zalinge, "A New Curvature-Corrected Bandgap Reference," *IEEE J. of Solid-State Circuits,* vol. 17, pp. 1139–1143, Dec. 1982.

6. M. Gunawan et al., "A Curvature-Corrected Low-Voltage Bandgap Reference," *IEEE J. of Solid-State Circuits,* vol. 28, pp. 667–670, June 1993.

7. T. Brooks and A. L. Westwisk, "A Low-Power Differential CMOS Bandgap Reference," *ISSCC Dig. of Tech. Papers,* pp. 248–249, Feb. 1994.

8. H. Banba et al., "A CMOS Bandgap Reference Circuit with Sub-1-V Operation," *IEEE J. of Solid-State Circuits,* vol. 34, pp. 670–674, May 1999.

9. B. Gilbert, "Monolithic Voltage and Current References: Themes and Variations," pp. 269–352 in *Analog Circuit Design,* J. H. Huijsing, R. J. van de Plassche, and W. M. C. Sansen, Editors, Boston: Kluwer Academic Publishers, 1996.

Chapter 12

Introduction to Switched-Capacitor Circuits

Our study of amplifiers in previous chapters has dealt only with cases where the input signal is continuously available and applied to the circuit and the output signal is continuously observed. Called "continuous-time" circuits, such amplifiers find wide application in audio, video, and high-speed analog systems. In many situations, however, we may sense the input only at periodic instants of time, ignoring its value at other times. The circuit then processes each "sample," producing a valid output at the end of each period. Such circuits are called "discrete-time" or "sampled-data" systems.

In this chapter, we study a common class of discrete-time systems called "switched capacitor (SC) circuits." Our objective is to provide the foundation for more advanced topics such as filters, comparators, ADCs, and DACs. Most of our study deals with switched-capacitor amplifiers but the concepts can be applied to other discrete-time circuits as well. Beginning with a general view of SC circuits, we describe sampling switches and their speed and precision issues. Next, we analyze switched-capacitor amplifiers, considering unity-gain, noninverting, and multiply-by-two topologies. Finally, we examine a switched-capacitor integrator.

12.1 General Considerations

In order to understand the motivation for sampled-data circuits, let us first consider the simple continuous-time amplifier shown in Fig. 12.1(a). Used extensively with bipolar op amps, this circuit presents a difficult issue if implemented in CMOS technology. Recall that, to achieve a high voltage gain, the open-loop output resistance of CMOS op amps is maximized, typically approaching hundreds of kilo-ohms. We therefore suspect that R_2 heavily drops the open-loop gain, degrading the precision of the circuit. In fact, with the aid of the simple equivalent circuit shown in Fig. 12.1(b), we can write

$$-A_v \left(\frac{V_{out} - V_{in}}{R_1 + R_2} R_1 + V_{in} \right) - R_{out} \frac{V_{out} - V_{in}}{R_1 + R_2} = V_{out}, \tag{12.1}$$

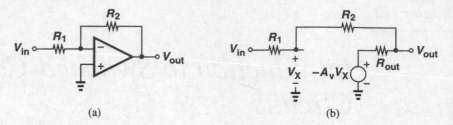

Figure 12.1   (a) Continuous-time feedback amplifier, (b) equivalent circuit of (a).

and hence

$$\frac{V_{out}}{V_{in}} = -\frac{R_2}{R_1} \cdot \frac{A_v - \dfrac{R_{out}}{R_2}}{1 + \dfrac{R_{out}}{R_1} + A_v + \dfrac{R_2}{R_1}}.$$ (12.2)

Equation (12.2) implies that, compared to the case where $R_{out} = 0$, the closed-loop gain suffers from inaccuracies in both the numerator and the denominator. Also, the input resistance of the amplifier, approximately equal to R_1, loads the preceding stage while introducing thermal noise.

Example 12.1

 Using the feedback techniques described in Chapter 8, calculate the closed-loop gain of the circuit of Fig. 12.1(a) and compare the result with Eq. (12.2).

Solution

 With the aid of the approach described in Example 8.9, the reader can prove that

$$\frac{V_{out}}{V_{in}} = \frac{-R_2^2 A_v}{R_2^2 + R_1 R_{out} + R_2 R_{out} + (1 + A_v) R_1 R_2}$$ (12.3)

$$= -\frac{R_2}{R_1} \cdot \frac{A_v}{\dfrac{R_2}{R_1} + \dfrac{R_{out}}{R_2} + \dfrac{R_{out}}{R_1} + 1 + A_v}.$$ (12.4)

The two results are approximately equal if $R_{out}/R_2 \ll A_v$, a condition required to ensure the transmission through R_2 is negligible.

In the circuit of Fig. 12.1(a), the closed-loop gain is set by the ratio of R_2 and R_1. In order to avoid reducing the open-loop gain of the op amp, we postulate that the resistors can be replaced by capacitors [Fig. 12.2(a)]. But, how is the bias voltage at node X set? We may add a large feedback resistor as in Fig. 12.2(b), providing dc feedback while negligibly affecting the ac behavior of the amplifier in the frequency band of interest. Such an arrangement is indeed practical if the circuit senses *only* high-frequency signals. But suppose, for example,

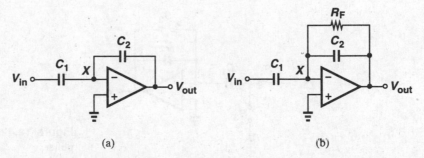

Figure 12.2 (a) Continuous-time feedback amplifier using capacitors, (b) use of resistor to define bias point.

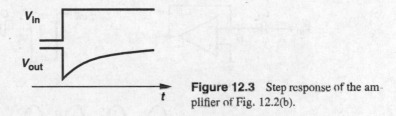

Figure 12.3 Step response of the amplifier of Fig. 12.2(b).

the circuit is to amplify a voltage step. Illustrated in Fig. 12.3, the response contains a step change due to the initial amplification by the circuit consisting of C_1, C_2, and the op amp, followed by a "tail" resulting from the loss of charge on C_2 through R_F. From another point of view, the circuit may not be suited to amplify *wideband* signals because it exhibits a high-pass transfer function. In fact, the transfer function is given by

$$\frac{V_{out}}{V_{in}}(s) \approx -\frac{R_F \dfrac{1}{C_2 s}}{R_F + \dfrac{1}{C_2 s}} \div \frac{1}{C_1 s} \tag{12.5}$$

$$= -\frac{R_F C_1 s}{R_F C_2 s + 1}, \tag{12.6}$$

indicating that $V_{out}/V_{in} \approx -C_1/C_2$ only if $\omega \gg (R_F C_2)^{-1}$.

The above difficulty can be remedied by increasing $R_F C_2$, but in many applications the required values of the two components become prohibitively large. We must therefore seek other methods of establishing the bias while utilizing capacitive feedback networks.

Let us now consider the switched-capacitor circuit depicted in Fig. 12.4, where three switches control the operation: S_1 and S_3 connect the left plate of C_1 to V_{in} and ground, respectively, and S_2 provides unity-gain feedback. We first assume the open-loop gain of the op amp is very large and study the circuit in two phases. First, S_1 and S_2 are on and S_3 is off, yielding the equivalent circuit of Fig. 12.5(a). For a high-gain op amp, $V_B = V_{out} \approx 0$, and hence the voltage across C_1 is approximately equal to V_{in}. Next, at $t = t_0$, S_1 and S_2

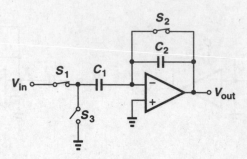

Figure 12.4 Switched-capacitor amplifier.

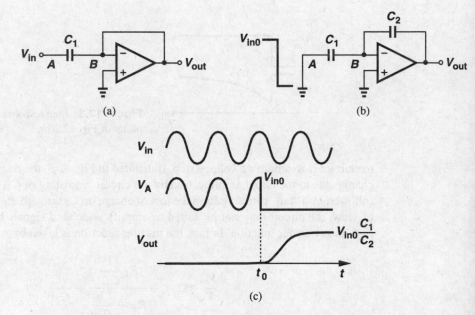

Figure 12.5 Circuit of Fig. 12.4 in (a) sampling mode, (b) amplification mode.

turn off and S_3 turns on, pulling node A to ground. Since V_A changes from V_{in0} to 0, the output voltage must change from zero to $V_{in0}C_1/C_2$.

The output voltage change can also be calculated by examining the transfer of charge. Note that the charge stored on C_1 just before t_0 is equal to $V_{in0}C_1$. After $t = t_0$, the negative feedback through C_2 drives the op amp input differential voltage and hence the voltage across C_1 to zero (Fig. 12.6). The charge stored on C_1 at $t = t_0$ must then be transferred to C_2, producing an output voltage equal to $V_{in0}C_1/C_2$. Thus, the circuit amplifies V_{in0} by a factor of C_1/C_2.

Several attributes of the circuit of Fig. 12.4 distinguish it from continuous-time implementations. First, the circuit devotes some time to "sample" the input, setting the output to zero and providing no amplification during this period. Second, after sampling, for $t > t_0$, the circuit ignores the input voltage V_{in}, amplifying the sampled voltage. Third, the circuit

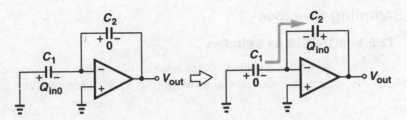

Figure 12.6   Transfer of charge from C_1 to C_2.

configuration changes considerably from one phase to another, as seen in Fig. 12.5(a) and (b), raising concern about its stability.

What is the advantage of the amplifier of Fig. 12.4 over that in Fig. 12.1? In addition to sampling capability, we note from the waveforms depicted in Fig. 12.5 that after V_{out} settles, the current through C_2 approaches zero. That is, the feedback capacitor does not reduce the open-loop gain of the amplifier if the output voltage is given enough time to settle. In Fig. 12.1, on the other hand, R_2 continuously loads the amplifier.

The switched-capacitor amplifier of Fig. 12.4 lends itself to implementation in CMOS technology much more easily than in other technologies. This is because discrete-time operations require switches to perform sampling as well as a high input impedance to sense the stored quantities with no corruption. For example, if the op amp of Fig. 12.4 incorporates bipolar transistors at its input, the base current drawn from the inverting input in the amplification phase [Fig. 12.5(b)] creates an error in the output voltage. The existence of simple switches and a high input impedance have made CMOS technology the dominant choice for sampled-data applications.

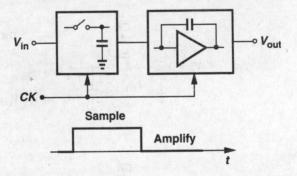

Figure 12.7   General view of switched-capacitor amplifier.

The foregoing discussion leads to the conceptual view illustrated in Fig. 12.7 for switched-capacitor amplifiers. In the simplest case, the operation takes place in two phases: sampling and amplification. Thus, in addition to the analog input, V_{in}, the circuit requires a clock to define each phase.

Our study of SC amplifiers proceeds according to these two phases. First, we analyze various sampling techniques. Second, we consider SC amplifier topologies.

12.2 Sampling Switches

12.2.1 MOSFETS as Switches

A simple sampling circuit consists of a switch and a capacitor [Fig. 12.8(a)]. A MOS transistor can serve as a switch [Fig. 12.8(b)] because (a) it can be on while carrying zero

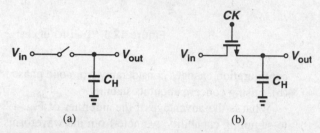

(a)   (b)

Figure 12.8 (a) Simple sampling circuit, (b) implementation of the switch by a MOS device.

current, and (b) its source and drain voltages are not "pinned" to the gate voltage, i.e., if the gate voltage varies, the source or drain voltage need not follow that variation. By contrast, bipolar transistors lack both of these properties, typically necessitating complex circuits to perform sampling.

To understand how the circuit of Fig. 12.8(b) samples the input, first consider the simple cases depicted in Fig. 12.9, where the gate command, CK, goes high at $t = t_0$. In Fig. 12.9(a) we assume that $V_{in} = 0$ for $t \geq t_0$ and the capacitor has an initial voltage equal to V_{DD}.

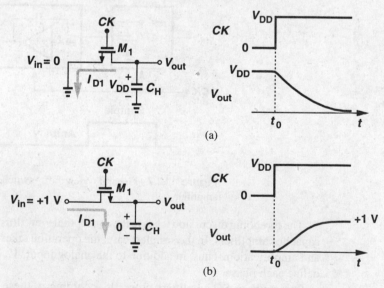

Figure 12.9 Response of a sampling circuit to different input levels and initial conditions.

Thus, at $t = t_0$, M_1 senses a gate-source voltage equal to V_{DD} while its drain voltage is also equal to V_{DD}. The transistor therefore operates in saturation, drawing a current of $I_{D1} = (\mu_n C_{ox}/2)(W/L)(V_{DD} - V_{TH})^2$ from the capacitor. As V_{out} falls, at some point $V_{out} = V_{DD} - V_{TH}$, driving M_1 into the triode region. The device nevertheless continues to discharge C_H until V_{out} approaches zero. We note that for $V_{out} \ll 2(V_{DD} - V_{TH})$, the transistor can be viewed as a resistor equal to $R_{on} = [\mu_n C_{ox}(W/L)(V_{DD} - V_{TH})]^{-1}$.

Now consider the case in Fig. 12.9(b), where $V_{in} = +1$ V, $V_{out}(t = t_0) = 0$ V, and $V_{DD} = 3$ V. Here, the terminal of M_1 connected to C_H acts as the source, and the transistor turns on with $V_{GS} = +3$ V, but $V_{DS} = +1$ V. Thus, M_1 operates in the triode region, charging C_H until V_{out} approaches $+1$ V. For $V_{out} \approx +1$ V, M_1 exhibits an on-resistance of $R_{on} = [\mu_n C_{ox}(W/L)(V_{DD} - V_{in} - V_{TH})]^{-1}$.

The above observations reveal two important points. First, a MOS switch can conduct current in either direction simply by exchanging the role of its source and drain terminals. Second, as shown in Fig. 12.10, when the switch is on, V_{out} follows V_{in} and when the switch

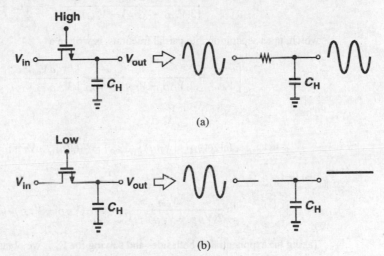

Figure 12.10   Track and hold capabilities of a sampling circuit.

is off, V_{out} remains constant. Thus, the circuit "tracks" the signal when CK is high and "freezes" the instantaneous value of V_{in} across C_H when CK goes low.

Example 12.2 ━━━

In the circuit of Fig. 12.9(a), calculate V_{out} as a function of time. Assume $\lambda = 0$.

Solution

Before V_{out} drops below $V_{DD} - V_{TH}$, M_1 is saturated and we have:

$$V_{out}(t) = V_{DD} - \frac{I_{D1}t}{C_H} \tag{12.7}$$

$$= V_{DD} - \frac{1}{2}\mu_n C_{ox}\frac{W}{L}(V_{DD} - V_{TH})^2 \frac{t}{C_H}. \tag{12.8}$$

After

$$t_1 = \frac{2V_{TH}C_H}{\mu_n C_{ox}\dfrac{W}{L}(V_{DD} - V_{TH})^2},$$ (12.9)

M_1 enters the triode region, yielding a time-dependent current. We therefore write:

$$C_H\frac{dV_{out}}{dt} = -I_{D1}$$ (12.10)

$$= -\frac{1}{2}\mu_n C_{ox}\frac{W}{L}\left[2(V_{DD} - V_{TH})V_{out} - V_{out}^2 \; big\right] \quad t > t_1.$$ (12.11)

Rearranging (12.11), we have

$$\frac{dV_{out}}{[2(V_{DD} - V_{TH}) - V_{out}]V_{out}} = -\frac{1}{2}\mu_n \frac{C_{ox}}{C_H}\frac{W}{L}dt,$$ (12.12)

which, upon separation into partial fractions, is written as

$$\left[\frac{1}{V_{out}} + \frac{1}{2(V_{DD} - V_{TH}) - V_{out}}\right]\frac{dV_{out}}{V_{DD} - V_{TH}} = -\mu_n \frac{C_{ox}}{C_H}\frac{W}{L}dt.$$ (12.13)

Thus,

$$\ln V_{out} - \ln[2(V_{DD} - V_{TH}) - V_{out}] = -(V_{DD} - V_{TH})\mu_n \frac{C_{ox}}{C_H}\frac{W}{L}(t - t_1),$$ (12.14)

that is,

$$\ln \frac{V_{out}}{2(V_{DD} - V_{TH}) - V_{out}} = -(V_{DD} - V_{TH})\mu_n \frac{C_{ox}}{C_H}\frac{W}{L}(t - t_1).$$ (12.15)

Taking the exponential of both sides and solving for V_{out}, we obtain

$$V_{out} = \frac{2(V_{DD} - V_{TH})\exp\left[-(V_{DD} - V_{TH})\mu_n \dfrac{C_{ox}}{C_H} \cdot \dfrac{W}{L}(t - t_1)\right]}{1 + \exp\left[-(V_{DD} - V_{TH})\mu_n \dfrac{C_{ox}}{C_H} \cdot \dfrac{W}{L}(t - t_1)\right]}.$$ (12.16)

In the circuit of Fig. 12.9(b), we assumed $V_{in} = +1$ V (Fig. 12.11). Now suppose $V_{in} = V_{DD}$. How does V_{out} vary with time? Since the gate and drain of M_1 are at the same potential, the transistor is saturated and we have:

$$C_H\frac{dV_{out}}{dt} = I_{D1}$$ (12.17)

$$= \frac{1}{2}\mu_n C_{ox}\frac{W}{L}(V_{DD} - V_{out} - V_{TH})^2,$$ (12.18)

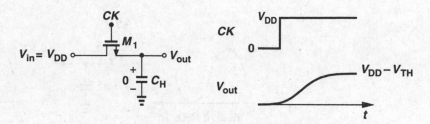

Figure 12.11   Maximum output level in an NMOS sampler.

where channel-length modulation is neglected. It follows that

$$\frac{dV_{out}}{(V_{DD} - V_{out} - V_{TH})^2} = \frac{1}{2}\mu_n\frac{C_{ox}}{C_H}\frac{W}{L}dt, \tag{12.19}$$

and hence

$$\frac{1}{V_{DD} - V_{out} - V_{TH}}\bigg|_0^{V_{out}} = \frac{1}{2}\mu_n\frac{C_{ox}}{C_H}\frac{W}{L}t\bigg|_0^t, \tag{12.20}$$

where body effect is neglected and $V_{out}(t = 0)$ is assumed zero. Thus,

$$V_{out} = V_{DD} - V_{TH} - \frac{1}{\dfrac{1}{2}\mu_n\dfrac{C_{ox}}{C_H}\dfrac{W}{L}t + \dfrac{1}{V_{DD} - V_{TH}}}. \tag{12.21}$$

Equation (12.21) implies that as $t \to \infty$, $V_{out} \to V_{DD} - V_{TH}$. This is because as V_{out} approaches $V_{DD} - V_{TH}$, the overdrive voltage of M_1 vanishes, reducing the current available for charging C_H to negligible values. Of course, even for $V_{out} = V_{DD} - V_{TH}$, the transistor conducts some subthreshold current and, given enough time, eventually brings V_{out} to V_{DD}. Nonetheless, as mentioned in Chapter 3, for typical operation speeds, it is reasonable to assume that V_{out} does not exceed $V_{DD} - V_{TH}$.

The foregoing analysis demonstrates a serious limitation of MOS switches: if the input signal level is close to V_{DD}, then the output provided by an NMOS switch cannot track the input. From another point of view, the on-resistance of the switch increases considerably as the input and output voltages approach $V_{DD} - V_{TH}$. We may then ask: what is the maximum input level that the switch can pass to the output faithfully? In Fig. 12.11, for $V_{out} \approx V_{in}$, the transistor must operate in deep triode region and hence the upper bound of V_{in} equals $V_{DD} - V_{TH}$. As explained below, in practice V_{in} must be quite lower than this value.

Example 12.3

In the circuit of Fig. 12.12, calculate the minimum and maximum on-resistance of M_1. Assume $\mu_n C_{ox} = 50\ \mu\text{A/V}^2$, $W/L = 10/1$, $V_{TH} = 0.7$ V, $V_{DD} = 3$ V, and $\gamma = 0$.

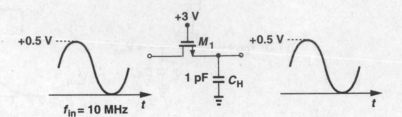

Figure 12.12

Solution

We note that in the steady state, M_1 remains in the triode region because the gate voltage is higher than both V_{in} and V_{out} by a value greater than V_{TH}. If $f_{in} = 10$ MHz, we predict that V_{out} tracks V_{in} with a negligible phase shift due to the on-resistance of M_1 and C_H. Assuming $V_{out} \approx V_{in}$, we need not distinguish between the source and drain terminals, obtaining

$$R_{on1} = \frac{1}{\mu_n C_{ox} \dfrac{W}{L} (V_{DD} - V_{in} - V_{TH})}. \tag{12.22}$$

Thus, $R_{on1,max} \approx 1.11$ kΩ and $R_{on1,min} \approx 870$ Ω. By contrast, if the maximum input level is raised to 1.5 V, then $R_{on1,max} = 2.5$ kΩ.

MOS devices operating in deep triode region are sometimes called "zero-offset" switches to emphasize that they exhibit no dc shift between the input and output voltages of the simple sampling circuit of Fig. 12.8(b).[1] This is evident from examples of Fig. 12.9, where the output eventually becomes equal to the input. Nonexistent in bipolar technology, the zero offset property proves crucial in precise sampling of analog signals.

We have thus far considered only NMOS switches. The reader can verify that the foregoing principles apply to PMOS switches as well. In particular, as shown in Fig. 12.13, a PMOS transistor fails to operate as a zero-offset switch if its gate is grounded and its drain terminal senses an input voltage of $|V_{THP}|$ or less. In other words, the on-resistance of the device rises rapidly as the input and output levels drop to $|V_{THP}|$ above ground.

12.2.2 Speed Considerations

What determines the speed of the sampling circuits of Fig. 12.8? We must first define the speed here. Illustrated in Fig. 12.14, a simple, but versatile measure of speed is the time required for the output voltage to go from zero to the maximum input level after the switch turns on. Since V_{out} would take infinite time to become equal to V_{in0}, we consider the output settled when it is within a certain "error band," ΔV, around the final value. For example, we say the output settles to 0.1% accuracy after t_S seconds, meaning that in Fig. 12.14, $\Delta V / V_{in0} = 0.1\%$. Thus, the speed specification must be accompanied by an accuracy

---

[1] We assume the circuit following the sampler draws no input dc current.

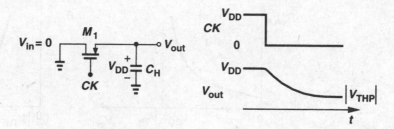

Figure 12.13 Sampling circuit using PMOS switch.

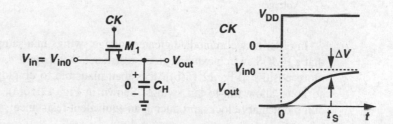

Figure 12.14 Definition of speed in a sampling circuit.

specification as well. Note that after $t = t_S$, we can consider the source and drain voltages to be approximately equal

From the circuit of Fig. 12.14, we surmise that the sampling speed is given by two factors: the on-resistance of the switch and the value of the sampling capacitor. Thus, to achieve a higher speed, a large aspect ratio and a small capacitor must be used. However, as illustrated in Fig. 12.12, the on-resistance also depends on the input level, yielding a greater time constant for more positive inputs (in the case of NMOS switches). From Eq. (12.22), we plot the on-resistance of the switch as a function of the input level [Fig. 12.15(a)], noting the sharp rise as V_{in} approaches $V_{DD} - V_{TH}$. For example, if we restrict the variation of R_{on} to a range of 4 to 1, then the maximum input level is given by

$$\frac{1}{\mu_n C_{ox} \dfrac{W}{L}(V_{DD} - V_{in,max} - V_{TH})} = \frac{4}{\mu_n C_{ox} \dfrac{W}{L}(V_{DD} - V_{TH})}. \tag{12.23}$$

That is,

$$V_{in,max} = \frac{3}{4}(V_{DD} - V_{TH}). \tag{12.24}$$

This value falls around $V_{DD}/2$, translating to severe swing limitations. Note that the device threshold voltage directly limits the voltage swings.[2]

---

[2]By contrast, the output swing of cascode stages is typically limited by overdrive voltages rather than the threshold voltage.

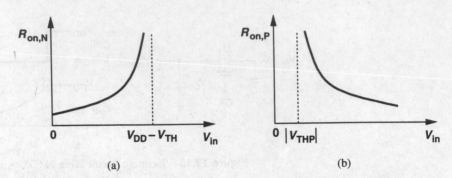

Figure 12.15   On-resistance of (a) NMOS and (b) PMOS devices as a function of input voltage.

In order to accommodate greater voltage swings in a sampling circuit, we first observe that a PMOS switch exhibits an on-resistance that *decreases* as the input voltage becomes more positive [Fig. 12.15(b)]. It is then plausible to employ "complementary" switches so as to allow rail-to-tail swings. Shown in Fig. 12.16(a), such a combination requires complementary clocks, producing an equivalent resistance:

$$R_{on,eq} = R_{on,N} \| R_{on,P} \tag{12.25}$$

$$= \frac{1}{\mu_n C_{ox}\left(\frac{W}{L}\right)_N (V_{DD} - V_{in} - V_{THN})} \left\| \frac{1}{\mu_p C_{ox}\left(\frac{W}{L}\right)_P (V_{in} - |V_{THP}|)} \right. \tag{12.26}$$

$$= \frac{1}{\mu_n C_{ox}\left(\frac{W}{L}\right)_N (V_{DD} - V_{THN}) - \left[\mu_n C_{ox}\left(\frac{W}{L}\right)_N - \mu_p C_{ox}\left(\frac{W}{L}\right)_P\right] V_{in} - \mu_p C_{ox}\left(\frac{W}{L}\right)_P |V_{THP}|}. \tag{12.27}$$

Interestingly, if $\mu_n C_{ox}(W/L)_N = \mu_p C_{ox}(W/L)_P$, then $R_{on,eq}$ is independent of the input level.[3] Figure 12.16(b) plots the behavior of $R_{on,eq}$ in the general case, revealing much less variation than that corresponding to each switch alone.

For high-speed input signals, it is critical that the NMOS and PMOS switches in Fig. 12.16(a) turn off simultaneously so as to avoid ambiguity in the sampled value. If, for example, the NMOS device turns off Δt seconds earlier than the PMOS device, then the output voltage tends to track the input for the remaining Δt seconds, but with a large, input-dependent time constant (Fig. 12.17). This effect gives rise to distortion in the sampled value. For moderate precision, the simple circuit shown in Fig. 12.18 provides complementary clocks by duplicating the delay of inverter I_1 through the gate G_2.

---

[3] In reality, V_{THN} and V_{THP} vary with V_{in} through body effect but we ignore this variation here.

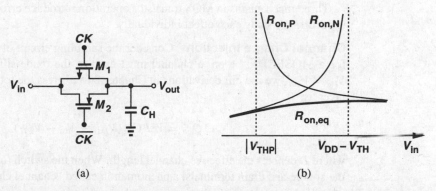

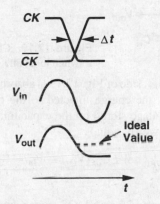

(a)

(b)

Figure 12.16 (a) Complementary switch, (b) on-resistance of the complementary switch.

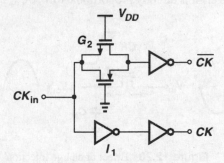

Figure 12.17 Distortion generated if complementary switches do not turn off simultaneously.

Figure 12.18 Simple circuit generating complementary clocks.

12.2.3 Precision Considerations

Our foregoing study of MOS switches indicates that a larger W/L or a smaller sampling capacitor results in a higher speed. In this section, we show that these methods of increasing the speed degrade the precision with which the signal is sampled.

Three mechanisms in MOS transistor operation introduce error at the instant the switch turns off. We study each effect individually.

Channel Charge Injection   Consider the sampling circuit of Fig. 12.19 and recall that for a MOSFET to be on, a channel must exist at the oxide-silicon interface. Assuming $V_{in} \approx V_{out}$, we use our derivations in Chapter 2 to express the total charge in the inversion layer as

$$Q_{ch} = WLC_{ox}(V_{DD} - V_{in} - V_{TH}), \tag{12.28}$$

where L denotes the effective channel length. When the switch turns off, Q_{ch} exits through the source and drain terminals, a phenomenon called "channel charge injection."

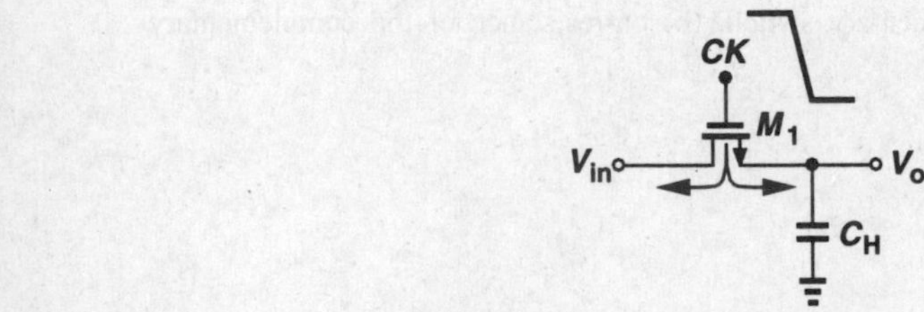

Figure 12.19   Charge injection when a switch turns off.

The charge injected to the left side of Fig. 12.19 is absorbed by the input source, creating no error. On the other hand, the charge injected to the right side is deposited on C_H, introducing an error in the voltage stored on the capacitor. For example, if half of Q_{ch} is injected onto C_H, the resulting error equals

$$\Delta V = \frac{WLC_{ox}(V_{DD} - V_{in} - V_{TH})}{2C_H}. \tag{12.29}$$

Illustrated in Fig. 12.20, the error for an NMOS switch appears as a negative "pedestal" at the output. Note that the error is directly proportional to WLC_{ox} and inversely proportional to C_H.

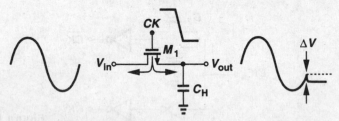

Figure 12.20   Effect of charge injection.

An important question that arises now is: why did we assume in arriving at (12.29) that exactly *half* of the channel charge is injected onto C_H? In reality, the fraction of charge that exits through the source and drain terminals is a relatively complex function of various parameters such as the impedance seen at each terminal to ground and the transition time of the clock [1, 2]. Investigations of this effect have not yielded any rule of thumb that can predict the charge splitting in terms of such parameters. Furthermore, in many cases, these parameters, e.g., the clock transition time, are poorly controlled. Also, most circuit simulation programs model charge injection quite inaccurately. As a worst-case estimate, we can assume that the entire channel charge is injected onto the sampling capacitor.

How does charge injection affect the precision? Assuming all of the charge is deposited on the capacitor, we express the sampled output voltage as

$$V_{out} \approx V_{in} - \frac{WLC_{ox}(V_{DD} - V_{in} - V_{TH})}{C_H}, \tag{12.30}$$

where the phase shift between the input and output is neglected. Thus,

$$V_{out} = V_{in}\left(1 + \frac{WLC_{ox}}{C_H}\right) - \frac{WLC_{ox}}{C_H}(V_{DD} - V_{TH}), \tag{12.31}$$

suggesting that the output deviates from the ideal value through two effects: a non-unity gain equal to $1 + WLC_{ox}/C_H$,[4] and a constant offset voltage $-WLC_{ox}(V_{DD} - V_{TH})/C_H$ (Fig. 12.21). In other words, since we have assumed channel charge is a *linear* function of the input voltage, the circuit exhibits only gain error and dc offset.

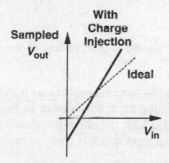

Figure 12.21 Input/output characteristic of sampling circuit in the presence of charge injection.

In the foregoing discussion, we tacitly assumed that V_{TH} is constant. However, for NMOS switches (in an n-well technology), body effect must be taken into account.[5] Since $V_{TH} = V_{TH0} + \gamma(\sqrt{2\phi_B + V_{SB}} - \sqrt{2\phi_B})$, and $V_{BS} \approx -V_{in}$, we have

$$V_{out} = V_{in} - \frac{WLC_{ox}}{C_H}\left(V_{DD} - V_{in} - V_{TH0} - \gamma\sqrt{2\phi_B + V_{in}} + \gamma\sqrt{2\phi_B}\right), \tag{12.32}$$

$$= V_{in}\left(1 + \frac{WLC_{ox}}{C_H}\right) + \gamma\frac{WLC_{ox}}{C_H}\sqrt{2\phi_B + V_{in}}$$

$$- \frac{WLC_{ox}}{C_H}\left(V_{DD} - V_{TH0} + \gamma\sqrt{2\phi_B}\right). \tag{12.33}$$

---

[4]The voltage gain is *greater* than unity because the pedestal becomes smaller as the input level rises.

[5]Even for PMOS switches, the n-well is connected to the most positive supply voltage because the source and drain terminals of the switch may interchange during sampling.

It follows that the nonlinear dependence of V_{TH} upon V_{in} introduces nonlinearity in the input/output characteristic.

In summary, charge injection contributes three types of errors in MOS sampling circuits: gain error, dc offsets, and nonlinearity. In many applications, the first two can be tolerated or corrected whereas the last cannot.

It is instructive to consider the speed-precision trade-off resulting from charge injection. Representing the speed by a simple time constant τ and the precision by the error ΔV due to charge injection, we define a figure of merit as $F = (\tau \Delta V)^{-1}$. Writing

$$\tau = R_{on} C_H \tag{12.34}$$

$$= \frac{1}{\mu_n C_{ox}(W/L)(V_{DD} - V_{in} - V_{TH})} C_H, \tag{12.35}$$

and

$$\Delta V = \frac{WLC_{ox}}{C_H}(V_{DD} - V_{in} - V_{TH}), \tag{12.36}$$

we have

$$F = \frac{\mu_n}{L^2}. \tag{12.37}$$

Thus, to the first order, the trade-off is independent of the switch width and the sampling capacitor.

Clock Feedthrough   In addition to channel charge injection, a MOS switch couples the clock transitions to the sampling capacitor through its gate-drain or gate-source overlap capacitance. Depicted in Fig. 12.22, the effect introduces an error in the sampled output voltage. Assuming the overlap capacitance is constant, we express the error as

$$\Delta V = V_{CK} \frac{WC_{ov}}{WC_{ov} + C_H}, \tag{12.38}$$

where C_{ov} is the overlap capacitance per unit width. The error ΔV is independent of the input level, manifesting itself as a constant offset in the input/output characteristic. As with charge injection, clock feedthrough leads to a trade-off between speed and precision as well.

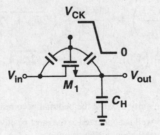

Figure 12.22 Clock feedthrough in a sampling circuit.

kT/C Noise Recall from Example 7.1 that a resistor charging a capacitor gives rise to a total rms noise voltage of $\sqrt{kT/C}$. As shown in Fig. 12.23, a similar effect occurs in sampling circuits. The on-resistance of the switch introduces thermal noise at the output and, when the switch turns off, this noise is stored on the capacitor along with the instantaneous value of the input voltage. It can be proved that the rms voltage of the sampled noise in this case is still approximately equal to $\sqrt{kT/C}$ [3, 4].

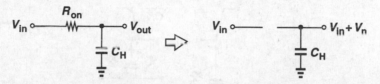

Figure 12.23 Thermal noise in a sampling circuit.

The problem of kT/C noise limits the performance in many high-precision applications. In order to achieve a low noise, the sampling capacitor must be sufficiently large, thus loading other circuits and degrading the speed.

12.2.4 Charge Injection Cancellation

The dependence of charge injection upon the input level and the trade-off expressed by (12.37) make it necessary to seek methods of cancelling the effect of charge injection so as to achieve a higher F. We consider a few such techniques here.

To arrive at the first technique, we postulate that the charge injected by the main transistor can be *removed* by means of a second transistor. As shown in Fig. 12.24, a "dummy" switch, M_2, driven by $\overline{CK}$ is added to the circuit such that after M_1 turns off and M_2 turns on, the channel charge deposited by the former on C_H is absorbed by the latter to create a channel. Note that both the source and drain of M_2 are connected to the output node.

How do we ensure that the charge injected by M_1, Δq_1, is equal to that absorbed by M_2, Δq_2? Suppose half of the channel charge of M_1 is injected onto C_H, i.e.,

$$\Delta q_1 = \frac{W_1 L_1 C_{ox}}{2}(V_{CK} - V_{in} - V_{TH1}). \tag{12.39}$$

Since $\Delta q_2 = W_2 L_2 C_{ox}(V_{CK} - V_{in} - V_{TH2})$, if we choose $W_2 = 0.5 W_1$ and $L_2 = L_1$, then $\Delta q_2 = \Delta q_1$. Unfortunately, the assumption of equal splitting of charge between source and drain is generally invalid, making this approach less attractive.

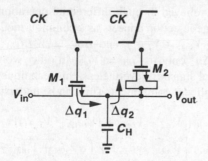

Figure 12.24 Addition of dummy device to reduce charge injection and clock feedthrough.

Interestingly, with the choice $W_2 = 0.5W_1$ and $L_2 = L_1$, the effect of clock feedthrough is suppressed. As depicted in Fig. 12.25, the total charge in V_{out} is zero because

$$-V_{CK}\frac{W_1 C_{ov}}{W_1 C_{ov} + C_H + 2W_2 C_{ov}} + V_{CK}\frac{2W_2 C_{ov}}{W_1 C_{ov} + C_H + 2W_2 C_{ov}} = 0. \qquad (12.40)$$

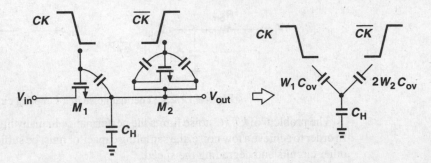

Figure 12.25 Clock feedthrough suppression by dummy switch.

Another approach to lowering the effect of charge injection incorporates both PMOS and NMOS devices such that the opposite charge packets injected by the two cancel each other (Fig. 12.26). For Δq_1 to cancel Δq_2, we must have $W_1 L_1 C_{ox}(V_{CK} - V_{in} - V_{THN}) = W_2 L_2 C_{ox}(V_{in} - |V_{THP}|)$. Thus, the cancellation occurs for only one input level. Even for clock feedthrough, the circuit does not provide complete cancellation because the gate-drain overlap capacitance of NFETs is not equal to that of PFETs.

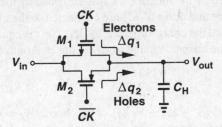

Figure 12.26 Use of complementary switches to reduce charge injection.

Our knowledge of the advantages of differential circuits suggests that the problem of charge injection may be relieved through differential operation. As shown in Fig. 12.27, we surmise that the charge injection appears as a common-mode disturbance. But, writing $\Delta q_1 = WLC_{ox}(V_{CK} - V_{in1} - V_{TH1})$, and $\Delta q_2 = WLC_{ox}(V_{CK} - V_{in2} - V_{TH2})$, we recognize that $\Delta q_1 = \Delta q_2$ only if $V_{in1} = V_{in2}$. In other words, the overall error is not suppressed for differential signals. Nevertheless, this technique both removes the constant offset and lowers the nonlinear component. This can be understood by writing

$$\Delta q_1 - \Delta q_2 = WLC_{ox}[(V_{in2} - V_{in1}) + (V_{TH2} - V_{TH1})] \qquad (12.41)$$

$$= WLC_{ox}\left[V_{in2} - V_{in1} + \gamma\left(\sqrt{2\phi_F + V_{in2}} - \sqrt{2\phi_F + V_{in1}}\right)\right]. \qquad (12.42)$$

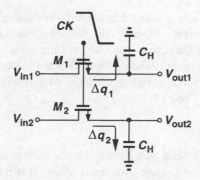

Figure 12.27 Differential sampling circuit.

Since for $V_{in1} = V_{in2}$, $\Delta q_1 - \Delta q_2 = 0$, the characteristic exhibits no offset. Also, the nonlinearity of body effect now appears in both square-root terms of (12.42), leading to only odd-order distortion (Chapter 13).

The problem of charge injection continues to limit the speed-precision envelope in sampled-data systems. Many cancellation techniques have been introduced but each leading to other trade-offs. One such technique, called "bottom-plate sampling," is widely used in switched-capacitor circuits and is described later in this chapter.

12.3 Switched-Capacitor Amplifiers

As mentioned in Section 12.1 and exemplified by the circuit of Fig. 12.4, CMOS feedback amplifiers are more easily implemented with a capacitive feedback network than a resistive one. Having examined sampling techniques, we are now ready to study a number of switched-capacitor amplifiers. Our objective is to understand the underlying principles as well as the speed-precision trade-offs encountered in the design of each circuit.

Before studying SC amplifiers, it is helpful to briefly look at the physical implementation of capacitors in CMOS technology. A simple capacitor structure is shown in Fig. 12.28(a), where the "top plate" is realized by a polysilicon layer and the "bottom plate" by a heavily

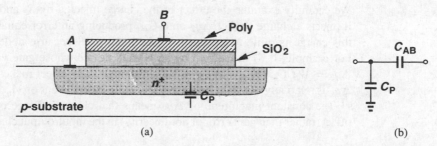

(a)   (b)

Figure 12.28 (a) Monolithic capacitor structure, (b) circuit model of (a) including parasitic capacitance to the substrate.

doped n^+ region. The dielectric is the thin oxide layer used in MOS devices as well.[6] An important concern in using this structure is the parasitic capacitance between each plate and the substrate. In particular, the bottom plate suffers from substantial junction capacitance to the underlying p region—typically about 10 to 20% of the oxide capacitance. For this reason, we usually model the capacitor as in Fig. 12.28(b). Monolithic capacitors are described in more detail in Chapters 17 and 18.

12.3.1 Unity-Gain Sampler/Buffer

While a unity-gain amplifier can be realized with no resistors or capacitors in the feedback network [Fig. 12.29(a)], for discrete-time applications, it still requires a sampling circuit. We may therefore conceive the circuit shown in Fig. 12.29(b) as a sampler/buffer. However, the input-dependent charge injected by S_1 onto C_H limits the accuracy here.

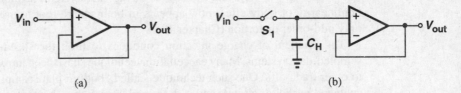

(a)   (b)

Figure 12.29   (a) Unity-gain buffer, (b) sampling circuit followed by unity-gain buffer.

Now consider the topology depicted in Fig. 12.30(a), where three switches control the sampling and amplification modes. In the sampling mode, S_1 and S_2 are on and S_3 is off, yielding the topology shown in Fig. 12.30(b). Thus, $V_{out} = V_X \approx 0$, and the voltage across C_H tracks V_{in}. At $t = t_0$, when $V_{in} = V_0$, S_1 and S_2 turn off and S_3 turns on, placing the capacitor around the op amp and entering the circuit into the amplification mode [Fig. 12.30(c)]. Since the op amp's high gain requires that node X still be a virtual ground and since the charge on the capacitor must be conserved, V_{out} rises to a value approximately equal to V_0. This voltage is therefore "frozen" and it can be processed by subsequent stages.

With proper timing, the circuit of Fig. 12.30(a) can substantially alleviate the problem of channel charge injection. As Fig. 12.31 illustrates in "slow motion," in the transition from the sampling mode to the amplification mode, S_2 turns off slightly *before* S_1 does. We carefully examine the effect of the charge injected by S_2 and S_1. When S_2 turns off, it injects a charge packet Δq_2 onto C_H, producing an error equal to $\Delta q_2/C_H$. However, this charge is quite independent of the input level because node X is a virtual ground. For example, if S_2 is realized by an NMOS device whose gate voltage equals V_{CK}, then $\Delta q_2 = WLC_{ox}(V_{CK} - V_{TH} - V_X)$. Although body effect makes V_{TH} a function of V_X, Δq_2 is relatively constant because V_X is quite independent of V_{in}.

The constant magnitude of Δq_2 means that channel charge of S_2 introduces only an offset (rather than gain error or nonlinearity) in the input/output characteristic. As described

---

[6] The oxide in this type of capacitor is typically thicker than the MOS gate oxide because silicon dioxide grows faster on a heavily-doped material.

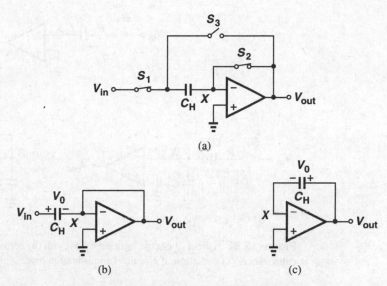

Figure 12.30 (a) Unity-gain sampler, (b) circuit of (a) in sampling mode, (c) circuit of (a) in amplification mode.

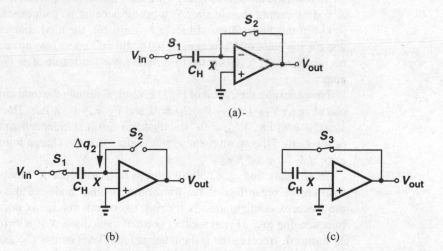

Figure 12.31 Operation of the unity-gain sampler in slow motion.

below, this offset can easily be removed by differential operation. But, how about the charge injected by S_1 onto C_H? Let us set V_{in} to zero and suppose S_1 injects a charge packet Δq_1 onto node P [Fig. 12.32(a)]. If the capacitance connected from X to ground (including the input capacitance of the op amp) is zero, V_P and V_X jump to infinity. To simplify the analysis, we assume a total capacitance equal to C_X from X to ground [Fig. 12.32(b)], and we will see shortly that its value does not affect the results. In Fig. 12.32(b), each of C_H

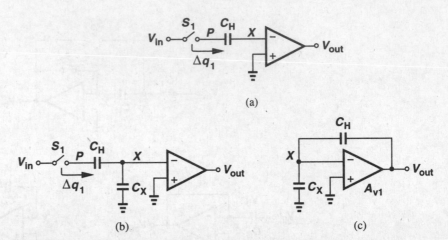

Figure 12.32 Effect of charge injected by S_1 with (a) zero and (b) finite op amp input capacitance, (c) transition of circuit to amplification mode.

and C_X carries a charge equal to Δq_1. Now, as shown in Fig. 12.32(c), we place C_H around the op amp, seeking to obtain the resulting output voltage.

To calculate the output voltage, we must make an important observation: the total charge at node X cannot change after S_2 turns off because no path exists for electrons to flow into or out of this node. Thus, if before S_1 turns off, the total charge on the right plate of C_H and the top plate of C_X is zero, it must still add up to zero after S_1 injects charge because no *resistive* path is connected to X. The same holds true after C_H is placed around the op amp.

Now consider the circuit of Fig. 12.32(c), assuming the total charge at node X is zero. We can write $C_X V_X - (V_{out} - V_X)C_H = 0$, and $V_X = -V_{out}/A_{v1}$. Thus, $-(C_X + C_H)V_{out}/A_{v1} - V_{out}C_H = 0$, i.e., $V_{out} = 0$. Note that this result is independent of Δq_1, capacitor values, or the gain of the op amp, thereby revealing that the charge injection by S_1 introduces no error *if S_2 turns off first*.

In summary, in Fig. 12.30(a), after S_2 turns off, node X "floats," maintaining a constant total charge regardless of the transitions at other nodes of the circuit. As a result, after the feedback configuration is formed, the output voltage is not influenced by the charge injection due to S_1. From another point of view, node X is a virtual ground at the moment S_2 turns off, freezing the instantaneous input level across C_H and yielding a charge equal to $V_0 C_H$ on the left plate of C_H. After settling with feedback, node X is again a virtual ground, forcing C_H to still carry $V_0 C_H$ and hence the output voltage to be approximately equal to V_0.

The effect of the charge injected by S_1 can be studied from yet another perspective. Suppose in Fig. 12.32(c), the output voltage is finite and positive. Then, since $V_X = V_{out}/(-A_{v1})$, V_X must be finite and negative, requiring negative charge on the top plate of C_X. For the total charge at X to be zero, the charge on the left plate of C_H must be positive and that on its right plate negative, giving $V_{out} \leq 0$. Thus, the only valid solution is $V_{out} = 0$.

The third switch in Fig. 12.30(a), S_3, also merits attention. In order to turn on, S_3 must establish an inversion layer at its oxide interface. Does the required channel charge come from C_H or from the op amp? We note from the foregoing analysis that after the feedback circuit has settled, the charge on C_H equals $V_0 C_H$, unaffected by S_3. The channel charge of this switch is therefore entirely supplied by the op amp, introducing no error.

Our study of Fig. 12.30(a) thus far suggests that, with proper timing, the charge injected by S_1 and S_3 is unimportant and the channel charge of S_2 results in a constant offset voltage. Fig. 12.33 depicts a simple realization of the clock edges to ensure S_1 turns off after S_2 does.

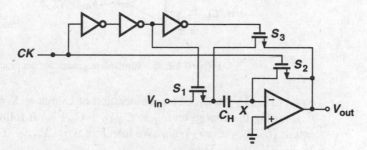

Figure 12.33 Generation of proper clock edges for unity-gain sampler.

The input-independent nature of the charge injected by the reset switch allows complete cancellation by differential operation. Illustrated in Fig. 12.34, such an approach employs a differential op amp along with two sampling capacitors so that the charge injected by S_2 and S_2' appears as a *common-mode* disturbance at nodes X and Y. This is in contrast to the behavior of the differential circuit shown in Fig. 12.27, where the input-dependent charge injection still leads to nonlinearity. In reality, S_2 and S_2' exhibit a finite charge injection mismatch, an issue resolved by adding another switch, S_{eq}, that turns off slightly after S_2 and S_2' (and before S_1 and S_1'), thereby equalizing the charge at nodes X and Y.

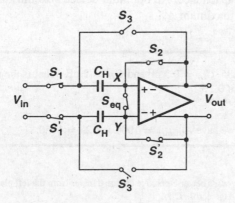

Figure 12.34 Differential realization of unity-gain sampler.

Precision Considerations   The circuit of Fig. 12.30(a) operates as a unity-gain buffer in the amplification mode, producing an output voltage approximately equal to the voltage stored across the capacitor. How close to unity is the gain here? As a general case, we assume the op amp exhibits a finite input capacitance C_{in} and calculate the output voltage when the circuit goes from the sampling mode to the amplification mode (Fig. 12.35). Owing to the finite gain of the op amp, $V_X \neq 0$ in the amplification mode, giving a charge

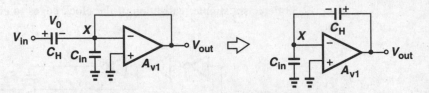

Figure 12.35   Equivalent circuit for accuracy calculations.

equal to $C_{in}V_X$ on C_{in}. The conservation of charge at X requires that $C_{in}V_X$ come from C_H, raising the charge on C_H to $C_H V_0 + C_{in}V_X$.[7] It follows that the voltage across C_H equals $(C_H V_0 + C_{in}V_X)/C_H$. We therefore write $V_{out} - (C_H V_0 + C_{in}V_X)/C_H = V_X$ and $V_X = -V_{out}/A_{v1}$. Thus,

$$V_{out} = \frac{V_0}{1 + \dfrac{1}{A_{v1}}\left(\dfrac{C_{in}}{C_H} + 1\right)} \tag{12.43}$$

$$\approx V_0\left[1 - \frac{1}{A_{v1}}\left(\frac{C_{in}}{C_H} + 1\right)\right]. \tag{12.44}$$

As expected, if $C_{in}/C_H \ll 1$, then $V_{out} \approx V_0/(1 + A_{v1}^{-1})$. In general, however, the circuit suffers from a gain error of approximately $-(C_{in}/C_H + 1)/A_{v1}$, suggesting that the input capacitance must be minimized even if speed is not critical. Recall from Chapter 9 that to increase A_{v1}, we may choose a large width for the input transistors of the op amp, but at the cost of higher input capacitance. An optimum device size must therefore yield minimum gain error rather than maximum A_{v1}.

Example 12.4 ───

In the circuit of Fig. 12.35, $C_{in} = 0.5$ pF and $C_H = 2$ pF. What is the minimum op amp gain that guarantees a gain error of 0.1%?

Solution

Since $C_{in}/C_H = 0.25$, we have $A_{v1,min} = 1000 \times 1.25 = 1250$.

───

[7]The charge on C_H *increases* because positive charge transfer from the left plate of C_H to the top plate of C_{in} leads to a more positive voltage across C_H.

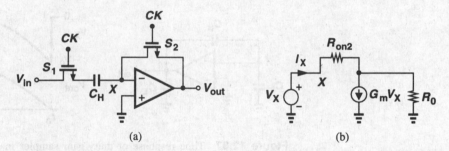

Figure 12.36   (a) Unity-gain sampler in sampling mode, (b) equivalent circuit of (a).

Speed Considerations   Let us first examine the circuit in the sampling mode
[Fig. 12.36(a)]. What is the time constant in this phase? The total resistance in series
with C_H is given by R_{on1} and the resistance between X and ground, R_X. Using the simple
op amp model shown in Fig. 12.36(b), where R_0 denotes the open-loop output impedance
of the op amp, we have

$$(I_X - G_m V_X)R_0 + I_X R_{on2} = V_X, \tag{12.45}$$

that is,

$$R_X = \frac{R_0 + R_{on2}}{1 + G_m R_0}. \tag{12.46}$$

Since typically $R_{on2} \ll R_0$ and $G_m R_0 \gg 1$, we have $R_X \approx 1/G_m$. For example, in
a telescopic op amp employing differential to single-ended conversion, G_m equals the
transconductance of each input transistor.

The time constant in the sampling mode is thus equal to

$$\tau_{sam} = \left(R_{on1} + \frac{1}{G_m} \right) C_H. \tag{12.47}$$

The magnitude of τ_{sam} must be sufficiently small to allow settling in the test case of Fig. 12.14
to the required precision.

Now let us consider the circuit as it enters the amplification mode. Shown in Fig. 12.37
along with both the op amp input capacitance and the load capacitance, the circuit must
begin with $V_{out} \approx 0$ and eventually produce $V_{out} \approx V_0$. If C_{in} is relatively small, we can
assume that the voltages across C_L and C_H do not change instantaneously, concluding that
if $V_{out} \approx 0$ and $V_{CH} \approx V_0$, then $V_X = -V_0$ at the beginning of the amplification mode.
In other words, the input difference sensed by the op amp initially jumps to a large value,
possibly causing the op amp to slew. But, let us first assume the op amp can be modeled by
a linear model and determine the output response.

To simplify the analysis, we represent the charge on C_H by an explicit series voltage
source, V_S, that goes from zero to V_0 at $t = t_0$ while C_H carries no charge itself (Fig. 12.38).
The objective is to obtain the transfer function $V_{out}(s)/V_S(s)$ and hence the step response.

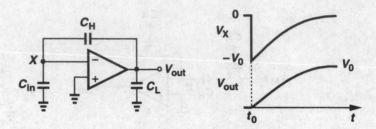

Figure 12.37 Time response of unity-gain sampler in amplification mode.

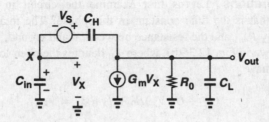

Figure 12.38 Equivalent circuit of unity-gain circuit in amplification mode.

We have

$$V_{out}\left(\frac{1}{R_0} + C_L s\right) + G_m V_X = (V_S + V_X - V_{out})C_H s. \tag{12.48}$$

Also, since the current through C_{in} equals $V_X C_{in} s$,

$$V_X \frac{C_{in} s}{C_H s} + V_X + V_S = V_{out}. \tag{12.49}$$

Calculating V_X from (12.49) and substituting in (12.48), we arrive at the transfer function

$$\frac{V_{out}}{V_S}(s) = R_0 \frac{(G_m + C_{in}s)C_H}{R_0(C_L C_{in} + C_{in}C_H + C_H C_L)s + G_m R_0 C_H + C_H + C_{in}}. \tag{12.50}$$

Note that for $s = 0$, (12.50) reduces to a form similar to (12.43). Since typically $G_m R_0 C_H \gg C_H, C_{in}$, we can simplify (12.50) as

$$\frac{V_{out}}{V_S}(s) = \frac{(G_m + C_{in}s)C_H}{(C_L C_{in} + C_{in}C_H + C_H C_L)s + G_m C_H}. \tag{12.51}$$

Thus, the response is characterized by a time constant equal to

$$\tau_{amp} = \frac{C_L C_{in} + C_{in}C_H + C_H C_L}{G_m C_H}, \tag{12.52}$$

which is independent of the op amp output resistance. This is because a higher R_0 leads to a greater loop gain, eventually yielding a constant closed-loop speed. If $C_{in} \ll C_L, C_H$, then (12.52) reduces to C_L/G_m, an expected result because with negligible C_{in}, the output resistance of the unity-gain buffer is equal to $1/G_m$.

We now study the slewing behavior of the circuit, considering a telescopic op amp as an example. Upon entering the amplification mode, the circuit may experience a large step at the inverting input (Fig. 12.37). As shown in Fig. 12.39, the tail current of the op amp's input differential pair is then steered to one side and its mirror current charges the capacitance seen at the output. Since M_2 is off during slewing, C_{in} is negligible and the slew rate is approximately equal to I_{SS}/C_L. The slewing continues until V_X is sufficiently close to the gate voltage of M_1, after which point the settling progresses with the time constant given in (12.52).

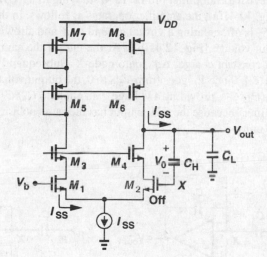

Figure 12.39 Unity-gain sampler during slewing.

Our foregoing studies reveal that the input capacitance of the op amp degrades both the speed and the precision of the unity-gain sampler/buffer. For this reason, the bottom plate of C_H in Fig. 12.30 is usually driven by the input signal or the output of the op amp and the top plate is connected to node X (Fig. 12.40), minimizing the parasitic capacitance seen from node X to ground. This technique is called "bottom-plate sampling." Driving the bottom plate by the input or the output also avoids the injection of substrate noise to node X (Chapter 18).

It is instructive to compare the performance of the sampling circuits shown in Figs. 12.29(b) and 12.30(a). In Fig. 12.29(b), the sampling time constant is smaller because it depends on only the on-resistance of the switch. More importantly, in Fig. 12.29(b), the amplification after the switch turns off is almost instantaneous, whereas in Fig. 12.30 it requires a finite settling time. However, the critical advantage of the unity-gain sampler is the input-independent charge injection.

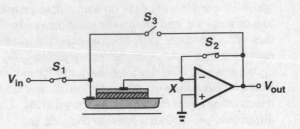

Figure 12.40 Connection of capacitor to the unity-gain sampler.

12.3.2 Noninverting Amplifier

In this section, we revisit the amplifier of Fig. 12.4, studying its speed and precision proper-ties. Repeated in Fig. 12.41(a), the amplifier operates as follows. In the sampling mode, S_1 and S_2 are on and S_3 is off, creating a virtual ground at X and allowing the voltage across C_1 to track the input voltage [Fig. 12.41(b)]. At the end of the sampling mode, S_2 turns off first, injecting a constant charge, Δq_2, onto node X. Subsequently, S_1 turns off and S_3 turns on [Fig. 12.41(c)]. Since V_P goes from V_{in0} to 0, the output voltage changes from 0 to approximately $V_{in0}(C_1/C_2)$, providing a voltage gain equal to C_1/C_2. We call the circuit a "noninverting amplifier" because the final output has the same polarity as V_{in0} and the gain can be greater than unity.

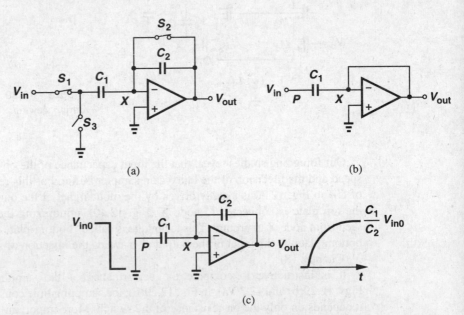

Figure 12.41 (a) Noninverting amplifier, (b) circuit of (a) in sampling mode, (c) transition of circuit to amplification mode.

As with the unity-gain circuit of Fig. 12.30(a), the noninverting amplifier avoids input-dependent charge injection by proper timing, namely, turning S_2 off before S_1 (Fig. 12.42). After S_2 is off, the total charge at node X remains constant, making the circuit insensitive to charge injection of S_1 or charge "absorption" of S_3. Let us first study the effect of S_1 carefully. As illustrated in Fig. 12.43, the charge injected by S_1, Δq_1, changes the voltage at node P by approximately $\Delta V_P = \Delta q_1/C_1$, and hence the output voltage by $-\Delta q_1/C_2$. However, after S_3 turns on, V_P drops to zero. Thus, the *overall* change in V_P is equal to $0 - V_{in0} = -V_{in0}$, producing an overall change in the output equal to $-V_{in0}(-C_1/C_2) = V_{in0}C_1/C_2$.

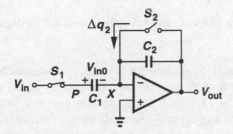

Figure 12.42   Transition of noninverting amplifier to amplification mode.

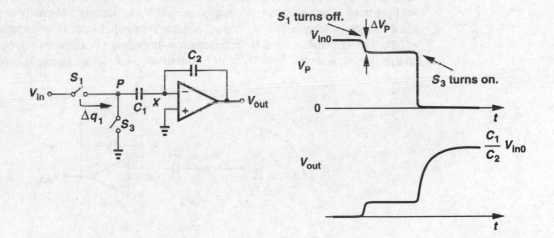

Figure 12.43   Effect of charge injected by S_1.

The key point here is that V_P goes from a fixed voltage, V_0, to another, 0, with an intermediate perturbation due to S_1. Since the output voltage of interest is measured after node P is connected to ground, the charge injected by S_1 does not affect the final output. From another perspective, as shown in Fig. 12.44, the charge on the right plate of C_1 at the instant S_2 turns off is approximately equal to $-V_{in0}C_1$. Also, the total charge at node X must remain constant after S_2 turns off. Thus, when node P is connected to ground and the circuit settles, the voltage across C_1 and hence its charge are nearly zero, and the charge $-V_{in0}C_1$ must reside on the left plate of C_2. In other words, the output voltage is approximately equal to $V_{in0}C_1/C_2$ regardless of the intermediate excursions at node P.

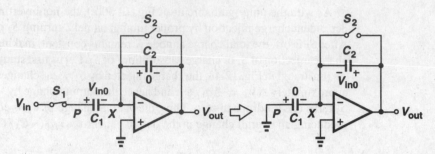

Figure 12.44   Charge redistribution in noninverting amplifier.

The foregoing discussion indicates that two other phenomena have no effect on the final output. First, from the time S_2 turns off until the time S_1 turns off, the input voltage may change significantly (Fig. 12.45) without introducing any error. In other words, the sampling instant is defined by the turn-off of S_2. Second, when S_3 turns on, it requires some channel charge but since the final value of V_P is zero, this charge is unimportant. Neither of these effects introduces error because the total charge at node X is conserved and V_P is eventually set by a fixed (zero) potential. To emphasize that V_P is initially and finally determined by fixed voltages, we say node P is "driven" or node P switches from a low-impedance node to another low-impedance node. Here the term low-impedance distinguishes node P, at which charge is not conserved, from "floating" nodes such as X, where charge is conserved.

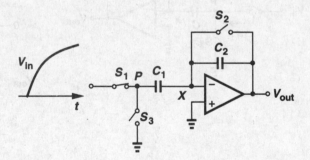

Figure 12.45   Effect of input change after S_2 turns off.

In summary, proper timing in Fig. 12.41(a) ensures that node X is perturbed by only the charge injection of S_2, making the final value of V_{out} free from errors due to S_1 and S_3. The constant offset due to S_2 can be suppressed by differential operation (Fig. 12.46).

Example 12.5

In the differential circuit of Fig. 12.46, suppose the equalizing switch is not used and S_2 and S_2' exhibit a threshold voltage mismatch of 10 mV. If $C_1 = 1$ pF, $C_2 = 0.5$ pF, $V_{TH} = 0.6$ V, and for all switches $WLC_{ox} = 50$ fF, calculate the dc offset measured at the output assuming all of the channel charge of S_2 and S_2' is injected onto X and Y, respectively.

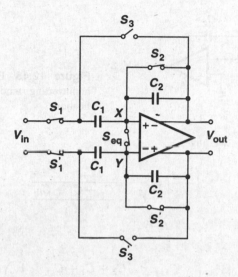

Figure 12.46 Differential realization of noninverting amplifier.

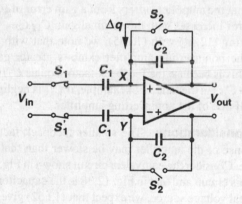

Figure 12.47

Solution

Simplifying the circuit as in Fig. 12.47, we have $V_{out} \approx \Delta q/C_2$, where $\Delta q = WLC_{ox}\Delta V_{TH}$. Note that C_1 does not appear in the result because X is a virtual ground, i.e., the voltage across C_1 changes only negligibly. Thus, the injected charge resides primarily on the left plate of C_2, giving an output error voltage equal to $\Delta V_{out} = WLC_{ox}\Delta V_{TH}/C_2 = 1$ mV.

Precision Considerations   As mentioned above, the circuit of Fig. 12.41(a) provides a nominal voltage gain of C_1/C_2. We now calculate the actual gain if the op amp exhibits a finite open-loop gain equal to A_{v1}. Depicted in Fig. 12.48 along with the input capacitance of the op amp, the circuit amplifies the input voltage change such that:

$$(V_{out} - V_X)C_2 s = V_X C_{in} s + (V_X - V_{in})C_1 s. \tag{12.53}$$

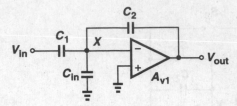

Figure 12.48 Equivalent circuit of noninverting amplifier during amplification.

Since $V_{out} = -A_{v1}V_X$, we have

$$\left| \frac{V_{out}}{V_{in}} \right| = \frac{C_1}{C_2 + \dfrac{C_2 + C_1 + C_{in}}{A_{v1}}}. \tag{12.54}$$

For large A_{v1},

$$\left| \frac{V_{out}}{V_{in}} \right| \approx \frac{C_1}{C_2} \left(1 - \frac{C_2 + C_1 + C_{in}}{C_2} \cdot \frac{1}{A_{v1}} \right), \tag{12.55}$$

implying that the amplifier suffers from a gain error of $(C_2 + C_1 + C_{in})/(C_2 A_{v1})$. Note that the gain error increases with the nominal gain C_1/C_2.

Comparing (12.44) with (12.55), we note that with $C_H = C_2$ and for a nominal gain of unity, the noninverting amplifier exhibits greater gain error than does the unity-gain sampler. This is because the feedback factor equals $C_2/(C_1 + C_{in} + C_2)$ in the former and $C_H/(C_H + C_{in})$ in the latter. For example, if C_{in} is negligible, the unity-gain sampler's gain error is half that of the noninverting amplifier.

Speed Considerations The smaller feedback factor in Fig. 12.48 suggests that the time response of the amplifier may be slower than that of the unity-gain sampler. This is indeed true. Consider the equivalent circuit shown in Fig. 12.49(a). Since the only difference between this circuit and that in Fig. 12.38 is the capacitor C_1, which is connected from node X to an ideal voltage source, we expect that (12.52) gives the time constant of this amplifier as well if C_{in} is replaced by $C_{in} + C_1$. But for a more rigorous analysis, we substitute V_{in}, C_1, and C_{in} in Fig. 12.49(a) by a Thevenin equivalent as in Fig. 12.49(b), where $\alpha = C_1/(C_1 + C_{in})$, and $C_{eq} = C_1 + C_{in}$, and note that

$$V_X = (\alpha V_{in} - V_{out}) \frac{C_{eq}}{C_{eq} + C_2} + V_{out}. \tag{12.56}$$

Thus,

$$\left[(\alpha V_{in} - V_{out}) \frac{C_{eq}}{C_{eq} + C_2} + V_{out} \right] G_m + V_{out} \left(\frac{1}{R_0} + C_L s \right) = (\alpha V_{in} - V_{out}) \frac{C_{eq} C_2}{C_{eq} + C_2} s, \tag{12.57}$$

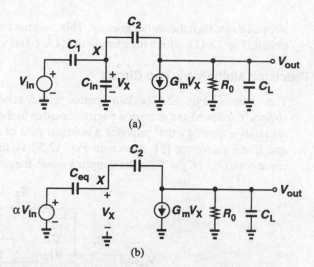

Figure 12.49 (a) Equivalent circuit of noninverting amplifier in amplification mode, (b) circuit of (a) with V_{in}, C_1, and C_{in} replaced by a Thevenin equivalent.

and hence

$$\frac{V_{out}}{V_{in}}(s) = \frac{-C_{eq}\dfrac{C_1}{C_1 + C_{in}}(G_m - C_2 s)R_0}{C_2 G_m R_0 + C_{eq} + C_2 + R_0[C_L(C_{eq} + C_2) + C_{eq}C_2]s} \qquad (12.58)$$

Note that for $s = 0$, (12.58) reduces to (12.54). For a large $G_m R_0$, we can simplify (12.58) to

$$\frac{V_{out}}{V_{in}}(s) \approx \frac{-C_{eq}\dfrac{C_1}{C_1 + C_{in}}(G_m - C_2 s)R_0}{R_0(C_L C_{eq} + C_L C_2 + C_{eq}C_2)s + G_m R_0 C_2}, \qquad (12.59)$$

obtaining a time constant of

$$\tau_{amp} = \frac{C_L C_{eq} + C_L C_2 + C_{eq}C_2}{G_m C_2}, \qquad (12.60)$$

which is the same as the time constant of Fig. 12.37 if C_{in} is replaced by $C_{in} + C_1$. Note the direct dependence of τ_{amp} upon the nominal gain, C_1/C_2.

It is instructive to examine the amplifier's time constant for the special case $C_L = 0$. Equation (12.60) yields $\tau_{amp} = (C_1 + C_{in})/G_m$, a value *independent* of the feedback capacitor. This is because, while a larger C_2 introduces heavier loading at the output, it also provides a greater feedback factor.

The reader may wonder why Eq. (12.58) yields a negative gain for the circuit that we have called a "noninverting" amplifier. This equation simply means if the left plate of C_1 is stepped *down*, then the output goes *up*. This does not contradict the operation of the original circuit (Fig. 12.41), where the *change* in V_P is equal to $-V_{in}$.

12.3.3 Precision Multiply-by-Two Circuit

The circuit of Fig. 12.41(a) can operate with a relatively high closed-loop gain, but it suffers from speed and precision degradation due to the low feedback factor. In this section, we study a topology that provides a nominal gain of two while achieving a higher speed and lower gain error [5]. Shown in Fig. 12.50(a), the amplifier incorporates two equal capacitors, $C_1 = C_2 = C$. In the sampling mode, the circuit is configured as in Fig. 12.50(b),

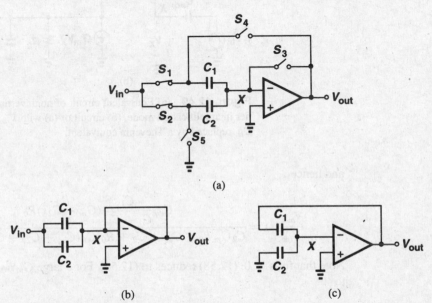

Figure 12.50 (a) Multiply-by-two circuit, (b) circuit of (a) in sampling mode, (c) circuit of (a) in amplification mode.

establishing a virtual ground at X and allowing the voltage across C_1 and C_2 to track V_{in}. In the transition to the amplification mode, S_3 turns off first, C_1 is placed around the op amp, and the left plate of C_2 is switched to ground [Fig. 12.50(c)]. Since at the moment S_3 turns off, the total charge on C_1 and C_2 equals $2V_{in0}C$ (if the charge injected by S_3 is neglected), and since the voltage across C_2 approaches zero in the amplification mode, the final voltage across C_1 and hence the output voltage are approximately equal to $2V_{in0}$. This can also be seen from the slow motion illustration of Fig. 12.51.

The reader can show that the charge injected by S_1 and S_2 and absorbed by S_4 and S_5 is unimportant and that injected by S_3 introduces a constant offset. The offset can be suppressed by differential operation.

The speed and precision of the multiply-by-two circuit are expressed by (12.60) and (12.55), respectively, but the advantage of the circuit is the higher feedback factor for a given closed-loop gain. Note, however, that the input capacitance of the multiply-by-two circuit in the sampling mode is higher.

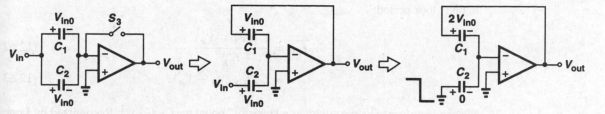

Figure 12.51 Transition of multiply-by-two circuit to amplification mode in slow motion.

2.4 Switched-Capacitor Integrator

Integrators are used in many analog systems. Examples include filters and oversampled analog-to-digital converters. Fig. 12.52 depicts a continuous-time integrator, whose output can be expressed as

$$V_{out} = -\frac{1}{RC_F}\int V_{in}dt, \qquad (12.61)$$

if the op amp gain is very large. For sampled-data systems, we must devise a discrete-time counterpart of this circuit.

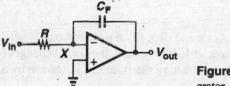

Figure 12.52 Continuous-time integrator.

Before studying SC integrators, let us first point out an interesting property. Consider a resistor connected between two nodes [Fig. 12.53(a)], carrying a current equal to $(V_A - V_B)/R$. The role of the resistor is to take a certain amount of charge from node A every second and move it to node B. Can we perform the same function by a capacitor? Suppose in the circuit of Fig. 12.53(b), capacitor C_S is alternately connected to nodes A and B at a clock rate f_{CK}. The *average* current flowing from A to B is then equal to the charge moved

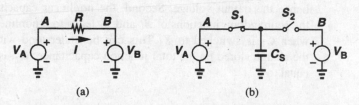

(a)   (b)

Figure 12.53 (a) Continuous-time and (b) discrete-time resistors.

in one clock period:

$$\overline{I_{AB}} = \frac{C_S(V_A - V_B)}{f_{CK}^{-1}} \tag{12.62}$$

$$= C_S f_{CK}(V_A - V_B). \tag{12.63}$$

We can therefore view the circuit as a "resistor" equal to $(C_S f_{CK})^{-1}$. Recognized by James Clark Maxwell, this property formed the foundation for many modern switched-capacitor circuits.

Let us now replace resistor R in Fig. 12.52 by its discrete-time equivalent, arriving at the integrator of Fig. 12.54(a). We note that in every clock cycle, C_1 absorbs a charge equal

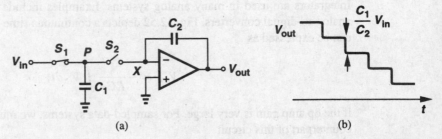

(a)   (b)

Figure 12.54 (a) Discrete-time integrator, (b) response of circuit to a constant input voltage.

to $C_1 V_{in}$ when S_1 is on and deposits the charge on C_2 when S_2 is on (node X is a virtual ground). For example, if V_{in} is constant, the output changes by $V_{in}C_1/C_2$ every clock cycle [Fig. 12.54(b)]. Approximating the staircase waveform by a ramp, we note that the circuit behaves as an integrator.

The final value of V_{out} in Fig. 12.54(a) after every clock cycle can be written as

$$V_{out}(kT_{CK}) = V_{out}[(k-1)T_{CK}] - V_{in}[(k-1)T_{CK}] \cdot \frac{C_1}{C_2}, \tag{12.64}$$

where the gain of the op amp is assumed large. Note that the small-signal settling time constant as charge is transferred from C_1 to C_2 is given by (12.52).

The integrator of Fig. 12.54(a) suffers from two important drawbacks. First, the input-dependent charge injection of S_1 introduces nonlinearity in the charge stored on C_1 and hence the output voltage. Second, the nonlinear capacitance at node P resulting from the source/drain junctions of S_1 and S_2 leads to a nonlinear charge-to-voltage conversion when C_1 is switched to X. This can be understood with the aid of Fig. 12.55, where the charge stored on the total junction capacitance, C_j, is *not* equal to $V_{in0}C_j$, but rather equal to

$$q_{cj} = \int_0^{V_{in0}} C_j dV. \tag{12.65}$$

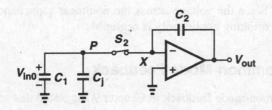

Figure 12.55 Effect of junction capacitance nonlinearity in SC integrator.

Since C_j is a function of voltage, q_{cj} exhibits a nonlinear dependence on V_{in0}, thereby creating a nonlinear component at the output after the charge is transferred to the integration capacitor.

An integrator topology that resolves both of the foregoing issues is shown in Fig. 12.56(a). We study the circuit's operation in the sampling and integration modes. As shown in Fig. 12.56(b), in the sampling mode S_1 and S_3 are on and S_2 and S_4 are off, allowing the voltage across C_1 to track V_{in} while the op amp and C_2 hold the previous value. In the transition to the integration mode, S_3 turns off first, injecting a constant charge onto C_1, S_1 turns off next, and subsequently S_2 and S_4 turn on [Fig. 12.56(c)]. The charge stored on C_1 is therefore transferred to C_2 through the virtual ground node.

Since S_3 turns off first, it introduces only a constant offset, which can be suppressed by differential operation. Moreover, because the left plate of C_1 is "driven" (Section 12.3.2), the charge injection or absorption of S_1 and S_2 contributes no error. Also, since node X is a virtual ground, the charge injected or absorbed by S_4 is constant and independent of V_{in}.

How about the nonlinear junction capacitance of S_3 and S_4? We observe that the voltage across this capacitance goes from near zero in the sampling mode to virtual ground in the

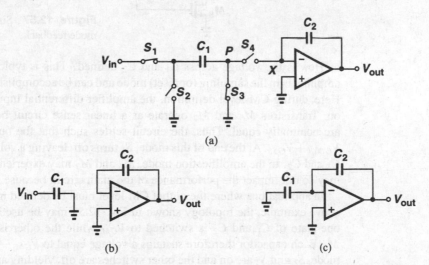

(a)

(b)   (c)

Figure 12.56   (a) Parasitic-insensitive integrator, (b) circuit of (a) in sampling mode, (c) circuit of (a) in integration mode.

integration mode. Since the voltage across the nonlinear capacitance changes by a very small amount, the resulting nonlinearity is negligible.

12.5 Switched-Capacitor Common-Mode Feedback

Our study of common-mode feedback in Chapter 9 suggested that sensing the output CM level by means of resistors lowers the differential voltage gain of the circuit considerably. We also observed that sensing techniques using MOSFETs that operate as source followers or variable resistors suffer from a limited linear range. Switched-capacitor CMFB networks provide an alternative that avoids both of these difficulties (but the circuit must be refreshed periodically.)

In switched-capacitor common-mode feedback, the outputs are sensed by capacitors rather than resistors. Figure 12.57 depicts a simple example, where equal capacitors C_1 and C_2 reproduce at node X the average of the changes in each output voltage. Thus, if V_{out1} and V_{out2} experience, say, a positive CM change, then V_X and hence I_{D5} increase, pulling V_{out1} and V_{out2} down. The output CM level is then equal to V_{GS2} plus the voltage across C_1 and C_2.

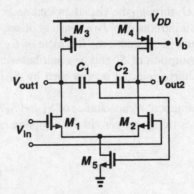

Figure 12.57 Simple SC common-mode feedback.

How is the voltage across C_1 and C_2 defined? This is typically carried out when the amplifier is in the sampling (or reset) mode and can be accomplished as shown in Fig. 12.58. Here, during CM level definition, the amplifier differential input is zero and switch S_1 is on. Transistors M_6 and M_7 operate as a linear sense circuit because their gate voltages are nominally equal. Thus, the circuit settles such that the ouput CM level is equal to $V_{GS6,7} + V_{GS5}$. At the end of this mode, S_1 turns off, leaving a voltage equal to $V_{GS6,7}$ across C_1 and C_2. In the amplification mode, M_6 and M_7 may experience a large nonlinearity but they do not impact the performance of the main circuit because S_1 is off.

In applications where the output CM level must be defined more accurately than in the above example, the topology shown in Fig. 12.59 may be used. Here, in the reset mode one plate of C_1 and C_2 is switched to V_{CM} while the other is connected to the gate of M_6. Each capacitor therefore sustains a voltage equal to $V_{CM} - V_{GS6}$. In the amplification mode, S_2 and S_3 are on and the other switches are off, yielding an output CM level equal to $V_{CM} - V_{GS6} + V_{GS5}$. Proper definition of I_{D3} and I_{D4} with respect to I_{REF} can guarantee that $V_{GS5} = V_{GS6}$ and hence the output CM level is equal to V_{CM}.

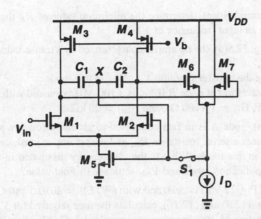

Figure 12.58 Definition of the voltage across C_1 and C_2.

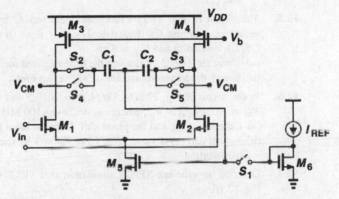

Figure 12.59 Alternative topology for definition of output CM level.

With large output swings, the speed of the CMFB loop may in fact influence the settling of the differential output [6]. For this reason, part of the tail current of the differential pairs in Figs. 12.58 and 12.59 can be provided by a *constant* current source so that M_5 makes only small adjustments to the circuit.

Problems

Unless otherwise stated, in the following problems, use the device data shown in Table 2.1 and assume $V_{DD} = 3$ V where necessary. Also, assume all transistors are in saturation.

12.1. The circuit of Fig. 12.2(b) is designed with $C_1 = 2$ pF and $C_2 = 0.5$ pF.

    (a) Assuming $R_F = \infty$ but the op amp has an output resistance R_{out}, derive the transfer function $V_{out}(s)/V_{in}(s)$.

(b) If the op amp is ideal, determine the minimum value of R_F that guarantees a gain error of 1% for an input frequency of 1 MHz.

12.2. Suppose in Fig. 12.5(a), the op amp is characterized by a transconductance G_m and an output resistance R_{out}.

(a) Determine the transfer function V_{out}/V_{in} in this mode.

(b) Plot the waveform at node B if V_{in} is a 100-MHz sinusoid with a peak amplitude of 1 V, $C_1 = 1$ pF, $G_m = 1/(100\ \Omega)$, and $R_{out} = 20$ kΩ.

12.3. In Fig. 12.5(b), node A is in fact connected to ground through a switch (Fig. 12.4). If the switch introduces a series resistance R_{on} and the op amp is ideal, calculate the time constant of the circuit in this mode. What is the total energy dissipated in the switch as the circuit enters the amplification mode and V_{out} settles to its final value?

12.4. The circuit of Fig. 12.9(a) is designed with $(W/L)_1 = 20/0.5$ and $C_H = 1$ pF.

(a) Using Eqs. (12.9) and (12.16), calculate the time required for V_{out} to drop to $+1$ mV.

(b) Approximating M_1 by a linear resistor equal to $[\mu_n C_{ox}(W/L)_1(V_{DD}-V_{TH})]^{-1}$, calculate the time required for V_{out} to drop to $+1$ mV and compare the result with that obtained in part (a).

12.5. The circuit of Fig. 12.11 cannot be characterized by a single time constant because the resistance charging C_H (equal to $1/g_{m1}$ if $\gamma = 0$) varies with the output level. Assume $(W/L)_1 = 20/0.5$ and $C_H = 1$ pF.

(a) Using Eq. (12.21), calculate the time required for V_{out} to reach 2.1 V.

(b) Sketch the transconductance of M_1 versus time.

12.6. In the circuit of Fig. 12.8(b), $(W/L)_1 = 20/0.5$ and $C_H = 1$ pF. Assume $\lambda = \gamma = 0$ and $V_{in} = V_0 \sin \omega_{in} t + V_m$, where $\omega_{in} = 2\pi \times (100\ \text{MHz})$.

(a) Calculate R_{on1} and the phase shift from the input to the output if $V_0 = V_m = 10$ mV.

(b) Repeat part (a) if $V_0 = 10$ mV but $V_m = 1$ V. The variation of the phase shift translates to distortion.

12.7. Describe an efficient SPICE simulation that yields the plot of $R_{on,eq}$ for the circuit of Fig. 12.16.

12.8. The sampling network of Fig. 12.16 is designed with $(W/L)_1 = 20/0.5$, $(W/L)_2 = 60/0.5$, and $C_H = 1$ pF. If $V_{in} = 0$ and the initial value of V_{out} is $+3$ V, estimate the time required for V_{out} to drop to $+1$ mV.

12.9. In the circuit of Fig. 12.19, $(W/L)_1 = 20/0.5$ and $C_H = 1$ pF. Calculate the maximum error at the output due to charge injection. Compare this error with that resulting from clock feedthrough.

12.10. The circuit of Fig. 12.60 samples the input on C_1 when CK is high and connects C_1 and C_2 when CK is low. Assume $(W/L)_1 = (W/L)_2$ and $C_1 = C_2$.

(a) If the initial voltages across C_1 and C_2 are zero and $V_{in} = 2$ V, plot V_{out} versus time for many clock cycles. Neglect charge injection and clock feedthrough.

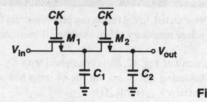

Figure 12.60

(b) What is the maximum error in V_{out} due to charge injection and clock feedthrough of M_1 and M_2? Assume the channel charge of M_2 splits equally between C_1 and C_2.

(c) Determine the sampled kT/C noise at the output after M_2 turns off.

12.11. For $V_{in} = V_0 \sin \omega_0 t + V_0$, where $V_0 = 0.5$ V and $\omega_0 = 2\pi \times (10 \text{ MHz})$, plot the output waveforms of the circuits shown in Fig. 12.29(b) and 12.30(a). Assume a clock frequency of 50 MHz.

12.12. In Fig. 12.45, S_1 turns off Δt seconds after S_2 and S_3 turns on Δt seconds after S_1 turns off. Plot the output waveform, taking into account the charge injection and clock feedthough of S_1-S_3. Assume all of the switches are NMOS devices.

12.13. The circuit of Fig. 12.48 is designed with $C_1 = 2$ pF, $C_{in} = 0.2$ pF and $A_v = 1000$. What is the maximum nominal gain, C_1/C_2, that the circuit can provide with a gain error of 1%?

12.14. In Problem 12.13, what is the maximum nominal gain if $G_m = 1/(100 \, \Omega)$ and the circuit must achieve a time constant of 2 ns in the amplification mode? Assume $C_{in} = 0.2$ pF and calculate C_1 and C_2.

12.15. The integrator of Fig. 12.54 is designed with $C_1 = C_2 = 1$ pF and a clock frequency of 100 MHz. Neglecting charge injection and clock feedthrough, sketch the output if the input is a 10-MHz sinusoid with a peak amplitude of 0.5 V. Approximating C_1, S_1, and S_2 by a resistor, estimate the output amplitude.

12.16. Consider the switched-capacitor amplifier depicted in Fig. 12.61, where the common-mode feedback is not shown. Assume $(W/L)_{1-4} = 50/0.5$, $I_{SS} = 1$ mA, $C_1 = C_2 = 2$ pF, $C_3 = C_4 = 0.5$ pF, and the output CM level is 1.5 V. Neglect the transistor capacitances.

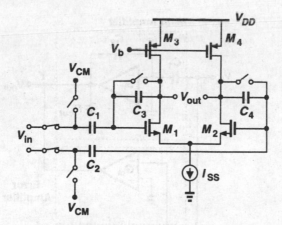

Figure 12.61

(a) What is the maximum allowable output voltage swing in the amplification mode?

(b) Determine the gain error of the amplifier.

(c) What is the small-signal time constant in the amplification mode?

12.17. Repeat Problem 12.16(c) if the gate-source capacitance of M_1 and M_2 is not neglected.

12.18. A differential circuit incorporating a well-designed common-mode feedback network exhibits the open-loop input-output characteristic shown in Fig. 12.62(a). In some circuits, however, the characteristic appears as in Fig. 12.62(b). Explain how this effect occurs.

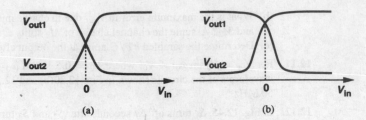

Figure 12.62

12.19. In the common-mode feedback network of Fig. 12.58, assume $W/L = 50/0.5$ for all transistors, $I_{D5} = 1$ mA, and $I_{D6,7} = 50$ μA. Determine the allowable range of the input common-mode level.

12.20. Repeat Problem 12.19 if $(W/L)_{6,7} = 10/0.5$.

12.21. Suppose in the common-mode feedback network of Fig. 12.58, S_1 injects a charge of Δq onto the gate of M_5. How much do the gate voltage of M_5 and the output common-mode level change due to this error?

12.22. In the circuit of Fig. 12.63, each op amp is represented by a Norton equivalent and characterized by G_m and R_{out}. The output currents of two op amps are summed at node Y [7]. (The circuit is shown in the amplification mode.) Note that the main amplifier and the auxiliary amplifier are identical and the error amplifier senses the voltage variation at node X and injects a

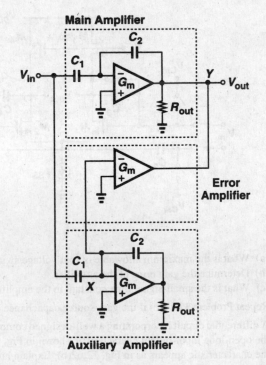

Figure 12.63

proportional current into node Y. The output impedance of the error amplifier is much greater than R_{out}. Assume $G_m R_{out} \gg 1$.

(a) Calculate the gain error of the circuit.

(b) Repeat part **(a)** if the auxiliary and error amplifiers are eliminated and compare the results.

References

1. G. Wegmann, E. A. Vittoz, and F. Rahali, "Charge Injection in Analog MOS Switches," *IEEE J. Solid-State Circuits*, vol. SC-22, pp. 1091–1097, Dec. 1987.

2. B. J. Sheu and C. Hu, "Switch-Induced Error Voltage on a Switched Capacitor," *IEEE J. Solid-State Circuits*, vol. SC-19, pp. 519–525, April 1984.

3. R. Gregorian and G. C. Temes, *Analog MOS Integrated Circuits for Signal Processing*, New York: John Wiley and Sons, 1986.

4. J. H. Fischer, "Noise Sources and Calculation Techniques for Switched Capacitor Filters," *IEEE J. Solid-State Circuits*, vol. 17, pp. 742–752, Aug. 1982.

5. B. S. Song, M. F. Tompsett, and K. R. Lakshmikumar, "A 12-Bit 1-Msample/s Capacitor-Averaging Pipelined A/D Converter," *IEEE J. Solid-State Circuits*, vol. SC-23, pp. 1324–1333, Dec. 1988.

6. B. Razavi, *Principles of Data Conversion System Design*, New York: IEEE Press, 1995.

7. P. C. Yu and H.-S. Lee, "A High-Swing 2-V CMOS Op Amp with Replica-Amp Gain Enhancement," *IEEE J. Solid-State Circuits*, vol. 28, pp. 1265–1272, Dec. 1993.

Chapter 13

Nonlinearity and Mismatch

In Chapters 6 and 7, we dealt with two types of nonidealities, namely, frequency response and noise, that limit the performance of analog circuits. In this chapter, we study two other imperfections that prove critical in high-precision analog design and trade with many other performance parameters. These effects are nonlinearity and mismatch.

We first define metrics for quantifying the effects of nonlinearity. Next, we study non-linearity in differential circuits and feedback systems and examine several linearization techniques. We then deal with the problem of mismatch and dc offsets in differential circuits. Finally, we consider a number of offset cancellation methods and describe the effect of offset cancellation on random noise.

13.1 Nonlinearity

13.1.1 General Considerations

As we have observed in the large-signal analysis of single-stage and differential amplifiers, circuits usually exhibit a nonlinear input/output characteristic. Depicted in Fig. 13.1, such a characteristic deviates from a straight line as the input swing increases. Two examples

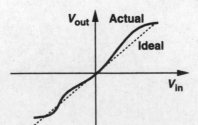

Figure 13.1 Input/output characteristic of a nonlinear system.

are shown in Fig. 13.2. In a common-source stage or a differential pair, the output variation becomes heavily nonlinear as the input level increases. In other words, for a small input swing, the output is a reasonable replica of the input but for large swings the output exhibits "saturated" levels.

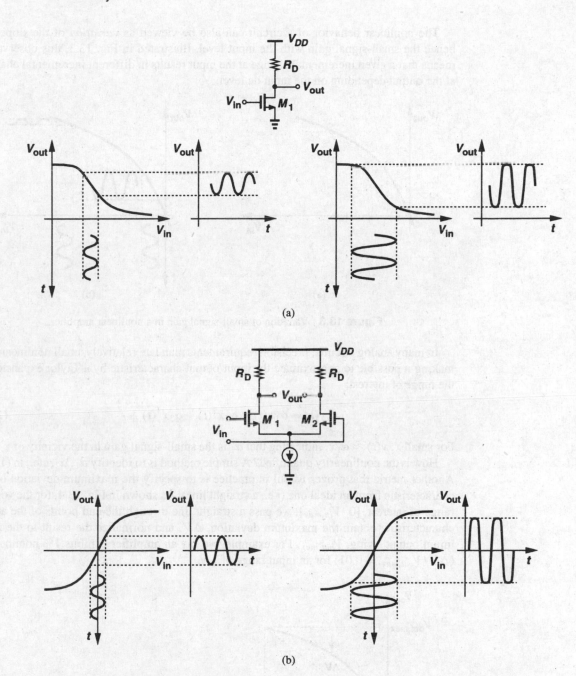

Figure 13.2   Distortion in (a) a common-source stage and (b) a differential pair.

The nonlinear behavior of a circuit can also be viewed as *variation* of the slope and hence the small-signal gain with the input level. Illustrated in Fig. 13.3, this observation means that a given incremental change at the input results in different incremental changes at the output depending on the input dc level.

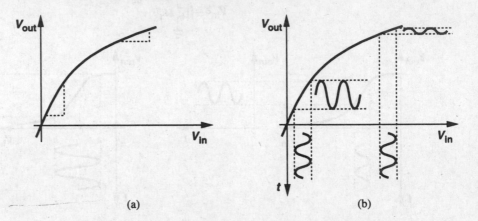

(a)   (b)

Figure 13.3   Variation of small-signal gain in a nonlinear amplifier.

In many analog circuits, precision requirements mandate relatively small nonlinearities, making it possible to approximate the input/output characteristic by a Taylor expansion in the range of interest:

$$y(t) = \alpha_1 x(t) + \alpha_2 x^2(t) + \alpha_3 x^3(t) + \cdots. \tag{13.1}$$

For small x, $y(t) \approx \alpha_1 x$, indicating that α_1 is the small-signal gain in the vicinity of $x \approx 0$.

How is the nonlinearity quantified? A simple method is to identify α_1, α_2, etc., in (13.1). Another metric that proves useful in practice is to specify the maximum deviation of the characteristic from an ideal one (i.e., a straight line). As shown in Fig. 13.4, for the voltage range of interest, $[0 \ V_{in,max}]$, we pass a straight line through the end points of the actual characteristic, obtain the maximum deviation, ΔV, and normalize the result to the maximum output swing, $V_{out,max}$. For example, we say an amplifier exhibits 1% nonlinearity ($\Delta V / V_{out,max} = 0.01$) for an input range of 1 V.

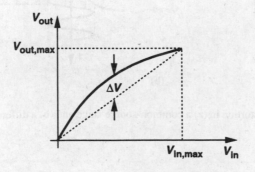

Figure 13.4 Definition of nonlinearity.

Example 13.1

The input/output characteristic of a differential amplifier is approximated as $y(t) = \alpha_1 x(t) + \alpha_3 x^3(t)$. Calculate the maximum nonlinearity if the input range is from $x = -x_{max}$ to $x = +x_{max}$.

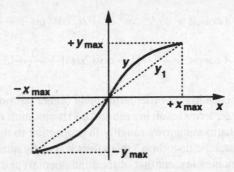

Figure 13.5

Solution

As depicted in Fig. 13.5, we can express the straight line passing through the end points as

$$y_1 = \frac{\alpha_1 x_{max} + \alpha_3 x_{max}^3}{x_{max}} x \tag{13.2}$$

$$= \left(\alpha_1 + \alpha_3 x_{max}^2\right)x. \tag{13.3}$$

The difference between y and y_1 is therefore equal to

$$\Delta y = y - y_1 \tag{13.4}$$

$$= \alpha_1 x + \alpha_3 x^3 - \left(\alpha_1 + \alpha_3 x_{max}^2\right)x. \tag{13.5}$$

Setting the derivative of Δy with respect to x to zero, we have $x = x_{max}/\sqrt{3}$ and the maximum deviation is equal to $2\alpha_3 x_{max}^3/(3\sqrt{3})$. Normalized to the maximum output, the nonlinearity is obtained as

$$\frac{\Delta y}{y_{max}} = \frac{2\alpha_3 x_{max}^3}{3\sqrt{3} \times 2\left(\alpha_1 x_{max} + \alpha_3 x_{max}^3\right)}. \tag{13.6}$$

Note that the factor of 2 in the denominator is included because the maximum peak-to-peak output swing is equal to $2(\alpha_1 x_{max} + \alpha_3 x_{max}^3)$. For small nonlinearities, we can neglect $\alpha_3 x_{max}^3$ with respect to $\alpha_1 x_{max}$, arriving at

$$\frac{\Delta y}{y_{max}} \approx \frac{\alpha_3}{3\sqrt{3}\alpha_1} x_{max}^2. \tag{13.7}$$

Note that the relative nonlinearity is proportional to the square of the maximum input swing in this example.

The nonlinearity of a circuit can also be characterized by applying a sinusoid at the input and measuring the harmonic content of the output. Specifically, if in (13.1), $x(t) = A\cos\omega t$, then

$$y(t) = \alpha_1 A\cos\omega t + \alpha_2 A^2\cos^2\omega t + \alpha_3\cos^3\omega t + \cdots \tag{13.8}$$

$$= \alpha_1 A\cos\omega t + \frac{\alpha_2 A^2}{2}[1+\cos(2\omega t)] + \frac{\alpha_3 A^3}{4}[3\cos\omega t + \cos(3\omega t)] + \cdots. \tag{13.9}$$

We observe that higher-order terms yield higher harmonics. In particular, even-order terms and odd-order terms result in even and odd harmonics, respectively. Note that the magnitude of the nth harmonic grows roughly in proportion to the nth power of the input amplitude. Called "harmonic distortion," this effect is usually quantified by summing the power of all of the harmonics (except that of the fundamental) and normalizing the result to the power of the fundamental. Such a metric is called the "total harmonic distortion" (THD). For a third-order nonlinearity:

$$THD = \frac{(\alpha_2 A^2/2)^2 + (\alpha_3 A^3/4)^2}{(\alpha_1 A + 3\alpha_3 A^3/4)^2}. \tag{13.10}$$

Harmonic distortion is undesirable in most signal processing applications, including audio and video systems. High-quality audio products such as compact disc (CD) players require a THD of about 0.01% (-80 dB) and video products, about 0.1% (-60 dB).

13.1.2 Nonlinearity of Differential Circuits

Differential circuits exhibit an "odd-symmetric" input/output characteristic, i.e., $f(-x) = -f(x)$. For the Taylor expansion of (13.1) to be an odd function, all of the even-order terms, α_{2j} must be zero:

$$y(t) = \alpha_1 x(t) + \alpha_3 x^3(t) + \alpha_5 x^5(t) + \cdots, \tag{13.11}$$

indicating that a differential circuit driven by a differential signal produces no even harmonics. This is another very important property of differential operation.

In order to appreciate the reduction of nonlinearity obtained by differential operation, let us consider the two amplifiers shown in Fig. 13.6, each of which is designed to provide a small-signal voltage gain of

$$|A_v| \approx g_m R_D \tag{13.12}$$

$$= \mu_n C_{ox}\frac{W}{L}(V_{GS}-V_{TH})R_D. \tag{13.13}$$

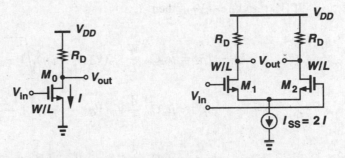

Figure 13.6 Single-ended and differential amplifiers providing the same voltage gain.

Suppose a signal $V_m \cos \omega t$ is applied to each circuit. Examining only the drain currents for simplicity, we can write for the common-source stage:

$$I_{D0} = \frac{1}{2}\mu_n C_{ox}\frac{W}{L}(V_{GS} - V_{TH} + V_m \cos\omega t)^2$$

$$= \frac{1}{2}\mu_n C_{ox}\frac{W}{L}(V_{GS} - V_{TH})^2 + \mu_n C_{ox}\frac{W}{L}(V_{GS} - V_{TH})V_m \cos \omega t$$

$$+ \frac{1}{2}\mu_n C_{ox}\frac{W}{L}V_m^2 \cos^2 \omega t$$

$$= I + \mu_n C_{ox}\frac{W}{L}(V_{GS} - V_{TH})V_m \cos \omega t + \frac{1}{4}\mu_n C_{ox}\frac{W}{L}V_m^2[1 + \cos(2\omega t)]. \quad (13.14)$$

Thus, the amplitude of the second harmonic, A_{HD2}, normalized to that of the fundamental, A_F, is

$$\frac{A_{HD2}}{A_F} = \frac{V_m}{4(V_{GS} - V_{TH})}. \quad (13.15)$$

On the other hand, for M_1 and M_2 in Fig. 13.6, we have from Chapter 4:

$$I_{D1} - I_{D2} = \frac{1}{2}\mu_n C_{ox}\frac{W}{L}V_{in}\sqrt{\frac{4I_{SS}}{\mu_n C_{ox}\dfrac{W}{L}} - V_{in}^2} \quad (13.16)$$

$$= \frac{1}{2}\mu_n C_{ox}\frac{W}{L}V_{in}\sqrt{4(V_{GS} - V_{TH})^2 - V_{in}^2}. \quad (13.17)$$

If $|V_{in}| \ll V_{GS} - V_{TH}$, then

$$I_{D1} - I_{D2} = \mu_n C_{ox} \frac{W}{L} V_{in}(V_{GS} - V_{TH}) \sqrt{1 - \frac{V_{in}^2}{4(V_{GS} - V_{TH})^2}} \qquad (13.18)$$

$$\approx \mu_n C_{ox} \frac{W}{L} V_{in}(V_{GS} - V_{TH}) \left[1 - \frac{V_{in}^2}{8(V_{GS} - V_{TH})^2} \right] \qquad (13.19)$$

$$= \mu_n C_{ox} \frac{W}{L} (V_{GS} - V_{TH}) \left[V_m \cos \omega t - \frac{V_m^3 \cos^3 \omega t}{8(V_{GS} - V_{TH})^2} \right]. \qquad (13.20)$$

Since $\cos^3 \omega t = [3 \cos \omega t + \cos(3\omega t)]/4$, we obtain,

$$I_{D1} - I_{D2} = g_m \left[V_m - \frac{3V_m^3}{32(V_{GS} - V_{TH})^2} \right] \cos \omega t - g_m \frac{V_m^3 \cos(3\omega t)}{32(V_{GS} - V_{TH})^2}. \qquad (13.21)$$

If $V_m \gg 3V_m^3/[8(V_{GS} - V_{TH})^2]$, then

$$\frac{A_{HD3}}{A_F} \approx \frac{V_m^2}{32(V_{GS} - V_{TH})^2}. \qquad (13.22)$$

Comparison of (13.15) and (13.22) indicates that the differential circuit exhibits much less distortion than its single-ended counterpart while providing the same voltage gain and output swing. For example, if $V_m = 0.2(V_{GS} - V_{TH})$, (13.15) and (13.22) yield a distortion of 5% and 0.125%, respectively.

While achieving a lower distortion, the differential pair consumes twice as much power as the CS stage because $I_{SS} = 2I$. The key point, however, is that even if the bias current of M_0 is raised to $2I$, (13.15) predicts that the distortion decreases by only a factor of $\sqrt{2}$ (with W/L maintained constant).

13.1.3 Effect of Negative Feedback on Nonlinearity

In Chapter 8, we observed that negative feedback makes the closed-loop gain relatively independent of the op amp's open-loop gain. Since nonlinearity can be viewed as variation of the small-signal gain with the input level, we expect that negative feedback suppresses this variation as well, yielding higher linearity for the closed-loop system.

Analysis of nonlinearity in a feedback system is quite complex. Here, we consider a simple, "mildly nonlinear" system to gain more insight. The reason is that, if properly designed, a feedback amplifier exhibits only small distortion components, lending itself to this type of analysis.

Let us assume that the core amplifier in the system of Fig. 13.7 has an input-output characteristic $y \approx \alpha_1 x + \alpha_2 x^2$. We apply a sinusoidal input $x(t) = V_m \cos \omega t$, postulating that the output contains a fundamental component and a second harmonic and hence can

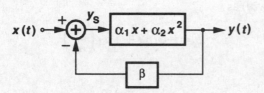

Figure 13.7 Feedback system incorporating a nonlinear feedforward amplifier.

be approximated as $y \approx a \cos \omega t + b \cos 2\omega t$.[1] Our objective is to determine a and b. The output of the subtractor can be written as

$$y_S = x(t) - \beta y(t) \tag{13.23}$$

$$= V_m \cos \omega t - \beta(a \cos \omega t + b \cos 2\omega t) \tag{13.24}$$

$$= (V_m - \beta a) \cos \omega t - \beta b \cos 2\omega t. \tag{13.25}$$

This signal experiences the nonlinearity of the feedforward amplifier, thereby producing an output given by:

$$y(t) = \alpha_1[(V_m - \beta) \cos \omega t - \beta b \cos 2\omega t]$$

$$+ \alpha_2[(V_m - \beta a) \cos \omega t - \beta b \cos 2\omega t]^2 \tag{13.26}$$

$$= [\alpha_1(V_m - \beta a) - \alpha_2(V_m - \beta a)\beta b] \cos \omega t$$

$$+ \left[-\alpha_1 \beta b + \frac{\alpha_2(V_m - \beta a)^2}{2} \right] \cos 2\omega t + \cdots. \tag{13.27}$$

The coefficients of $\cos \omega t$ and $\cos 2\omega t$ in (13.27) must be equal to a and b, respectively:

$$a = (\alpha_1 - \alpha_2 \beta b)(V_m - \beta a) \tag{13.28}$$

$$b = -\alpha_1 \beta b + \frac{\alpha_2(V_m - \beta a)^2}{2}. \tag{13.29}$$

The assumption of small nonlinearity implies that both α_2 and b are small quantities, yielding $a \approx \alpha_1(V_m - \beta a)$ and hence

$$a = \frac{\alpha_1}{1 + \beta \alpha_1} V_m, \tag{13.30}$$

which is to be expected because $\beta \alpha_1$ is the loop gain. To calculate b, we write

$$V_m - \beta a \approx \frac{a}{\alpha_1}, \tag{13.31}$$

---

[1] Note that higher harmonics and phase shifts through the system are neglected.

thus expressing (13.29) as

$$b = -\alpha_1 \beta b + \frac{1}{2}\alpha_2 \left(\frac{a}{\alpha_1}\right)^2. \tag{13.32}$$

That is,

$$b(1 + \alpha_1 \beta) = \frac{\alpha_2}{2}\left(\frac{a}{\alpha_1}\right)^2 \tag{13.33}$$

$$= \frac{\alpha_2}{2\alpha_1^2}\frac{\alpha_1^2}{(1 + \beta\alpha_1)^2}V_m^2. \tag{13.34}$$

It follows that

$$b = \frac{\alpha_2 V_m^2}{2}\frac{1}{(1 + \beta\alpha_1)^3}. \tag{13.35}$$

For a meaningful comparison, we normalize the amplitude of the second harmonic to that of the fundamental:

$$\frac{b}{a} = \frac{\alpha_2 V_m}{2}\frac{1}{\alpha_1}\frac{1}{(1 + \beta\alpha_1)^2}. \tag{13.36}$$

Without feedback, on the other hand, such a ratio would be equal to $(\alpha_2 V_m^2/2)/\alpha_1 V_m = \alpha_2 V_m/(2\alpha_1)$. Thus, the relative magnitude of the second harmonic has dropped by a factor of $(1 + \beta\alpha_1)^2$.

As described in Chapter 8, a feedback circuit employing a feedforward amplifier with a finite gain suffers from gain error. For a feedforward gain of A_0 and a feedback factor of β, the relative gain error is approximately equal to $1/(\beta A_0)$. If the feedforward amplifier exhibits nonlinearity, it is possible to derive a simple relationship between the gain error and maximum nonlinearity of the overall feedback circuit. As illustrated in Fig. 13.8, we draw two straight lines, one representing the ideal characteristic (with a slope $1/\beta$) and another passing through the end points of the actual characteristic. We note that with this construction, the nonlinearity, Δy_2, is always smaller than the gain error, Δy_1. This is of course true only if the small-signal gain drops monotonically as x goes from 0 to x_{max}, a

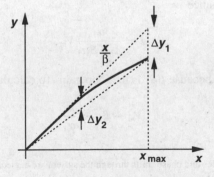

Figure 13.8 Gain error and nonlinearity in a feedback system.

typical behavior in most analog circuits. Thus, a sufficient condition to ensure $\Delta y_2 < \epsilon$ is to guarantee that $\Delta y_1 < \epsilon$ by choosing a high open-loop gain for the amplifier.

The above condition is often applied in analog design because it is much easier to predict the open-loop gain than its nonlinearity. Of course, this simplification is obtained at the cost of a pessimistic choice of the amplifier's gain, an issue that becomes more serious as short-channel devices limit the voltage gain that can be achieved.

13.1.4 Capacitor Nonlinearity

In switched-capacitor circuits, the voltage dependence of capacitors may introduce substantial distortion. While for a linear capacitor we have $Q = CV$, for a voltage-dependent capacitor we must write $dQ = C \, dV$. Thus, the total charge on a capacitor sustaining a voltage V_1 is

$$Q(V_1) = \int_0^{V_1} C \, dV. \tag{13.37}$$

To study the effect of capacitor nonlinearity, we express each capacitor as $C = C_0(1 + \alpha_1 V + \alpha_2 V^2 + \cdots)$.

Let us consider the noninverting amplifier of Fig. 12.41(a), repeated in Fig. 13.9, as an example. At the beginning of the amplification mode, C_1 has a voltage equal to V_{in0} and C_2 a voltage of zero. Assuming $C_1 \approx MC_0(1 + \alpha_1 V)$, where M is the nominal closed-loop gain ($C_1 = MC_2$), we obtain the charge across C_1 as

$$Q_1 = \int_0^{V_{in0}} C_1 \, dV \tag{13.38}$$

$$= \int_0^{V_{in0}} MC_0(1 + \alpha_1 V) \, dV \tag{13.39}$$

$$= MC_0 V_{in0} + MC_0 \frac{\alpha_1}{2} V^2. \tag{13.40}$$

Similarly, if $C_2 \approx C_0(1 + \alpha_1 V)$, then the charge on this capacitor at the end of the amplification mode is

$$Q_2 = \int_0^{V_{out}} C_2 \, dV \tag{13.41}$$

$$= C_0 V_{out} + C_0 \frac{\alpha_1}{2} V_{out}^2. \tag{13.42}$$

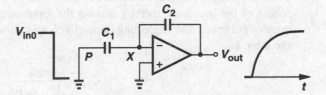

Figure 13.9   Effect of capacitor nonlinearity.

Equating Q_1 and Q_2 and solving for V_{out}, we have

$$V_{out} = \frac{1}{\alpha_1}\left(-1 + \sqrt{1 + M\alpha_1^2 V_{in0}^2 + 2M\alpha_1 V_{in0}}\right). \tag{13.43}$$

The last two terms under the square root are usually much less than unity and, since for $\epsilon \ll 1$, $\sqrt{1+\epsilon} \approx 1 + \epsilon/2 - \epsilon^2/8$, we can write

$$V_{out} \approx MV_{in0} + (1 - M)\frac{M\alpha_1}{2}V_{in0}^2. \tag{13.44}$$

The second term in the above equation represents the nonlinearity resulting from the voltage dependence of the capacitor.

13.1.5 Linearization Techniques

While amplifiers using "global" feedback (e.g., the switched-capacitor topologies of Chapter 12) can achieve a high linearity, stability and settling issues of feedback circuits limit their usage in high-speed applications. For this reason, many other techniques have been invented to linearize amplifiers with less compromise in speed.

The principle behind linearization is to reduce the dependence of the gain of the circuit upon the input level. This usually translates into making the gain relatively independent of the transistor bias currents.

The simplest linearization method is source degeneration by means of a linear resistor. As shown in Fig. 13.10 for a common-source stage and revealed by the observations in the

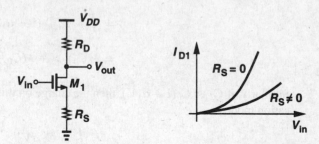

Figure 13.10   Common-source stage with resistive degeneration.

previous section, degeneration reduces the signal swing applied between the gate and the source of the transistor, thereby making the input/output characteristic more linear. From another point of view, neglecting body effect, we can write the overall transconductance of the stage as

$$G_m = \frac{g_m}{1 + g_m R_S}, \tag{13.45}$$

which for large $g_m R_S$ approaches $1/R_S$, an input-independent value.

Note that the amount of linearization depends on $g_m R_S$ rather on R_S alone. With a relatively constant G_m, the voltage gain, $G_m R_D$, is also relatively independent of the input and the amplifier is linearized.

Example 13.2

A common-source stage biased at a current I_1 experiences an input voltage swing that varies the drain current from $0.75 I_1$ to $1.25 I_1$. Calculate the variation of the small-signal voltage gain (a) with no degeneration, (b) with degeneration such that $g_m R_S = 2$, where g_m denotes the transconductance at $I_D = I_1$.

Solution

Assuming square-law behavior, we have $g_m \propto \sqrt{I_D}$. For the case of no degeneration:

$$\frac{g_{m,high}}{g_{m,low}} = \sqrt{\frac{1.25}{0.75}}. \tag{13.46}$$

With $g_m R_S = 2$,

$$\frac{G_{m,high}}{G_{m,low}} = \frac{\dfrac{\sqrt{1.25}\, g_m}{1 + \sqrt{1.25}\, g_m R_S}}{\dfrac{\sqrt{0.75}\, g_m}{1 + \sqrt{0.75}\, g_m R_S}} \tag{13.47}$$

$$= \sqrt{\frac{1.25}{0.75}} \cdot \frac{1 + 2\sqrt{0.75}}{1 + 2\sqrt{1.25}} \tag{13.48}$$

$$= 0.84 \sqrt{\frac{1.25}{0.75}}. \tag{13.49}$$

Thus, degeneration decreases the variation of the small-signal gain by approximately 16% in this case.

Resistive degeneration presents trade-offs between linearity, noise, power dissipation, and gain. For reasonable input voltage swings (e.g., 1 V_{pp}), it may be quite difficult to achieve even a voltage gain of 2 in a common-source stage if the nonlinearity is to remain below 1%.

A differential pair can be degenerated as shown in Figs. 13.11(a) and (b). In Fig. 13.11(a), I_{SS} flows through the degeneration resistors, thereby consuming a voltage headroom of $I_{SS} R_S/2$, an important issue if a high level of degeneration is required. The circuit of Fig. 13.11(b), on the other hand, does not involve this issue but it suffers from a slightly higher noise (and offset voltage) because the two tail current sources introduce some differential error. The reader can prove that if the output noise current of each current source is equal to $\overline{I_n^2}$, then the input-referred noise voltage of the circuit of Fig. 13.11(b) is higher than that of Fig. 13.11(a) by $2\overline{I_n^2} R_S^2$.

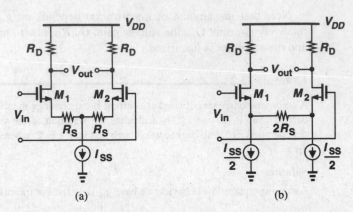

Figure 13.11   Source degeneration applied to a differential pair.

Resistive degeneration requires high-quality resistors, a commodity unavailable in many of today's CMOS technologies (Chapter 17). As depicted in Fig. 13.12, the resistor can be replaced by a MOSFET operating in deep triode region. However, for large input swings, M_3 may not remain in deep triode region, thereby experiencing substantial change in its on-resistance. Furthermore, V_b must track the input common-mode level so that R_{on3} can be defined accurately.

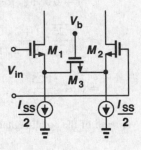

Figure 13.12   Differential pair degenerated by a MOSFET operating in deep triode region.

Another linearization technique is illustrated in Fig. 13.13 [1]. Here, M_3 and M_4 are in deep triode region if $V_{in} = 0$. As the gate voltage of M_1 becomes more positive than the gate voltage of M_2, transistor M_3 stays in the triode region because $V_{D3} = V_{G3} - V_{GS1}$ whereas

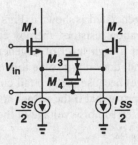

Figure 13.13   Differential pair degenerated by two MOSFETs operating in the triode region.

M_4 eventually enters the saturation region because its drain voltage rises and its gate and source voltages fall. Thus, the circuit remains relatively linear even if one degeneration device goes into saturation. For the widest linear region, [1] suggests $(W/L)_{1,2} \approx 7(W/L)_{3,4}$.

A linearization technique avoiding the use of resistors is based on the observation that a MOSFET operating in the triode region can provide a linear I_D/V_{GS} characteristic if its drain-source voltage is held constant: $I_D = (1/2)\mu C_{ox}(W/L)[2(V_{GS} - V_{TH})V_{DS} - V_{DS}^2]$. Illustrated in Fig. 13.14, the technique employs amplifiers A_1 and A_2 along with cascode devices M_3 and M_4 to force V_X and V_Y to be equal to V_b for varying input levels.

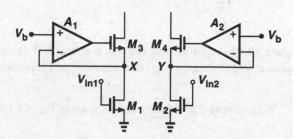

Figure 13.14 Differential pair using input devices operating in the triode region.

This circuit suffers from several drawbacks. First, the transconductance of M_1 and M_2, equal to $\mu_n C_{ox}(W/L)V_{DS}$, is relatively small because V_{DS} must be low enough to ensure each input transistor remains in the triode region. Second, the input common-mode level must be tightly controlled and it must track V_b so as to define I_{D1} and I_{D2}. Third, M_3, M_4, and the two auxiliary amplifiers contribute substantial noise to the output.

Another approach to linearizing voltage amplifiers is to perform "post-correction." Illustrated in Fig. 13.15, the idea is to view the amplifier as a voltage-to-current (V/I) converter followed by a current-to-voltage (I/V) converter. If the V/I converter can be described as $I_{out} = f(V_{in})$ and the I/V converter as $V_{out} = f^{-1}(I_{in})$, then V_{out} is a linear function of V_{in}. That is, the second stage corrects the nonlinearity introduced by the first stage. As an example, recall from Chapter 4 that for the circuit shown in Fig. 13.16(a), we have

$$V_{in1} - V_{in2} = V_{GS1} - V_{GS2} \tag{13.50}$$

$$= \sqrt{\frac{2I_{D1}}{\mu_n C_{ox}\left(\dfrac{W}{L}\right)_{1,2}}} - \sqrt{\frac{2I_{D2}}{\mu_n C_{ox}\left(\dfrac{W}{L}\right)_{1,2}}}. \tag{13.51}$$

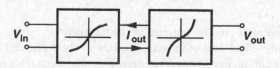

Figure 13.15 Voltage amplifier viewed as a cascade of two nonlinear stages.

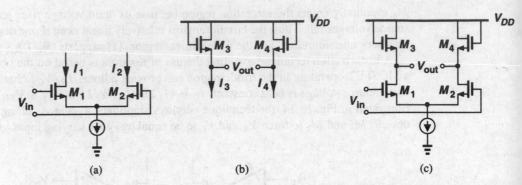

Figure 13.16 (a) Differential pair with nonlinear I/V characteristic, (b) diode-connected devices with nonlinear V/I characteristic, (c) circuit having linear input/output characteristic.

We also note that for the circuit shown in Fig. 13.16(b),

$$V_{out} = V_{GS3} - V_{GS4} \tag{13.52}$$

$$= \sqrt{\frac{2I_3}{\mu_n C_{ox}\left(\dfrac{W}{L}\right)_{3,4}}} - \sqrt{\frac{2I_4}{\mu_n C_{ox}\left(\dfrac{W}{L}\right)_{3,4}}}, \tag{13.53}$$

where channel-length modulation and body effect are neglected. It follows that for the circuit shown in Fig. 13.16(c),

$$V_{out} = \sqrt{\frac{2I_{D1}}{\mu_n C_{ox}\left(\dfrac{W}{L}\right)_{3,4}}} - \sqrt{\frac{2I_{D2}}{\mu_n C_{ox}\left(\dfrac{W}{L}\right)_{3,4}}} \tag{13.54}$$

$$= \frac{1}{\sqrt{\left(\dfrac{W}{L}\right)_{3,4}}}(V_{in1} - V_{in2})\, sqrt\left(\frac{W}{L}\right)_{1,2}. \tag{13.55}$$

Thus, as derived in Chapter 4, the voltage gain is equal to

$$A_v = \sqrt{\frac{\left(\dfrac{W}{L}\right)_{1,2}}{\left(\dfrac{W}{L}\right)_{3,4}}}, \tag{13.56}$$

a quantity independent of the bias currents of the transistors.

In practice, body effect and other nonidealities in short-channel devices give rise to nonlinearity in this circuit. Furthermore, as the differential input level increases, driving M

or M_2 into the subthreshold region, Eqs. (13.51) and (13.53) no longer hold and the gain drops sharply.

13.2 Mismatch

Our study of amplifiers in the previous chapters has mostly assumed that the circuits are perfectly symmetric, i.e., the two sides exhibit identical properties and bias currents. In reality, however, nominally-identical devices suffer from a finite mismatch due to uncertainties in each step of the manufacturing process. For example, as illustrated in Fig. 13.17, the gate dimensions of MOSFETs suffer from random, microscopic variations and hence mismatches between the equivalent lengths and widths of two transistors that are identically laid out. Also, MOS devices exhibit threshold voltage mismatch because, from (2.1), V_{TH} is a function of the doping levels in the channel and the gate, and these levels vary randomly from one device to another.

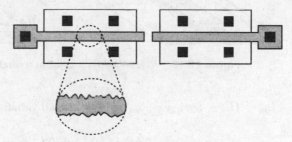

Figure 13.17 Random mismatches due to microscopic variations in device dimensions.

Study of mismatch consists of two steps: (1) identify and formulate the mechanisms that lead to mismatch between devices; (2) analyze the effect of device mismatches upon the performance of circuits. Unfortunately, the first step is quite complex and heavily dependent on the fabrication technology and the layout, often requiring actual measurements of mismatches. For example, the achievable mismatch between capacitors is typically quoted to be 0.1%, but this value is not derived from any fundamental quantities. We therefore consider only some basic trends and intuitive results. Layout techniques for minimum mismatch are described in Chapter 18.

Expressing the characteristics of a MOSFET in saturation as $I_D = (1/2)\mu C_{ox}(W/L)$ $(V_{GS} - V_{TH})^2$, we observe that mismatches between μ, C_{ox}, W, L, and V_{TH} result in mismatches between drain currents (for a given V_{GS}) or gate-source voltages (for a given drain current) of two nominally-identical transistors. Intuitively, we expect that as W and L increase, their relative mismatches, $\Delta W/W$ and $\Delta L/L$, respectively, decrease, i.e., larger devices exhibit smaller mismatches. A more important observation is that all of the mismatches decrease as the *area* of the transistor, WL, increases. For example, increasing W reduces both $\Delta W/W$ *and* $\Delta L/L$. This is because as WL increases, random variations experience greater "averaging," thereby falling in magnitude. For the case depicted in Fig. 13.18, $\Delta L_2 < \Delta L_1$ because, if the device is viewed as many small parallel transistors (Fig. 13.19), each having a width W_0, then we can write the equivalent length as

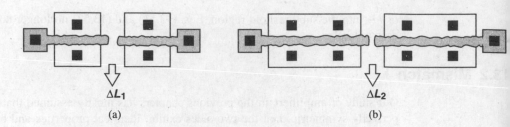

Figure 13.18 Reduction of length mismatch as a result of increasing the width.

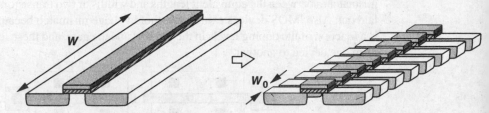

Figure 13.19 Wide MOSFET viewed as a parallel combination of narrow devices.

$L_{eq} \approx (L_1 + L_2 + \cdots + L_n)/n$. The overall variation is therefore given by

$$\Delta L_{eq} \approx \left(\Delta L_1^2 + \Delta L_2^2 + \cdots + \Delta L_n^2\right)^{1/2}/n \qquad (13.57)$$

$$= \frac{\left(n\Delta L_0^2\right)^{1/2}}{n} \qquad (13.58)$$

$$= \frac{\Delta L_0}{\sqrt{n}}, \qquad (13.59)$$

where ΔL_0 is the statistical variation of the length for a transistor with width W_0. Equation (13.59) reveals that for a given W_0, as n increases, the variation of L_{eq} decreases.

The above result can be extended to other device parameters as well. For example, we postulate that μC_{ox} and V_{TH} suffer from less mismatch if the device area increases. Illustrated in Fig. 13.20, the reason is that a large transistor can be decomposed into a series and

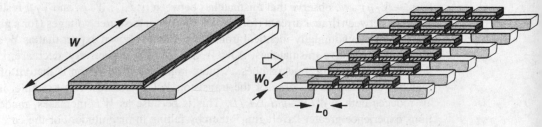

Figure 13.20 Large MOSFET viewed as a combination of small devices.

parallel combination of small unit transistors with dimensions W_0 and L_0, each exhibiting $(\mu C_{ox})_j$ and V_{THj}. For given W_0 and L_0, as the number of unit transistors increases, μC_{ox} and V_{TH} experience greater averaging, leading to smaller mismatch between two large transistors.

The foregoing qualitative observations have been verified mathematically and experimentally [2, 3]. Here, we state without proof that

$$\Delta V_{TH} = \frac{A_{VTH}}{\sqrt{WL}} \tag{13.60}$$

$$\Delta \left(\mu C_{OX} \frac{W}{L} \right) = \frac{A_K}{\sqrt{WL}}, \tag{13.61}$$

where A_{VTH} and A_K are proportionality factors.

Interestingly, A_{VTH} has been observed to scale down with the gate oxide thickness [3]. From the data in [4], $A_{VTH} \approx 10$ mV·μm for $t_{ox} \approx 100$ Å. Thus, in a 0.6-μm technology with $t_{ox} = 100$ Å, two 100 μm/0.6 μm devices ($L_{eff} \approx 0.5\ \mu$m) exhibit a threshold mismatch of 1.4 mV. With this information, we can write

$$\Delta V_{TH} = \frac{0.1 t_{ox}}{\sqrt{WL}}\ \text{mV}, \tag{13.62}$$

where t_{ox} is expressed in angstroms and W and L in microns. Since the channel capacitance is proportional to WLC_{ox}, we note that ΔV_{TH} and the channel capacitance bear a trade-off.

We now study the effect of device mismatch upon the performance of circuits. Mismatches lead to three significant phenomena: dc offsets, finite even-order distortion, and lower common-mode rejection. The last phenomenon was studied in Chapter 4.

DC Offsets   Consider the differential pair shown in Fig. 13.21(a). With $V_{in} = 0$ and perfect symmetry, $V_{out} = 0$, but in the presence of mismatches, $V_{out} \neq 0$. We say the circuit suffers from a dc "offset" equal to the observed value of V_{out} when V_{in} is set to

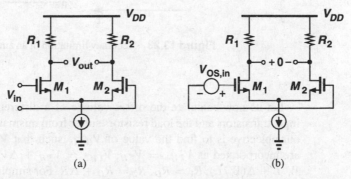

(a)   (b)

Figure 13.21   (a) Differential pair with offset measured at the output, (b) circuit of (a) with its offset referred to the input.

zero. In practice, it is more meaningful to specify the input-referred offset voltage, defined as the input level that forces the output voltage to go to zero [Fig. 13.21(b)]. Note that $|V_{OS,in}| = |V_{OS,out}|/A_v$. As with random noise, the polarity of random offsets is unimportant.

How does offset limit the performance? Suppose the differential pair of Fig. 13.21 is to amplify a small input voltage. Then, as depicted in Fig. 13.22, the output contains amplified replicas of both the signal and the offset. In a cascade of direct-coupled amplifiers, the dc offset may experience so much gain that it drives the latter stages into nonlinear operation.

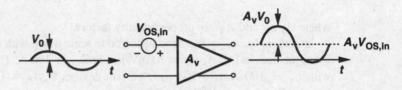

Figure 13.22   Effect of offset in an amplifier.

A more important effect of offset is the limitation on the precision with which signals can be measured. For example, if an amplifier is used to determine whether the input signal is greater or less than a reference, V_{REF} (Fig. 13.23), then the input-referred offset imposes a lower bound on the minimum $V_{in} - V_{REF}$ that can be detected reliably.

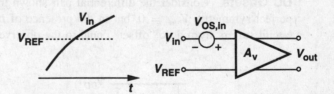

Figure 13.23   Accuracy limitation of an amplifier due to offset.

Let us now calculate the offset voltage of a differential pair, assuming that both the input transistors and the load resistors suffer from mismatch. As illustrated in Fig. 13.21(b), our objective is to find the value of $V_{OS,in}$ such that $V_{out} = 0$. The device mismatches are incorporated as $V_{TH1} = V_{TH}, V_{TH2} = V_{TH} + \Delta V_{TH}; (W/L)_1 = W/L, (W/L)_2 = W/L + \Delta(W/L); R_1 = R_D, R_2 = R_D + \Delta R$. For simplicity, $\lambda = \gamma = 0$, and mismatches in $\mu_n C_{ox}$ are neglected. For $V_{out} = 0$, we must have $I_{D1} R_1 = I_{D2} R_2$, concluding that I_D cannot be equal to I_{D2}. Thus, we assume $I_{D1} = I_D, I_{D2} = I_D + \Delta I_D$.

Since $V_{OS,in} = V_{GS1} - V_{GS2}$, we have

$$V_{OS,in} = \sqrt{\frac{2I_{D1}}{\mu_n C_{ox}\left(\dfrac{W}{L}\right)_1} + V_{TH1}} - \sqrt{\frac{2I_{D2}}{\mu_n C_{ox}\left(\dfrac{W}{L}\right)_2} - V_{TH2}} \qquad (13.63)$$

$$= \sqrt{\frac{2}{\mu_n C_{ox}}}\left[\sqrt{\frac{I_D}{\dfrac{W}{L}}} - \sqrt{\frac{I_D + \Delta I_D}{\dfrac{W}{L} + \Delta\left(\dfrac{W}{L}\right)}}\right] - \Delta V_{TH} \qquad (13.64)$$

$$= \sqrt{\frac{2}{\mu_n C_{ox}}}\sqrt{\frac{I_D}{W/L}}\left[1 - \sqrt{\frac{1 + \dfrac{\Delta I_D}{I_D}}{1 + \Delta\left(\dfrac{W}{L}\right)\Big/\left(\dfrac{W}{L}\right)}}\right] - \Delta V_{TH}. \quad (13.65)$$

Assuming $\Delta I_D/I_D$ and $\Delta(W/L)/(W/L) \ll 1$, and noting that for $\epsilon \ll 1$ we can write $\sqrt{1+\epsilon} \approx 1 + \epsilon/2$ and $(\sqrt{1+\epsilon})^{-1} \approx 1 - \epsilon/2$, we reduce (13.65) to

$$V_{OS,in} = \sqrt{\frac{2I_D}{\mu_n C_{ox}\left(\dfrac{W}{L}\right)}}\left\{1 - \left(1 + \frac{\Delta I_D}{2I_D}\right)\left[1 - \frac{\Delta(W/L)}{2(W/L)}\right]\right\} - \Delta V_{TH} \quad (13.66)$$

$$= \sqrt{\frac{2I_D}{\mu_n C_{ox}\left(\dfrac{W}{L}\right)}}\left[\frac{-\Delta I_D}{2I_D} + \frac{\Delta(W/L)}{2(W/L)}\right] - \Delta V_{TH}, \qquad (13.67)$$

where the product of two small quantities is neglected. Recall that $I_{D1}R_1 = I_{D2}R_2$ and hence $I_D R_D = (I_D + \Delta I_D)(R_D + \Delta R_D) \approx I_D R_D + R_D \Delta I_D + I_D \Delta R_D$. Consequently, $\Delta I_D/I_D \approx -\Delta R_D/R_D$, and

$$V_{OS,in} = \frac{1}{2}\sqrt{\frac{2I_D}{\mu_n C_{ox}\left(\dfrac{W}{L}\right)}}\left[\frac{\Delta R_D}{R_D} + \frac{\Delta(W/L)}{(W/L)}\right] - \Delta V_{TH}. \qquad (13.68)$$

We also recognize that the square-root quantity is approximately equal to the equilibrium overdrive voltage of each transistor, $V_{GS} - V_{TH}$, and

$$V_{OS,in} = \frac{V_{GS} - V_{TH}}{2}\left[\frac{\Delta R_D}{R_D} + \frac{\Delta(W/L)}{(W/L)}\right] - \Delta V_{TH}. \qquad (13.69)$$

Equation (13.69) is an important result, revealing the dependence of $V_{OS,in}$ on device mismatches and bias conditions. We note that (1) the contribution of load resistor mismatch and transistor dimension mismatch *increases* with the equilibrium overdrive, and (2) the threshold voltage mismatch is directly referred to the input. Thus, it is desirable to minimize $V_{GS} - V_{TH}$ by lowering the tail current or increasing the transistor widths. In reality, since mismatches are independent statistical variables, we express (13.69) as[2]

$$V_{OS,in}^2 = \left(\frac{V_{GS} - V_{TH}}{2}\right)^2 \left\{\left(\frac{\Delta R_D}{R_D}\right)^2 + \left[\frac{\Delta(W/L)}{(W/L)}\right]^2\right\} + \Delta V_{TH}^2, \tag{13.70}$$

where squared quantities represent standard deviations.

To gain more insight into the effect of offset, let us establish an analogy between offset and *noise*. If the two inputs of a differential pair are shorted, the output voltage exhibits a finite noise, that is, a voltage that varies with time. We may therefore say that the offset voltage of a differential pair resembles a very low-frequency noise component, varying so slowly that it appears constant in our measurements. Viewed as such, offsets can be incorporated as noise sources, allowing us to utilize analysis techniques developed in Chapter 7. To this end, we represent the offset of two nominally-identical transistors by a voltage source equal to (13.70) in series with the gate of one of the transistors.

Example 13.3

Calculate the input-referred offset voltage of the circuit shown in Fig. 13.24(a). Assume all of the transistors operate in saturation.

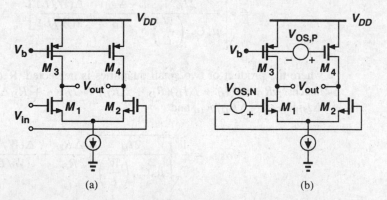

(a)   (b)

Figure 13.24

---

[2]As mentioned earlier, ΔV_{TH} does depend on W, an effect that can be added as a cross-correlation term. We neglect this term here for simplicity.

Solution

We insert the offsets of the NMOS and PMOS pairs as in Fig. 13.24(b). To obtain $I_{D1} = I_{D2}$ and $I_{D3} = I_{D4}$, we have from (13.69),

$$V_{OS,N} = \frac{(V_{GS} - V_{TH})_N}{2}\left[\frac{\Delta(W/L)}{W/L}\right]_N + \Delta V_{TH,N} \tag{13.71}$$

$$V_{OS,P} = \frac{|V_{GS} - V_{TH}|_P}{2}\left[\frac{\Delta(W/L)}{W/L}\right]_P + \Delta V_{TH,P}. \tag{13.72}$$

From the noise analysis in Chapter 7, $V_{OS,P}$ is amplified by a gain of $g_{mP}(r_{ON}\|r_{OP})$ and divided by $g_{mN}(r_{ON}\|r_{OP})$ when referred to the main input. As a result,

$$V_{OS,in} = \left\{\frac{|V_{GS} - V_{TH}|_P}{2}\left[\frac{\Delta(W/L)}{W/L}\right]_P + \Delta V_{TH,P}\right\}\frac{g_{mP}}{g_{mN}}$$

$$+ \frac{(V_{GS} - V_{TH})_N}{2}\left[\frac{\Delta(W/L)}{W/L}\right]_N + \Delta V_{TH,N}. \tag{13.73}$$

In practice, we add the "power" of these terms, as exemplified by (13.70). Note that, as with noise, the contribution of the offset of the PMOS pair is proportional to g_{mP}/g_{mN}.

The foregoing example can be better understood if we study the offset behavior of current sources. Consider the nominally-identical current sources M_1 and M_2 in Fig. 13.25. Neglecting channel-length modulation, we determine the total mismatch between I_{D1} and I_{D2} by calculating the total differential. Recall from calculus that if $y = f(x_1, x_2, \ldots)$, then the total differential is given by

$$\Delta y = \frac{\partial f}{\partial x_1}\Delta x_1 + \frac{\partial f}{\partial x_2}\Delta x_2 + \cdots. \tag{13.74}$$

Equation (13.74) simply means that each mismatch component Δx_j is weighted by the corresponding sensitivity $\partial f/\partial x_j$ as it contributes to the total mismatch. Since $I_D = (1/2)\mu_n C_{ox}(W/L)(V_{GS} - V_{TH})^2$, we have

$$\Delta I_D = \frac{\partial I_D}{\partial(W/L)}\Delta\left(\frac{W}{L}\right) + \frac{\partial I_D}{\partial(V_{GS} - V_{TH})}\Delta(V_{GS} - V_{TH}), \tag{13.75}$$

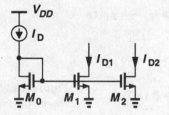

Figure 13.25 Mismatch between two current sources.

where mismatches in $\mu_n C_{ox}$ are neglected. It follows that

$$\Delta I_D = \frac{1}{2}\mu_n C_{ox}(V_{GS}-V_{TH})^2 \Delta\left(\frac{W}{L}\right) - \mu_n C_{ox}\frac{W}{L}(V_{GS}-V_{TH})\Delta V_{TH}. \quad (13.76)$$

Unlike the input-referred offset *voltage*, current mismatch is usually normalized to the average value to allow a meaningful comparison:

$$\frac{\Delta I_D}{I_D} = \frac{\Delta(W/L)}{W/L} - 2\frac{\Delta V_{TH}}{V_{GS}-V_{TH}}. \quad (13.77)$$

This result suggests that, to minimize current mismatch, the overdrive voltage must be *maximized*, a trend opposite of that in (13.69). This is because as $V_{GS} - V_{TH}$ increases, threshold mismatch has lesser effect on the device currents.

The dependence of offset voltage and current mismatches upon the overdrive voltage is similar to our observations in Chapter 7 for corresponding noise quantities. For a given current, the input noise voltage of a differential pair increases as the overdrive increases because $g_m = 2I_D/(V_{GS} - V_{TH})$. Also, the output noise current of current sources is proportional to g_m and hence proportional to $V_{GS} - V_{TH}$.

Even-Order Distortion   Our study of nonlinearity in Section 13.1 implies that, by virtue of odd symmetry, differential circuits are free from even-order distortion. In reality, however, mismatches degrade the symmetry, thereby introducing a finite even-order nonlinearity.

Analysis of the even-order distortion in the presence of mismatches is generally quite complex, often necessitating simulations. Here, we consider a simple case to gain some insight. Suppose the two signal paths in a differential circuit are represented by $y_1 \approx \alpha_1 x_1 + \alpha_2 x_1^2 + \alpha_3 x_1^3$ and $y_2 \approx \beta_1 x_2 + \beta_2 x_2^2 + \beta_3 x_2^3$ (Fig. 13.26). The differential output is given by

$$y_1 - y_2 = (\alpha_1 x_1 - \beta_2 x_2) + (\alpha_2 x_1^2 - \beta_2 x_2^2) + (\alpha_3 x_1^3 - \beta_3 x_2^3), \quad (13.78)$$

which, for $x_1 = -x_2$, reduces to

$$y_1 - y_2 = (\alpha_1 + \beta_1)x_1 + (\alpha_2 - \beta_2)x_1^2 + (\alpha_3 + \beta_3)x_1^3. \quad (13.79)$$

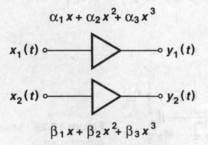

Figure 13.26 Effect of mismatch on second-order distortion.

If $x_1(t) = A \cos \omega t$, then the second harmonic has an amplitude equal to $(\alpha_2 - \beta_2)A^2/2$, i.e., proportional to the mismatch between the second-order coefficients of the input/output characteristic.

We should also mention that since at high frequencies, signals experience considerable phase shift, even-order distortion may arise from *phase* mismatch. This point is considered in Problem 13.1.

In circuits dissipating a high power, thermal gradients across the chip may create asymmetries. For example, if one transistor of a differential pair is closer to a high-power output stage than the other transistor, then mismatches arise between the threshold voltages and the mobilities of the two transistors.

13.2.1 Offset Cancellation Techniques

As mentioned above, the threshold voltage mismatch of MOSFETS trades with the channel capacitance. For example, a threshold mismatch of 1 mV translates to roughly 300 fF of channel capacitance for each transistor in a 0.6-μm technology. If many differential pairs are connected in parallel (e.g., in an A/D converter), the input capacitance becomes prohibitively large, severely degrading the speed and/or demanding high power dissipation in the preceding stage. Another difficulty is that mechanical stress may increase the offset voltages after a circuit is packaged. For these reasons, many high-precision systems require electronic cancellation of the offsets. As explained below, offset cancellation can also reduce $1/f$ noise of amplifiers considerably.

As our first step toward understanding the principle of offset cancellation, let us consider the circuit of Fig. 13.27(a), where a differential amplifier having an input-referred offset

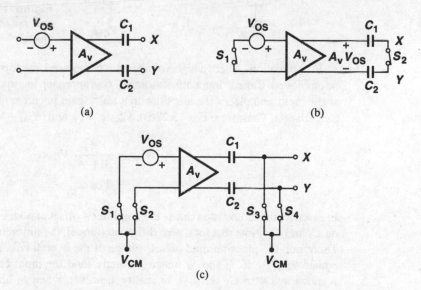

(a)   (b)

(c)

Figure 13.27   (a) Simple amplifier with capacitive coupling at the output, (b) circuit of (a) with its inputs and outputs shorted, (c) proper setting of the common-mode level during offset cancellation.

voltage V_{OS} is followed by two series capacitors. Now suppose, as shown in Fig. 13.27(b), the inputs are shorted together, driving the amplifier output to $V_{out} = A_v V_{OS}$. Furthermore, assume that during this period, nodes X and Y are shorted together as well. We note that when all of the node voltages are settled and $A_v V_{OS}$ is stored across C_1 and C_2, a zero differential input results in a zero difference between V_X and V_Y. Thus, after S_1 and S_2 turn off, the circuit consisting of the amplifier and C_1 and C_2 exhibits a zero offset voltage, amplifying only *changes* in the differential input voltage. In practice, the inputs and outputs must be shorted to proper common-mode voltages [Fig. 13.27(c)].

In summary, this type of offset cancellation "measures" the offset by setting the differential input to zero and stores the result on capacitors in series with the output. The circuit therefore requires a dedicated offset cancellation period, during which the actual input is disabled. Fig. 13.28 depicts the final topology, where CK denotes the offset cancellation command. Called "output offset storage," this technique reduces the overall offset to zero if S_3-S_4 exhibit no charge injection mismatch. Note, however, that if A_v is large, $A_v V_{OS}$ may "saturate" the amplifier output. For this reason, A_v is typically chosen to be less than roughly 10.

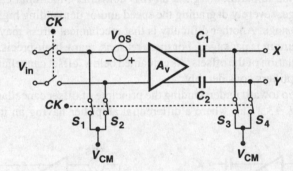

Figure 13.28 Control of amplification and offset cancellation modes by a clock.

In applications where a high voltage gain is required, the topology of Fig. 13.29(a) may be employed. Called "input offset storage," this approach incorporates two series capacitors at the input and places the amplifier in a unity-gain negative-feedback loop during offset cancellation. Thus, from Fig. 13.29(b), $V_{out} = V_{XY}$ and $(V_{out} - V_{OS})(-A_v) = V_{out}$. That is,

$$V_{out} = \frac{A_v}{1 + A_v} V_{OS} \tag{13.80}$$

$$\approx V_{OS}. \tag{13.81}$$

In essence, the circuit reproduces the amplifier's offset at nodes X and Y, storing the result on C_1 and C_2. Note that for a zero differential input, the differential output is equal to V_{OS}. Therefore, the input-referred offset voltage of the overall circuit (after S_3 and S_4 turn off) equals V_{OS}/A_v if S_3 and S_4 match perfectly (and the input capacitance of the amplifier is much less than C_1 and C_2). In reality, however, when S_3 and S_4 turn off, their charge injection mismatch may saturate the amplifier if A_v is very large.

The general drawback of input and output storage techniques is that they introduce capacitors in the signal path, a particularly serious issue in op amps and feedback systems

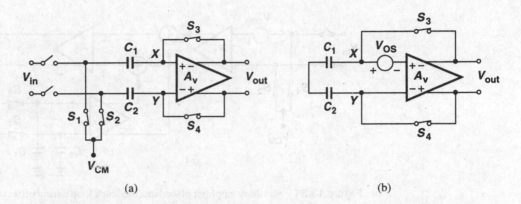

(a)   (b)

Figure 13.29 (a) Input offset storage, (b) circuit of (a) in the offset cancellation mode.

The bottom-plate parasitic of the capacitors may reduce the magnitude of the poles in the circuit, thereby degrading the phase margin. Even in open-loop amplifiers, this parasitic may limit the settling speed, intensifying the speed-power trade-off.

To resolve the above issues, the offset cancellation scheme can isolate the signal path from the offset storage capacitors though the use of an "auxiliary" amplifier. Consider the topology shown in Fig. 13.30, where A_{aux} amplifies the differential voltage V_1 stored across C_1 and C_2 and subtracts the result from the output of A_1. We note that if $V_{OS1}A_1 = V_1A_{aux}$, then for $V_{in} = 0$, $V_{out} = 0$, and the circuit is free from offsets. The key point here is that C_1 and C_2 do not appear in the signal path.

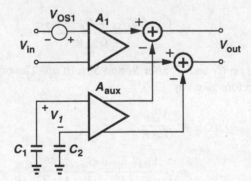

Figure 13.30 Addition of an auxiliary stage to remove the offset of an amplifier.

How is V_1 generated in Fig. 13.30? This is accomplished as illustrated in Fig. 13.31. Here, a second stage, A_2, is added and its output is sensed by A_{aux} during offset cancellation. To understand the operation, suppose that first only S_1 and S_2 are on, yielding $V_{out} = V_{OS1}A_1A_2$. Now, assume S_3 and S_4 turn on, placing A_2 and A_{aux} in a negative feedback loop. The reader can show that V_{out} then drops by a factor approximately equal to the loop gain: $V_{OS1}A_1A_2/(A_2A_{aux}) = V_{OS1}A_1/A_{aux}$. Stored across C_1 and C_2, this value is indeed the required V_1 in Fig. 13.30 because $(V_{OS1}A_1/A_{aux})A_{ux} = V_{OS1}A_1$.

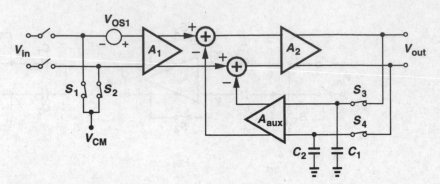

Figure 13.31 Auxiliary amplifier placed in a feedback loop during offset cancellation.

The topology of Fig. 13.31 suffers from two drawbacks. First, two voltage gain stages in the signal path may not be desirable in a high-speed op amp. Second, addition of the output voltages of A_1 and A_2 is quite difficult. For these reasons, the technique is usually realized as shown in Fig. 13.32(a), where each G_m stage is simply a differential pair and the R stage represents a transimpedance amplifier. As exemplified by Fig. 13.32(b), G_{m1} and R may in fact constitute a one-stage op amp while G_{m2} adds an offset correction current at the low impedance nodes X and Y.

Let us now examine the offset cancellation in Fig. 13.32(a) carefully, taking the offset voltage of G_{m2} into account as well. As depicted in Fig. 13.33, we can write:

$$[G_{m1}V_{OS1} - G_{m2}(V_{out} - V_{OS2})]R = V_{out}. \tag{13.82}$$

Thus,

$$V_{out} = \frac{G_{m1}RV_{OS1} + G_{m2}RV_{OS2}}{1 + G_{m2}R}. \tag{13.83}$$

This voltage is stored on C_1 and C_2 after S_3 and S_4 turn off. The offset voltage referred to the main input is therefore given by

$$V_{OS,tot} = \frac{V_{out}}{G_{m1}R} \tag{13.84}$$

$$= \frac{V_{OS1}}{1 + G_{m2}R} + \frac{G_{m2}}{G_{m1}}\frac{V_{OS2}}{1 + G_{m2}R} \tag{13.85}$$

$$\approx \frac{V_{OS1}}{G_{m2}R} + \frac{V_{OS2}}{G_{m1}R}, \tag{13.86}$$

where we have assumed $G_{m2}R \gg 1$. If $G_{m2}R$ and $G_{m1}R$ are large, as in the op amp of Fig. 13.32(b), then $V_{OS,tot}$ is very small.

The offset cancellation of Fig. 13.32 warrants a cautionary note. Upon turning off, S_3 and S_4 may inject slightly unequal charges onto C_1 and C_2, respectively, creating an error

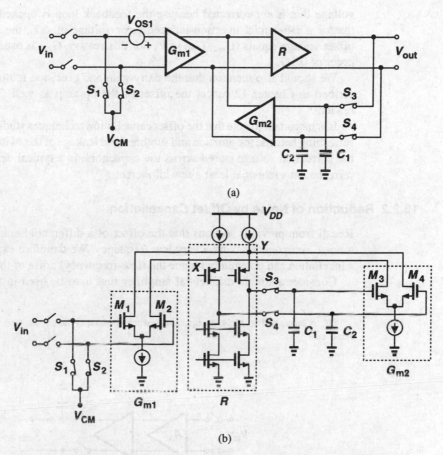

(a)

(b)

Figure 13.32 (a) Circuit of Fig. 13.31 using G_m and R stages, (b) realization of (a) in a folded-cascode op amp.

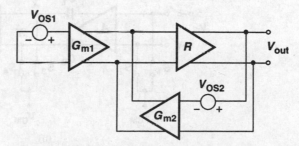

Figure 13.33 Circuit of Fig. 13.32(a) including offset of G_{m2}.

voltage that is *not* corrected because the feedback loop is opened. The reader can prove thatfor a differential injection-induced error voltage of ΔV, the resulting input-referred offset voltage equals $(G_{m2}/G_{m1})\Delta V$. For this reason, G_{m2} is usually chosen to be on the order of $0.1G_{m1}$.

We should also mention that the unity-gain and precision multiply-by-two circuits described in Chapter 12 cancel the offset of the op amp as well. The proof is left to the reader.[3]

It is important to note that the offset cancellation techniques studied here require periodic refreshing because the junction and subthreshold leakage of the switches eventually corrupts the correction voltage stored across the capacitors. In a typical design, the offset must be refreshed at a rate of at least a few kilohertz.

13.2.2 Reduction of Noise by Offset Cancellation

Recall from previous sections that the offset of a differential amplifier can be viewed as a noise component having a very low frequency. We therefore expect that periodic offset cancellation can potentially reduce the (low-frequency) noise of the circuit as well.

Consider a simple differential amplifier that is to be used in the front-end of a sam-

---

[3]If, as shown in Fig. 12.34, an equalizing switch is added to the circuit, then the op amp offset may not be removed.

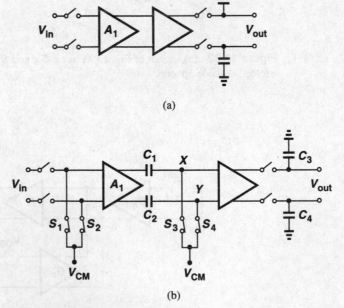

(a)

(b)

Figure 13.34 (a) Front end of a sampler, (b) circuit of (a) with offset cancellation applied to the first stage.

pling system [Fig. 13.34(a)]. Here, the noise of A_1 directly corrupts V_{in}. The $1/f$ noise of A_1 proves especially problematic if the signal spectrum extends from zero to only a few megahertz, because the $1/f$ noise corner frequency is typically around 500 kHz to 1 MHz.

Now suppose the amplifier undergoes offset cancellation before *every* sampling operation [Fig. 13.34(b)]. That is, as depicted in Fig. 13.35, the input is disabled; the offset of A_1 is stored on C_1 and C_2; the input is enabled and amplified by A_1 and A_2 and stored on C_3 and C_4; and finally the sampling switches are turned off. How does the noise of A_1 affect the final output? Denoting the time elapsed from the end of offset cancellation to the end of sampling by $\Delta t = t_2 - t_1$, we recall that at $t = t_1$, $V_{XY} = 0$. Thus, from t_1 to t_2, only *high-frequency* noise components of A_1, on the order of $> 1/\Delta t$, change V_{XY} significantly. In other words, offset cancellation suppresses noise frequencies below roughly $1/\Delta t$.

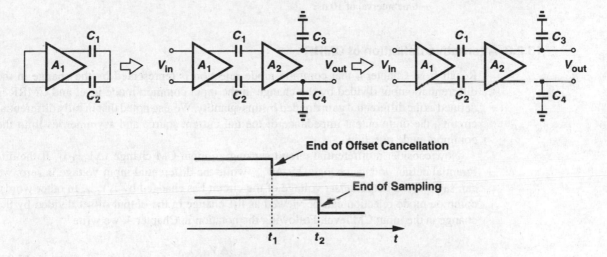

Figure 13.35 Sequence of operations in the sampler.

To better understand this concept, let us consider a numerical example. Assuming $\Delta t = 10$ ns, we examine two noise components, one at 1 MHz and another at 10 MHz, approximating each with a sinusoid (Fig. 13.36). For a sinusoid of amplitude A and frequency f, the maximum slew rate is equal to $2\pi f A$ and hence the maximum variation in Δt seconds is $2\pi f A \Delta t$. Normalizing this value to the amplitude, we obtain the change for 1-MHz and 10-MHz components as $\Delta V_1/A = 6.3\%$ and $\Delta V_2/A = 63\%$, respectively. We therefore conclude that noise frequencies below a few megahertz do not have sufficient time to change if the sampling occurs only 10 ns after the end of offset cancellation.

Originally utilized in charge-coupled devices (CCDs), the foregoing property of offset cancellation is called "correlated double sampling" (CDS) because it involves two consecutive sampling operations (the first being offset storage) that are so tightly spaced in time that they do not allow (low-frequency) noise components to vary significantly. A powerful technique, CDS finds wide usage in suppressing the $1/f$ noise of MOS circuits. Nonetheless, it leads to aliasing of wideband noise [5].

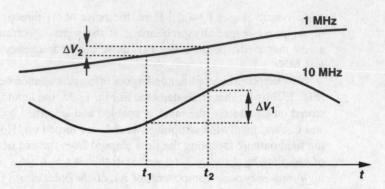

Figure 13.36 Variation of 1-MHz and 10-MHz noise components in a time interval of 10 ns.

13.2.3 Alternative Definition of CMRR

Recall from Chapter 4 that common-mode rejection is represented by the change in the differential output divided by the change in the input common-mode level and CMRR is defined as the differential gain divided by this quantity. We also noted that in fully differential circuits, the finite output impedance of the tail current source and asymmetries limit the common-mode rejection.

Now consider a differential circuit sensing an input CM change, $\Delta V_{in,CM}$. If the differential output voltage changes by ΔV_{out} while the differential input voltage is zero, we can say that the output *offset* voltage of the circuit has changed by ΔV_{out}. In other words, common-mode rejection can be viewed as the change in the output offset divided by the change in the input CM level. Following the notation in Chapter 4, we write

$$A_{CM-DM} = \frac{\Delta V_{OS,out}}{\Delta V_{CM,in}}. \tag{13.87}$$

Since $CMRR = A_{DM}/A_{CM-DM}$, we have

$$CMRR = \frac{A_{DM}}{\dfrac{\Delta V_{OS,out}}{\Delta V_{CM,in}}} \tag{13.88}$$

$$= \frac{\dfrac{\Delta V_{CM,in}}{\Delta V_{OS,out}}}{A_{DM}}. \tag{13.89}$$

Noting that $\Delta V_{OS,out}/A_{DM}$ is in fact the input-referred offset voltage, we have

$$CMRR = \frac{\Delta V_{CM,in}}{\Delta V_{OS,in}}. \tag{13.90}$$

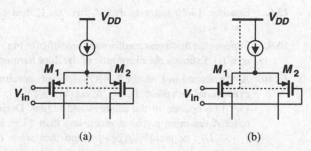

Figure 13.37 PMOS differential pair (a) without and (b) with body effect.

The above result proves useful in analyzing the behavior of circuits. For example, suppose an op amp incorporates a PMOS differential pair at the input. Which one of the topologies shown in Fig. 13.37 yields a higher CMRR? In Fig. 13.37(a), body effect is eliminated and the threshold voltages of M_1 and M_2 are independent of the input CM level. In Fig. 13.37(b), on the other hand, M_1 and M_2 experience body effect and, if they suffer from mismatches in their body effect coefficients, then the difference between V_{TH1} and V_{TH2}, i.e., the input offset voltage, *varies* with the input CM level, degrading the common-mode rejection.

Problems

Unless otherwise stated, in the following problems, use the device data shown in Table 2.1 and assume $V_{DD} = 3$ V where necessary. Also, assume all transistors are in saturation.

13.1. The input-output characteristic of an amplifier is approximated as $y(t) = \alpha_1 x(t) + \alpha_2 x^2(t)$ in the range $x = [0 \ x_{max}]$.
(a) What is the maximum nonlinearity?
(b) What is the THD for $x(t) = (x_{max} \cos \omega t + x_{max})/2$?

13.2. In the circuits of Fig. 13.6, $W/L = 20/0.5$ and $I = 0.5$ mA. Calculate the harmonic distortion in each circuit if the input signal has a peak amplitude of 100 mV. How do the results change if we double W/L or I?

13.3. For the circuits of Fig. 13.6(a), plot the THD and the input-referred thermal noise as a function of (a) W/L, (b) I. Identify the trade-offs between noise, linearity, and power dissipation.

13.4. In Fig. 13.6, *two* effects lead to a trade-off between nonlinearity and voltage gain. Describe these effects.

13.5. The circuit of Fig. 13.6(a) is designed with $W/L = 50/0.5$, $I = 1$ mA, and $R_D = 2$ kΩ. The circuit is placed in a feedback loop similar to that of Fig. 13.7 with $\beta = 0.2$ and senses an input sinusoid with a peak amplitude of 10 mV. Calculate the THD at the output.

13.6. Suppose in Fig. 13.14, A_1 and A_2 have an input-referred noise voltage V_n. Neglecting other sources of noise, calculate the input-referred noise voltage of the overall circuit.

13.7. Equation 13.36 suggests that if the open-loop gain, α_1, increases while other parameters remain constant, then the harmonic distortion drops sharply. Repeat Problem 13.5 with $W/L = 200/0.5$ to achieve a higher open-loop gain and explain the results.

13.8. Equation 13.36 suggests that if $\beta\alpha_1 \gg 1$, then $b/a \propto \beta^{-2}$. Repeat Problem 13.5 with $\beta = 0.4$.

13.9. Suppose the nonlinear feedforward amplifier in Fig. 13.7 is characterized by $y(t) = \alpha_1 x(t) + \alpha_3 x^3(t)$. Estimate the magnitude of the third harmonic at the output of the overall system.

13.10. As mentioned in Chapter 2, MOS devices operating in the subthreshold region exhibit an exponential behavior: $I_D = I_0 \exp[V_{GS}/(\zeta V_T)]$. Suppose both of the circuits shown in Fig. 13.6 operate in the subthreshold region. Derive expressions for the harmonic amplitudes if the input signal is much less than ζV_T. For the differential pair, first prove that $I_{D1} - I_{D2} \propto \tanh[V_{in}/(2\zeta V_T)]$ and then write the Taylor expansion of the hyperbolic tangent.

13.11. The mobility of MOSFETs is in fact a function of the gate-source voltage and expressed as $\mu = \mu_0/[1 + \theta(V_{GS} - V_{TH})]$, where θ is an empirical factor (Chapter 16). Assuming $\theta(V_{GS} - V_{TH}) \ll 1$ and using the relationship $(1 + \epsilon)^{-1} \approx 1 - \epsilon$ for $\epsilon \ll 1$, calculate the third harmonic in the circuit of Fig. 13.6(a).

13.12. The input devices of a differential pair have an effective length of 0.5 μm.
  (a) Assuming $\Delta V_{TH} = 0.1 t_{ox}/\sqrt{WL}$ and neglecting other mismatches, determine the minimum width of the transistors such that $V_{OS} \leq 5$ mV.
  (b) If the tail current is 1 mA, what is the maximum input swing that gives a THD of 1%?

13.13. Repeat Problem 13.12(b) if the tolerable input offset is 2 mV and compare the results.

13.14. Determine the dimensions of M_1 and M_2 in Fig. 13.25 such that $I_{D1} \approx I_{D2} = 0.5$ mA, $\Delta I_D/I_D = 2\%$, and $V_{GS} - V_{TH} = 0.5$ V. Assume $\Delta V_{TH} = 0.1 t_{ox}/\sqrt{WL}$ and neglect other mismatches.

13.15. Source degeneration can improve the matching between current sources if resistor mismatches are small. Prove that in the circuit of Fig. 13.38,

$$\frac{\Delta I_D}{I_D} = \frac{1}{1 + g_m R_S}\left[\frac{\Delta(\mu_n C_{ox})}{\mu_n C_{ox}} + \frac{\Delta(W/L)}{(W/L)} - \frac{2\Delta V_{TH}}{V_{GS} - V_{TH}} - g_m\Delta R_S\right], \qquad (13.91)$$

where ΔR_S denotes the mismatch between R_{S1} and R_{S2}. Note that for an appreciable reduction of $\Delta I/I_D$, R_S must be greater than $1/g_m$.

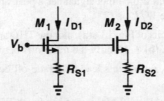

Figure 13.38

13.16. In the circuit of Fig. 13.26, assume $\alpha_j = \beta_j$ but $x_1(t) = A\cos\omega t$ and $x_2(t) = A\cos(\omega t + \theta)$, where θ denotes a small phase mismatch. Calculate the magnitude of the second harmonic at the output.

13.17. In the circuit of Fig. 13.39, M_3 and M_4 suffer from a threshold mismatch of ΔV_{TH} and the circuit is otherwise symmetric. Assuming $\lambda \neq 0$ but $\gamma = 0$, calculate the input-referred offset voltage. What happens as $R_D \to \infty$?

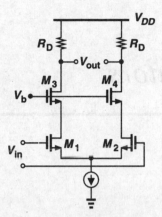

Figure 13.39

13.18. In the circuit of Fig. 13.29, the amplifier has an input capacitance (between X and Y) equal to C_{in}. Calculate the input offset voltage after offset compensation.

13.1. The circuit of Fig. 13.29 is designed for an input offset voltage of 1 mV. If the width of the transistors in the input differential pair of the amplifier is doubled, what is the overall input offset voltage? (Neglect the input capacitance of the amplifier.)

13.20. Explain why the circuit of Fig. 13.24 suffers from a trade-off between the input offset and the output voltage swing (for a given tail current).

References

1. F. Krummenacher and N. Joehl, "A 4-MHz CMOS Continuous-Time Filter with On-Chip Automatic Tuning," *IEEE J. Solid-State Circuits*, vol. 23, pp. 750–758, June 1988.
2. K. R. Lakshmikumar, R. A. Hadaway, and M. A. Copeland, "Characterization and Modeling of Mismatches in MOS Transistors for Precision Analog Design," *IEEE J. Solid-State Circuits*, vol. 21, pp. 1057–1066, Dec. 1986.
3. M. J. M. Pelgrom, A. C. J. Duinmaiger, and A. P. G. Welbers, "Matching Properties of MOS Transistors," *IEEE J. Solid-State Circuits*, vol. SC-24, pp. 1433–1439, Oct. 1989.
4. M. J. M. Pelgrom, H. P. Tuinhout, and M. Vertregt, "Transistor Matching in Analog CMOS Applications," *IEDM Dig. of Tech. Papers*, pp. 34.1.1–34.1.4, Dec. 1998.
5. C. C. Enz and G. C. Temes, "Circuit Techniques for Reducing the Effects of Op-Amp Imperfections: Autozeroing, Correlated Double Sampling, and Chopper Stabilization," *Proc. IEEE*, vol. 84, pp. 1584–1614, Nov. 1996.

Chapter 14

Oscillators

Oscillators are an integral part of many electronic systems. Applications range from clock generation in microprocessors to carrier synthesis in cellular telephones, requiring vastly different oscillator topologies and performance parameters. Robust, high-performance oscillator design in CMOS technology continues to pose interesting challenges. As described in Chapter 15, oscillators are usually embedded in a phase-locked system.

This chapter deals with the analysis and design of CMOS oscillators, more specifically, voltage-controlled oscillators (VCOs). Beginning with a general study of oscillation in feedback systems, we introduce ring oscillators and LC oscillators along with methods of varying the frequency of oscillation. We then describe a mathematical model of VCOs that will be used in the analysis of PLLs in Chapter 15.

14.1 General Considerations

A simple oscillator produces a periodic output, usually in the form of voltage. As such, the circuit has no input while sustaining the output indefinitely. How can a circuit oscillate? Recall from Chapter 10 that negative feedback systems may oscillate, i.e., an oscillator is a badly-designed feedback amplifier![1] Consider the unity-gain negative feedback circuit shown in Fig. 14.1, where

$$\frac{V_{out}}{V_{in}}(s) = \frac{H(s)}{1 + H(s)}. \tag{14.1}$$

As mentioned in Chapter 10, if the amplifier itself experiences so much phase shift at high frequencies that the overall feedback becomes positive, then oscillation may occur. More accurately, if for $s = j\omega_0$, $H(j\omega_0) = -1$, then the closed-loop gain approaches infinity at ω_0. Under this condition, the circuit amplifies its own noise components at ω_0 indefinitely. In fact, as conceptually illustrated in Fig. 14.2, a noise component at ω_0 experiences a total gain of unity and a phase shift of 180°, returning to the subtractor as a negative replica

---

[1] It is said, "In the high-frequency world, amplifiers oscillate and oscillators don't."

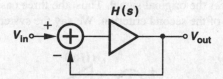

Figure 14.1   Feedback system.

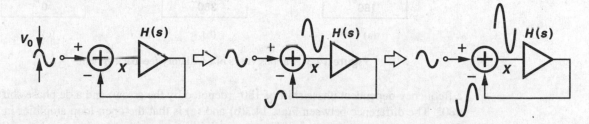

Figure 14.2   Evolution of oscillatory system with time.

of the input. Upon subtraction, the input and the feedback signals give a larger difference. Thus, the circuit continues to "regenerate," allowing the component at ω_0 to grow.

For the oscillation to begin, a loop gain of unity or greater is necessary. This can be seen by following the signal around the loop over many cycles and expressing the amplitude of the subtractor's output in Fig. 14.2 as a geometric series (if $\angle H(j\omega_0) = 180°$):

$$V_X = V_0 + |H(j\omega_0)|V_0 + |H(j\omega_0)|^2 V_0 + |H(j\omega_0)|^3 V_0 + \cdots. \qquad (14.2)$$

If $|H(j\omega_0)| > 1$, the above summation diverges whereas if $|H(j\omega_0)| < 1$, then

$$V_X = \frac{V_0}{1 - |H(j\omega_0)|} < \infty. \qquad (14.3)$$

In summary, if a negative-feedback circuit has a loop gain that satisfies two conditions:

$$|H(j\omega_0)| \geq 1 \qquad (14.4)$$

$$\angle H(j\omega_0) = 180°, \qquad (14.5)$$

then the circuit may oscillate at ω_0. Called "Barkhausen criteria," these conditions are necessary but not sufficient [1]. In order to ensure oscillation in the presence of temperature and process variations, we typically choose the loop gain to be at least twice or three times the required value.

We may state the second Barkhausen criterion as $\angle H(j\omega) = 180°$ or a *total* phase shift of 360°. This should not be confusing: if the system is designed to have a low-frequency negative feedback, it already produces 180° of phase shift in the signal traveling around the loop (as represented by the subtractor in Fig. 14.1), and $\angle H(j\omega) = 180°$ denotes an additional *frequency-dependent* phase shift that, as illustrated in Fig. 14.2, ensures the feedback signal *enhances* the original signal. Thus, the three cases illustrated in Fig. 14.3 are equivalent in terms of the second criterion. We say the system of Fig. 14.3(a) exhibits

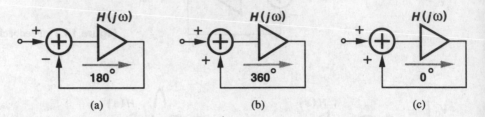

Figure 14.3 Various views of oscillatory feedback system.

a frequency-dependent phase shift of 180° (denoted by the arrow) and a dc phase shift of 180°. The difference between Figs. 14.3(b) and (c) is that the open-loop amplifier in the former contains enough stages with proper polarities to provide a total phase shift of 360° at ω_0 whereas that in the latter produces *no* phase shift at ω_0. Examples of these topologies are presented later in this chapter.

CMOS oscillators in today's technology are typically implemented as "ring oscillators" or "LC oscillators." We study each type in the following sections.

14.2 Ring Oscillators

A ring oscillator consists of a number of gain stages in a loop. To arrive at the actual implementation, we begin by attempting to make a single-stage feedback circuit oscillate.

Example 14.1 _____

Explain why a single common-source stage does not oscillate if it is placed in a unity-gain loop.

Solution

From Fig. 14.4, it is seen that the open-loop circuit contains only one pole, thereby providing a maximum frequency-dependent phase shift of 90° (at a frequency of infinity). Since the common-source stage exhibits a dc phase shift of 180° due to the signal inversion from the gate to the drain, the maximum total phase shift is 270°. The loop therefore fails to sustain oscillation growth.

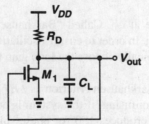

Figure 14.4

The above example suggests that oscillation may occur if the circuit contains multiple stages and hence multiple poles. Indeed, such a topology was considered *undesirable* in Chapter 10 because it led to inadequate phase margin in op amps. We therefore surmise that if the circuit of Fig. 14.4 is modified as shown in Fig. 14.5, then two significant poles appear in the signal path, allowing the frequency-dependent phase shift to approach 180°.

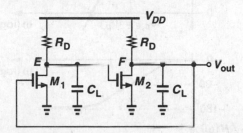

Figure 14.5 Two-pole feedback system.

Unfortunately, this circuit exhibits *positive* feedback near zero frequency due to the signal inversion through each common-source stage. As a result, it simply "latches up" rather than oscillates. That is, if V_E rises, V_F falls, thereby turning M_1 off and allowing V_E to rise further. This may continue until V_E reaches V_{DD} and V_F drops to near zero, a state that will remain indefinitely.

To gain more insight into the oscillation conditions, let us assume an ideal inverting stage (with zero phase shift at all frequencies) is inserted in the loop of Fig. 14.5, providing *negative* feedback near zero frequency and eliminating the problem of latch-up (Fig. 14.6). Does this circuit oscillate? We note that the loop contains only two poles: one at E and another

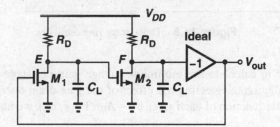

Figure 14.6 Two-pole feedback system with additional signal inversion.

at F. The frequency-dependent phase shift can therefore reach 180°, but at a frequency of infinity. Since the loop gain vanishes at very high frequencies, we observe that the circuit does not satisfy both of Barkhausen's criteria at the same frequency (Fig. 14.7), failing to oscillate.

The foregoing discussion points to the need for greater phase shift around the loop, suggesting the possibility of oscillation if the third inverting stage in Fig. 14.6 contains a pole that contributes significant phase. We then arrive at the topology depicted in Fig. 14.8. If the three stages are identical, the total phase shift around the loop, ϕ, reaches $-135°$ at $\omega = \omega_{p,E}(= \omega_{p,F} = \omega_{p,G})$ and $-270°$ at $\omega = \infty$. Consequently, ϕ equals $-180°$ at $\omega < \infty$, where the loop gain can be still greater than or equal to unity. This circuit indeed oscillates if the loop gain is sufficient and it is an example of a ring oscillator.

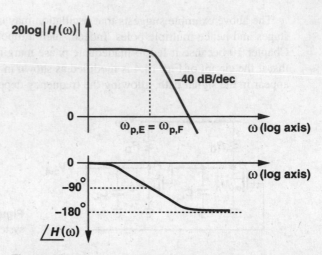

Figure 14.7   Loop gain characteristics of a two-pole system.

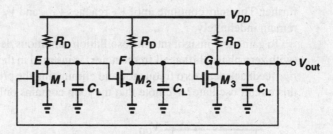

Figure 14.8   Three-stage ring oscillator.

It is instructive to calculate the minimum voltage gain per stage in Fig. 14.8 that is necessary for oscillation. Neglecting the effect of the gate-drain overlap capacitance and denoting the transfer function of each stage by $-A_0/(1 + s/\omega_0)$, we have for the loop gain:

$$H(s) = -\frac{A_0^3}{(1 + \dfrac{s}{\omega_0})^3}. \tag{14.6}$$

The circuit oscillates only if the frequency-dependent phase shift equals 180°, i.e., if each stage contributes 60°. The frequency at which this occurs is given by

$$\tan^{-1}\frac{\omega_{osc}}{\omega_0} = 60° \tag{14.7}$$

and hence:

$$\omega_{osc} = \sqrt{3}\omega_0. \tag{14.8}$$

The minimum voltage gain per stage must be such that the magnitude of the loop gain at ω_{osc} is equal to unity:

$$\frac{A_0^3}{\left[\sqrt{1 + (\frac{\omega_{osc}}{\omega_0})^2}\right]^3} = 1. \tag{14.9}$$

It follows from (14.8) and (14.9) that

$$A_0 = 2. \tag{14.10}$$

In summary, a three-stage ring oscillator requires a low-frequency gain of 2 per stage, and it oscillates at a frequency of $\sqrt{3}\omega_0$, where ω_0 is the 3-dB bandwidth of each stage.

Let us now examine the waveforms at the three nodes of the oscillator of Fig. 14.8. Since each stage contributes a frequency-dependent phase shift of 60° as well as a low-frequency signal inversion, the waveform at each node is 240° (or 120°) out of phase with respect to its neighboring nodes (Fig. 14.9). The ability to generate multiple phases is a very useful property of ring oscillators.

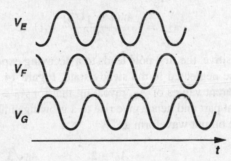

Figure 14.9 Waveforms of a three-stage ring oscillator.

Amplitude Limiting   The natural question at this point is: what happens if in the three-stage ring of Fig. 14.8, $A_0 \neq 2$? We know from Barkhausen's criteria that if $A_0 < 2$, the circuit fails to oscillate, but what if $A_0 > 2$? To answer this question, we first model the oscillator by a linear feedback system, as depicted in Fig. 14.10. Note that the feedback is positive (i.e., V_{out} is *added* to V_{in}) because $H(s)$ in Eq. (14.6) already includes the negative polarity resulting from three inversions in the signal path. The closed-loop transfer function is:

$$\frac{V_{out}(s)}{V_{in}(s)} = \frac{\dfrac{-A_0^3}{(1 + s/\omega_0)^3}}{1 + \dfrac{A_0^3}{(1 + s/\omega_0)^3}} \tag{14.11}$$

$$= \frac{-A_0^3}{(1 + s/\omega_0)^3 + A_0^3}. \tag{14.12}$$

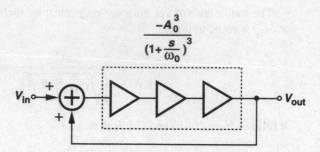

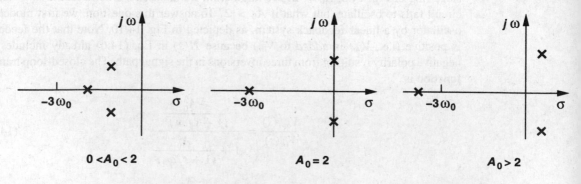

Figure 14.10   Linear model of three-stage ring oscillator.

The denominator of (14.12) can be expanded as:

$$(1 + \frac{s}{\omega_0})^3 + A_0^3 = (1 + \frac{s}{\omega_0} + A_0) \left[(1 + \frac{s}{\omega_0})^2 - (1 + \frac{s}{\omega_0})A_0 + A_0^2 \right]. \tag{14.13}$$

Thus, the closed-loop system exhibits three poles:

$$s_1 = (-A_0 - 1)\omega_0 \tag{14.14}$$

$$s_{2,3} = [\frac{A_0(1 \pm j\sqrt{3})}{2} - 1]\omega_0. \tag{14.15}$$

Since A_0 itself is positive, the first pole leads to a decaying exponential term: $\exp[(-A_0 - 1)\omega_0 t]$, which can be neglected in the steady state. Figure 14.11 illustrates the locations of the poles for different values of A_0, revealing that for $A_0 > 2$, the two complex poles exhibit a positive real part and hence give rise to a growing sinusoid. Neglecting the effect of s_1, we express the output waveform as

$$V_{out}(t) = a \exp(\frac{A_0 - 2}{2}\omega_0 t) \cos(\frac{A_0\sqrt{3}}{2}\omega_0 t). \tag{14.16}$$

Thus, if $A_0 > 2$, the exponential envelope grows to infinity.

Figure 14.11   Poles of three-stage ring oscillator for various values of gain.

In practice, as the oscillation amplitude increases, the stages in the signal path experience nonlinearity and eventually "saturation," limiting the maximum amplitude. We may say the poles begin in the right half plane and eventually move to the imaginary axis to stop the growth. If the small-signal loop gain is greater than unity, the circuit must spend enough time in saturation so that the "average" loop gain is still equal to unity.[2]

Example 14.2

Shown in Fig.14.12 is a differential implementation of the oscillator of Fig. 14.8. What is the maximum voltage swing of each stage?

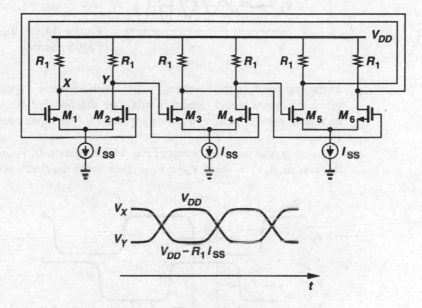

Figure 14.12

Solution

If the gain per stage is well above 2, then the amplitude grows until each differential pair experiences complete switching, that is, until I_{SS} is completely steered to one side every half cycle. As a result, the swing at each node is equal to $I_{SS}R_1$. From the waveforms shown in Fig. 14.12, we also observe that each stage is in its high-gain region for only a fraction of the period, (e.g., when $|V_X - V_Y|$ is small).

A simple implementation of ring oscillators that does not require resistors is depicted in Fig. 14.13. Suppose the circuit is released with an initial voltage at each node equal

---

[2]While intuitive, these statements are not rigorous. The concepts of transfer function, poles, and loop gain are difficult to apply to a nonlinear circuit.

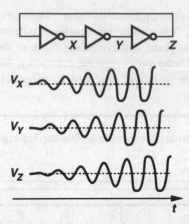

V_X

V_Y

V_Z

Figure 14.13 Ring oscillator using CMOS inverters.

to the trip point of the inverters, V_{trip}.[3] With identical stages and no noise in the devices, the circuit would remain in this state indefinitely,[4] but noise components disturb each node voltage, yielding a growing waveform. The signal eventually exhibits rail-to-rail swings.

Let us now assume the circuit of Fig. 14.13 begins with $V_X = V_{DD}$ (Fig. 14.14). Under this condition, $V_Y = 0$ and $V_Z = V_{DD}$. Thus, when the circuit is released, V_X begins to fall

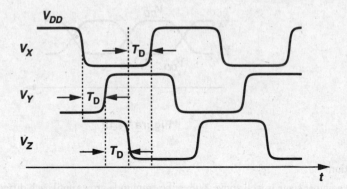

Figure 14.14 Waveforms of ring oscillator when one node is initialized at V_{DD}.

to zero (because the first inverter senses a high input), forcing V_Y to rise to V_{DD} after one inverter delay, T_D, and V_Z to fall to zero after another inverter delay. The circuit therefore oscillates with a delay of T_D between consecutive node voltages, yielding a period of $6T_D$.

The above small-signal and large-signal analyses raise an interesting question. While the small-signal oscillation frequency is given by $A_0\sqrt{3}\omega_0/2$ [from Eq. (14.16)], the large-signal

[3]The trip point of an inverter is the input voltage that results in an equal output voltage.

[4]This is indeed how SPICE predicts the circuit's behavior. To start the oscillation in SPICE, one of the nodes must be initialized at a different voltage.

value is $1/(6T_D)$. Are these two values equal? Not necessarily. After all, ω_0 is determined by the small-signal output resistance and capacitance of each inverter near the trip point whereas T_D results from the large-signal, nonlinear current drive and capacitances of each stage. In other words, when the circuit is released with all inverters at their trip point, the oscillation begins with a frequency of $\sqrt{3}A_0\omega_0/2$ but, as the amplitude grows and the circuit becomes nonlinear, the frequency shifts to $1/(6T_D)$ (which is a lower value).

Ring oscillators employing more than three stages are also feasible. The total number of inversions in the loop must be odd so that the circuit does not latch up. For example, as shown in Fig. 14.15(a), a ring can incorporate five inverters, providing a frequency of

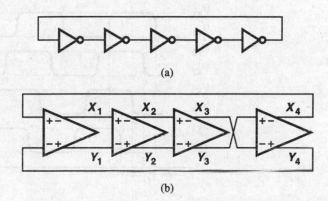

(a)

(b)

Figure 14.15 (a) Five-stage single-ended ring oscillator, (b) four-stage differential ring oscillator.

$1/(10T_D)$. On the other hand, the differential implementation can utilize an *even* number of stages by simply configuring one stage such that it does not invert. Illustrated in Fig. 14.15(b), this flexibility demonstrates another advantage of differential circuits over their single-ended counterparts.

Example 14.3

What is the minimum required voltage gain per stage in the four-stage oscillator of Fig. 14.15(b)? How many signal phases are provided by the circuit?

Solution

Using a notation similar to that for Fig. 14.8, we have:

$$H(s) = -\frac{A_0^4}{(1 + \dfrac{s}{\omega_0})^4}. \tag{14.17}$$

For the circuit to oscillate, each stage must contribute a frequency-dependent phase shift of $180°/4 = 45°$. The frequency at which this occurs is given by $\tan^{-1}\omega_{osc}/\omega_0 = 45°$ and hence $\omega_{osc} = \omega_0$. The minimum voltage gain is therefore derived as

$$\frac{A_0}{\sqrt{1 + (\frac{\omega_{osc}}{\omega_0})^2}} = 1. \tag{14.18}$$

That is, $A_0 = \sqrt{2}$. As expected, this value is lower than that required in a three-stage ring.

With $45°$ of phase shift per stage, the oscillator provides four phases and their complements. This is illustrated in Fig. 14.16.

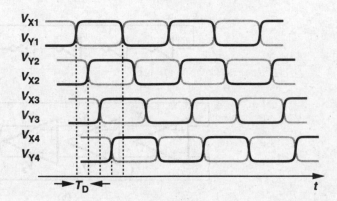

Figure 14.16

The number of stages in a ring oscillator is determined by various requirements, including speed, power dissipation, noise immunity, etc. In most applications, three to five stages provide optimum performance (for differential implementations).

Example 14.4

Determine the maximum voltage swings and the minimum supply voltage of a ring oscillator incorporating differential pairs with resistive loads (e.g., as in Fig. 14.12) if no transistor must enter the triode region. Assume each stage experiences complete switching.

Solution

Figure 14.17(a) shows two stages in cascade. If each stage experiences complete switching, then each drain voltage, e.g., V_X or V_Y, varies between V_{DD} and $V_{DD} - I_{SS}R_P$. Thus, when M_1 is fully on, its gate and drain voltages are equal to V_{DD} and $V_{DD} - I_{SS}R_P$, respectively. For this transistor to remain in saturation, we have $I_{SS}R_P \leq V_{TH}$, i.e., the peak-to-peak swing at each drain must not exceed V_{TH}.

How is the minimum supply voltage determined? If V_{DD} is lowered, the voltage at the common source node of each differential pair, e.g., V_P in Fig. 14.17(a), falls, eventually driving the tail transistor into the triode region. We must therefore calculate V_P for the worst case, noting that V_P does vary with time because M_1 and M_2 carry unequal currents when the input difference becomes large.

Now consider the stand-alone circuit of Fig. 14.17(b), assuming the inputs vary between V_{DD} and $V_{DD} - I_{SS}R_P$. How does V_P vary? When the gate voltage of M_1, V_1, is equal to V_{DD} and M_1 carries

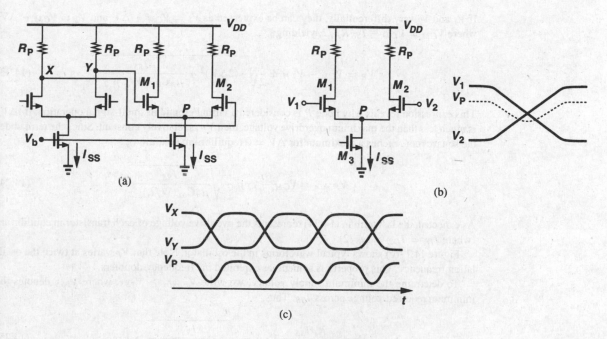

Figure 14.17

all of I_{SS},

$$V_P = V_{DD} - \sqrt{\frac{2I_{SS}}{\mu_n C_{ox}(W/L)_{1,2}}} - V_{TH}. \tag{14.19}$$

As V_1 falls and V_2 rises, so does V_P because, so long as M_2 is off, M_1 operates as a source follower. When the difference between V_1 and V_2 reaches $\sqrt{2}(V_{GS,eq} - V_{TH})$, where $V_{GS,eq}$ denotes the equilibrium overdrive of each transistor, M_2 turns on. To calculate V_P after this point, we note that $I_{D1} + I_{D2} = I_{SS}$, $V_{GS1} \doteq V_1 - V_P$, and $V_{GS2} = V_2 - V_P$. Thus,

$$\frac{1}{2}\mu_n C_{ox}(\frac{W}{L})_{1,2}(V_1 - V_P - V_{TH})^2 + \frac{1}{2}\mu_n C_{ox}(\frac{W}{L})_{1,2}(V_2 - V_P - V_{TH})^2 = I_{SS}. \tag{14.20}$$

Expanding the quadratic terms and rearranging the result, we have

$$2V_P^2 - 2(V_1 - V_{TH} + V_2 - V_{TH})V_P + (V_1 - V_{TH})^2 + (V_2 - V_{TH})^2 - \frac{2I_{SS}}{\mu_n C_{ox}(W/L)_{1,2}} = 0. \tag{14.21}$$

It follows that

$$V_P = \frac{1}{2}[V_1 + V_2 - 2V_{TH} \pm \sqrt{-(V_1 - V_2)^2 + \frac{4I_{SS}}{\mu_n C_{ox}(W/L)_{1,2}}}]. \tag{14.22}$$

If V_1 and V_2 vary differentially, they can be expressed as $V_1 = V_{CM} + \Delta V$ and $V_2 = V_{CM} - \Delta V$, where $V_{CM} = V_{DD} - I_{SS}R_P/2$, yielding

$$V_P = V_{CM} - V_{TH} \pm \frac{1}{2}\sqrt{-(2\Delta V)^2 + \frac{4I_{SS}}{\mu_n C_{ox}(W/L)_{1,2}}}. \tag{14.23}$$

This expression reveals why node P is considered a virtual ground in small-signal operation: if $|\Delta V|$ is much less than the maximum overdrive voltage, then V_P is relatively constant. Since the term under the square root reaches a maximum for $\Delta V = 0$ (equilibrium condition),

$$V_{P,min} = V_{CM} - V_{TH} - \sqrt{\frac{I_{SS}}{\mu_n C_{ox}(W/L)_{1,2}}}. \tag{14.24}$$

As expected, the last term in (14.24) represents the overdrive voltage of each transistor in equilibrium (where $I_{D1} = I_{D2} = I_{SS}/2$).

Figure 14.17(c) shows typical waveforms in the oscillator. Note that V_P varies at twice the oscillation frequency. This property is sometimes exploited in "frequency doublers."

To determine the minimum supply voltage, we write $V_{P,min} \geq V_{ISS}$, where V_{ISS} denotes the minimum required voltage across I_{SS}. Thus,

$$V_{DD} - \frac{R_P I_{SS}}{2} - V_{TH} - \sqrt{\frac{I_{SS}}{\mu_n C_{ox}(W/L)_{1,2}}} \geq V_{ISS}, \tag{14.25}$$

and

$$V_{DD} \geq V_{ISS} + V_{TH} + \sqrt{\frac{I_{SS}}{\mu_n C_{ox}(W/L)_{1,2}}} + \frac{R_P I_{SS}}{2}. \tag{14.26}$$

The terms on the right are: the voltage headroom consumed by a current source, one threshold voltage, the equilibrium overdrive, and half of the swing at each node.

In CMOS technologies lacking high-quality resistors, the implementation of Fig. 14.17(a) must be modified. While a PMOS transistor operating in the deep triode region can serve as the load [Fig. 14.18(a)], the gate voltage must be set so as to define the on-resistance accurately. Alternatively, a diode-connected load can be utilized [Fig. 14.18(b)] but at the cost of one threshold voltage in the headroom. Figure 14.18(c) shows a more efficient load where an NMOS source follower is inserted between the drain and gate of each PMOS transistor. With the output sensed at nodes X and Y, M_3 and M_4 consume only a voltage headroom equal to $|V_{DS3,4}|$. If $V_{GS5} \approx V_{TH3}$, then M_3 operates at the edge of the triode region and the small-signal resistance of the load is roughly equal to $1/g_{m3}$ (with the assumption $\lambda = \gamma = 0$) (Problem 14.4).

The load of Fig. 14.18(c) exhibits another interesting property as well. Since the gate-source capacitance of M_3 is driven by the source follower, the time constant associated with the load is smaller than that of a diode-connected transistor. Also, the finite output resistance of the follower may yield an inductive behavior for the load (Problem 14.5).

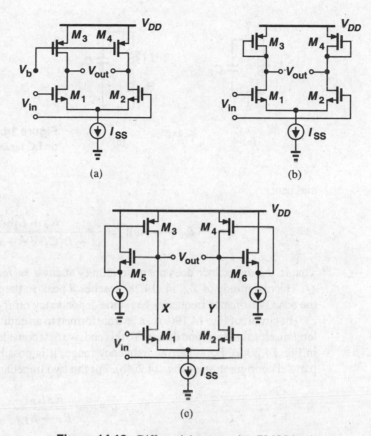

Figure 14.18 Differential stages using PMOS loads.

4.3 LC Oscillators

Monolithic inductors have gradually appeared in bipolar and CMOS technologies in the past 10 years, making it possible to design oscillators based on passive resonant circuits. Before delving into such oscillators, it is instructive to review basic properties of RLC circuits.

As shown in Fig. 14.19(a), an inductor L_1 placed in parallel with a capacitor C_1 resonates at a frequency $\omega_{res} = 1/\sqrt{L_1 C_1}$. At this frequency, the impedances of the inductor, $jL_1\omega_{res}$, and the capacitor, $1/(jC_1\omega_{res})$, are equal and opposite, thereby yielding an infinite impedance. We say the circuit has an infinite quality factor, Q. In practice, inductors (and capacitors) suffer from resistive components. For example, the series resistance of the metal wire used in the inductor can be modeled as shown in Fig. 14.19(b). We define the Q of the inductor as $L_1\omega/R_S$. For this circuit, the reader can show that the equivalent impedance is given by

$$Z_{eq}(s) = \frac{R_S + L_1 s}{1 + L_1 C_1 s^2 + R_S C_1 s},$$
(14.27)

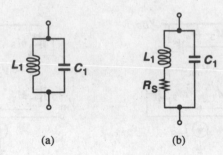

(a)   (b)

Figure 14.19 (a) Ideal and (b) realistic LC tanks.

and hence,

$$|Z_{eq}(s = j\omega)|^2 = \frac{R_S^2 + L_1^2\omega^2}{(1 - L_1 C_1 \omega^2)^2 + R_S^2 C_1^2 \omega^2}. \tag{14.28}$$

That is, the impedance does not go to infinity at any $s = j\omega$. We say the circuit has a finite Q. The magnitude of Z_{eq} in (14.28) reaches a peak in the vicinity of $\omega = 1/\sqrt{L_1 C_1}$, but the actual resonance frequency has some dependency on R_S.

The circuit of Fig. 14.19(b) can be transformed to an equivalent topology that more easily lends itself to analysis and design. To this end, we first consider the series combination shown in Fig. 14.20(a). For a narrow frequency range, it is possible to convert the circuit to the parallel configuration of Fig. 14.20(b). For the two impedances to be equivalent:

$$L_1 s + R_S = \frac{R_P L_P s}{R_P + L_P s}. \tag{14.29}$$

Considering only the steady state response, we assume $s = j\omega$ and rewrite (14.29) as

$$(L_1 R_P + L_P R_S) j\omega + R_S R_P - L_1 L_P \omega^2 = R_P L_P j\omega. \tag{14.30}$$

This relationship must hold for all values of ω (in a narrow range), mandating that

$$L_1 R_P + L_P R_S = R_P L_P \tag{14.31}$$

$$R_S R_P - L_1 L_P \omega^2 = 0. \tag{14.32}$$

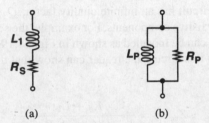

(a)   (b)

Figure 14.20 Conversion of a series combination to a parallel combination.

Calculating R_P from the latter and substituting in the former, we have

$$L_P = L_1(1 + \frac{R_S^2}{L_1^2\omega^2}). \tag{14.33}$$

Recall that $L_1\omega/R_S = Q$, a value typically greater than 3 for monolithic inductors. Thus,

$$L_P \approx L_1 \tag{14.34}$$

and

$$R_P \approx \frac{L_1^2\omega^2}{R_S} \tag{14.35}$$

$$\approx Q^2 R_S. \tag{14.36}$$

In other words, the parallel network has the same reactance but a resistance Q^2 times the series resistance. This concept holds valid for a first-order RC network as well if the Q of the series combination is defined as $1/(C\omega)/R_S$.

The above transformation allows the conversion illustrated in Fig. 14.21, where $C_P = C_1$. The equivalence of course breaks down as ω departs susbtantially from the resonance

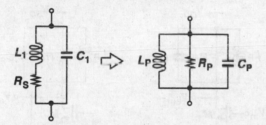

Figure 14.21   Conversion of a tank to three parallel components.

frequency. The insight gained from the parallel combination is that at $\omega_1 = 1/\sqrt{L_pC_p}$, the tank reduces to a simple resistor; i.e., the phase difference between the voltage and current of the tank drops to zero. Plotting the magnitude of the tank impedance versus frequency [Fig. 14.22(a)], we note that the behavior is inductive for $\omega < \omega_1$ and capacitive for $\omega > \omega_1$. We then surmise that the phase of the impedance is positive for $\omega < \omega_1$ and negative for $\omega > \omega_1$ [Fig. 14.22(b)]. These observations prove useful in studying LC oscillators. (Why do we expect the phase shift to approach $+90°$ at very low frequencies and $-90°$ at very high frequencies?)

Let us now consider the "tuned" stage of Fig. 14.23(a), where an LC tank operates as the load. At resonance, $jL_p\omega = 1/(jC_p\omega)$ and the voltage gain equals $-g_{m1}R_P$. (Note that the gain of the circuit is very small at frequencies near zero.) Does this circuit oscillate if the output is connected to the input [Fig. 14.23(b)]? At resonance, the total phase shift around the loop is equal to $180°$ (rather than $360°$). Also, from Fig. 14.22(b), the frequency-dependent phase shift of the tank never reaches $180°$. Thus, the circuit does not oscillate.

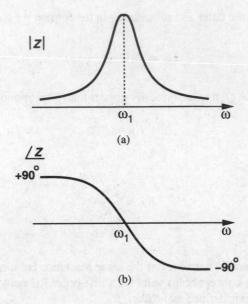

Figure 14.22 (a) Magnitude and (b) phase of the impedance of an LC tank as a function of frequency.

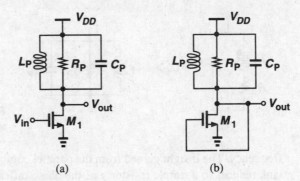

Figure 14.23 (a) Tuned gain stage, (b) stage of (a) in feedback.

Before modifying the circuit for oscillatory behavior, let us observe another interesting property of the gain stage of Fig. 14.23(a) that distinguishes it from a common-source topology using a resistive load. Suppose, as shown in Fig. 14.24, the stage is biased at drain current I_1. If the series resistance of L_p is small, the dc level of V_{out} is close to V_{DD}. How does V_{out} vary if a small sinusoidal voltage at the resonance frequency is applied to the input? We expect V_{out} to be an inverted sinusoid with an average value near V_{DD} because the inductor cannot sustain a large dc drop. In other words, if the average value of V_{out} deviates significantly from V_{DD}, then the inductor series resistance must carry an average current greater than I_1. Thus, the peak output level in fact *exceeds* the supply voltage, a

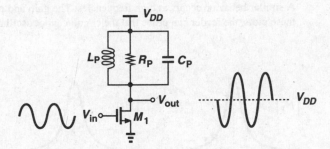

Figure 14.24 Output signal levels in a tuned stage.

important and often useful attribute of the LC load. For example, with proper design, the output peak-to-peak swing can be larger than V_{DD}.

We now study two types of LC oscillators.

14.3.1 Crossed-Coupled Oscillator

Suppose we place two stages of Fig. 14.23(a) in a cascade, as depicted in Fig. 14.25. While similar to the topology of Fig. 14.5, this configuration does not latch up because its low-

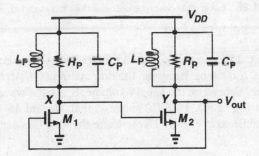

Figure 14.25 Two tuned stages in a feedback loop.

frequency gain is very small. Furthermore, at resonance, the total phase shift around the loop is zero because each stage contributes zero frequency-dependent phase shift. That is, if $g_{m1}R_P g_{m2}R_P \geq 1$, then the loop oscillates. Note that V_X and V_Y are differential waveforms. (Why?)

Example 14.5

Sketch the open-loop voltage gain and phase of the circuit shown in Fig. 14.25. Neglect transistor capacitances.

Solution

The magnitude of the transfer function has a shape similar to that in Fig. 14.22(a) but with sharper rise and fall because it results from the *product* of those of the two stages. The total phase at low frequencies is given by signal inversion by each common-source stage plus a 90° phase shift due to each tank.

A similar behavior occurs at high frequencies. The gain and phase are sketched in Fig. 14.26. From these plots, the reader can prove that the circuit cannot oscillate at any other frequency.

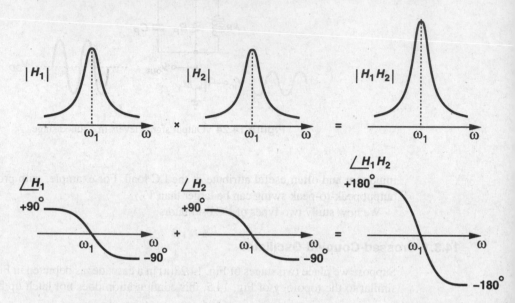

Figure 14.26   Loop gain characteristics of the circuit shown in Fig. 14.25.

The circuit of Fig. 14.25 serves as the core of many LC oscillators and is sometimes drawn as in Fig. 14.27(a) or (b). However, the drain currents of M_1 and M_2 and hence the output swings heavily depend on the supply voltage. Since the waveforms at X and Y are differential, the drawing in Fig. 14.27(b) suggests that M_1 and M_2 can be converted to a differential pair as depicted in Fig. 14.27(c), where the total bias current is defined by I_{SS}.

Example 14.6

For the circuit of Fig. 14.27(c), plot V_X and V_Y and I_{D1} and I_{D2} as the oscillation begins.

Solution

If the circuit begins with zero difference between V_X and V_Y, then $V_X = V_Y \approx V_{DD}$. The two transistors share the tail current equally. If $(g_{m1,2}R_P)^2 \geq 1$, where R_P is the equivalent parallel resistance of the tank at resonance, then noise components at the resonance frequency are continually amplified by M_1 and M_2, allowing the oscillation to grow. The drain currents of M_1 and M_2 vary according to the instantaneous value of $V_X - V_Y$ (as in a differential pair).

As shown in Fig. 14.28, the oscillation amplitude grows until the loop gain drops at the peaks. In fact, if $g_{m1,2}R_P$ is large enough, the difference between $V_X - V_Y$ reaches a level that steers the entire tail current to one transistor, turning the other off. Thus, in the steady state, I_{D1} and I_{D2} vary between zero and I_{SS}.

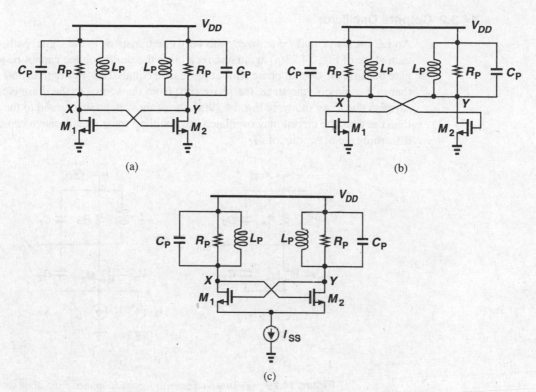

Figure 14.27 (a) Redrawing of the oscillator shown in Fig. 14.25, (b) another redrawing of the circuit, (c) addition of tail current source to lower supply sensitivity.

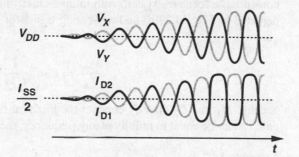

Figure 14.28

The oscillator of Fig. 14.27(c) is constructed in fully differential form. The supply sensitivity of the circuit, however, is nonzero even with perfect symmetry. This is because the drain junction capacitances of M_1 and M_2 vary with the supply voltage. We return to this issue in Example 14.9.

14.3.2 Colpitts Oscillator

An LC oscillator may be realized with only one transistor in the signal path. Consider the gain stage of Fig. 14.23(a) again and recall that the drain voltage cannot be applied to the gate because the overall phase shift at resonance equals 180° rather than 360°. Also, recall that in a common-gate stage, the phase shift from the source to the drain is zero. We then surmise that if, as shown in Fig. 14.29(a), the drain voltage is returned to the source rather than the gate, the circuit may oscillate. The coupling must incorporate a capacitor to avoid disturbing the bias point of M_1.

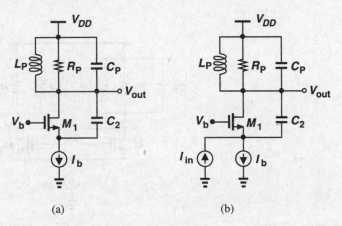

(a)   (b)

Figure 14.29   (a) Tuned stage with feedback applied from drain to source, (b) addition of input current to calculate closed-loop gain.

Unfortunately, owing to insufficient loop gain, the circuit of Fig. 14.29(a) does not oscillate. To prove this point, we invoke the view of Fig. 14.1, where an oscillator is considered a feedback system with infinite closed-loop gain. Applying an input current as depicted in Fig. 14.29(b) and neglecting transistor parasitics, we obtain the closed-loop gain as:

$$\frac{V_{out}}{I_{in}} = L_P s || \frac{1}{C_P s} || R_P \tag{14.37}$$

because M_1 and C_2 directly conduct the input current to the tank. Since the closed-loop gain cannot be equal to infinity at any frequency, the circuit fails to oscillate.

Example 14.7

The reader may wonder why the input to the feedback system is realized as a current source applied to the source of the transistor rather than a voltage source applied to its gate. Perform the analysis with the latter stimulus.

Solution

From Fig. 14.30, we note that with a finite variation of V_{in}, the change in I_b is still zero if the bias current source is ideal. Thus, if the source-bulk junction capacitance of M_1 is neglected, the

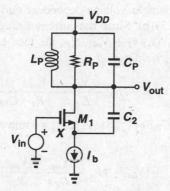

Figure 14.30

change in the tank current is zero, yielding $V_{out}/V_{in} = 0$. Interestingly, V_X does vary with V_{in}, but M_1 generates a small-signal current that cancels that through C_2. The reader can prove that $V_X/V_{in} = g_m/(g_m + C_2 s)$.

The above example reveals two important points. First, to excite a circuit into oscillation, the stimulus can be applied at different points. (That is, the noise of any device in the loop can initiate the oscillation.[5]) Second, in Fig. 14.30, V_{out}/V_{in} is zero because the impedance connected between the source of M_1 and ground is infinity. We then add a capacitor from this node to ground as shown in Fig. 14.31(a), seeking conditions of oscillation. Note that the capacitor in parallel with L_P is removed. The reason will become clear later.

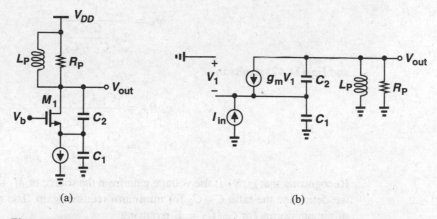

Figure 14.31   (a) Colpitts oscillator, (b) equivalent circuit of (a) with input stimulus.

---

[5]This is because the natural frequencies of a linear (observable) system do not depend on the location of the stimulus. Of course, the type of stimulus (voltage or current) must be chosen such that when it is set to zero, the circuit returns to its original topology. For example, driving the gate of M_1 in Fig. 14.30 by a current changes the natural frequencies of the circuit.

Approximating M_1 by a single voltage-dependent current source, we construct the equivalent circuit of Fig. 14.31(b). Since the current through the parallel combination of L_P and R_P is given by $V_{out}/(L_P s) + V_{out}/R_P$, the total current through C_1 is equal to $I_{in} - V_{out}/(L_P s) - V_{out}/R_P$, yielding

$$V_1 = -(I_{in} - \frac{V_{out}}{L_P s} - \frac{V_{out}}{R_P})\frac{1}{C_1 s}. \tag{14.38}$$

Writing the current through C_2 as $(V_{out} + V_1)C_2 s$, we sum all of the currents at the output node:

$$-g_m(I_{in} - \frac{V_{out}}{L_P s} - \frac{V_{out}}{R_P})\frac{1}{C_1 s} + [V_{out} - (I_{in} - \frac{V_{out}}{L_P s} - \frac{V_{out}}{R_P})\frac{1}{C_1 s}]C_2 s + \frac{V_{out}}{L_P s} + \frac{V_{out}}{R_P} = 0. \tag{14.39}$$

It follows that

$$\frac{V_{out}}{I_{in}} = \frac{R_P L_P s(g_m + C_2 s)}{R_P C_1 C_2 L_P s^3 + (C_1 + C_2)L_P s^2 + [g_m L_P + R_P(C_1 + C_2)]s + g_m R_P}. \tag{14.40}$$

Note that, as expected, (14.40) reduces to $(L_P s \| R_P)$ if $C_1 = 0$. The circuit oscillates if the closed-loop transfer function goes to infinity at an imaginary value of s, $s_R = j\omega_R$. Consequently, both the real and imaginary parts of the denominator must drop to zero at this frequency:

$$-R_P C_1 C_2 L_P \omega_R^3 + [g_m L_P + R_P(C_1 + C_2)]\omega_R = 0 \tag{14.41}$$

$$-(C_1 + C_2)L_P \omega_R^2 + g_m R_P = 0. \tag{14.42}$$

Since with typical values, $g_m L_P \ll R_P(C_1 + C_2)$, Eq. (14.41) yields:

$$\omega_R^2 = \frac{1}{L_P \dfrac{C_1 C_2}{C_1 + C_2}}, \tag{14.43}$$

and Eq. (14.42) results in

$$g_m R_P = \frac{(C_1 + C_2)^2}{C_1 C_2} \tag{14.44}$$

$$= \frac{C_1}{C_2}(1 + \frac{C_2}{C_1})^2. \tag{14.45}$$

Recognizing that $g_m R_P$ is the voltage gain from the source of M_1 to the output (if $g_{mb} = 0$), we determine the ratio C_1/C_2 for minimum required gain. The reader can prove that the minimum occurs for $C_1/C_2 = 1$, requiring

$$g_m R_P \geq 4. \tag{14.46}$$

Equation (14.46) demonstrates an important disadvantage of the Colpitts oscillator with respect to the cross-coupled topology of Fig. 14.27(c). The former demands a voltage gain of at least 4 at resonance and the latter, only unity. This issue is critical if the inductor suffers from a low Q and hence a small R_P, a common situation in CMOS technologies. As a consequence, the cross-coupled scheme is used more widely.

The foregoing analysis neglected the capacitance that appears in parallel with the inductor. As suggested in Problem 14.10, if this capacitance, C_P, is included in the equivalent circuit, Eq. (14.43) is modified as:

$$\omega_R^2 = \cfrac{1}{L_P(C_P + \cfrac{C_1 C_2}{C_1 + C_2})}, \tag{14.47}$$

whereas (14.46) remains unchanged. Thus, C_P is simply included in parallel with the series combination of C_1 and C_2.

14.3.3 One-Port Oscillators

Our development of oscillators thus far has been based on feedback systems. An alternative view that provides more insight into the oscillation phenomenon employs the concept of "negative resistance." To arrive at this view, let us first consider a simple tank that is stimulated by a current impulse [Fig. 14.32(a)]. The tank responds with a decaying oscillatory behavior because, in every cycle, some of the energy that reciprocates between

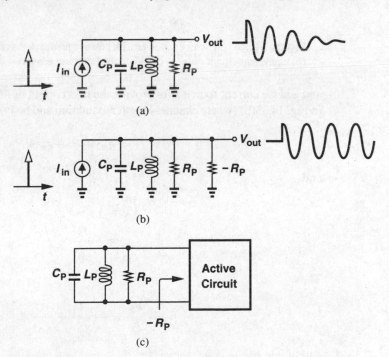

Figure 14.32   (a) Decaying impulse response of a tank, (b) addition of negative resistance to cancel loss in R_P, (c) use of an active circuit to provide negative resistance.

the capacitor and the inductor is lost in the form of heat in the resistor. Now suppose a resistor equal to $-R_P$ is placed in parallel with R_P and the experiment is repeated [Fig. 14.32(b)]. Since $R_P||(-R_P) = \infty$, the tank oscillates indefinitely. Thus, if a one-port circuit exhibiting a negative resistance is placed in parallel with a tank [Fig. 14.32(c)], the combination may oscillate. Such a topology is called a one-port oscillator.

How can a circuit provide a negative resistance? Recall that feedback multiplies or divides the input and output impedances of circuits by a factor equal to one plus the loop gain. Thus, if the loop gain is sufficiently *negative*, (i.e., the feedback is sufficiently positive), a negative resistance is achieved. As a simple example, let us apply positive feedback around a source follower. The follower introduces no signal inversion and neither must the feedback network. As depicted in Fig. 14.33(a), we implement the feedback by a common-gate stage

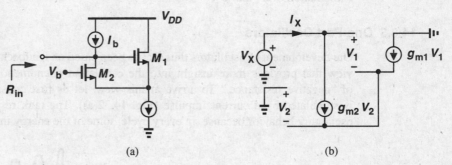

(a)   (b)

Figure 14.33 (a) Source follower with positive feedback to create negative input impedance, (b) equivalent circuit of (a) to calculate the input impedance.

and add the current source I_b to provide the bias current of M_2. From the equivalent circuit in Fig. 14.33(b) (where channel-length modulation and body effect are neglected), we have

$$I_X = g_{m2}V_2 = -g_{m1}V_1 \qquad (14.48)$$

and

$$V_X = V_1 - V_2 \qquad (14.49)$$

$$= -\frac{I_X}{g_{m1}} - \frac{I_X}{g_{m2}}. \qquad (14.50)$$

Thus,

$$\frac{V_X}{I_X} = -(\frac{1}{g_{m1}} + \frac{1}{g_{m2}}), \qquad (14.51)$$

and, if $g_{m1} = g_{m2} = g_m$, then

$$\frac{V_X}{I_X} = \frac{-2}{g_m}. \qquad (14.52)$$

Negative resistance becomes more intuitive if we bear in mind that it is an *incremental* quantity, that is, negative resistance indicates that if the applied voltage *increases*, the current drawn by the circuit *decreases*. In Fig. 14.33(a), for example, if the input voltage increases, so does the source voltage of M_1, decreasing the drain current of M_2 and allowing part of I_b to flow to the input source.

With a negative resistance available, we can now construct an oscillator as illustrated in Fig. 14.34. Here, R_P denotes the equivalent parallel resistance of the tank and, for oscillation

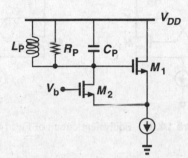

Figure 14.34 Oscillator using negative input resistance of a source follower with positive feedback.

build-up, $R_P - 2/g_m \geq 0$. Note that the inductor provides the bias current of M_2, obviating the need for a current source. If the small-signal resistance presented by M_1 and M_2 to the tank is less negative than $-R_P$, then the circuit experiences large swings such that each transistor is nearly off for part of the period, thereby yielding an "average" resistance of $-R_P$.

The circuit of Fig. 14.34 is similar to the stage of Fig. 14.29(a) but with the feedback capacitor replaced by a source follower. More interestingly, the circuit can be redrawn as in Fig. 14.35(a), bearing a resemblance to Fig. 14.27(c). In fact, if the drain current of M_1 flows

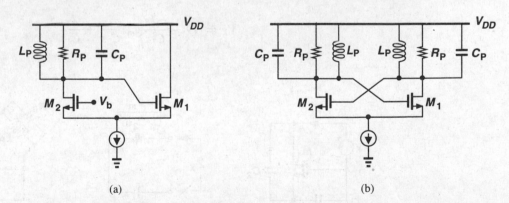

(a)   (b)

Figure 14.35   (a) Redrawing of the topology shown in Fig. 14.34, (b) differential version of (a).

through a tank and the resulting voltage is applied to the gate of M_2, the topology of Fig. 14.35(b) is obtained. Ignoring bias paths and merging the two tanks into one (Fig. 14.36), we note that the cross-coupled pair must provide a negative resistance of $-R_P$ between

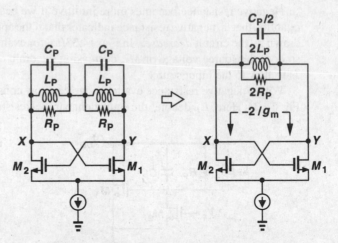

Figure 14.36   Equivalent circuit of Fig. 14.35(b).

nodes X and Y to enable oscillation. The reader can prove that this resistance is equal to $-2/g_m$ and hence it is necessary that $R_P \geq 1/g_m$. Thus, the circuit can be viewed as either a feedback system or a negative resistance in parallel with a lossy tank. This topology is also called a "negative-G_m oscillator."

As another method of creating negative resistance, consider the topology depicted in Fig. 14.37(a), where none of the nodes is grounded and channel-length modulation, body effect, and transistor capacitances are neglected. Since the drain current of M_1 is equal to $(-I_X/C_1 s)g_m$, we have

$$V_X = \left(I_X - \frac{-I_X}{C_1 s}g_m\right)\frac{1}{C_2 s} + \frac{I_X}{C_1 s} \tag{14.53}$$

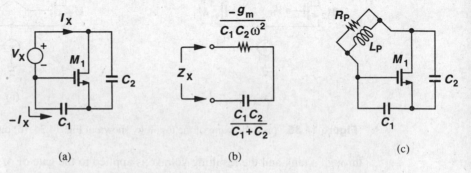

(a)   (b)   (c)

Figure 14.37   (a) Circuit topology providing negative resistance, (b) equivalent circuit of (a), (c) oscillator using (a).

and hence

$$\frac{V_X}{I_X} = \frac{g_m}{C_1 C_2 s^2} + \frac{1}{C_2 s} + \frac{1}{C_1 s}. \tag{14.54}$$

For $s = j\omega$, this impedance consists of a negative resistance equal to $-g_m/(C_1 C_2 \omega^2)$ in series with the series combination of C_1 and C_2 [Fig. 14.37(b)]. Thus, as shown in Fig. 14.37(c), if an inductor is placed between the gate and drain of M_1, the circuit may oscillate. Of the three nodes in the circuit, one can be an ac ground, resulting in the three different topologies illustrated in Fig. 14.38. The circuit of Fig. 14.38(a) is in fact based on a source follower, whose input impedance was found in Chapter 6 to contain a negative real part. The configuration of Fig. 14.38(b) is a Colpitts oscillator.

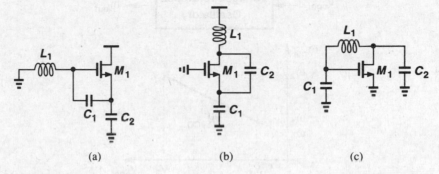

(a)   (b)   (c)

Figure 14.38   Oscillator topologies derived from the circuit of Fig. 14.37(c).

Example 14.8

Redraw the circuits of Fig. 14.38 with proper biasing.

Solution

The circuits are redrawn in Fig. 14.39.

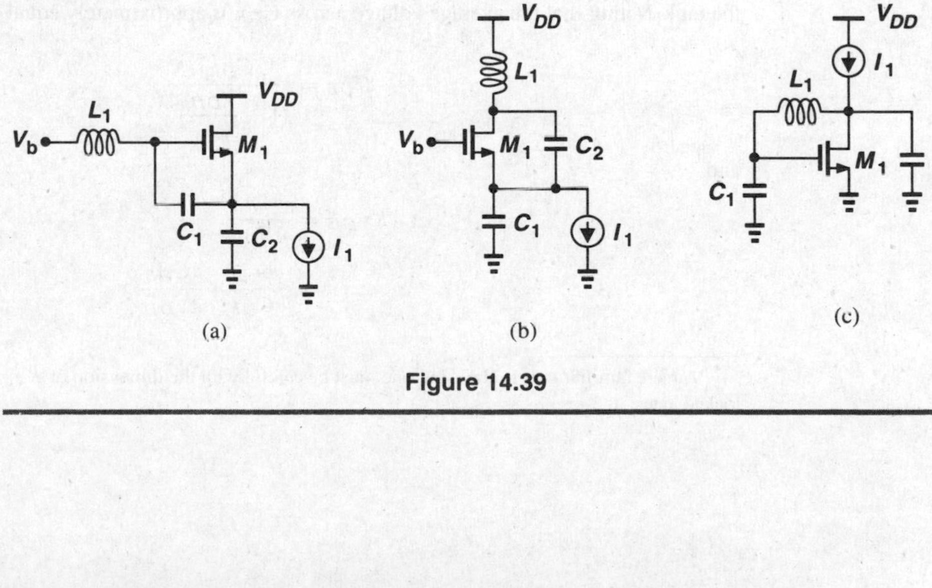

(a)   (b)   (c)

Figure 14.39

14.4 Voltage-Controlled Oscillators

Most applications require that oscillators be "tunable," i.e., their output frequency be a function of a control input, usually a voltage. An ideal voltage-controlled oscillator is a circuit whose output frequency is a linear function of its control voltage (Fig. 14.40):

$$\omega_{out} = \omega_0 + K_{VCO} V_{cont}. \tag{14.55}$$

Here, ω_0 represents the intercept corresponding to $V_{cont} = 0$ and K_{VCO} denotes the "gain" or "sensitivity" of the circuit (expressed in rad/s/V).[6] The achievable range, $\omega_2 - \omega_1$, is called the "tuning range."

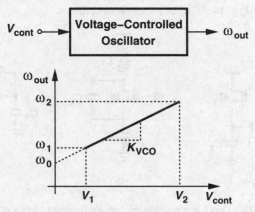

Figure 14.40   Definition of a VCO.

Example 14.9

In the negative-G_m oscillator of Fig. 14.27(c), assume $C_P = 0$, consider only the drain junction capacitance, C_{DB}, of M_1 and M_2, and explain why V_{DD} can be viewed as the control voltage. Calculate the gain of the VCO.

Solution

Since C_{DB} varies with the drain-bulk voltage, if V_{DD} changes, so does the resonance frequency of the tank. Noting that the average voltage across C_{DB} is approximately equal to V_{DD}, we write

$$C_{DB} = \frac{C_{DB0}}{(1 + \dfrac{V_{DD}}{\phi_B})^m}, \tag{14.56}$$

and

$$K_{VCO} = \frac{\partial \omega_{out}}{\partial V_{DD}} \tag{14.57}$$

$$= \frac{\partial \omega_{out}}{\partial C_{DB}} \cdot \frac{\partial C_{DB}}{\partial V_{DD}}. \tag{14.58}$$

---

[6]A more familiar unit is Hz/V but one must be careful with the dimension of K_{VCO} in the context of phase-locked loops.

With $\omega_{out} = 1/\sqrt{L_P C_{DB}}$, we have

$$K_{VCO} = \frac{-1}{2\sqrt{L_P C_{DB}} C_{DB}} \cdot \frac{-m C_{DB}}{\phi_B(1 + \frac{V_{DD}}{\phi_B})} \tag{14.59}$$

$$= \frac{m}{2\phi_B(1 + \frac{V_{DD}}{\phi_B})} \cdot \omega_{out}. \tag{14.60}$$

Note that the relationship between ω_{out} and V_{cont} is nonlinear because K_{VCO} varies with V_{DD} and ω_{out}.

Before modifying the oscillators studied in the previous sections for tunability, we summarize the important performance parameters of VCOs.

Center Frequency   The center frequency (i.e., the midrange value in Fig. 14.40) is determined by the environment in which the VCO is used. For example, in the clock generation network of a microprocessor, the VCO may be required to run at the clock rate or even twice that. Today's CMOS VCOs achieve center frequencies as high as 10 GHz.

Tuning Range   The required tuning range is dictated by two parameters: (1) the variation of the VCO center frequency with process and temperature and (2) the frequency range necessary for the application. The center frequency of some CMOS oscillators may vary by a factor of two at the extremes of process and temperature, thus mandating a sufficiently wide ($\geq 2\times$) tuning range to guarantee that the VCO output frequency can be driven to the desired value. Also, some applications incorporate clock frequencies that must vary by one to two orders of magnitude depending on the mode of operation, demanding a proportionally wide tuning range.

An important concern in the design of VCOs is the variation of the output phase and frequency as a result of noise on the control line. For a given noise amplitude, the noise in the output frequency is proportional to K_{VCO} because $\omega_{out} = \omega_0 + K_{VCO} V_{cont}$. Thus, to minimize the effect of noise in V_{cont}, the VCO gain must be *minimized*, a constraint in direct conflict with the required tuning range. In fact, if, as shown in Fig. 14.40, the allowable range of V_{cont} is from V_1 to V_2 (e.g., from 0 to V_{DD}) and the tuning range must span at least ω_1 to ω_2, then K_{VCO} must satisfy the following requirement:

$$K_{VCO} \geq \frac{\omega_2 - \omega_1}{V_2 - V_1}. \tag{14.61}$$

Note that, for a given tuning range, K_{VCO} increases as the supply voltage decreases, making the oscillator more sensitive to noise on the control line.

Tuning Linearity   As exemplified by Eq. (14.60), the tuning characteristics of VCOs exhibit nonlinearity, i.e., their gain, K_{VCO}, is not constant. As explained in Chapter 15, such nonlinearity degrades the settling behavior of phase-locked loops. For this reason, it is desirable to minimize the variation of K_{VCO} across the tuning range.

Actual oscillator characteristics typically exhibit a high gain region in the middle of the range and a low gain at the two extremes (Fig. 14.41). Compared to a linear characteristic (the gray line), the actual behavior displays a maximum gain *greater* than that predicted

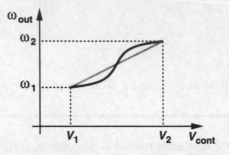

Figure 14.41 Nonlinear VCO characteristic.

by (14.61), implying that, for a given tuning range, nonlinearity inevitably leads to higher sensitivity for some region of the characteristic.

Output Amplitude   It is desirable to achieve a large output oscillation amplitude, thus making the waveform less sensitive to noise. The amplitude trades with power dissipation, supply voltage, and (as explained in Section 14.4.2) even the tuning range. Also, the amplitude may vary across the tuning range, an undesirable effect.

Power Dissipation   As with other analog circuits, oscillators suffer from trade-offs between speed, power dissipation, and noise. Typical oscillators drain 1 to 10 mW of power.

Supply and Common-Mode Rejection   Oscillators are quite sensitive to noise, especially if they are realized in single-ended form. As seen in Example 14.9, even differential oscillators exhibit supply sensitivity. The design of oscillators for high noise immunity is a difficult challenge. Note that noise may be coupled to the control line of a VCO as well. For these reasons, it is preferable (but not always possible) to employ differential paths for both the oscillation signal and the control line.

Output Signal Purity   Even with a constant control voltage, the output waveform of a VCO is not perfectly periodic. The electronic noise of the devices in the oscillator and supply noise lead to noise in the output phase and frequency. These effects are quantified by "jitter" and "phase noise" and determined by the requirements of each application.

14.4.1 Tuning in Ring Oscillators

Recall from Section 14.2 that the oscillation frequency, f_{osc}, of an N-stage ring equals $(2NT_D)^{-1}$, where T_D denotes the large-signal delay of each stage. Thus, to vary the frequency, T_D can be adjusted.

As a simple example, consider the differential pair of Fig. 14.42 as one stage of a ring oscillator. Here, M_3 and M_4 operate in the triode region, each acting as a variable resistor controlled by V_{cont}. As V_{cont} becomes more positive, the on-resistance of M_3 and M_4 increases, thus raising the time constant at the output, τ_1, and lowering f_{osc}. If M_3 and M_4

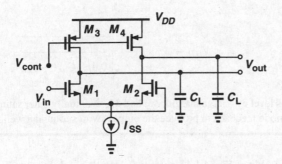

Figure 14.42 Differential pair with variable output time constant.

remain in deep triode region,

$$\tau_1 = R_{on3,4}C_L \tag{14.62}$$

$$= \frac{C_L}{\mu_p C_{ox}(\frac{W}{L})_{3,4}(V_{DD} - V_{cont} - |V_{THP}|)}. \tag{14.63}$$

In the above equation, C_L denotes the total capacitance seen at each output to ground (including the input capacitance of the following stage). The delay of the circuit is roughly proportional to τ_1, yielding

$$f_{osc} \propto \frac{1}{T_D} \tag{14.64}$$

$$\propto \frac{\mu_p C_{ox}(\frac{W}{L})_{3,4}(V_{DD} - V_{cont} - |V_{THP}|)}{C_L}. \tag{14.65}$$

Interestingly, f_{osc} is linearly proportional to V_{cont}.

Example 14.10 ————————————————————————————————

For given device dimensions and bias currents in Fig. 14.42, determine the maximum allowable value of V_{cont}. What happens if M_3 and M_4 enter saturation?

Solution

Let us assume (somewhat arbitrarily) that M_3 and M_4 remain in deep triode region if $|V_{DS3,4}| \leq 0.2 \times 2|V_{GS3,4} - V_{THP}|$. If each stage in the ring experiences complete switching, then the maximum drain current of M_3 and M_4 is equal to I_{SS}. To satisfy the above condition, we must have $I_{SS}R_{on3,4} \leq 0.4(V_{DD} - V_{cont} - |V_{THP}|)$, and hence

$$\frac{I_{SS}}{\mu_p C_{ox}(\frac{W}{L})_{3,4}(V_{DD} - V_{cont} - |V_{THP}|)} \leq 0.4(V_{DD} - V_{cont} - |V_{THP}|). \tag{14.66}$$

It follows that

$$V_{cont} \leq V_{DD} - |V_{THP}| - \sqrt{\frac{I_{SS}}{0.4\mu_p C_{ox}(\frac{W}{L})_{3,4}}}. \tag{14.67}$$

If V_{cont} exceeds this level by a large margin, M_3 and M_4 eventually enter saturation. Each stage then requires common-mode feedback to produce the output swings around a well-defined CM level.

The differential pair of Fig. 14.42 suffers from a critical drawback: the output swing of the circuit varies considerably across the tuning range. With complete switching, each stage provides a differential output swing of $2I_{SS}R_{on3,4}$. Thus, a tuning range of, say, two to one translates to a twofold variation in the swing.

In order to minimize the swing variation, the tail current can be adjusted by V_{cont} as well such that, as V_{cont} becomes more positive, I_{SS} decreases. The circuit nonetheless requires a means of maintaining $I_{SS}R_{on3,4}$ relatively constant. To this end, let us consider the circuit in Fig. 14.43(a), where M_5 operates in the deep triode region and amplifier A_1 applies

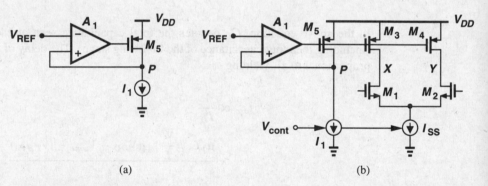

(a)   (b)

Figure 14.43   (a) Simple feedback circuit defining V_P, (b) replica biasing to define voltage swings in a ring oscillator.

negative feedback to the gate of M_5. If the loop gain is sufficiently large, the differential input voltage of A_1 must be small, giving $V_P \approx V_{REF}$ and $|V_{DS5}| \approx V_{DD} - V_{REF}$. Thus, the feedback ensures a relatively constant drain-source voltage even if I_1 varies. In fact, as I_1, say, decreases, A_1 raises the gate voltage of M_5 such that $R_{on5}I_1 \approx V_{DD} - V_{REF}$.

The topology of Fig. 14.43(a) can serve as a "replica circuit" for the stages of a ring oscillator, thereby defining the oscillation amplitude. Illustrated in Fig. 14.43(b), the idea is to "servo" the on-resistance of M_3 and M_4 to that of M_5 and vary the frequency by adjusting I_1 and I_{SS} simultaneously [2]. If M_3 and M_4 are identical to M_5 and I_{SS} to I_1, then V_X and V_Y vary from V_{DD} to $V_{DD} - V_{REF}$ as M_1 and M_2 steer the tail current to one side or the other. Thus, if process and temperature variations, say, decrease, I_1 and I_{SS}, then A_1 increases the on-resistance of M_3-M_5, forcing V_P and hence V_X and V_Y (when M_1 or M_2 is fully on) equal to V_{REF}.

The bandwidth of the op amp A_1 in Fig. 14.43(b) is of some concern. If a change in V_{cont} takes a long time to change ω_{out}, then the settling speed of a PLL using this VCO degrades significantly (Chapter 15).

Example 14.11

How does the oscillation frequency depend on I_{SS} for a VCO incorporating the stage of Fig. 14.43(b).

Solution

Noting that $R_{on3,4}I_{SS} \approx V_{DD} - V_{REF}$, we have $R_{on3,4} \approx (V_{DD} - V_{REF})/I_{SS}$ and hence

$$f_{osc} \propto \frac{1}{R_{on3,4}C_L} \tag{14.68}$$

$$\propto \frac{I_{SS}}{(V_{DD} - V_{REF})C_L}. \tag{14.69}$$

Thus, the characteristic is relatively linear.

Delay Variation by Positive Feedback   To arrive at another tuning technique, recall that a cross-coupled transistor pair such as that of Fig. 14.36 exhibits a negative resistance of $-2/g_m$, a value that can be controlled by the bias current. A negative resistance $-R_N$ placed in parallel with a positive resistance $+R_P$ gives an equivalent value $+R_N R_P/(R_N - R_P)$, which is more positive if $|-R_N| > |+R_P|$. This idea can be applied to each stage of a ring oscillator as illustrated in Fig. 14.44(a). Here, the load of the differential pair consists of

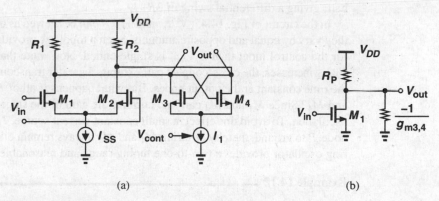

(a)   (b)

Figure 14.44   (a) Differential stage with variable negative-resistance load, (b) half-circuit equivalent of (a).

resistors R_1 and R_2 ($R_1 = R_2 = R_P$) and the cross-coupled pair M_3-M_4. As I_1 increases, the small-signal differential resistance $-2/g_{m3,4}$ becomes less negative and, from the half circuit of Fig. 14.44(b), the equivalent resistance $R_P||(-1/g_{m3,4}) = R_P/(1 - g_{m3,4}R_P)$ increases, thereby lowering the frequency of oscillation.

An important issue in the circuit of Fig. 14.44(a) is that as I_1 varies, so do the currents steered by M_3 and M_4 to R_1 and R_2. Thus, the output voltage swing is not constant across the tuning range. To minimize this effect, I_{SS} can be varied in the *opposite* direction such that the total current steered between R_1 and R_2 remains constant. In other words, it is desirable to vary I_1 and I_{SS} *differentially* while their sum is fixed, a characteristic provided by a differential pair. Illustrated in Fig. 14.45, the idea is to employ a differential pair M_5-M_6 to steer I_T to M_1-M_2 or M_3-M_4 so that $I_{SS} + I_1 = I_T$. Since I_T must flow through R_1 and

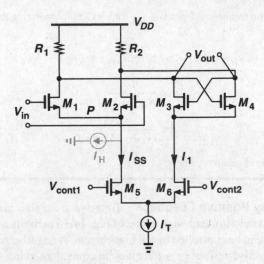

Figure 14.45   Use of a differential pair to steer current between M_1-M_2 and M_3-M_4.

R_2, if M_1-M_4 experience complete switching in each cycle of oscillation, then I_T is steered to R_1 (through M_1 and M_3) in half a period and to R_2 (through M_2 and M_4) in the other half, giving a differential swing of $2R_P I_T$.

In the circuit of Fig. 14.45, V_{cont1} and V_{cont2} can be viewed as differential control lines if they vary by equal and opposite amounts. Such a topology provides higher noise immunity for the control input than if V_{cont} is single-ended. Now, note that as V_{cont1} decreases and V_{cont2} increases, the cross-coupled pair exhibits a greater transconductance, thereby raising the time constant at the output nodes. But what happens if all of I_T is steered by M_6 to M_3 and M_4? Since M_1 and M_2 carry no current, the gain of the stage falls to zero, prohibiting oscillation. To avoid this effect, a small constant current source, I_H, can be connected from node P to ground, thereby ensuring M_1 and M_2 always remain on. With typical values, this ring oscillator provides a two-to-one tuning range and reasonable linearity.

Example 14.12

Calculate the minimum value of I_H in Fig. 14.45 to guarantee a low-frequency gain of 2 when all of I_T is steered to the cross-coupled pair.

Solution

The small-signal voltage gain of the circuit equals $g_{m1,2} R_P / (1 - g_{m3,4} R_P)$. Assuming square-law devices, we have

$$\sqrt{\mu_n C_{ox} (\frac{W}{L})_{1,2} I_H} \; \frac{R_P}{1 - \sqrt{\mu_n C_{ox} (\frac{W}{L})_{3,4} I_T R_P}} \geq 2. \tag{14.70}$$

That is,

$$I_H \geq \frac{4\left[1 - \sqrt{\mu_n C_{ox}(\frac{W}{L})_{3,4} I_T R_P}\right]^2}{\mu_n C_{ox}(\frac{W}{L})_{1,2} R_P^2}. \tag{14.71}$$

An important drawback of using the differential pair M_5-M_6 in the circuit of Fig. 14.45 is the additional voltage headroom that it consumes. As depicted in Fig. 14.46, for M_5 to remain in saturation, V_P must be sufficiently higher than V_N. When $V_{cont1} = V_{cont2}$,

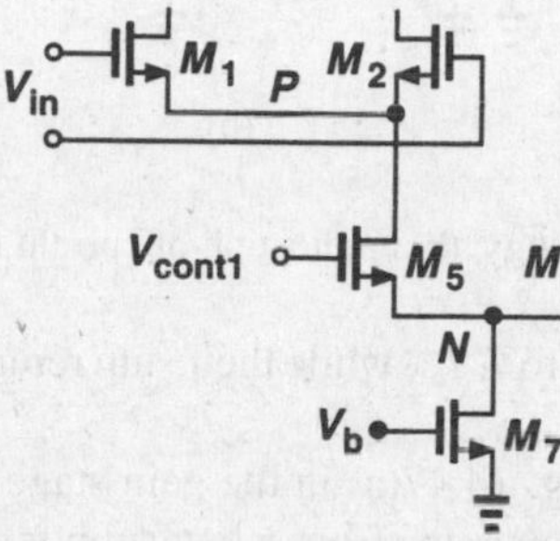

Figure 14.46 Headroom calculation for a current-steering topology.

the minimum allowable drain-source voltage of M_5 is equal to its equilibrium overdrive voltage, implying that, compared to that calculated in Example 14.4, the supply voltage must be higher by this value. Note also that if V_{cont1} or V_{cont2} is allowed to vary above its equilibrium value by more than V_{TH}, then M_5 or M_6 enters the triode region.

The above observation reveals a trade-off between voltage headroom and the *sensitivity* of the VCO. In order to minimize the sensitivity with a given tuning range, the transconductance of M_5-M_6 must be *minimized*. (That is, to steer all of the tail current, the differential pair must require a *large* $V_{cont1} - V_{cont2}$.) However, for a given tail current, $g_m = 2I_D/(V_{GS} - V_{TH})$, indicating a large equilibrium overdrive for M_5-M_6 and a correspondingly higher value for the minimum required supply voltage.

We should mention that the pair M_5-M_6 need not remain in complete saturation. If the drain voltages are low enough to drive these transistors into the triode region, then the equivalent transconductance of the differential pair drops, thus demanding a greater $V_{cont1} - V_{cont2}$ to steer the tail current. This phenomenon in fact translates to a *lower* VCO sensitivity. In practice, careful simulations are required to ensure the VCO characteristic remains relatively linear across the range of interest.[7]

At low supply voltages, it is desirable to avoid the voltage headroom consumed by M_5-M_6 in Fig. 14.45. The issue can be resolved by means of "current folding." Suppose, as illustrated in Fig. 14.47(a), a differential pair drives two current mirrors, generating I_{out1} and I_{out2}. Since $I_1 + I_2 = I_{SS}$, $I_{out1} = KI_1$, and $I_{out2} = KI_2$, we have $I_{out1} + I_{out2} = KI_{SS}$. Thus, as $V_{in1} - V_{in2}$ goes from a very negative value to a very positive value, I_{out1} varies

---

[7] If both M_5 and M_6 are in the triode region and $V_{cont1} \neq V_{cont2}$, then supply voltage variations affect the current steered between the two transistors, introducing noise in the frequency of oscillation.

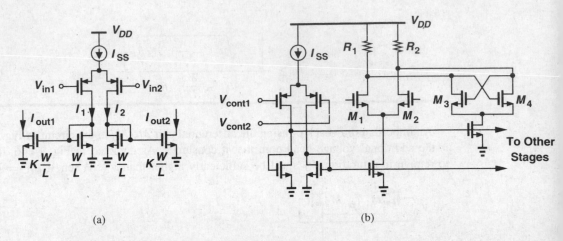

Figure 14.47 (a) Current folding topology, (b) application of current folding to current steering.

from KI_{SS} to zero and I_{out2} from zero to KI_{SS} while their sum remains constant - a behavior similar to that of a differential pair.

We now utilize the topology of Fig. 14.47(a) in the gain stage of Fig. 14.44(a). Shown in Fig. 14.47(b), the resulting circuit operates from a low supply voltage.

Delay Variation by Interpolation Another approach to tuning ring oscillators is based on "interpolation" [3, 4]. As illustrated in Fig. 14.48(a), each stage consists of a fast path and

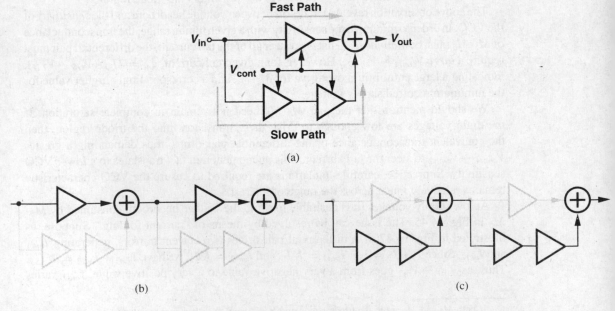

Figure 14.48 (a) Interpolating delay stage, (b) smallest delay, (b) largest delay.

a slow path whose outputs are summed and whose gains are adjusted by V_{cont} in opposite directions. At one extreme of the control voltage, only the fast path is on and the slow path is disabled, yielding the maximum oscillation frequency [Fig. 14.48(b)]. Conversely, at the other extreme, only the slow path is on and the fast path is off, providing the minimum oscillation frequency [Fig. 14.48(c)]. If V_{cont} lies between the two extremes, each path is partially on and the total delay is a weighted sum of their delays.

To better understand the concept of interpolation, let us implement the topology of Fig. 14.48(a) at the transistor level. Each stage can be simply realized as a differential pair whose gain is controlled by its tail current. But how are the two outputs summed? Since the two transistors in a differential pair provide output *currents*, the outputs of the two pairs can be added in the current domain. As depicted in Fig. 14.49(a), simply shorting the outputs of two pairs performs the current addition, e.g., for small signals, $I_{out} = g_{m1,2}V_{in1} + g_{m3,4}V_{in2}$. The overall interpolating stage therefore assumes the configuration shown in Fig. 14.49(b), where V_{cont}^+ and V_{cont}^- denote voltages that vary in opposite directions (so that when one path turns on, the other turns off). The output currents of M_1-M_2 and M_3-M_4 are summed at X and Y and flow through R_1 and R_2, producing V_{out}.

In the circuit of Fig. 14.49(b), the gain of each stage is varied by the tail current to achieve interpolation. But it is desirable to maintain constant voltage swings. We also recognize that the gain of the differential pair M_5-M_6 need not be varied because even if only the gain of M_3-M_4 drops to zero, the slow path is fully disabled. We then surmise that if the tail currents of M_1-M_2 and M_3-M_4 vary in opposite directions such that their sum remains constant, we achieve both interpolation between the two paths and constant output swings.

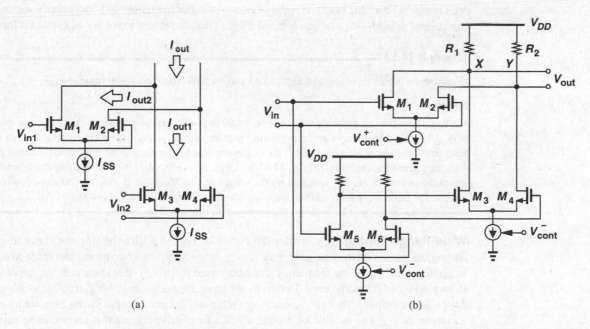

Figure 14.49 (a) Addition of currents of two differential pairs, (b) interpolating delay stage.

Illustrated in Fig. 14.50, the resulting circuit employs the differential pair M_7-M_8 to steer I_{SS} between M_1-M_2 and M_3-M_4. If V_{cont} is very negative, M_8 is off and only the fast path

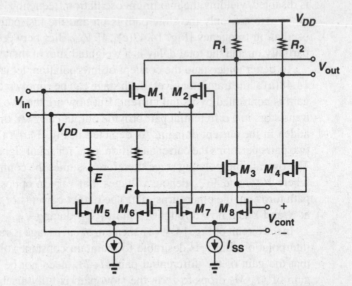

Figure 14.50 Interpolating delay stage with current steering.

amplifies the input. Conversely, if V_{cont} is very positive, M_7 is off and only the slow path is enabled. Since the slow path in this case employs one more stage than the fast path, the VCO achieves a tuning range of roughly two to one. For operation with low supply voltages, the control pair M_7-M_8 can be replaced by the current-folding topology of Fig. 14.47(a).

Example 14.13

Combine the tuning techniques of Figs. 14.45 and 14.50 to achieve a wider tuning range.

Solution

We begin with the interpolating stage of Fig. 14.50 and add a cross-coupled pair to the output nodes [Fig. 14.51(a)]. However, in order to obtain constant voltage swings, the total current through the load resistors must remain constant. This is accomplished by replacing the control differential pair with the current-folding circuit of Fig. 14.47(a). Depicted in Fig. 14.51(b), the resulting configuration steers the current to M_1-M_2 to speed up the circuit and to M_3-M_4 and M_{10}-M_{11} to slow down the circuit. The tail current source dimensions are chosen such that $I_{SS1} = I_{SS2} + I_{SS3}$.

Wide-Range Tuning Except for the circuit of Fig. 14.43(b), the ring oscillator tuning techniques presented thus far achieve a tuning range of typically no more than three to one. In applications where the frequency must be varied by orders of magnitude, the topology shown in Fig. 14.52 can be used. Driven by the input, the additional PMOS transistors M_5 and M_6 pull each output node to V_{DD}, creating a relatively constant output swing even with large variations in I_{SS}. The oscillation frequency of a ring incorporating this stage can be varied by more than four orders of magnitude with less than a twofold variation in the amplitude.

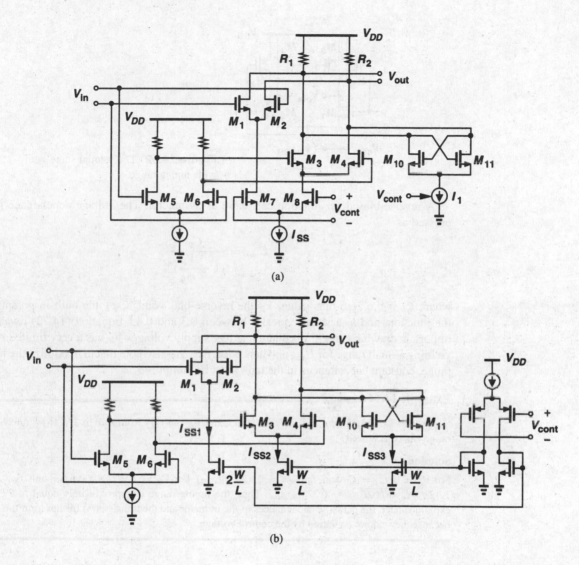

(a)

(b)

Figure 14.51

14.4.2 Tuning in LC Oscillators

The oscillation frequency of LC topologies is equal to $f_{osc} = 1/(2\pi\sqrt{LC})$, suggesting that only the inductor and capacitor values can be varied to tune the frequency and other parameters such as bias currents and transistor transconductances affect f_{osc} negligibly. Since it is difficult to vary the value of monolithic inductors, we simply change the tank capacitance to tune the oscillator. Voltage-dependent capacitors are called "varactors."[8]

---

[8]The term "varicap" is also used.

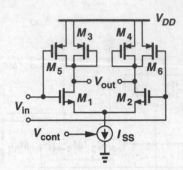

Figure 14.52 Differential stage with wide tuning range.

A reverse-biased *pn* junction can serve as a varactor. The voltage dependence is expressed as

$$C_{var} = \frac{C_0}{(1 + \frac{V_R}{\phi_B})^m},$$ (14.72)

where C_0 is the zero-bias value, V_R the reverse-bias voltage, ϕ_B the built-in potential of the junction, and m a value typically between 0.3 and 0.4.[9] Equation (14.72) reveals an important drawback of LC oscillators: at low supply voltages V_R has a very limited range, yielding a small range for C_{var} and hence for f_{osc}. We also note that to maximize the tuning range, constant capacitances in the tank must be *minimized*.

Example 14.14

Suppose in Eq. (14.72), $\phi_B = 0.7$ V, $m = 0.35$, and V_R can vary from zero to 2 V. How much tuning range can be achieved?

Solution

For $V_R = 0$, $C_j = C_0$ and $f_{osc,min} = 1/(2\pi\sqrt{LC_0})$. For $V_R = 2$ V, $C_j \approx 0.62C_0$ and $f_{osc,max} = 1/(2\pi\sqrt{L \times 0.62C_0}) \approx 1.27 f_{osc,min}$. Thus, the tuning range is approximately equal to 27%. As explained later, the parasitic capacitances of the inductor and the transistor(s) further limit this range because they cannot be varied by the control voltage.

Let us now add varactor diodes to a cross-coupled LC oscillator (Fig. 14.53). To avoid forward-biasing D_1 and D_2 significantly, V_{cont} must not exceed V_X or V_Y by more than a few hundred millivolts. Thus, if the peak amplitude at each node is A, then $0 < V_{cont} < V_{DD} - A + 300$ mV, where it is assumed a forward bias of 300 mV creates negligible current. Interestingly, the circuit suffers from a trade-off between the output swing and the tuning range. This effect appears in most LC oscillators.

Note that, since the swings at X and Y are typically large (e.g., 1 V_{pp} at each node), the capacitance of D_1 and D_2 *varies* with time. Nonetheless, the "average" value of the capacitance is still a function of V_{cont}, providing the tuning range.

---

[9]Note that $m = 0.5$ for an abrupt junction, but *pn* junctions in CMOS technology are not abrupt.

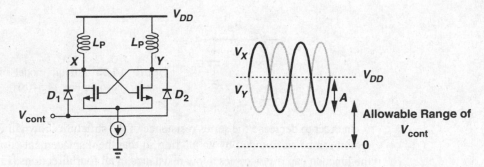

Figure 14.53 LC oscillator using varactor diodes.

How are varactor diodes realized in CMOS technology? Illustrated in Fig. 14.54 are two types of *pn* junctions. In Fig. 14.54(a), the anode is inevitably grounded whereas in Fig.

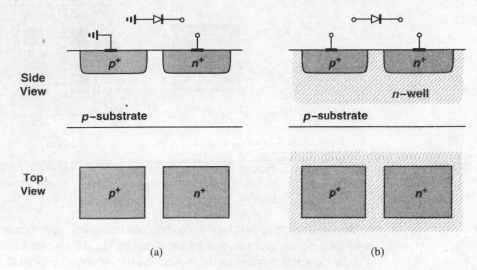

Figure 14.54 Diodes realized in CMOS technology.

14.54(b), both terminals are floating. For the circuit of Fig. 14.53, only the floating diode can be used. To increase the capacitance of the junction, the p^+ and n^+ areas (and hence the n-well) are enlarged.

Upon closer examination, the structure of Fig. 14.54(b) suffers from a number of drawbacks. First, the n-well material has a high resistivity, creating a resistance in series with the reverse-biased diode and lowering the quality factor of the capacitance. Second, the n-well displays substantial capacitance to the substrate, contributing a constant capacitance to the tank and limiting the tuning range. The diode is therefore represented as shown in Fig. 14.55, where C_n represents the (voltage-dependent) capacitance between the n-well and the substrate.[10]

---

[10]In circuit simulations, C_n is replaced by a diode having proper junction capacitance.

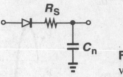

Figure 14.55 Circuit model of the varactor shown in Fig. 14.54(b).

In order to decrease the series resistance of the structure shown in Fig. 14.54(b), the p^+ region can be surrounded by an n^+ ring so that the displacement current flowing through the junction capacitance sees a low resistance in all four directions [Fig. 14.56(a)]. Since a single minimum-size p^+ area has a small capacitance, many of these units can be placed in parallel [Fig. 14.56(b)]. The n-well, however, must accommodate the entire set, exhibiting a large capacitance to the substrate.

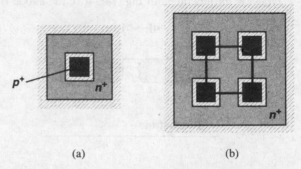

(a)   (b)

Figure 14.56 (a) Reduction of series resistance by surrounding the p^+ region by an n^+ ring, (b) several diodes in parallel.

It is instructive at this point to examine the unwanted capacitances in the circuit of Fig. 14.53, i.e., the components that are not varied by V_{cont}. We identify three such capacitances: (1) the capacitance between the n-well and the substrate associated with D_1 and D_2; (2) the capacitances contributed by the transistors to each node, i.e., C_{GD}, $2C_{GD}$ (the factor of 2 arising from Miller effect[11]), and C_{DB}; (3) the parasitic capacitance of the inductor itself. Monolithic inductors are typically implemented as metal spiral structures (Fig. 14.57) having relatively large dimensions ($S \approx 100\text{-}200 \ \mu m$). Their capacitance to the substrate is therefore quite large.

In Fig. 14.53, it is desirable to connect the anode of the diodes to nodes X and Y, thereby eliminating the parasitic n-well capacitances from the tank. Shown in Fig. 14.58 is a topology allowing such a modification. Here, the cross-coupled pair incorporates PMOS devices, providing swings around the ground potential. However, owing to their lower mobility, the PMOS transistors must be wider than their NMOS counterparts so as to exhibit the same transconductance. This increases the second component mentioned above.

---

[11]If the gate and drain voltages vary by equal and opposite amounts, the Miller multiplication factor is equal to 2 regardless of the small-signal gain.

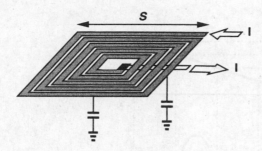

Figure 14.57 Spiral inductor structure.

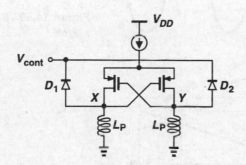

Figure 14.58 Negative-G_m oscillator using PMOS devices to eliminate n-well capacitance from the tanks.

The design of low-noise CMOS LC oscillators with acceptable tuning range is still a topic of active research. Issues such as phase noise and inductor and varactor design continue to intrigue researchers. MOS varactors have also been investigated as an alternative to *pn* junctions.

14.5 Mathematical Model of VCOs

The definition of the voltage-controlled oscillator given by Eq. (14.55) specifies the relationship between the control voltage and the output frequency. The dependence is "memoryless" because a change in V_{cont} immediately results in a change in ω_{out}. But how is the output signal of the VCO expressed as a function of time? To answer this question, we must review the concepts of phase and frequency.

Consider the waveform $V_0(t) = V_m \sin \omega_0 t$. The argument of the sinusoid is called the "total phase" of the signal. In this example, the phase varies linearly with time, exhibiting a slope equal to ω_0. Note that, as depicted in Fig. 14.59, every time $\omega_0 t$ crosses an integer multiple of π, $V_0(t)$ crosses zero.

Now consider two waveforms $V_1(t) = V_m \sin[\phi_1(t)]$ and $V_2(t) = V_m \sin[\phi_2(t)]$. where $\phi_1(t) = \omega_1 t$, $\phi_2(t) = \omega_2 t$, and $\omega_1 < \omega_2$. As illustrated in Fig. 14.60, $\phi_2(t)$ crosses integer multiples of π faster than $\phi_1(t)$ does, yielding faster variations in $V_2(t)$. We say $V_2(t)$ accumulates phase faster.

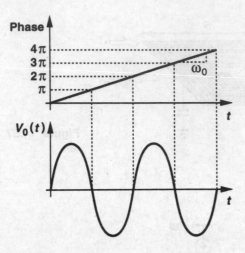

Figure 14.59 Illustration of phase of a signal.

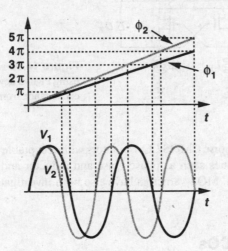

Figure 14.60 Variation of phase for two signals.

The above study reveals that the faster the phase of a waveform varies, the higher the frequency of the waveform, suggesting that the frequency[12] can be defined as the derivative of the phase with respect to time:

$$\omega = \frac{d\phi}{dt}. \tag{14.73}$$

Example 14.15

Figure 14.61(a) shows the phase of a sinusoidal waveform with constant amplitude as a function of time. Plot the waveform in the time domain.

---

[12]The quantity $\omega = 2\pi f$ is called the "radian frequency" (and expressed in rad/s) to distinguish it from f (expressed in Hz). In this book, we call both the frequency, but use ω more often to avoid the factor 2π.

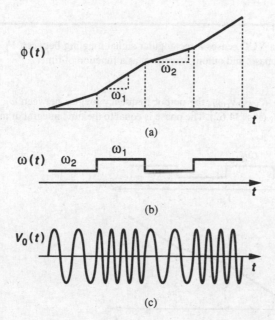

(a)

(b)

(c)

Figure 14.61

Solution

Taking the time derivative of $\phi(t)$, we obtain the behavior illustrated in Fig. 14.61(b). The frequency therefore periodically toggles between ω_1 and ω_2, yielding the waveform shown in Fig. 14.61(c). (This is a simple example of binary frequency modulation, called "frequency shift keying" and utilized in wireless pagers and many other communication systems.)

Equation (14.73) indicates that, if the frequency of a waveform is known as a function of time, then the phase can be computed as

$$\phi = \int \omega \, dt + \phi_0. \tag{14.74}$$

In particular, since for a VCO, $\omega_{out} = \omega_0 + K_{VCO} V_{cont}$, we have

$$V_{out}(t) = V_m \cos(\int \omega_{out} dt + \phi_0) \tag{14.75}$$

$$= V_m \cos(\omega_0 t + K_{VCO} \int V_{cont} dt + \phi_0). \tag{14.76}$$

Equation (14.76) proves essential in the analysis of VCOs and PLLs.[13] The initial phase ϕ_0 is usually unimportant and is assumed zero hereafter.

---

[13]Note that K_{VCO} cannot be brought out of the integral if the characteristic is nonlinear.

Example 14.16

The control line of a VCO senses a rectangular signal toggling between V_1 and V_2 at a period T_m. Plot the frequency, phase, and output waveform as a function of time.

Solution

Since $\omega_{out} = \omega_0 + K_{VCO} V_{cont}$, the output frequency toggles between $\omega_1 = \omega_0 + K_{VCO} V_1$ and $\omega_2 = \omega_0 + K_{VCO} V_2$ (Fig. 14.62). The phase is equal to the time integral of this result, rising linearly

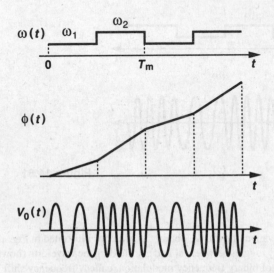

Figure 14.62

with time at a slope of ω_1 for half the input period and ω_2 for the other half. The output waveform of the VCO is similar to that shown in Fig. 14.61. Thus, a VCO can operate as a frequency modulator.

As explained in Chapter 15, if a VCO is placed in a phase-locked loop, then only the second term of the total phase in Eq. (14.76) is of interest. This term, $K_{VCO} \int V_{cont} dt$, is called the "excess phase," ϕ_{ex}. In fact, in the analysis of PLLs, we view the VCO as a system whose input and output are the control voltage and the excess phase, respectively:

$$\phi_{ex} = K_{VCO} \int V_{cont} dt. \qquad (14.77)$$

That is, the VCO operates as an *ideal* integrator, providing a transfer function:

$$\frac{\Phi_{ex}}{V_{cont}}(s) = \frac{K_{VCO}}{s}. \qquad (14.78)$$

Example 14.17

A VCO senses a small sinusoidal control voltage $V_{cont} = V_m \cos \omega_m t$. Determine the output waveform and its spectrum.

Solution

The output is expressed as

$$V_{out}(t) = V_0 \cos(\omega_0 t + K_{VCO} \int V_{cont} dt) \tag{14.79}$$

$$= V_0 \cos(\omega_0 t + K_{VCO} \frac{V_m}{\omega_m} \sin \omega_m t) \tag{14.80}$$

$$= V_0 \cos \omega_0 t \cos(K_{VCO} \frac{V_m}{\omega_m} \sin \omega_m t) \tag{14.81}$$

$$- V_0 \sin \omega_0 t \sin(K_{VCO} \frac{V_m}{\omega_m} \sin \omega_m t).$$

If V_m is small enough that $K_{VCO} V_m/\omega_m \ll 1$ rad, then

$$V_{out}(t) \approx V_0 \cos \omega_0 t - V_0(\sin \omega_0 t)(K_{VCO} \frac{V_m}{\omega_m} \sin \omega_m t) \tag{14.82}$$

$$= V_0 \cos \omega_0 t - \frac{K_{VCO} V_m V_0}{2\omega_m}[\cos(\omega_0 - \omega_m)t - \cos(\omega_0 + \omega_m)t]. \tag{14.83}$$

The output therefore consists of three sinusoids having frequencies of ω_0, $\omega_0 - \omega_m$, and $\omega_0 + \omega_m$. The spectrum is shown in Fig. 14.63. The components at $\omega_0 \pm \omega_m$ are called "sidebands."

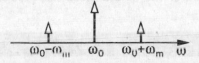

Figure 14.63

The above example reveals that variation of the control voltage with time may create unwanted components at the output. Indeed, when a VCO operates in the steady state, the control voltage must experience very little variation.[14] This issue is studied in Chapter 15.

A common mistake in expressing the phase of signals arises from the familiar form $V_m \cos \omega_0 t$. Here, the phase is equal to the product of frequency and time, creating the impression that such equality holds in all conditions. We may even deduce that, since the output frequency of a VCO is given by $\omega_0 + K_{VCO} V_{cont}$, the output waveform can be written as $V_m \cos[(\omega_0 + K_{VCO} V_{cont})t]$. To understand why this is incorrect, let us compute the frequency as the derivative of the phase:

$$\omega = \frac{d}{dt}[(\omega_0 + K_{VCO} V_{cont})t] \tag{14.84}$$

$$= K_{VCO} \frac{dV_{cont}}{dt}t + \omega_0 + K_{VCO} V_{cont}. \tag{14.85}$$

---

[14]Except when the VCO senses a signal to perform frequency modulation.

The first term in this expression is redundant, vanishing only if $dV_{cont}/dt = 0$. Thus, in the general case, the phase cannot be written as the product of time and frequency.

Our study of VCOs in this section has assumed sinusoidal output waveforms. In practice, depending on the type and speed of the oscillator, the output may contain significant harmonics, even approaching a rectangular waveform. How should Eq. (14.76) be modified in this case? We expect that $V_{out}(t)$ can be expressed as a Fourier series:

$$V_{out}(t) = V_1 \cos(\omega_0 t + \phi_1) + V_2 \cos(2\omega_0 t + \phi_2) + \cdots. \tag{14.86}$$

We also note that if the (fundamental) frequency of a rectagular waveform is changed by Δf, the frequency of its second harmonic must change by $2\Delta f$, etc. Thus, if V_{cont} varies by ΔV, then the frequency of the first harmonic varies by $K_{VCO}\Delta V$, the frequency of the second harmonic by $2K_{VCO}\Delta V$, etc. That is,

$$V_{out}(t) = V_1 \cos\left(\omega_0 t + K_{VCO}\int V_{cont}dt + \theta_1\right) + V_2 \cos\left(2\omega_0 t + 2K_{VCO}\int V_{cont}dt + \theta_2\right) + \cdots, \tag{14.87}$$

where $\theta_1, \theta_2, \cdots$ are constant phases necessary for the representation of each harmonic in the Fourier series expansion.

Equation (14.87) suggests that the harmonics of an oscillator output can be readily taken into account. For this reason, we often limit our calculations to the first harmonic even though we may draw the waveforms in rectangular shape rather than sinusoidal shape.

Problems

Unless otherwise stated, in the following problems, use the device data shown in Table 2.1 and assume $V_{DD} = 3$ V where necessary. Also, assume all transistors are in saturation.

14.1. For the circuit of Fig. 14.6, determine the open-loop tranfer function and calculate the phase margin. Assume $g_{m1} = g_{m2} = g_m$ and neglect other capacitances.

14.2. In the circuit of Fig. 14.8, assume $g_{m1} = g_{m2} = g_{m3} = (200 \ \Omega)^{-1}$.
  (a) What is the minimum value of R_D that ensures oscillation?
  (b) Determine the value of C_L for an oscillation frequency of 1 GHz and a total low-frequency loop gain of 16.

14.3. For the circuit of Fig. 14.12, determine the minimum value of I_{SS} that guarantees oscillation. (Hint: if the circuit is at the edge of oscillation, the swings are quite small.)

14.4. Prove that the small-signal resistance of the composite load in Fig. 14.18(c) is roughly equal to $1/g_{m3}$.

14.5. Including only the gate-source capacitance of M_3 in Fig. 14.18(c), explain under what condition the impedance of the composite load (seen at the drain of M_3) becomes inductive.

14.6. If each inductor in Fig. 14.25 exhibits a series resistance of R_S, how low must R_S be to ensure the low-frequency loop gain is less than unity? (This condition is necessary to avoid latchup.)

14.7. Explain why the V_X and V_Y waveforms in Fig. 14.28 are closer to sinusoids (i.e., they contain smaller harmonics) than the I_{D1} and I_{D2} waveforms.

14.8. Determine the minimum value of I_{SS} in Fig. 14.27(c) that guarantees oscillation. Estimate the maximum value of I_{SS} that guarantees M_1 and M_2 do not enter the triode region.

14.9. Repeat Example 14.7 by applying a current stimulus to the drain of M_1.

14.10. Prove that if a capacitor C_P is placed in parallel with L_P in Fig. 14.31(a), then Eq. (14.47) results.

14.11. The Colpitts oscillator of Fig. 14.31(a) was analyzed and its oscillation conditions were derived by applying a current stimulus to the source. Repeat the analysis by applying a voltage stimulus to the gate of M_1.

14.12. Repeat the analysis of the Colpitts oscillator for the topologies in Figs. 14.38(a) and (c). Determine the oscillation condition and the frequency of oscillation.

14.13. The stage of Fig. 14.45 is designed with $I_T = 1$ mA and $(W/L)_{1,2} = 50/0.5$. Assume $I_H \ll I_1$.
  (a) Determine the minimum value of $R_1 = R_2 = R$ to ensure oscillation in a three-stage ring.
  (b) Determine $(W/L)_{3,4}$ such that $g_{m3,4}R = 0.5$ when each of M_3 and M_4 carries $I_T/2$.
  (c) Calculate the minimum value of I_H to guarantee oscillation.
  (d) If the common-mode level of V_{cont1} and V_{cont2} is 1.5 V, calculate $(W/L)_{5,6}$ such that I_T sustains 0.5 V when $V_{cont1} = V_{cont2}$.

14.14. Repeat Example 14.14 if each inductor in the circuit contributes a constant capacitance equal to C_1.

14.15. The VCO of Fig. 14.53 is designed for operation at 1 GHz.
  (a) If $L_P = 5$ nH and the total (fixed) parasitic capacitance seen at X (and Y) to ground is 500 fF, determine the maximum capacitance that D_1 and D_2 can add to the circuit.
  (b) If the tail current is equal to 1 mA and the Q of each inductor at 1 GHz is equal to 4, estimate the output voltage swing.

References

1. N. M. Nguyen and R. G. Meyer, "Start-up and Frequency Stability in High-Frequency Oscillators," *IEEE Journal of Solid-State Circuits,* vol. 27, pp. 810–820, May 1992.
2. I. A. Young, J. K. Greason, and K. L. Wong, "A PLL Clock Generator with 5 to 110 MHz of Lock Range for Microprocessors," *IEEE Journal of Solid-State Circuits,* vol. SC-27, pp.1599–1607, Nov. 1992.
3. B. Lai and R. C. Walker, "A Monolithic 622 Mb/sec Clock Extraction and Data Retiming Circuit," *ISSCC Dig. Tech. Papers,* pp. 144–145, Feb. 1991.
4. S. K. Enam and A. A. Abidi, "NMOS ICs for Clock and Data Regeneration in Gigabit-per-Second Optical-Fiber Receivers," *IEEE Journal of Solid-State Circuits,* vol. SC-27, pp. 1763–1774, Dec. 1992.

Chapter 15

Phase-Locked Loops

The concept of phase locking was invented in the 1930s and swiftly found wide usage in electronics and communication. While the basic phase-locked loop has remained nearly the same since then, its implementation in different technologies and for different applications continues to challenge designers. A PLL serving the task of clock generation in a microprocessor appears quite similar to a frequency synthesizer used in a cellphone, but the actual circuits are designed quite differently.

This chapter deals with the analysis and design of PLLs with particular attention to implementations in VLSI technologies. A thorough study of PLLs would require an entire book by itself, but our objective here is to lay the foundation for more advanced work. Beginning with a simple PLL architecture, we study the phenomenon of phase locking and analyze the behavior of PLLs in the time and frequency domains. We then address the problem of lock acquisition and describe charge-pump PLLs (CPPLLs) and their nonidealities. Finally, we examine jitter in PLLs, study delay-locked loops (DLLs), and present a number of PLL applications.

15.1 Simple PLL

A PLL is a feedback system that compares the output phase with the input phase. The comparison is performed by a "phase comparator" or "phase detector" (PD). It is therefore beneficial to define the PD rigorously.

15.1.1 Phase Detector

A phase detector is a circuit whose average output, $\overline{V_{out}}$, is linearly proportional to the phase difference, $\Delta\phi$, between its two inputs (Fig. 15.1). In the ideal case, the relationship between $\overline{V_{out}}$ and $\Delta\phi$ is linear, crossing the origin for $\Delta\phi = 0$. Called the "gain" of the PD, the slope of the line, K_{PD}, is expressed in V/rad.

A familiar example of phase detector is the exclusive OR (XOR) gate. As shown in Fig. 15.2, as the phase difference between the inputs varies, so does the width of the output pulses, thereby providing a dc level proportional to $\Delta\phi$. While the XOR circuit produces

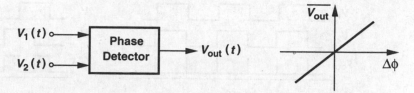

Figure 15.1   Definition of phase detector.

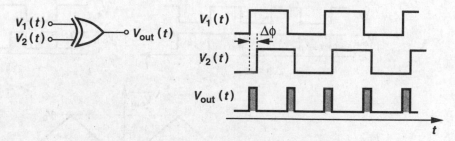

Figure 15.2   Exclusive OR gate as phase detector.

error pulses on both rising and falling edges, other types of PD may respond only to positive or negative transitions.

Example 15.1

If the output swing of the XOR in Fig. 15.2 is V_0 volts, what is the gain of the circuit as a phase detector? Plot the input-output characteristic of the PD.

Solution

If the phase difference increases from zero to $\Delta\phi$ radians, the area under each pulse increases by $V_0 \cdot \Delta\phi$. Since each period contains *two* pulses, the average value rises by $2[V_0 \cdot \Delta\phi/(2\pi)]$, yielding a gain of V_0/π. Note that the gain is independent of the input frequency.

To construct the input-output characteristic, we examine the circuit's response to various input phase differences. As illustrated in Fig. 15.3, the average output voltage rises to $[V_0/\pi] \times \pi/2 = V_0/2$ for $\Delta\phi = \pi/2$ and V_0 for $\Delta\phi = \pi$. For $\Delta\phi > \pi$, the average begins to *drop*, falling to $V_0/2$ for $\Delta\phi = 3\pi/2$ and zero for $\Delta\phi = 2\pi$. The characteristic is therefore periodic, exhibiting both negative and positive gains.

The operation of phase detectors is similar to that of differential amplifiers in that both sense the *difference* between the two inputs, generating a proportional output.

15.1.2 Basic PLL Topology

To arrive at the concept of phase locking, let us consider the problem of aligning the output phase of a VCO with the phase of a reference clock. As illustrated in Fig. 15.4(a), the rising

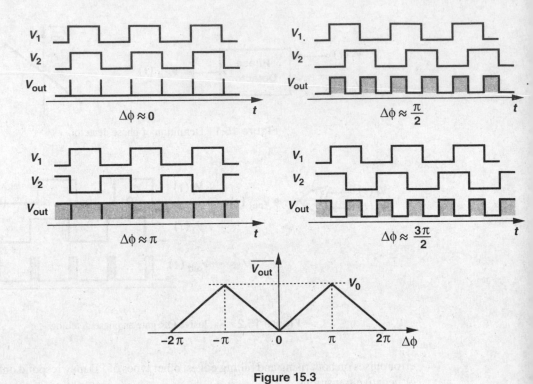

Figure 15.3

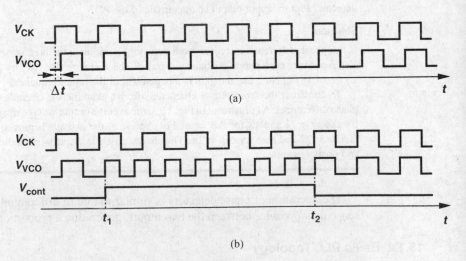

Figure 15.4 (a) Two waveforms with a skew, (b) change of VCO frequency to eliminate the skew.

edges of V_{VCO} are "skewed" by Δt seconds with respect to V_{CK}, and we wish to eliminate this error. Assuming that the VCO has a single control input, V_{cont}, we note that to vary the phase, we *must* vary the frequency and allow the integration $\phi = \int (\omega_0 + K_{VCO} V_{cont}) \, dt$ to take place. For example, suppose as shown in Fig. 15.4(b), the VCO frequency is stepped to a higher value at $t = t_1$. The circuit then accumulates phase faster, gradually decreasing the phase error. At $t = t_2$, the phase error drops to zero and, if V_{cont} returns to its original value, V_{VCO} and V_{CK} remain aligned. Interestingly, the alignment can be accomplished by stepping the VCO frequency to a *lower* value for a certain time interval as well (Problem 15.2). Thus, phase alignment can be achieved only by a (temporary) frequency change.

The foregoing experiment suggests that the output phase of a VCO can be aligned with the phase of a reference if (1) the frequency of the VCO is changed momentarily, (2) a means of comparing the two phases, i.e., a phase detector, is used to determine when the VCO and reference signals are aligned. The task of aligning the output phase of the VCO with the phase of the reference is called "phase locking."

From the above observations, we surmise that a PLL simply consists of a PD and a VCO in a feedback loop [Fig. 15.5(a)]. The PD compares the phases of V_{out} and V_{in}, generating an error that varies the VCO frequency until the phases are aligned, i.e., the loop is locked.

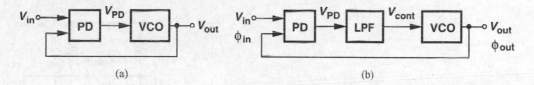

(a)   (b)

Figure 15.5 (a) Feedback loop comparing input and output phases, (b) simple PLL.

This topology, however, must be modified because (1) as exemplified by the waveforms of Fig. 15.2, the PD output, V_{PD}, consists of a dc component (desirable) and high-frequency components (undesirable), and (2) as mentioned in Chapter 14, the control voltage of the oscillator must remain quiet in the steady state, i.e., the PD output must be filtered. We therefore interpose a low-pass filter (LPF) between the PD and the VCO [Fig. 15.5(b)], suppressing the high-frequency components of the PD output and presenting the dc level to the oscillator. This forms the basic PLL topology. For now, we assume the LPF has a gain of unity at low frequencies (e.g., as in a first-order RC section).

It is important to bear in mind that the feedback loop of Fig. 15.5(b) compares the *phases* of the input and output. Unlike the feedback topologies studied in the previous chapters, PLLs typically require no knowledge of voltages or currents in their feedback operation. If the loop gain is large enough, the difference between the input phase, ϕ_{in}, and the output phase, ϕ_{out}, falls to a small value in the steady state, providing phase alignment.

For subsequent analyses of PLLs, we must define the phase lock condition carefully. If the loop of Fig. 15.5(b) is locked, we postulate that $\phi_{out} - \phi_{in}$ is constant and preferably small. We therefore define the loop to be locked if $\phi_{out} - \phi_{in}$ does not change with time.

An important corollary of this definition is that

$$\frac{d\phi_{out}}{dt} - \frac{d\phi_{in}}{dt} = 0 \qquad (15.1$$

and hence

$$\omega_{out} = \omega_{in}. \qquad (15.2$$

This is a unique property of PLLs and will be revisited more closely later.

In summary, when locked, a PLL produces an output that has a small phase error wit respect to the input but exactly the same frequency. The reader may then wonder why PLL is used at all. A short piece of wire would seem to perform the task even better! W answer this question in Section 15.5.

Example 15.2

Implement a simple PLL in CMOS technology.

Solution

Figure 15.6 illustrates an implementation utilizing an XOR gate as the phase detector. The VCO

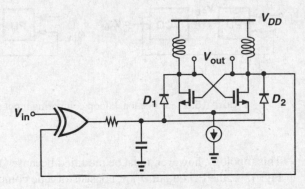

Figure 15.6

configured as a negative-G_m LC oscillator whose frequency is tuned by varactor diodes.

PLL Waveforms in Locked Condition

In order to familiarize ourselves with th behavior of PLLs, we begin with the simplest case: the circuit is locked and we wish examine the waveforms at each point around the loop. As illustrated in Fig. 15.7(a), V and V_{out} exhibit a small phase difference but equal frequencies. The PD therefore generate pulses as wide as the skew between the input and the output[1] and the low-pass filter extrac the dc component of V_{PD}, applying the result to the VCO. We assume the LPF has a ga of unity at low frequencies. The small pulses in V_{LPF} are called "ripple."

---

[1]In this example, the PD produces pulses only on the rising transitions.

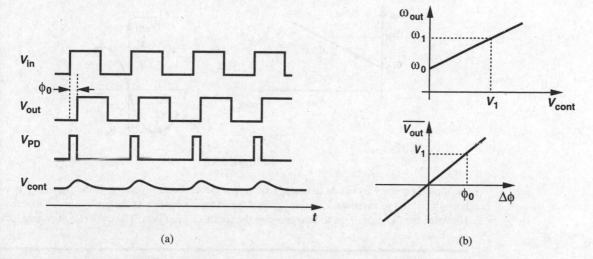

Figure 15.7   (a) Waveforms in a PLL in locked condition, (b) calculation of phase error.

In the waveforms of Fig. 15.7(a), two quantities are unknown: ϕ_0 and the dc level of V_{cont}. To determine these values, we construct the VCO and PD characteristics [Fig. 15.7(b)]. If the input and output frequencies are equal to ω_1, then the required oscillator control voltage is unique and equal to V_1. This voltage must be produced by the phase detector, demanding a phase error determined by the PD characteristic. More specifically, since $\omega_{out} = \omega_0 + K_{VCO}V_{cont}$ and $\overline{V_{PD}} = K_{PD}\Delta\phi$, we can write

$$V_1 = \frac{\omega_1 - \omega_0}{K_{VCO}}, \tag{15.3}$$

and

$$\phi_0 = \frac{V_1}{K_{PD}} \tag{15.4}$$

$$= \frac{\omega_1 - \omega_0}{K_{PD}K_{VCO}}. \tag{15.5}$$

Equation (15.5) reveals two important points: (1) as the input frequency of the PLL varies, so does the phase error; (2) to minimize the phase error, $K_{PD}K_{VCO}$ must be maximized.

Example 15.3 _____

A PLL incorporates a VCO and a PD having the characteristics shown in Fig. 15.8. Explain what happens as the input frequency varies in the locked condition.

Solution

The PD characteristic is relatively linear near the origin but exhibits a small-signal gain of zero if the phase difference equals $\pm\pi/2$, at which point the average output is equal to $\pm V_0$. Now suppose

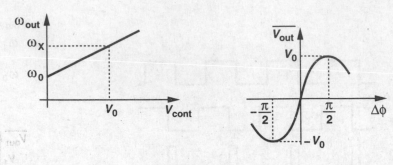

Figure 15.8

the input frequency increases from ω_0, requiring a greater control voltage. If the frequency is high enough $(= \omega_x)$ to mandate $V_{cont} = V_0$, then the PD must operate at the peak of its characteristic. However, the PD gain drops to zero here and the feedback loop fails. Thus, the circuit cannot lock if $\omega_{in} = \omega_X$.

With the basic understanding of PLLs developed thus far, we now return to Eq. (15.2). The exact equality of the input and output frequencies of a PLL in the locked condition is a critical attribute. The significance of this property can be seen from two observations. First, in many applications, even a very small (deterministic) frequency error may prove unacceptable. For example, if a data stream is to be processed synchronously by a clocked system, even a slight difference between the data rate and the clock frequency results in "drift," creating errors (Fig. 15.9). Second, the equality would *not* exist if the PLL compared the input and output frequencies rather than phases. As illustrated in Fig. 15.10(a), a loop employing a frequency detector (FD) would suffer from a finite difference between ω_i and ω_{out} due to various mismatches and other nonidealities. This can be understood by an analogy with the unity-gain feedback circuit of Fig. 15.10(b). Even if the op amp

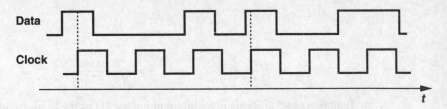

Figure 15.9   Drift of data with respect to clock in the presence of small frequency error.

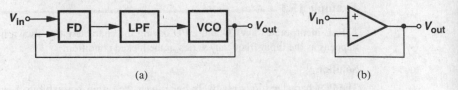

Figure 15.10   (a) Frequency-locked loop, (b) unity-gain feedback amplifier.

open-loop gain is infinity, the input-referred offset voltage leads to a finite error between V_{in} and V_{out}.

Small Transients in Locked Condition   Let us now analyze the response of a PLL in locked condition to small phase or frequency transients at the input.

Consider a PLL in the locked condition and assume the input and output waveforms can be expressed as

$$V_{in}(t) = V_A \cos \omega_1 t \tag{15.6}$$

$$V_{out}(t) = V_B \cos(\omega_1 t + \phi_0), \tag{15.7}$$

where higher harmonics are neglected and ϕ_0 is the static phase error. Suppose, as shown in Fig. 15.11, the input experiences a phase step of ϕ_1 at $t = t_1$, i.e., $\phi_{in} = \omega_1 t + \phi_1 u(t - t_1)$.[2] Since the output of the LPF does not change instantaneously, the VCO initially continues to

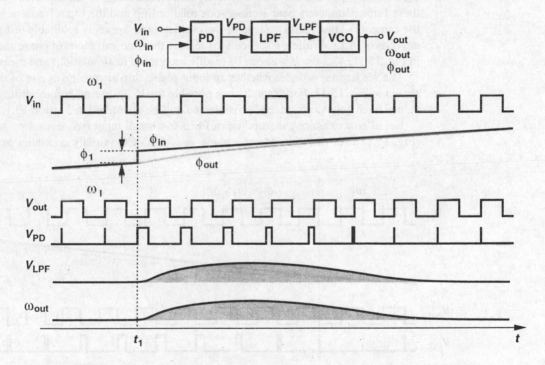

Figure 15.11   Response of a PLL to a phase step.

oscillate at ω_1. The growing phase difference between the input and the output then creates wide pulses at the output of the PD, forcing V_{LPF} to rise gradually. As a result, the VCO frequency begins to change, attempting to minimize the phase error. Note that the loop is not locked during the transient because the phase error varies with time.

---

[2]In this example, ϕ_{in} and ϕ_{out} denote the *total* phases of the input and output, respectively.

What happens after the VCO frequency begins to change? If the loop is to return to lock, ω_{out} must eventually go back to ω_1, requiring that V_{LPF} and hence $\phi_{out} - \phi_{in}$ also return to their original values. Since ϕ_{in} has changed by ϕ_1, the variation in the VCO frequency is such that the *area* under ω_{out} provides an additional phase of ϕ_1 in ϕ_{out}:

$$\int_{t1}^{\infty} \omega_{out}dt = \phi_1. \tag{15.8}$$

Thus, when the loop settles, the output becomes equal to

$$V_{out}(t) = V_B \cos[\omega_1 t + \phi_0 + \phi_1 u(t - t_1)]. \tag{15.9}$$

Consequently, as shown in Fig. 15.11, ϕ_{out} gradually "catches up" with ϕ_{in}.

It is important to make two observations. (1) After the loop returns to lock, *all* of the parameters (except for the total input and output phases) assume their original values. That is, $\phi_{in} - \phi_{out}$, V_{LPF}, and the VCO frequency remain unchanged—an expected result because these three parameters bear a one-to-one relationship and the input frequency has stayed the same. (2) The control voltage of the oscillator can serve as a suitable test point in the analysis of PLLs. While it is difficult to measure the time variations of phase and frequency in Fig. 15.11, $V_{cont}(= V_{LPF})$ can be readily monitored in simulations and measurements.

The reader may wonder whether an input phase step always gives rise to the response shown in Fig. 15.11. For example, is it possible for V_{LPF} to ring before settling to its final value? Such behavior is indeed possible and will be quantified in Section 15.1.3.

Let us now examine the response of PLLs to a small input frequency step $\Delta\omega$ at $t = t_1$ (Fig. 15.12). As with the case of a phase step, the VCO initially continues to oscillate at

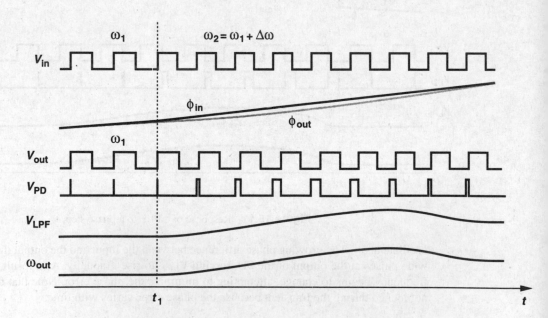

Figure 15.12   Response of a PLL to a small frequency step.

ω_1. Thus, the PD generates increasingly wider pulses, and V_{LPF} rises with time. As ω_{out} approaches $\omega_1 + \Delta\omega$, the width of the pulses generated by the PD decreases, eventually settling to a value that produces a dc component equal to $(\omega_1 + \Delta\omega - \omega_0)/K_{VCO}$. In contrast to the case of phase step, the response of a PLL to a frequency step entails a permanent change in both the control voltage and the phase error. If the input frequency is varied slowly, ω_{out} simply "tracks" ω_{in}.

The exact settling behavior of PLLs depends on the various loop parameters and will be studied in Section 15.1.3. But, to arrive at an important observation, we consider the phase step response depicted in Fig. 15.13, where V_{cont} rings before settling to its final value.

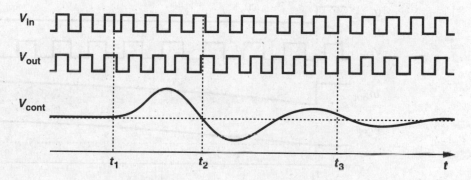

Figure 15.13 Example of phase step response.

Consider the state of the loop at $t = t_2$. At this point, the output frequency is equal to its final value (because V_{cont} is equal to its final value) but the loop continues the transient because the phase error deviates from the required value. Similarly, at $t = t_3$, the phase error is equal to its final value but the output frequency is not. In other words, for the loop to settle, both the phase and the frequency must settle to proper values.

Example 15.4

Consider the PLL shown in Fig. 15.14, where an external voltage V_{ex} is added to the output of the low-pass filter.[3] (a) Determine the phase error and V_{LPF} if the loop is locked and $V_{ex} = V_1$. (b) Suppose V_{ex} steps from V_1 to V_2 at $t = t_1$. How does the loop respond?

Solution

(a) If the loop is locked, $\omega_{out} = \omega_{in}$ and $V_{cont} = (\omega_{in} - \omega_0)/K_{VCO}$. Thus, $V_{LPF} = (\omega_{in} - \omega_0)/K_{VCO} - V_1$ and $\Delta\phi = V_{LPF}/K_{PD} = (\omega_{in} - \omega_0)/(K_{PD}K_{VCO}) - V_1/K_{PD}$.

(b) When V_{ex} steps from V_1 to V_2, V_{cont} immediately goes from $(\omega_{in} - \omega_0)/K_{VCO}$ to $(\omega_{in} - \omega_0)/K_{VCO} + (V_2 - V_1)$, changing the VCO frequency to $\omega_{in} - K_{VCO}(V_1 - V_2)$. Since V_{LPF} cannot change instantaneously, the PD begins to generate increasingly wider pulses, raising V_{LPF} and increasing ω_{out}. When the loop returns to lock, ω_{out} becomes equal to ω_{in} and $V_{LPF} = (\omega_{in} - \omega_0)/K_{VCO} - V_2$. The phase error also changes to $(\omega_{in} - \omega_0)/(K_{PD}K_{VCO}) - V_2/K_{PD}$. Note that the

---

[3]This topology is used for some types of frequency modulation in wireless communication.

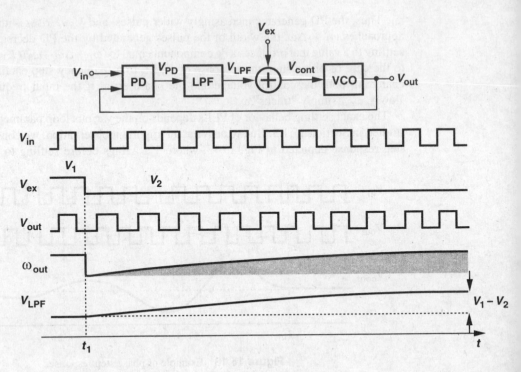

Figure 15.14

area under ω_{out} during the transient is equal to the change in the output phase and hence the change in the phase error:

$$\int_{t1}^{\infty} \omega_{out} dt = \frac{V_1 - V_2}{K_{PD}}. \tag{15.10}$$

From our study thus far, we conclude that phase-locked loops are "dynamic" systems, i.e., their response depends on the past values of the input and output. This is to be expected because the low-pass filter and the VCO introduce poles (and possibly zeros) in the loop transfer function. Moreover, we note that, so long as the input and the output remain perfectly periodic (i.e., $\phi_{in} = \omega_{in}t$ and $\phi_{out} = \omega_{in}t + \phi_0$), the loop operates in the steady state, exhibiting no transient. Thus, the PLL only responds to variations in the *excess* phase of the input or output. For example, in Fig. 15.11, $\phi_{in} = \omega_1 t + \phi_1 u(t - t_1)$ and in Fig. 15.12, $\phi_{in} = \omega_1 t + \Delta\omega \cdot t u(t - t_1)$.

15.1.3 Dynamics of Simple PLL

With the qualitative analysis of PLLs in the previous section, we can now study their transient behavior more rigorously. Assuming the loop is initially locked, we treat the PLL as a feedback system but recognize that the output quantity in this analysis must be the (excess) phase of the VCO because the "error amplifier" can only compare phases. Our objective is to determine the transfer function $\Phi_{out}(s)/\Phi_{in}(s)$ for both open-loop and closed-loop systems and subsequently study the time-domain response. Note that the dimensions change from phase to voltage through the PD and from voltage to phase through the VCO.

What does $\Phi_{out}(s)/\Phi_{in}(s)$ signify? An analogy with more familiar transfer functions proves useful here. A circuit having a transfer function $V_{out}(s)/V_{in}(s) = 1/(1 + s/\omega_0)$ is considered a low-pass filter because if V_{in} varies rapidly, V_{out} cannot fully track the input variations. Similarly, $\Phi_{out}(s)/\Phi_{in}(s)$ reveals how the output phase tracks the input phase if the latter changes slowly or rapidly.

To visualize the variation of the excess phase with time, consider the waveforms in Fig. 15.15. The period varies slowly in Fig. 15.15(a) and rapidly in Fig. 15.15(b). Thus, $y_2(t)$ experiences faster phase variation than does $y_2(t)$.

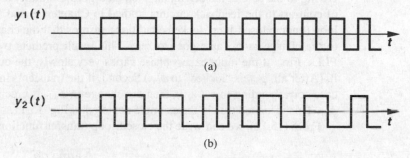

Figure 15.15   Slow and fast variation of the excess phase.

Let us construct a linear model of the PLL, assuming a first-order low-pass filter for simplicity. The PD output contains a dc component equal to $K_{PD}(\phi_{out} - \phi_{in})$ as well as high-frequency components. Since the latter are suppressed by the LPF, we simply model the PD by a subtractor whose output is "amplified" by K_{PD}. Illustrated in Fig. 15.16, the overall PLL model consists of the phase subtractor, the LPF transfer function $1/(1 + s/\omega_{LPF})$,

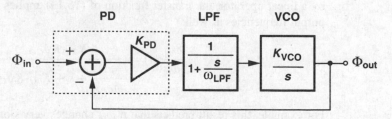

Figure 15.16   Linear model of type I PLL.

where ω_{LPF} denotes the -3-dB bandwidth, and the VCO transfer function K_{VCO}/s. Here, Φ_{in} and Φ_{out} denote the excess phases of the input and output waveforms, respectively. For example, if the total input phase experiences a step change, $\phi_1 u(t)$, then $\Phi_{in}(s) = \phi_1/s$.

The open-loop transfer function is given by

$$H(s)|_{\text{open}} = \frac{\Phi_{out}}{\Phi_{in}}(s)|_{\text{open}} \tag{15.11}$$

$$= K_{PD} \cdot \frac{1}{1 + \dfrac{s}{\omega_{LPF}}} \cdot \frac{K_{VCO}}{s}, \tag{15.12}$$

revealing one pole at $s = -\omega_{LPF}$ and another at $s = 0$. Note that the loop gain is equal to $H(s)|_{\text{open}}$ because of the unity feedback factor. Since the loop gain contains a pole at the origin, the system is called "type I."

Before computing the closed-loop transfer function, let us make an important observation. What is the loop gain if s is very small, i.e., if the input excess phase varies very slowly? Owing to the pole at the origin, the loop gain goes to infinity as s approaches zero, a point of contrast to the feedback circuits studied in Chapters 8 and 10. Thus, the phase-locked loop (under closed-loop, locked condition) ensures that the change in ϕ_{out} is *exactly* equal to the change in ϕ_{in} as s goes to zero. This result predicts two interesting properties of PLLs. First, if the input excess phase varies very slowly, the output excess phase "tracks" it. (After all, ϕ_{out} is "locked" to ϕ_{in}.) Second, if the transients in ϕ_{in} have decayed (another case corresponding to $s \to 0$), then the change in ϕ_{out} is precisely equal to the change in ϕ_{in}. This is indeed true in the example depicted in Fig. 15.11.

From (15.12), we can write the closed-loop transfer function as:

$$H(s)|_{\text{closed}} = \frac{K_{PD}K_{VCO}}{\dfrac{s^2}{\omega_{LPF}} + s + K_{PD}K_{VCO}}. \tag{15.13}$$

For the sake of brevity, we hereafter denote $H(s)|_{\text{closed}}$ simply by $H(s)$ or Φ_{out}/Φ_{in}. As expected, if $s \to 0$, $H(s) \to 1$ because of the infinite loop gain.

In order to analyze $H(s)$ further, we derive a relationship that allows a more intuitive understanding of the system. Recall that the instantaneous frequency of a waveform is equal to the time derivative of the phase: $\omega = d\phi/dt$. Since the frequency and the phase are related by a linear operator, the transfer function of (15.13) applies to variations in the input and output frequencies as well:

$$\frac{\omega_{out}}{\omega_{in}}(s) = \frac{K_{PD}K_{VCO}}{\dfrac{s^2}{\omega_{LPF}} + s + K_{PD}K_{VCO}}. \tag{15.14}$$

For example, this result predicts that if ω_{in} changes very slowly ($s \to 0$), then ω_{out} tracks ω_{in}, again an expected result because the loop is assumed locked. Equation (15.14) also indicates that if ω_{in} changes abruptly but the system is given enough time to settle ($s \to 0$), then the change in ω_{out} equals that in ω_{in} (as illustrated in the example of Fig. 15.12).

The above observation aids the analysis in two directions. First, some transient responses of the closed-loop system may be simpler to visualize in terms of changes in the frequency quantities rather than phase quantities. Second, since a change in ω_{out} must be accompanied by a change in V_{cont}, we have

$$H(s) = K_{VCO} \cdot \frac{V_{cont}}{\omega_{in}}(s). \tag{15.15}$$

That is, monitoring the response of V_{cont} to variations in ω_{in} indeed yields the response of the closed-loop system.

The second-order transfer function of (15.13) suggests that the step response of the type I system can be overdamped, critically damped, or underdamped. To derive the condition for each case, we rewrite the denominator in a familiar form used in control theory, $s^2 + 2\zeta\omega_n s + \omega_n^2$, where ζ is the "damping ratio" and ω_n is the "natural frequency." That is,

$$H(s) = \frac{\omega_n^2}{s^2 + 2\zeta\omega_n s + \omega_n^2}, \tag{15.16}$$

where

$$\omega_n = \sqrt{\omega_{LPF}K_{PD}K_{VCO}} \tag{15.17}$$

$$\zeta = \frac{1}{2}\sqrt{\frac{\omega_{LPF}}{K_{PD}K_{VCO}}}. \tag{15.18}$$

The two poles of the closed-loop system are given by

$$s_{1,2} = -\zeta\omega_n \pm \sqrt{(\zeta^2 - 1)\omega_n^2} \tag{15.19}$$

$$= (-\zeta \pm \sqrt{\zeta^2 - 1})\omega_n. \tag{15.20}$$

Thus, if $\zeta > 1$, both poles are real, the system is overdamped, and the transient response contains two exponentials with time constants $1/s_1$ and $1/s_2$. On the other hand, if $\zeta < 1$, the poles are complex and the response to an input frequency step $\omega_{in} = \Delta\omega u(t)$ is equal to

$$\omega_{out}(t) = \left\{1 - e^{-\zeta\omega_n t}[\cos(\omega_n\sqrt{1-\zeta^2}t) + \frac{\zeta}{\sqrt{1-\zeta^2}}\sin(\omega_n\sqrt{1-\zeta^2}t)]\right\}\Delta\omega u(t) \tag{15.21}$$

$$= [1 - \frac{1}{\sqrt{1-\zeta^2}}e^{-\zeta\omega_n t}\sin(\omega_n\sqrt{1-\zeta^2}t + \theta)]\Delta\omega u(t), \tag{15.22}$$

where ω_{out} denotes the change in the output frequency and $\theta = \sin^{-1}\sqrt{1-\zeta^2}$. Thus, as shown in Fig. 15.17, the step response contains a sinusoidal component with a frequency $\omega_n\sqrt{1-\zeta^2}$ that decays with a time constant $(\zeta\omega_n)^{-1}$. Note that the system exhibits the same response if a phase step is applied to the input and the output phase is observed.

The settling speed of PLLs is of great concern in most applications. Equation (15.22) indicates that the exponential decay determines how fast the output approaches its final

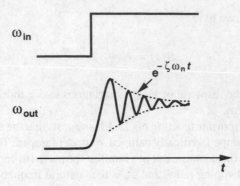

Figure 15.17   Underdamped response of PLL to a frequency step.

value, implying that $\zeta \omega_n$ must be maximized. For the type I PLL under study here, (15.17) and (15.18) yield

$$\zeta \omega_n = \frac{1}{2} \omega_{LPF}.$$

(15.23)

This result reveals a critical trade-off between the settling speed and the ripple on the VCO control line: the lower ω_{LPF}, the greater the suppression of the high-frequency components produced by the PD but the longer the settling time constant.

Example 15.5

A cellular telephone incorporates a 900-MHz phase-locked loop to generate the carrier frequencies. If $\omega_{LPF} = 2\pi \times (20 \text{ kHz})$ and the output frequency is to be changed from 901 MHz to 901.2 MHz, how long does the PLL output frequency take to settle within 100 Hz of its final value?

Solution

Since the step size is 200 kHz, we have

$$[1 - e^{-\zeta \omega_n t_s} \sin(\omega_n \sqrt{1 - \zeta^2} t_s + \theta)] \times 200 \text{ kHz} = 200 \text{ kHz} - 100 \text{ Hz}.$$

(15.24)

Thus,

$$e^{-\zeta \omega_n t_s} \sin(\omega_n \sqrt{1 - \zeta^2} t_s + \theta) = \frac{100 \text{ Hz}}{200 \text{ kHz}}.$$

(15.25)

In the worst case, the sinusoid is equal to unity and

$$e^{-\zeta \omega_n t_s} = 0.0005.$$

(15.26)

That is,

$$t_s = \frac{7.6}{\zeta \omega_n}$$

(15.27)

$$= \frac{15.2}{\omega_{LPF}}$$

(15.28)

$$= 0.12 \text{ ms}.$$

(15.29)

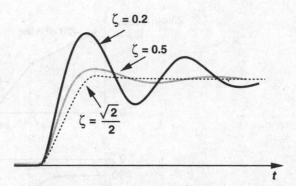

Figure 15.18 Underdamped response of a second-order system for various values of ζ.

In addition to the product $\zeta\omega_n$, the value of ζ itself is also important. Illustrated in Fig. 15.18 for several values of ζ and a constant ω_n, the step response exhibits severe ringing for $\zeta < 0.5$. In view of process and temperature variation of the loop parameters, ζ is usually chosen to be greater than $\sqrt{2}/2$ or even 1 to avoid excessive ringing.[4]

The choice of ζ entails other trade-offs as well. First, (15.18) implies that as ω_{LPF} is reduced to minimize the ripple on the control voltage, the stability degrades. Second, (15.5) and (15.18) indicate that both the phase error and ζ are inversely proportional to $K_{PD}K_{VCO}$; lowering the phase error inevitably makes the system less stable. In summary, the type I PLL suffers from trade-offs between the settling speed, the ripple on the control voltage (i.e., the quality of the output signal), the phase error, and the stability.

The stability behavior of PLLs can also be analyzed graphically, providing more insight. Recall from Chapter 10 that the Bode plots of the magnitude and phase of the loop gain readily yield the phase margin. Let us utilize (15.12) to construct such plots. As shown in Fig. 15.19, the loop gain begins from infinity at $\omega = 0$ and falls at a rate of 20 dB/dec for $\omega < \omega_{LPF}$ and at a rate of 40 dB/dec thereafter. The phase begins at $-90°$ and asymptotically reaches $-180°$.

What happens if a higher $K_{PD}K_{VCO}$ is chosen so as to minimize $\phi_{out} - \phi_{in}$? Since the entire gain plot in Fig. 15.19 is shifted up, the gain crossover moves to the right, thus degrading the phase margin. This is consistent with the dependence of ζ upon $K_{PD}K_{VCO}$.

As observed thus far, $K_{PD}K_{VCO}$ impacts many important parameters of PLLs. This quantity is sometimes called the loop gain (even though it is not dimensionless) due to the resemblance of $\Delta\phi = (\omega_{out} - \omega_0)/(K_{PD}K_{VCO})$ to the error equation in a feedback system.

The stability behavior of type I PLLs can also be analyzed by the locus of their poles in the complex plane as the parameter $K_{PD}K_{VCO}$ varies (Fig. 15.20). With $K_{PD}K_{VCO} = 0$,

---

[4]The value of ζ may also yield peaking in the transfer function. Thus, some applications require a ζ of 5 to 10 to avoid peaking in the presence of higher order poles.

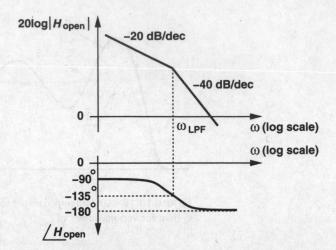

Figure 15.19 Bode plots of type I PLL.

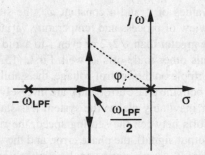

Figure 15.20 Root locus of type I PLL.

the loop is open, $\zeta = \infty$, and the two poles are given by $s_1 = -\omega_{LPF}$ and $s_2 = 0$. As $K_{PD}K_{VCO}$ increases (i.e., the feedback becomes stronger), ζ drops and the two poles, given by $s_{1,2} = (-\zeta \pm \sqrt{\zeta^2 - 1})\omega_n$, move toward each other on the real axis. For $\zeta = 1$ (i.e., $K_{PD}K_{VCO} = \omega_{LPF}/4$), $s_1 = s_2 = -\zeta\omega_n = -\omega_{LPF}/2$. As $K_{PD}K_{VCO}$ increases further, the two poles become complex, with a real part equal to $-\zeta\omega_n = -\omega_{LPF}/2$, moving in parallel with the $j\omega$ axis.

We recognize from Fig. 15.20 that, as s_1 and s_2 move away from the real axis, the system becomes less stable. In fact, the reader can prove that $\cos\varphi = \zeta$ (Problem 15.8), concluding that as φ approaches $90°$, ζ drops to zero.

Another transfer function that reveals the settling behavior of PLLs is that of the error at the output of the phase subtractor in Fig. 15.16. Defined as $H_e(s) = (\phi_{in} - \phi_{out})/\phi_{in}$, this transfer function can be obtained by noting that $\phi_{out}/\phi_{in} = H(s)$ and, from (15.13),

$$H_e(s) = 1 - H(s) \tag{15.30}$$

$$= \frac{s^2 + 2\zeta\omega_n s}{s^2 + 2\zeta\omega_n s + \omega_n^2}. \tag{15.31}$$

As expected, $H_e(s) \to 0$ if $s \to 0$ because the output tracks the input when the input varies very slowly or the transient has settled.

Example 15.6

Suppose a type I PLL experiences a frequency step $\Delta\omega$ at $t = 0$. Calculate the change in the phase error.

Solution

The Laplace transform of the frequency step equals $\Delta\omega/s$. Since $H_e(s)$ relates the phase error to the input phase, we write $\Phi_{in}(s) = (\Delta\omega/s)/s = \Delta\omega/s^2$. Thus, the Laplace transform of the phase error is

$$\Phi_e(s) = H_e(s) \cdot \frac{\Delta\omega}{s^2} \tag{15.32}$$

$$= \frac{s^2 + 2\zeta\omega_n s}{s^2 + 2\zeta\omega_n s + \omega_n^2} \cdot \frac{\Delta\omega}{s^2}. \tag{15.33}$$

From the final value theorem,

$$\phi_e(t = \infty) = \lim_{s \to 0} s\Phi_e(s) \tag{15.34}$$

$$= \frac{2\zeta}{\omega_n} \Delta\omega \tag{15.35}$$

$$- \frac{\Delta\omega}{K_{PD}K_{VCO}}, \tag{15.36}$$

which agrees with (15.5).

15.2 Charge-Pump PLLs

While type I PLLs have been realized widely in discrete form, their shortcomings often prohibit usage in high-performance integrated circuits. In addition to the trade-offs between ζ, ω_{LPF}, and the phase error, type I PLLs suffer from another critical drawback: limited acquisition range.

15.2.1 Problem of Lock Acquisition

Suppose when a PLL circuit is turned on, its oscillator operates at a frequency far from the input frequency, i.e., the loop is not locked. Under what conditions does the loop "acquire" lock? The transition of the loop from unlocked to locked condition is a very nonlinear phenomenon because the phase detector senses unequal frequencies. The problem of lock acquisition in type I PLLs has been studied extensively [1, 2], but we state without proof that the "acquisition range"[5] is on the order of ω_{LPF}, that is, the loop locks only if the difference between ω_{in} and ω_{out} is less than roughly ω_{LPF}.[6]

The problem of lock acquisition further tightens the trade-offs in type I PLLs. If ω_{LPF} is reduced to suppress the ripple on the control voltage, the acquisition range decreases. Note that even if the input frequency has a precisely controlled value, a wide acquisition range is often necessary because the VCO center frequency may vary considerably with process and temperature. In most of today's applications, the acquisition range of the simple PLL studied thus far proves inadequate.

In order to remedy the acquisition problem, modern PLLs incorporate frequency detection in addition to phase detection. Called "aided acquisition" and illustrated in Fig. 15.21, the idea is to compare ω_{in} and ω_{out} by means of a frequency detector, generate a dc com-

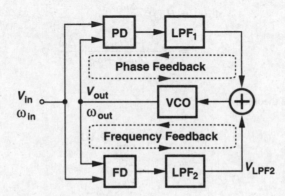

Figure 15.21 Addition of frequency detection to increase the acquisition range.

ponent V_{LPF2} proportional to $\omega_{in} - \omega_{out}$, and apply the result to the VCO in a negative-feedback loop. At the beginning, the FD drives ω_{out} toward ω_{in} while the PD output remains "quiet." When $|\omega_{out} - \omega_{in}|$ is sufficiently small, the phase-locked loop takes over, acquiring lock. Such a scheme increases the acquisition range to the tuning range of the VCO.[7]

15.2.2 Phase/Frequency Detector and Charge Pump

For periodic signals, it is possible to merge the two loops of Fig. 15.21 by devising a circuit that can detect both phase and frequency differences. Called a phase/frequency detector (PFD) and illustrated conceptually in Fig. 15.22, the circuit employs sequential logic to create three states and respond to the rising (or falling) edges of the two inputs. If initially $Q_A = Q_B = 0$, then a rising transition on A leads to $Q_A = 1$, $Q_B = 0$. The circuit remains

---

[5] Acquisition range, tracking range, lock range, capture range, and pull-in range are often used to describe the behavior of PLLs in the presence of input or VCO frequency variation. For our purposes, the acquisition range, the capture range, and the pull-in range are the same. The tracking range refers to the input frequency range across which a locked PLL can track the input. With the addition of frequency detection, the acquisition range becomes equal to the tracking range (for periodic signals).

[6] This is a very rough estimate. In practice, the acquisition range may be several times narrower or wider. It is also assumed that the tuning range of the VCO is large enough not to limit the acquisition range.

[7] This may not be true if the input is not periodic.

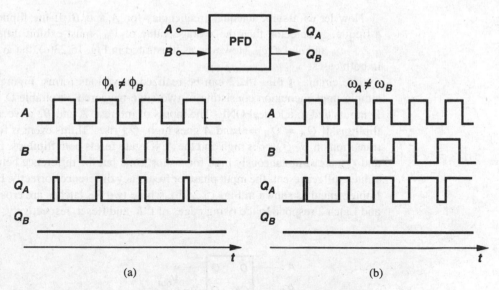

Figure 15.22 Conceptual operation of a PFD.

in this state until B goes high, at which point Q_A returns to zero. The behavior is similar for the B input.

In Fig. 15.22(a), the two inputs have equal frequencies but A leads B. The output Q_A continues to produce pulses whose width is proportional to $\phi_A - \phi_B$ while Q_B remains at zero. In Fig. 15.22(b), A has a higher frequency than B and Q_A generates pulses while Q_B does not. By symmetry, if A lags B or has a lower frequency than B, then Q_B produces pulses and Q_A remains quiet. Thus, the dc contents of Q_A and Q_B provide information about $\phi_A - \phi_B$ or $\omega_A - \omega_B$. The outputs Q_A and Q_B are called the "UP" and "DOWN" pulses, respectively.

Example 15.7

Explain whether a master-slave D flipflop can operate as a phase detector or a frequency detector. Assume the flipflop provides differential outputs.

Solution

As shown in Fig. 15.23(a), we first apply inputs having equal frequencies and a finite phase difference, assuming the output changes on the rising edge of the clock input. If A leads B, then V_{out} remains at a logical ONE indefinitely because the flipflop continues to sample the high levels of A. Conversely, if A lags B, then V_{out} remains low. Plotted in Fig. 15.23(b), the input-output characteristic of the circuit displays a very high gain at $\Delta\phi = 0, \pm\pi, \cdots$ and a zero gain at other values of $\Delta\phi$. The D flipflop is sometimes called a "bang-bang" phase detector to emphasize that the average value of V_{out} jumps from $-V_1$ to $+V_1$ as $\Delta\phi$ varies from slightly below zero to slightly above zero.

Now let us assume unequal frequencies for A and B. If the flipflop is to behave as a frequency detector, then the average value of V_{out} must exhibit different polarities for $\omega_A > \omega_B$ and $\omega_A < \omega_B$. However, as illustrated in Fig. 15.23(c), the average value is zero in both cases.

The circuit of Fig. 15.22 can be realized in various forms. Figure 15.24(a) shows a simple implementation consisting of two edge-triggered, resettable D flipflops with their D inputs tied to a logical ONE. The inputs of interest, A and B, serve as the clocks of the flipflops. If $Q_A = Q_B = 0$ and A goes high, Q_A rises. If this event is followed by a rising transition on B, Q_B goes high and the AND gate resets both flipflops. In other words, Q_A and Q_B are simultaneously high for a short time but the difference between their average values still represents the input phase or frequency difference correctly. Each flipflop can be implemented as shown in Fig. 15.24(b), where two RS latches are cross-coupled. Latch 1 and Latch 2 respond to the rising edges of CK and Reset, respectively.

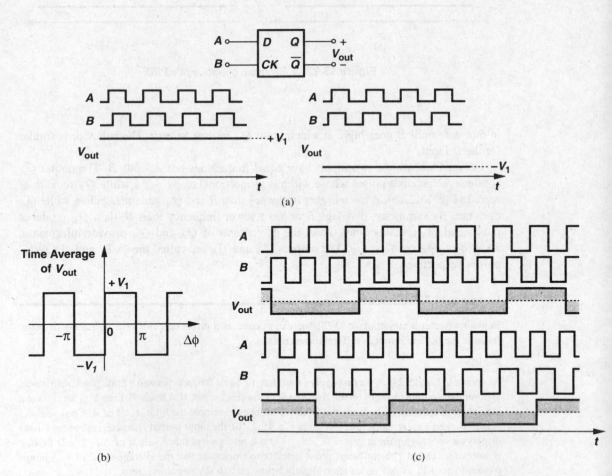

Figure 15.23 (a) D flipflop as a phase detector, (b) input/output characteristic, (c) response of D flipflop to unequal input frequencies.

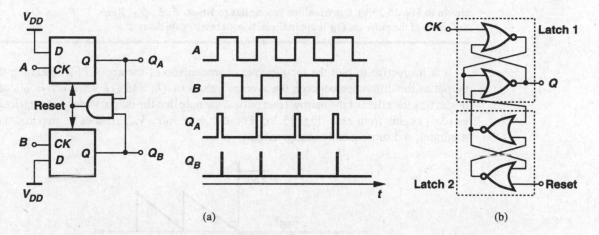

(a)   (b)

Figure 15.24   (a) Implementation of PFD, (b) implementation of D flipflop.

Example 15.8

Determine the width of the narrow reset pulses that appear in the Q_B waveform in Fig. 15.24(a).

Solution

Figure 15.25(a) illustrates the overall PFD at the gate level. If the circuit begins with $A = 1$, $Q_A = 1$, and $Q_B = 0$, a rising edge on B forces $\overline{Q}_B$ to go low and, one gate delay later, Q_B to go high. As

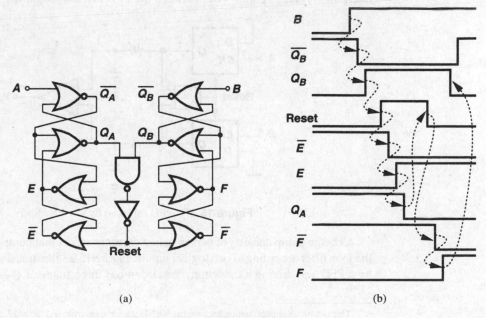

(a)   (b)

Figure 15.25

shown in Fig. 15.25(b), this transition propagates to Reset, $\overline{E}$, E, Q_A, Reset, $\overline{F}$, F, and Q_B. Thus, the width of the pulse on Q_B is approximately equal to 10 gate delays.[8]

It is instructive to plot the input-output characteristic of the above PFD. Defining the output as the difference between the average values of Q_A and Q_B when $\omega_A = \omega_B$ and neglecting the effect of the narrow reset pulses, we note that the output varies symmetrically as $|\Delta\phi|$ begins from zero (Fig. 15.26). For $\Delta\phi = \pm 360°$, V_{out} reaches its maximum or minimum and subsequently changes sign.

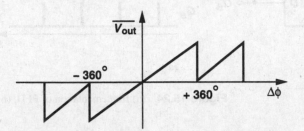

Figure 15.26 Input-output characteristic of the three-state PFD.

How is the PFD of Fig. 15.24(a) utilized in a phase-locked loop? Since the difference between the average values of Q_A and Q_B is of interest, the two outputs can be low-pass filtered and sensed differentially (Fig. 15.27). However, a more common approach is to interpose a "charge pump" (CP) between the PFD and the loop filter.

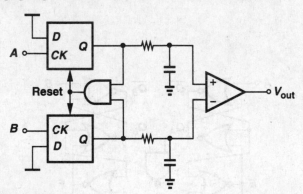

Figure 15.27 PFD followed by low-pass filters.

A charge pump consists of two switched current sources that pump charge into or out of the loop filter according to two logical inputs. Figure 15.28 illustrates a charge pump driven by a PFD and driving a capacitor. The circuit has three states. If $Q_A = Q_B = 0$, then S_1

---

[8]This is a rough approximation because the NAND gate, the inverter, and the NOR gates have different delays and fanouts.

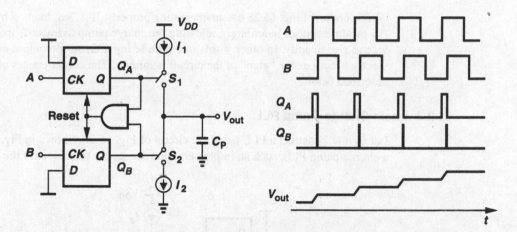

Figure 15.28 PFD with charge pump.

and S_2 are off and V_{out} remains constant. If Q_A is high and Q_B is low, then I_1 charges C_P. Conversely, if Q_A is low and Q_B is high, then I_2 discharges C_P. Thus, if, for example, A leads B, then Q_A continues to produce pulses and V_{out} rises steadily. Called UP and DOWN currents, respectively, I_1 and I_2 are nominally equal.

Example 15.9

What is the effect of the narrow pulses that appear in the Q_B waveform in Fig. 15.28?

Solution

Since Q_A and Q_B are simultaneously high for a finite period (approximately 10 gate delays from Example 15.8), the current supplied by the charge pump to C_P is affected. In fact, if $I_1 = I_2$, the current through S_1 simply flows through S_2 during the narrow reset pulse, leaving no current to charge C_P. Thus, as shown in Fig. 15.29, V_{out} remains constant after Q_B goes high.

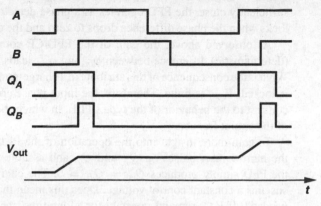

Figure 15.29

The circuit of Fig. 15.28 has an interesting property. If A, say, leads B by a finite amount, Q_A produces pulses indefinitely, allowing the charge pump to inject I_1 into C_P and forcing V_{out} to rise steadily. In other words, for a finite input error, the output eventually goes to $+\infty$ or $-\infty$, i.e., the "gain" of the circuit is infinity. The consequences of infinite gain are described below.

15.2.3 Basic Charge-Pump PLL

Let us now construct a PLL using the circuit of Fig. 15.28. Shown in Fig. 15.30 and called a charge-pump PLL, such an implementation senses the transitions at the input and output,

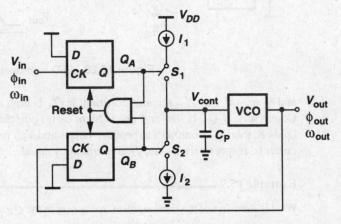

Figure 15.30 Simple charge-pump PLL.

detects phase or frequency differences, and activates the charge pump accordingly. When the loop is turned on, ω_{out} may be far from ω_{in}, and the PFD and the charge pump vary the control voltage such that ω_{out} approaches ω_{in}. When the input and output frequencies are sufficiently close, the PFD operates as a phase detector, performing phase lock. The loop locks when the phase difference drops to zero and the charge pump remains relatively idle.

As observed above, the gain of the PFD/CP combination is infinite, i.e., a nonzero (deterministic) difference between ϕ_{in} and ϕ_{out} leads to indefinite charge buildup on C_P. What is the consequence of this attribute in a charge-pump PLL? When the loop of Fig. 15.30 is locked, V_{cont} is finite. Therefore, the input phase error must be exactly *zero*.[9] This is in contrast to the behavior of the type I PLL, in which the phase error is finite and a function of the output frequency.

To gain more insight into the operation of the PLL shown in Fig. 15.30, let us ignore the narrow reset pulses on Q_A and Q_B and assume that after $\phi_{out} - \phi_{in}$ drops to zero, the PFD simply produces $Q_A = Q_B = 0$. The charge pump thus remains idle and C_P sustains a constant control voltage. Does this mean that the PFD and the CP are no longer needed?! If V_{cont} remains constant for a long time, the VCO frequency and phase begin to

[9]As explained in Section 15.3.1, mismatches still yield a finite phase error.

drift. In particular, the noise sources in the VCO create random variations in the oscillation frequency that can result in a large accumulation of phase error. The PFD then detects the phase difference, producing a corrective pulse on Q_A or Q_B that adjusts the VCO frequency through the charge pump and the filter. This is why we stated earlier that the PLL responds only to the *excess* phase of waveforms. We also note that, since in Fig. 15.30 phase comparison is performed in every cycle, the VCO phase and frequency cannot drift substantially.

Dynamics of CPPLL   In order to quantify the behavior of charge-pump PLLs, we must develop a linear model for the combination of the PFD, the charge pump, and the low-pass filter, thereby obtaining the transfer function. We therefore raise two questions: (1) Is the PFD/CP/LPF combination in Fig. 15.28 a linear system? (2) If so, how can its transfer function be computed?

To answer the first question, we test the system for linearity. For example, as illustrated in Fig. 15.31(a), we double the input phase difference and see if V_{out} exactly doubles.

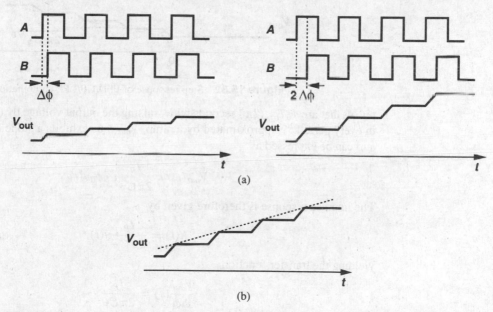

(a)

(b)

Figure 15.31   (a) Test of linearity of PFD/CP/LPF combination, (b) ramp approximation of the response.

Interestingly, the flat sections of V_{out} double but not the ramp sections. After all, the current charging or discharging C_P is constant, yielding a constant slope for the ramp—an effect similar to slewing in op amps. Thus, the system is not linear in the strict sense. To overcome this quandary, we approximate the output waveform by a ramp [Fig. 15.31(b)], arriving at a linear relationship between V_{out} and $\Delta\phi$. In a sense, we approximate a discrete-time system by a continuous-time model.

To answer the second question, we recall that the transfer function is the Laplace transform of the impulse response, requiring that we apply a phase difference impulse and compute V_{out} in the time domain. Since a phase difference impulse is difficult to visualize we apply a phase difference step, obtain V_{out}, and differentiate the result with respect to time

Let us assume the input period is T_{in} and the charge pump provides a current of $\pm I_P$ t the capacitor. As shown in Fig. 15.32, we begin with a zero phase difference and, at $t = ($ step the phase of B by ϕ_0, i.e., $\Delta\phi = \phi_0 u(t)$. As a result, Q_A or Q_B continues to produc

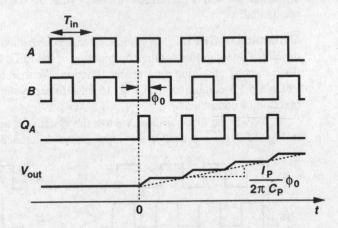

Figure 15.32   Step response of PFD/CP/LPF combination.

pulses that are $\phi_0 T_{in}/(2\pi)$ seconds wide, raising the output voltage by $(I_P/C_P)\phi_0 T_{in}/(2\pi$ in every period.[10] Approximated by a ramp, V_{out} thus exhibits a slope of $(I_P/C_P)\phi_0/(2\pi$ and can be expressed as

$$V_{out}(t) = \frac{I_P}{2\pi C_P}t \cdot \phi_0 u(t). \tag{15.3}$$

The impulse response is therefore given by

$$h(t) = \frac{I_P}{2\pi C_P}u(t), \tag{15.3}$$

yielding the transfer function

$$\frac{V_{out}}{\Delta\phi}(s) = \frac{I_P}{2\pi C_P} \cdot \frac{1}{s}. \tag{15.3}$$

Consequently, the PFD/CP/LPF combination contains a pole at the origin, a point of contra to the PD/LPF circuit used in the type I PLL. In analogy with the expression K_{VCO}/s, w call $I_P/(2\pi C_P)$ the "gain" of the PFD and denote it by K_{PFD}.

Example 15.10

Suppose the output quantity of interest in the circuit of Fig. 15.28 is the current injected by the char pump into the capacitor. Determine the transfer function from $\Delta\phi$ to this current, I_{out}.

---

[10]We neglect the effect of the narrow reset pulses that appear in the other output.

Solution

Since $V_{out}(s) = I_{out}/(C_P s)$, we have

$$\frac{I_{out}}{\Delta\phi}(s) = \frac{I_P}{2\pi}. \tag{15.40}$$

Let us now construct a linear model of charge-pump PLLs. Shown in Fig. 15.33, the model gives an open-loop transfer function

$$\frac{\Phi_{out}}{\Phi_{in}}(s)\big|_{\text{open}} = \frac{I_P}{2\pi C_P}\frac{K_{VCO}}{s^2}. \tag{15.41}$$

Since the loop gain has two poles at the origin, this topology is called a "type II" PLL. The closed-loop transfer function, denoted by $H(s)$ for the sake of brevity, is thus equal to

$$H(s) = \frac{\dfrac{I_P K_{VCO}}{2\pi C_P}}{s^2 + \dfrac{I_P K_{VCO}}{2\pi C_P}}. \tag{15.42}$$

This result is alarming because the closed-loop system contains two imaginary poles at $s_{1,2} = \pm j\sqrt{I_P K_{VCO}/(2\pi C_P)}$ and is therefore unstable. The instability arises because the loop gain has only two poles at the origin, (i.e., two ideal integrators). As shown in Fig. 15.34(a), each integrator contributes a constant phase shift of 90°, allowing the system to oscillate at the gain crossover frequency.

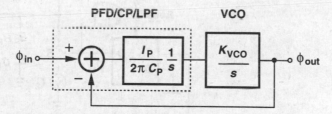

Figure 15.33   Linear model of simple charge-pump PLL.

In order to stabilize the system, we must modify the phase characteristic such that the phase shift is less than 180° at the gain crossover. As shown in Fig. 15.34(b), this is accomplished by introducing a zero in the loop gain, i.e., by adding a resistor in series with the loop filter capacitor (Fig. 15.35). Using the result of Example 15.10, the reader can prove (Problem 15.11) that the PFD/CP/LPF now has a transfer function

$$\frac{V_{out}}{\Delta\phi}(s) = \frac{I_P}{2\pi}\left(R_P + \frac{1}{C_P s}\right). \tag{15.43}$$

It follows that the PLL open-loop transfer function is equal to

$$\frac{\Phi_{out}}{\Phi_{in}}(s)\big|_{\text{open}} = \frac{I_P}{2\pi}\left(R_P + \frac{1}{C_P s}\right)\frac{K_{VCO}}{s}, \tag{15.44}$$

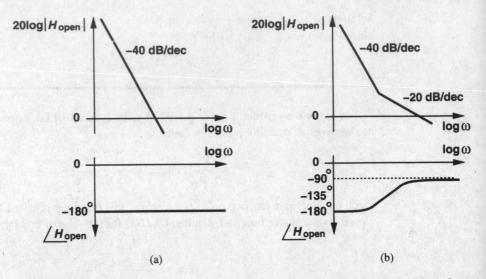

Figure 15.34 (a) Loop gain characteristics of simple charge-pump PLL, (b) addition of zero.

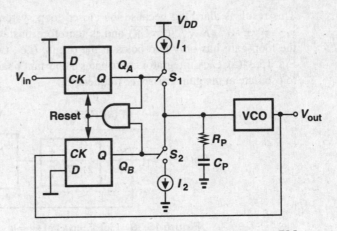

Figure 15.35 Addition of zero to charge-pump PLL.

and hence

$$H(s) = \frac{\dfrac{I_P K_{VCO}}{2\pi C_P}(R_P C_P s + 1)}{s^2 + \dfrac{I_P}{2\pi} K_{VCO} R_P s + \dfrac{I_P}{2\pi C_P} K_{VCO}}. \tag{15.45}$$

The closed-loop system contains a zero at $s_z = -1/(R_P C_P)$. Using the same notation as that for the type I PLL, we have

$$\omega_n = \sqrt{\frac{I_P K_{VCO}}{2\pi C_P}} \tag{15.46}$$

$$\zeta = \frac{R_P}{2}\sqrt{\frac{I_P C_P K_{VCO}}{2\pi}}. \tag{15.47}$$

As expected, if $R_P = 0$, then $\zeta = 0$. With complex poles, the decay time constant is given by $1/(\zeta \omega_n) = 4\pi/(R_P I_P K_{VCO})$.

Stability Issues   The stability behavior of type II PLLs is quite different from that of type I PLLs. We begin the analysis with the Bode plots of the loop gain [Eq. (15.44)]. Shown in Fig. 15.36, these plots suggest that if $I_P K_{VCO}$ decreases, the gain crossover frequency

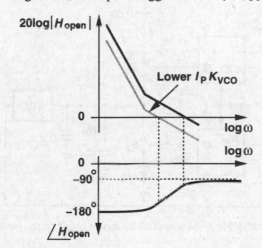

Figure 15.36 Stability degradation of charge-pump PLL as $I_P K_{VCO}$ decreases.

moves toward the origin, *degrading* the phase margin. Predicted by (15.47), this trend is in sharp contrast to that expressed by (15.18) and illustrated in Fig. 15.19.

It is also possible to construct the root locus of the closed-loop system in the complex plane. For $I_P K_{VCO} = 0$ (e.g., $I_P = 0$), the loop is open and both poles lie at the origin. For $I_P K_{VCO} > 0$, we have, $s_{1,2} = -\zeta \omega_n \pm \omega_n \sqrt{\zeta^2 - 1}$, and, since $\zeta \propto \sqrt{I_P K_{VCO}}$, the poles are complex if $I_P K_{VCO}$ is small. The reader can prove (Problem 15.14) that as $I_P K_{VCO}$ increases, s_1 and s_2 move on a circle centered at $\sigma = -1/(R_P C_P)$ with a radius $1/(R_P C_P)$ (Fig. 15.37). The poles return to the real axis at $\zeta = 1$, assuming a value of $-2/(R_P C_P)$. For $\zeta > 1$, the poles remain real, one approaching $-1/(R_P C_P)$ and the other going to $-\infty$ as $I_P K_{VCO} \to +\infty$. Since for complex s_1 and s_2, $\zeta = \cos\varphi$, we observe that as $I_P K_{VCO}$ exceeds zero, the system becomes more stable.

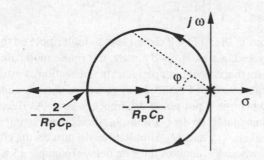

Figure 15.37   Root locus of type II PLL.

The compensated type II PLL of Fig. 15.35 suffers from a critical drawback. Since the charge pump drives the series combination of R_P and C_P, each time a current is injected into the loop filter, the control voltage experiences a large jump. Even in the locked condition, the mismatches between I_1 and I_2 and the charge injection and clock feedthrough of S_1 and S_2 introduce voltage jumps in V_{cont}. The resulting ripple severely disturbs the VCO, corrupting the output phase. To relax this issue, a second capacitor is usually added in parallel with R_P and C_P (Fig. 15.38), suppressing the initial step. The loop filter now is of

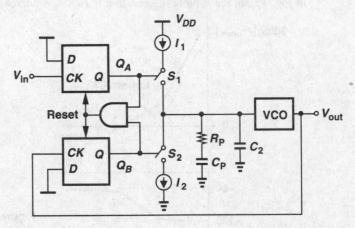

Figure 15.38   Addition of C_2 to reduce ripple on the control line.

second order, yielding a third-order PLL and creating stability difficulties [4]. Nonetheless, if C_2 is about one-fifth to one-tenth of C_P, the closed-loop time and frequency responses remain relatively unchanged.

Equation (15.47) implies that the loop becomes more stable as R_P increases. In reality, as R_P becomes very large, the stability degrades again. This effect is not predicted by the foregoing derivations because we have approximated the discrete-time system by a continuous-time loop. A more accurate analysis is given in [2], but simulations are often necessary to determine the stability bounds of CPPLLs.

15.3 Nonideal Effects in PLLs

15.3.1 PFD/CP Nonidealities

Several imperfections in the PFD/CP circuit lead to high ripple on the control voltage even when the loop is locked. As mentioned earlier, the ripple modulates the VCO frequency, producing a waveform that is no longer periodic. In this section, we study these nonidealities.

The PFD implementation of Fig. 15.24(a) generates narrow, coincident pulses on both Q_A and Q_B even when the input phase difference is zero. As illustrated in Fig. 15.39, if A and B rise simultaneously, so do Q_A and Q_B, thereby activating the reset. That is, even when the PLL is locked, Q_A and Q_B simultaneously turn on the charge pump for a finite period $T_P \approx 10T_D$, where T_D denotes the gate delay (Example 15.8).

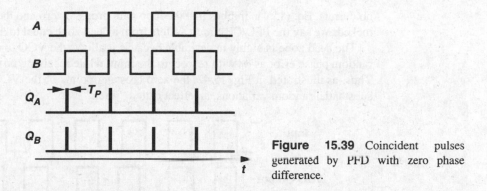

Figure 15.39 Coincident pulses generated by PFD with zero phase difference.

What are the consequences of the reset pulses on Q_A and Q_B? To understand why these pulses are *desirable*, we consider a hypothetical PFD that produces no pulses for a zero input phase difference [Fig. 15.40(a)]. How does such a PFD respond to a small phase error? As shown in Fig. 15.40(b), the circuit generates very narrow pulses on Q_A or Q_B.

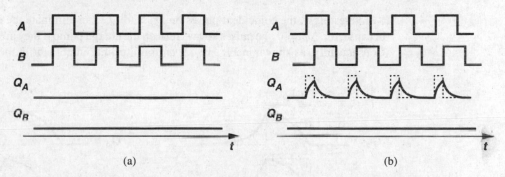

(a)   (b)

Figure 15.40 Output waveforms of a hypothetical PD with (a) zero input phase difference, and (b) a small input phase difference.

However, owing to the finite risetime and falltime resulting from the capacitance seen at these nodes, the pulse may not find enough time to reach a logical high level, failing to turn on the charge pump switches. In other words, if the input phase difference, $\Delta\phi$, falls below a certain value ϕ_0, then the output voltage of the PFD/CP/LPF combination is no longer a function of $\Delta\phi$. Since, as depicted in Fig. 15.41, for $|\Delta\phi| < \phi_0$ the charge pump injects

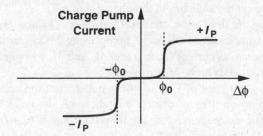

Figure 15.41 Dead zone in the charge pump current.

no current, Eq. (15.41) implies that the loop gain drops to zero and the output phase is not locked. We say the PFD/CP circuit suffers from a dead zone equal to $\pm \phi_0$ around $\Delta \phi = 0$.

The dead zone is highly undesirable because it allows the VCO to accumulate as much random phase error as ϕ_0 with respect to the input while receiving no corrective feedback. Thus, as illustrated in Fig. 15.42, the zero crossing points of the VCO output experience substantial random variations, an effect called "jitter."

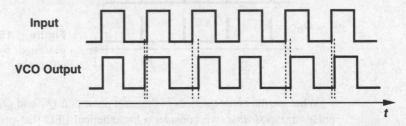

Figure 15.42 Jitter resulting from the dead zone.

Interestingly, the coincident pulses on Q_A and Q_B can eliminate the dead zone. This is because, for $\Delta \phi = 0$, the pulses always turn on the charge pump if they are sufficiently wide. Consequently, as shown in Fig. 15.43, an infinitesimal increment in the phase difference

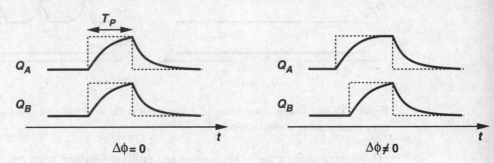

Figure 15.43 Response of actual PD to a small input phase difference.

results in a proportional increase in the net current produced by the charge pump. In other words, the dead zone vanishes if T_P is long enough to allow Q_A and Q_B to reach a valid logical level and turn on the switches in the charge pump.

While eliminating the dead zone, the reset pulses on Q_A and Q_B introduce other difficulties. Let us first implement the charge pump using MOS transistors [Fig. 15.44(a)]. Here, M_1 and M_2 operate as current sources and M_3 and M_4 as switches. The output Q_A is inverted so that when it goes high, M_4 turns on.

The first issue in the circuit of Fig. 15.44(a) stems from the delay difference between $\overline{Q_A}$ and Q_B in turning on their respective switches. As shown in Fig. 15.44(b), the net current injected by the charge pump into the loop filter jumps to $+I_P$ and $-I_P$, disturbing the oscillator control voltage periodically even if the loop is locked. To suppress this effect

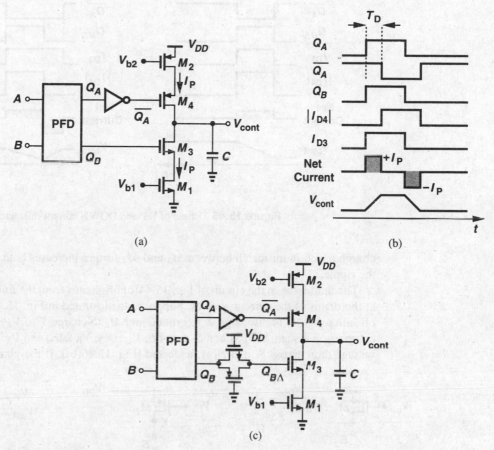

(a)   (b)

(c)

Figure 15.44 (a) Implementation of charge pump, (b) effect of skew between $\overline{Q_A}$ and Q_B, (c) suppression of skew by a pass gate.

a complementary pass gate can be interposed between Q_B and the gate of M_3, equalizing the delays [Fig. 15.44(c)].

The second issue in the CP of Fig. 15.44(c) relates to the mismatch between the drain currents of M_1 and M_2. As depicted in Fig. 15.45(a), even with perfect alignment of the UP and DOWN pulses, the net current produced by the charge pump is nonzero, changing V_{cont} by a constant increment at each phase comparison instant. How does the PLL respond to this error? For the loop to remain locked, the average value of the control voltage must remain constant. The PLL therefore creates a phase error between the input and the output such that the net current injected by the CP in every cycle is zero [Fig. 15.45(b)]. The relationship between the current mismatch and the phase error is determined in Problem 15.12. It is important to note that (1) the control voltage still experiences a periodic ripple, (2) owing to the low output impedance of short-channel MOSFETs, the current mismatch *varies* with the output voltage (i.e., with the VCO frequency), and (3) the clock feedthrough and

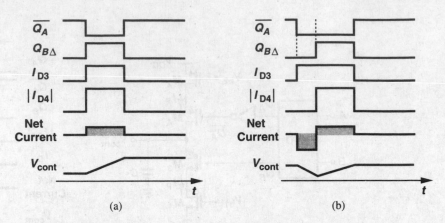

Figure 15.45 Effect of UP and DOWN current mismatch.

charge injection mismatch between M_3 and M_4 further increases both the phase error and the ripple.

The third issue in the circuit of Fig. 15.44(c) originates from the finite capacitance seen at the drains of the current sources. Suppose, as illustrated in Fig. 15.46(a), S_1 and S_2 are off, allowing M_1 to discharge X to ground and M_2 to charge Y to V_{DD}. At the next phase comparison instant, both S_1 and S_2 turn on, V_X rises, V_Y falls, and $V_X \approx V_Y \approx V_{cont}$ if the voltage drop across S_1 and S_2 is neglected [Fig. 15.46(b)]. If the phase error is zero and

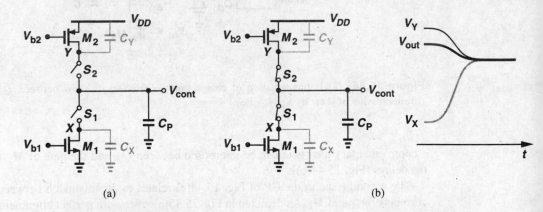

Figure 15.46 Charge sharing between C_P and capacitances at X and Y.

$I_{D1} = |I_{D2}|$, does V_{cont} remain constant after the switches turn on? Even if $C_X = C_Y$, the change in V_X is not equal to that in V_Y. For example, if V_{cont} is relatively high, V_X changes by a large amount and V_Y by a small amount. The difference between the two changes must therefore be supplied by C_P, leading to a jump in V_{cont}.

The above charge sharing phenomenon can be suppressed by "bootstrapping." Illustrated in Fig. 15.47 [3], the idea is to "pin" V_X and V_Y to V_{cont} after phase comparison is finished. When S_1 and S_2 turn off, S_3 and S_4 turn on, allowing the unity-gain amplifier to hold nodes

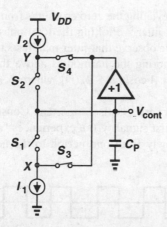

Figure 15.47 Bootstrapping X and Y to minimize charge sharing.

X and Y at a potential equal to V_{cont}. Note that the amplifier need not provide much current because $I_1 \approx I_2$. At the next phase comparison instant, S_1 and S_2 turn on, S_3 and S_4 turn off, and V_X and V_Y begin with a value equal to V_{cont}. Thus, no charge sharing occurs between C_P and the capacitances at X and Y.

15.3.2 Jitter in PLLs

The response of phase-locked loops to jitter is of extreme importance in most applications. We first describe the concepts of jitter and the rate of change of jitter.

As shown in Fig. 15.48, a strictly periodic waveform, $x_1(t)$, contains zero crossings that are evenly spaced in time. Now consider the nearly periodic signal $x_2(t)$, whose period

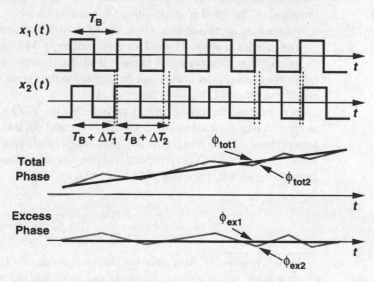

Figure 15.48 Ideal and jittery waveforms.

experiences small changes, deviating the zero crossings from their ideal points. We say the latter waveform suffers from jitter.[11] Plotting the total phase, ϕ_{tot}, and the excess phase, ϕ_{ex}, of the two waveforms, we observe that jitter manifests itself as variation of the excess phase with time. In fact, ignoring the harmonics above the fundamental, we can write $x_1(t) = A \cos \omega t$ and $x_2(t) = A \cos[\omega t + \phi_n(t)]$, where $\phi_n(t)$ models the variation of the period.[12]

The rate at which the jitter varies is also important. Consider the two jittery waveforms depicted in Fig. 15.49. The first signal, $y_1(t)$, experiences "slow jitter" because its instantaneous frequency varies slowly from one period to the next. The second signal, $y_2(t)$,

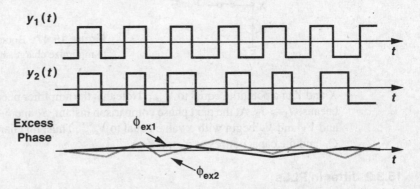

Figure 15.49   Illustration of slow and fast jitter.

experiences "fast jitter." The rate of change is also evident from the excess phase plots of the two waveforms.

Two jitter phenomena in phase-locked loops are of great interest: (a) the input exhibits jitter, and (b) the VCO produces jitter. Let us study each case, assuming the input and output waveforms are expressed as $x_{in}(t) = A \cos[\omega t + \phi_{in}(t)]$ and $x_{out}(t) = A \cos[\omega t + \phi_{out}(t)]$.

The transfer functions derived for type I and type II PLLs have a low-pass characteristic, suggesting that if $\phi_{in}(t)$ varies rapidly, then $\phi_{out}(t)$ does not fully track the variations. In other words, slow jitter at the input propagates to the output unattenuated but fast jitter does not. We say the PLL low-pass filters $\phi_{in}(t)$.

Now suppose the input is strictly periodic but the VCO suffers from jitter. Viewing jitter as random phase variations, we construct the model depicted in Fig. 15.50, where the input excess phase is set to zero [i.e., $x_{in}(t) = A \cos \omega t$] and a random component Φ_{VCO} is added to the output of the VCO to represent its jitter. The reader can show that the transfer function from Φ_{VCO} to Φ_{out} for a type II PLL is equal to

$$\frac{\Phi_{out}}{\Phi_{VCO}}(s) = \frac{s^2}{s^2 + 2\zeta\omega_n s + \omega_n^2}. \tag{15.48}$$

---

[11]Jitter is quantified by several different mathematical definitions, e.g., as in [5].

[12]The quantity $\phi_n(t)$ (or more commonly its spectrum) is called the "phase noise." In this book, we assume the jitter is uniquely represented by $\phi_n(t)$.

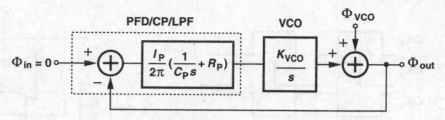

Figure 15.50   Effect of VCO jitter.

Interestingly, the characteristic has a high-pass nature, indicating that slow jitter components generated by the VCO are suppressed but fast jitter components are not. This can be understood with the aid of Fig. 15.50: If $\phi_{VCO}(t)$ changes slowly (e.g., the oscillation period drifts with temperature), then the comparison with $\phi_{in} = 0$ (i.e., a perfectly periodic signal) generates a slowly varying error that propagates through the LPF and adjusts the VCO frequency, thereby counteracting the change in ϕ_{VCO}. On the other hand, if ϕ_{VCO} varies rapidly, (e.g., high-frequency noise modulates the oscillation period), then the error produced by the phase detector is heavily attenuated by the poles in the loop, failing to correct for the change.

Figure 15.51 conceptually summarizes the response of PLLs to input jitter and VCO jitter. Depending on the application and the environment, one or both sources may be significant, requiring an optimum choice of the loop bandwidth.

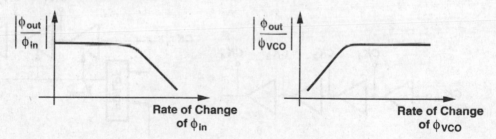

Figure 15.51   Transfer functions of jitter from input and VCO to the output.

15.4 Delay-Locked Loops

A variant of PLLs that has become popular in the past ten years is the delay-locked loop. To arrive at the concept, let us begin with an example. Suppose an application requires four clock phases with a precise spacing of $\Delta T = 1$ ns between consecutive edges [Fig. 15.52(a)]. How should these phases be generated? We can use a two-stage differential ring oscillator[13] to produce the four phases, but how do we guarantee that $\Delta T = 1$ ns

---

[13] As explained in Chapter 14, a simple two-stage CMOS ring oscillator may not oscillate. This example is merely for illustration purposes.

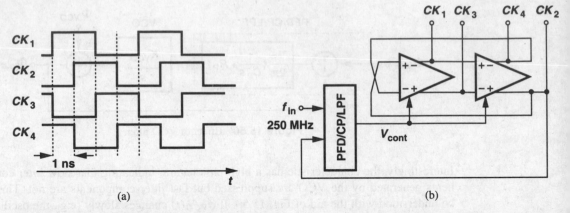

(a)   (b)

Figure 15.52 (a) Clock phases with edge-to-edge delay of 1 ns, (b) use of a phase-locked ring oscillator to generate the clock phases.

despite process and temperature variations? This requires that the oscillator be locked to a 250-MHz reference so that the output period is exactly equal to 4 ns [Fig. 15.52(b)].

An alternative approach to generating the clock phases of Fig. 15.52(a) is to apply the input clock to four delay stages in a cascade. Illustrated in Fig. 15.53(a), this technique nonetheless does not produce a well-defined edge spacing because the delay of each stage

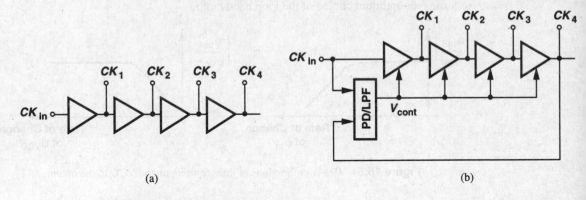

(a)   (b)

Figure 15.53 (a) Generation of clock edges by delay stages, (b) simple delay-locked loop.

varies with process and temperature. Now consider the circuit shown in Fig. 15.53(b), where the phase difference between CK_{in} and CK_4 is sensed by a phase detector, a proportional average voltage, V_{cont}, is generated, and the delay of the stages is adjusted with negative feedback. For a large loop gain, the phase difference between CK_{in} and CK_4 is small, that is, the four stages delay the clock by almost exactly one period, thereby establishing precise edge spacing.[14] This topology is called a delay-locked loop to emphasize that it incorporates a voltage-controlled delay line (VCDL) rather than a VCO. In practice, a charge pump is

---

[14] The total delay through the four stages may be equal to two or more periods. We return to this issue later.

interposed between the PD and the LPF to achieve an infinite loop gain. Each delay stage may be based on one of the ring oscillator stages described in Chapter 14.

The reader may wonder about the advantages of DLLs over PLLs. First, delay lines are generally less susceptible to noise than oscillators are because corrupted zero crossings of a waveform disappear at the end of a delay line whereas they are recirculated in an oscillator, thereby experiencing more corruption. Second, in the VCDL of Fig. 15.53(b), a change in the control voltage immediately changes the delay, that is, the transfer function $\Phi_{out}(s)/V_{cont}(s)$ is simply equal to the gain of the VCDL, K_{VCDL}. Thus, the feedback system of Fig. 15.53(b) has the same order as the LPF and its stability and settling issues are more relaxed than those of a PLL.

Example 15.11

Determine the closed-loop transfer function of the DLL shown in Fig. 15.54.

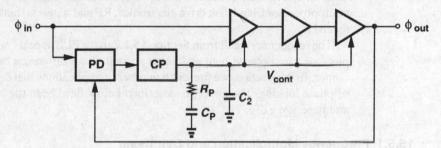

Figure 15.54

Solution

From Example 15.10, we write the transfer function of the PD/CP/LPF combination as

$$\frac{V_{cont}}{\Delta\Phi}(s) = \frac{I_P}{2\pi}\left[\left(R_P + \frac{1}{C_P s}\right)\middle\| \frac{1}{C_2 s}\right] \tag{15.49}$$

$$= \frac{I_P}{2\pi}\frac{R_P C_P s + 1}{(R_P C_P C_2 s + C_P + C_2)s}. \tag{15.50}$$

The closed-loop transfer function is thus equal to

$$\frac{\Phi_{out}}{\Phi_{in}}(s)\bigg|_{closed} = \frac{\dfrac{I_P K_{VCDL}}{2\pi}(R_P C_P s + 1)}{R_P C_P C_2 s^2 + [C_P + C_2 + I_P K_{VCDL} R_P C_P/(2\pi)]s + I_P K_{VCDL}/(2\pi)}. \tag{15.51}$$

This transfer function can be used to determine how ϕ_{out} settles if ϕ_{in} experiences a change. Note that in practice R_P may not be needed because the loop contains only one pole at the origin.

The principal drawback of DLLs is that they cannot generate a variable output frequency. This issue becomes clearer when we study the frequency synthesis capabilities of PLLs in Section 15.5.1. DLLs may also suffer from locked delay ambiguity. That is, if the total delay of the four stages in Fig. 15.53(b) can vary from below T_{in} to above $2T_{in}$, then the loop may lock with a CK_{in}-to-CK_4 delay equal to either T_{in} or $2T_{in}$. This ambiguity proves detrimental if the DLL must provide precisely-spaced clock edges because the edge-to-edge delay may settle to $2T_{in}/4$ rather than $T_{in}/4$. In such cases, additional circuitry is necessary to avoid the ambiguity. Also, mismatches between the delay stages and their load capacitances introduce error in the edge spacing, requiring large devices and careful layout.

15.5 Applications

After nearly 70 years since its invention, phase locking continues to find new applications in electronics, communication, and instrumentation. Examples include memories, microprocessors, hard disk drive electronics, RF and wireless transceivers, and optical fiber receivers.

The reader may recall from Section 15.1.2 that a PLL appears no more useful than a short piece of wire because both guarantee a small phase difference between the input and the output. In this section, we present a number of applications that demonstrate the versatility of phase locking. The concepts described below have been the topic of numerous books and papers, e.g., [6, 7].

15.5.1 Frequency Multiplication and Synthesis

Frequency Multiplication   A PLL can be modified such that it multiplies its input frequency by a factor of M. To arrive at the implementation, we exploit an analogy with voltage multiplication. As depicted in Fig. 15.55(a), a feedback system amplifies the input

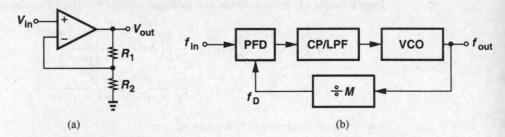

(a)   (b)

Figure 15.55   (a) Voltage amplification and (b) frequency multiplication.

voltage by a factor of M if the output voltage is divided by M [i.e., if $R_2/(R_1 + R_2) = 1/M$] and the result is compared with the input. Thus, as shown in Fig. 15.55(b), if the output *frequency* of a PLL is divided by M and applied to the phase detector, we have $f_{out} = Mf_{in}$. From another point of view, since $f_D = f_{out}/M$ and f_D and f_{in} must be equal in the locked condition, the PLL multiplies f_{in} by M. The $\div M$ circuit is realized as a counter that produces one output pulse for every M input pulses.

As with voltage division in Fig. 15.55(a), the feedback divider in the loop of Fig. 15.55(b) alters the system characteristics. Using (15.44), we rewrite (15.45) as

$$H(s) = \frac{\dfrac{I_P}{2\pi}\left(R_P + \dfrac{1}{C_P s}\right)\dfrac{K_{VCO}}{s}}{1 + \dfrac{1}{M}\dfrac{I_P}{2\pi}\left(R_P + \dfrac{1}{C_P s}\right)\dfrac{K_{VCO}}{s}} \tag{15.52}$$

$$= \frac{\dfrac{I_P K_{VCO}}{2\pi C_P}(R_P C_P s + 1)}{s^2 + \dfrac{I_P}{2\pi}\dfrac{K_{VCO}}{M}R_P s + \dfrac{I_P}{2\pi C_P}\dfrac{K_{VCO}}{M}}. \tag{15.53}$$

Note that $H(s) \to M$ as $s \to 0$, i.e., phase or frequency changes at the input result in an M-fold change in the corresponding output quantity. Comparing the denominators of (15.45) and (15.53), we observe that frequency division in the loop manifests itself as division of K_{VCO} by M. In other words, as far as the poles of the closed-loop system are concerned, we can assume the oscillator and the divider form a VCO with an equivalent gain of K_{VCO}/M. This is of course to be expected because, for the VCO/divider cascade shown Fig. 15.56, we have

$$\omega_{out} = \frac{\omega_0 + K_{VCO}V_{cont}}{M} \tag{15.54}$$

$$= \frac{\omega_0}{M} + \frac{K_{VCO}}{M}V_{cont}. \tag{15.55}$$

Thus, the combination cannot be distinguished from a VCO having an intercept frequency of ω_0/M and a gain of K_{VCO}/M.

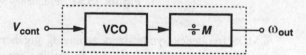

Figure 15.56 Equivalency of VCO/divider combination to a single VCO.

The foregoing discussion suggests that (15.46) and (15.47) can be respectively rewritten as

$$\omega_n = \sqrt{\frac{I_P}{2\pi C_P}\frac{K_{VCO}}{M}} \tag{15.56}$$

$$\zeta = \frac{R_P}{2}\sqrt{\frac{I_P C_P}{2\pi}\frac{K_{VCO}}{M}}. \tag{15.57}$$

Also, the decay time constant is modified to $(\zeta\omega_n)^{-1} = 4\pi M/(R_P I_P K_{VCO})$. It follows that inserting a divider in a type II loop degrades both the stability and the settling speed, requiring a proportional increase in the charge pump current.

The frequency-multiplying loop of Fig. 15.55(b) exhibits two interesting properties. First, unlike the voltage amplifier of Fig. 15.55(a), the PLL provides a multiplication factor *exactly* equal to M, an attribute resulting from the infinite loop gain and expressed by Eq. (15.53). Second, the output frequency can be varied by changing the divide ratio M, an extremely useful property in synthesizing frequencies. Note that DLLs cannot perform such synthesis.

Frequency Synthesis   Some systems require a periodic waveform whose frequency (a) must be very accurate (e.g., exhibit an error less than 10 ppm), and (b) can be varied in very fine steps (e.g., in steps of 30 kHz from 900 MHz to 925 MHz). Commonly encountered in wireless transceivers, such requirements can be met through frequency multiplication by PLLs.

Figure 15.57 shows the architecture of a phase-locked frequency synthesizer. The channel control input is a digital word that varies the value of M. Since $f_{out} = M f_{REF}$, the relative accuracy of f_{out} is equal to that of f_{REF}. For this reason, f_{REF} is derived from a stable, low-noise crystal oscillator. Note that f_{out} varies in steps equal to f_{REF} if M changes by one each time.

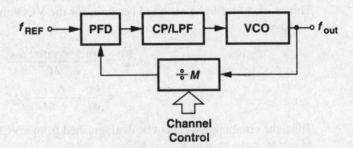

Figure 15.57   Frequency synthesizer.

CMOS frequency synthesizers achieving gigahertz output frequencies have been reported. Issues such as noise, sidebands, settling speed, frequency range, and power dissipation continue to challenge synthesizer designers.

15.5.2 Skew Reduction

The earliest usage of phase locking in digital systems was for skew reduction. Suppose a synchronous pair of data and clock lines enter a large digital chip as shown in Fig. 15.58. Since the clock typically drives a large number of transistors and long interconnects, it is first applied to a large buffer. Thus, the clock distributed on the chip may suffer from substantial skew with respect to the data, an undesirable effect because it reduces the timing budget for on-chip operations.

Now consider the circuit shown in Fig. 15.59, where CK_{in} is applied to an on-chip PLL and the buffer is placed *inside* the loop. Since the PLL guarantees a nominally-zero phase difference between CK_{in} and CK_B, the skew is eliminated. From another point of view, the constant phase shift introduced by the buffer is divided by the infinite loop gain of

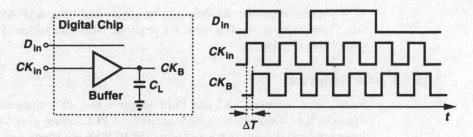

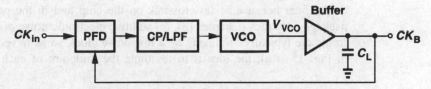

Figure 15.58 Skew between data and buffered clock.

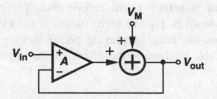

Figure 15.59 Use of a PLL to eliminate skew.

the feedback system. Note that the VCO output, V_{VCO}, may not be aligned with CK_{in}, a nonetheless unimportant issue because V_{VCO} is not used.

Example 15.12 ————————————————————————————————

Construct the voltage-domain counterpart of the loop shown in Fig. 15.59.

Solution

The buffer creates a constant phase shift in the signal generated by the VCO. The voltage-domain counterpart therefore assumes the topology shown in Fig. 15.60. We have

$$(V_{in} - V_{out})A + V_M = V_{out} \tag{15.58}$$

and hence

$$V_{out} = \frac{AV_{in} + V_M}{1 + A}. \tag{15.59}$$

As $A \to \infty$, $V_{out} \to V_{in}$.

Figure 15.60

We should note that the skew can be suppressed by a delay-locked loop as well. In fact, if frequency multiplication is not required, DLLs are preferred because they are less susceptible to noise.

15.5.3 Jitter Reduction

Recall from Section 15.3.2 that PLLs suppress fast jitter components at the input. For example, if a 1-GHz jittery signal is applied to a PLL having a bandwidth of 10 MHz, then input jitter components that vary faster than 10 MHz are attenuated. In a sense, the phase-locked loop operates as a narrowband filter centered around 1 GHz with a total bandwidth of 20 MHz. This is another important and useful property of PLLs.

Many applications must deal with jittery waveforms. Random binary signals experience jitter because of (a) crosstalk on the chip and in the package (Chapter 18), (b) package parasitics (Chapter 18), (c) additive electronic noise of devices, etc. Such waveforms are typically "retimed" by a low-noise clock so as to reduce the jitter. Illustrated in Fig. 15.61(a), the idea is to resample the midpoint of each bit by a D flipflop that

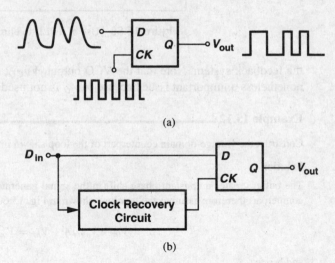

(a)

(b)

Figure 15.61 (a) Retiming data with D flipflop driven by a low-noise clock, (b) use of a phase-locked clock recovery circuit to generate the clock.

is driven by the clock. However, in many applications, the clock may not be available independently. For example, an optical fiber carries only the random data stream, providing no separate clock waveform at the receive end. The circuit of Fig. 15.61(a) is therefore modified as shown in Fig. 15.61(b), where a "clock recovery circuit" (CRC) produces the clock from the data. Employing phase locking with a relatively narrow loop bandwidth, the CRC minimizes the effect of the input jitter on the recovered clock.

Problems

Unless otherwise stated, in the following problems, use the device data shown in Table 2.1 and assume $V_{DD} = 3$ V where necessary. Also, assume all transistors are in saturation.

15.1. The Gilbert cell (Chapter 4) operates as an XOR gate with large input swings and as an analog multiplier with small input swings. Prove that an analog multiplier can be used to detect the phase difference between two sinusoids. Is the input-output characteristic of such a phase detector linear?

15.2. Redraw the waveforms of Fig. 15.4(b) if the VCO frequency is lowered at $t = t_1$. If the phase error between V_{CK} and V_{VCO} before $t = t_1$ is equal to ϕ_0 and f_{VCO} is lowered from f_H to f_L, determine the minimum $t_2 - t_1$ that is sufficient for phase alignment.

15.3. Explain why the low-pass filter in Fig. 15.5(b) cannot be replaced by a high-pass filter.

15.4. A PLL using an XOR gate as a phase detector locks with $\phi_{in} - \phi_{out} \approx 90°$ if $K_{PD}K_{VCO}$ is large. Explain why?

15.5. Using the characteristic of Fig. 15.3 as an example, explain why the polarity of feedback in a PLL (without frequency detection) is unimportant. (Hint: prove that the loop locks regardless of whether the initial phase difference falls in the positive-slope region or the negative-slope region.)

15.6. Assuming a first-order LPF in Fig. 15.14, determine the transfer function Φ_{out}/Φ_{ex}, where Φ_{out} denotes the excess phase of V_{out}.

15.7. A VCO used in a type I PLL exhibits nonlinearity in its input-output characteristic, i.e., K_{VCO} varies across the tuning range. If the damping ratio must remain between 1 and 1.5, how much variation can be tolerated in K_{VCO}?

15.8. Prove that in the root locus of Fig. 15.20, $\cos\theta = \zeta$.

15.9. A type I PLL incorporates a VCO with $K_{VCO} = 100$ MHz/V, a PD with $K_{PD} = 1$ V/rad, and an LPF with $\omega_{LPF} = 2\pi(1$ MHz$)$. Determine the step response of the PLL.

15.10. Explain why in the charge-pump PLL of Fig. 15.35, the control voltage of the VCO cannot be connected to the top plate of C_P.

15.11. Prove that the transfer function of the PFD/CP/LPF circuit in Fig. 15.35 is given by Eq. (15.43).

15.12. As illustrated in Fig. 15.45, mismatches between the UP and DOWN currents translate to phase offset at the input of a CPPLL. With the aid of the waveforms in Fig. 15.45, calculate the phase offset in terms of current mismatch.

15.13. For a VCO, we have $\omega_{out} = \omega_0 + K_{VCO}V_{cont}$. The control line experiences a small sinusoidal ripple, $V_{cont} = V_m \cos\omega_m t$. If the VCO is followed by a $\div M$ circuit, determine the output spectrum of the divider. Consider two cases: $\omega_0/M > \omega_m$ and $\omega_0/M < \omega_m$.

15.14. Prove that the root locus of a type II PLL is as shown in Fig. 15.37.

15.15. Determine the transfer function Φ_{out}/Φ_{ex} for the circuit of Fig. 15.14 if the PLL is modified to the architecture of Fig. 15.35.

15.16. When a charge-pump PLL incorporating a PFD is turned on, the VCO frequency may be far from the input frequency. Explain why the order of the PLL transfer function is lower by one while the PFD operates as a frequency detector.

References

1. R. E. Best, *Phase-Locked Loops,* Second Ed., New York: McGraw-Hill, 1993.

2. F. M. Gardner, *Phaselock Techniques,* Second Ed., New York: Wiley & Sons, 1979.

3. M. G. Johnson and E. L. Hudson, "A Variable Delay Line PLL for CPU-Coprocessor Synchronization," *IEEE Journal of Solid-State Circuits,* vol. 23, pp. 1218–1223, Oct. 1988.

4. F. M. Gardner, "Charge-Pump Phase-Locked Loops," *IEEE Trans. Comm.,* vol. COM-28, pp.1849–1858, Nov. 1980.

5. F. Herzel and B. Razavi, "A Study of Oscillator Jitter Due to Supply and Substrate Noise," *IEEE Transactions on Circuits and Systems, Part II,* vol.46, pp.56–62, Jan. 1999.

6. W. F. Egan, *Frequency Synthesis by Phase Lock,* New York: Wiley & Sons, 1981.

7. J. A. Crawford, *Frequency Synthesizer Design Handbook,* New York: Artech House, 1994.

Chapter 16

Short-Channel Effects and
Device Models

The square-law characteristics derived for MOSFETs in Chapter 2 provide moderate accuracies for devices with minimum channel lengths of greater than 4 μm, a value corresponding to technologies in production in the early 1980s. As device dimensions continue to scale down, reaching below 0.2 μm by the year 2000, higher order effects necessitate more complex models so as to attain enough accuracy in simulations.

The problem of device models in CMOS technology has constantly haunted analog designers, manifesting itself as substantial discrepancies between simulated and measured results. A number of comprehensive books [1, 2, 3] and hundreds of papers deal with the subject in great detail, but our objective here is to provide a basic understanding of short-channel effects and review some of the SPICE models developed to reflect such phenomena. Knowledge of these issues also proves useful in interpreting the anomalies that the designer may encounter in SPICE simulations.

We first describe the ideal scaling theory of MOS transistors. Next, we study short-channel effects such as threshold voltage variation, velocity saturation, and the dependence of the output impedance on the drain-source voltage. We then review MOS device models, including Levels 1–3 and the BSIM series. Finally, we discuss charge and capacitance modeling, temperature dependence, and process corners.

16.1 Scaling Theory

The two principal reasons for the dominance of CMOS technology in today's semiconductor industry are the zero static power dissipation of CMOS logic and the scalability of MOSFETs. In a paper published in 1974 [4], Dennard et al. recognized the tremendous potential of scaling MOS transistors, making predictions about speed and power dissipation of digital CMOS circuits as devices are shrunk.

The ideal scaling theory follows three rules: (1) reduce all lateral and vertical dimensions by $\alpha(> 1)$; (2) reduce the threshold voltage and the supply voltage by α; (3) increase all of the doping levels by α (Fig. 16.1). Since the dimensions and voltages scale together, all electric fields in the transistor remain constant, hence the name "constant-field scaling." Note that

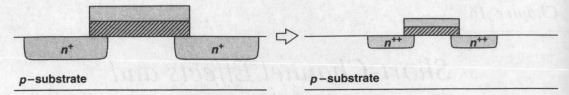

Figure 16.1 Ideal scaling of MOS transistor.

$W, L, t_{ox}, V_{DD}, V_{TH}$, and the depth and perimeter of the source and drain junctions scale down by α.

Let us examine the saturation drain current of a square-law device after scaling. Writing

$$I_{D,scaled} = \frac{1}{2}\mu_n(\alpha C_{ox})\left(\frac{W/\alpha}{L/\alpha}\right)\left(\frac{V_{GS}}{\alpha} - \frac{V_{TH}}{\alpha}\right)^2 \qquad (16.1)$$

$$= \frac{1}{2}\mu_n C_{ox}\frac{W}{L}(V_{GS} - V_{TH})^2\frac{1}{\alpha}, \qquad (16.2)$$

we observe that the current capability of the transistor *drops* by a factor of α. Note that the same result applies for the drain current in the triode region. The advantage of scaling, however, lies in the reduction of capacitances and power dissipation. The total channel capacitance is

$$C_{ch,scaled} = \frac{W}{\alpha}\frac{L}{\alpha}(\alpha C_{ox}) \qquad (16.3)$$

$$= \frac{1}{\alpha}WLC_{ox}. \qquad (16.4)$$

To calculate the source/drain junction capacitance, we first analyze the effect of ideal scaling on the total width of the depletion region. Recall that this width is given by

$$W_d = \sqrt{\frac{2\epsilon_{si}}{q}\left(\frac{1}{N_A} + \frac{1}{N_D}\right)(\phi_B + V_R)}, \qquad (16.5)$$

where N_A and N_D denote the doping levels of the two sides of the junction, $\phi_B = V_T \ln(N_A N_D/n_i^2)$, and V_R is the reverse-bias voltage. The built-in potential, ϕ_B, is a weak function of $N_A N_D$ and in fact it *increases* if $N_A N_D$ is scaled up by α^2. For now, we assume $V_R \gg \phi_B$ so that

$$W_{d,scaled} \approx \sqrt{\frac{2\epsilon_{si}}{q}\left(\frac{1}{\alpha N_A} + \frac{1}{\alpha N_D}\right)\frac{V_R}{\alpha}} \qquad (16.6)$$

$$\approx \frac{1}{\alpha}\sqrt{\frac{2\epsilon_{si}}{q}\left(\frac{1}{N_A} + \frac{1}{N_D}\right)V_R}. \qquad (16.7)$$

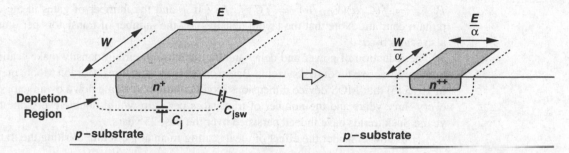

Figure 16.2   Scaling of S/D junction capacitances.

Thus, as with other dimensions, the width of each depletion region scales down by α, increasing the depletion region capacitance per unit area by the same factor.

As illustrated in Fig. 16.2, the bottom-plate capacitance of the S/D junction (per unit area), C_j, increases by a factor of α. The sidewall capacitance (per unit width), C_{jsw}, on the other hand, remains constant because the depth of the junction is reduced by α. It follows that

$$C_{S/D,scaled} = \frac{W}{\alpha}\frac{E}{\alpha}(\alpha C_j) + 2\left(\frac{W}{\alpha} + \frac{E}{\alpha}\right)(C_{jsw}) \tag{16.8}$$

$$= [WEC_j + 2(W + E)C_{jsw}]\frac{1}{\alpha}. \tag{16.9}$$

All of the capacitances therefore decrease by the scaling factor.

In digital applications, the scaling of the gate delay and power dissipation is of interest. Approximating the delay of a CMOS inverter by $T_d = (C/I)V_{DD}$ (Fig. 16.3), we have

$$T_{d,scaled} = \frac{C/\alpha}{I/\alpha}\frac{V_{DD}}{\alpha} \tag{16.10}$$

$$= \left(\frac{C}{I}V_{DD}\right)\frac{1}{\alpha}. \tag{16.11}$$

We conclude that the speed of digital circuits can potentially increase by the scaling factor. For power dissipation, we write $P = fCV_{DD}^2$, where f is the operating frequency. Thus,

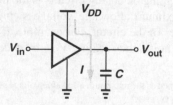

Figure 16.3   CMOS inverter.

$P_{scaled} = f(C/\alpha)(V_{DD}/\alpha)^2 = fCV_{DD}^2/\alpha^3$, if f and the number of gates in the circuit remain constant. Note that the layout density, i.e., the number of transistors per unit area, also scales by α^2.

The reduction of power and delay and the increase in circuit density make scaling extremely attractive for digital systems. Based on these observations, Gordon Moore predicted in 1975 [5] that MOS device dimensions would continue to scale down by a factor of two every three years and the number of transistors per chip would double every one to two years. Such trends have indeed persisted over the past 25 years.

Let us now consider the effect of ideal scaling in analog circuits. Writing the transconductance as

$$g_{m,scaled} = \mu(\alpha C_{ox})\frac{W/\alpha}{L/\alpha}\frac{V_{GS} - V_{TH}}{\alpha} \tag{16.12}$$

$$= \mu C_{ox}\frac{W}{L}(V_{GS} - V_{TH}), \tag{16.13}$$

we note that the transconductance remains constant if all of the dimensions and voltages (and currents) scale down. To calculate the output impedance in saturation, we first observe from Fig. 16.4 and Eq. (16.7) that the width of the depletion region around the drain decreases by α, and hence $\Delta L/L$ remains constant. Since $\lambda = (\Delta L/L)/V_{DS}$ (Chapter 2), λ increases by α and

$$r_{O,scaled} = \frac{1}{\alpha\lambda\dfrac{I_D}{\alpha}} \tag{16.14}$$

$$= \frac{1}{\lambda I_D}. \tag{16.15}$$

Thus, the intrinsic gain, $g_m r_O$, remains constant.

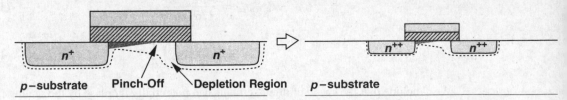

Figure 16.4   Effect of scaling on pinch-off.

The greatest impact of scaling on analog circuits is the reduction of the supply voltage. With ideal scaling, the maximum allowable voltage swings decrease by a factor of α, lowering the dynamic range[1] of the circuit. For example, if the lower end of the dynamic range is limited by thermal noise, then scaling V_{DD} by α decreases the dynamic range by

---

[1]Dynamic range is loosely defined as the maximum allowable voltage swing divided by the total noise voltage in the band of interest.

the same factor because g_m and hence thermal noise remain constant. Of course, since for analog circuits $(V_{DD}/\alpha)(I_{DD}/\alpha) = (V_{DD}I_{DD}/\alpha)^2$, the power dissipation drops by α^2.

In order to restore the dynamic range, the transconductance of the transistors must be increased by a factor of α^2 because thermal noise voltages and currents scale with $\sqrt{g_m}$. Thus, since voltage scaling requires that $V_{GS} - V_{TH}$ decrease by a factor of α, we note from $g_m = 2I_D/(V_{GS} - V_{TH})$ that I_D must increase by the same factor, leading to a power dissipation of $(V_{DD}/\alpha)(\alpha I_D) = V_{DD}I_D$. Also, from $g_m = \mu C_{ox}(W/L)(V_{GS} - V_{TH})$, we conclude that if C_{ox} is scaled up by α and L and $V_{GS} - V_{TH}$ are scaled down by α, then W must *increase* by α (whereas in ideal scaling it would decrease by this factor). That is, for a constant (thermal-noise limited) dynamic range, ideal scaling of linear analog circuits requires a *constant* power dissipation and a *higher* device capacitance, e.g., $(\alpha W)(L/\alpha)(\alpha C_{ox}) = \alpha WLC_{ox}$. Interestingly, if the lower end of the dynamic range is determined by kT/C noise, then to maintain a constant slew rate in switched-capacitor circuits, the bias current must scale up by a factor of α^2, resulting in an increase in the power dissipation [Problem 16.3(d)].

In practice, technology scaling has deviated from the ideal, constant-field scenario considerably. The supply voltage and MOS threshold voltage have not scaled as rapidly as device dimensions. For example, V_{DD} has decreased from 5 V to 2.5 V and V_{TH} from 0.8 V to 0.4 V as minimum channel length has dropped from 1 μm to 0.25 μm. Furthermore, many "short-channel" effects have plagued the transistors, making it difficult to obtain all of the benefits that would accrue with ideal scaling.

The reluctance of circuit designers to use a lower supply voltage and the fundamental limitations in decreasing the MOS threshold voltage have led to another scaling scenario: constant voltage scaling. In this case, the device dimensions shrink by α, the doping levels increase by α, and the voltages remain constant, thereby increasing the electric fields by α. Such high electric fields both raise the possibility of device breakdown and exacerbate short-channel effects. In reality, technology scaling has followed a mixture of constant-field and constant-voltage trends, thus demanding innovative device design so as to achieve reliability and performance.

16.2 Short-Channel Effects

In order to appreciate the need for sophisticated device models, we briefly study some of the phenomena that manifest themselves for channel lengths below approximately 3 μm. As we will see, a basic understanding of these effects also proves essential to the design of analog (and digital) circuits.

Small-geometry effects arise because five factors deviate the scaling from the ideal scenario: (1) the electric fields tend to increase because the supply voltage has not scaled proportionally; (2) the built-in potential term in Eq. (16.5) is neither scalable nor negligible; (3) the depth of S/D junctions cannot be reduced easily; (4) the mobility decreases as the substrate doping increases; (5) the subthreshold slope (described below) is not scalable.

16.2.1 Threshold Voltage Variation

The choice of the threshold voltage is based on the device performance in typical circuit applications. The upper bound is roughly equal to $V_{DD}/4$ to avoid degrading the speed of digital CMOS gates. The lower bound is determined by several factors: the subthreshold behavior, variation with temperature and process, and dependence upon the channel length [6].

Let us first consider the subthreshold behavior. For long-channel devices, the subthreshold drain current can be expressed as

$$I_D = \mu C_d \frac{W}{L} V_T^2 \left(\exp \frac{V_{GS} - V_{TH}}{\zeta V_T} \right) \left(1 - \exp \frac{-V_{DS}}{V_T} \right), \tag{16.16}$$

where $C_d = \sqrt{\epsilon_{si} q N_{sub}/(4\phi_B)}$ denotes the capacitance of the depletion region under the gate area, $V_T = kT/q$, and $\zeta = 1 + C_d/C_{ox}$ [6]. Equation (16.16) reveals two interesting properties. First, as V_{DS} exceeds a few V_T, I_D becomes independent of the drain-source voltage and the relationship reduces to Eq. (2.30). Second, under this condition the slope of I_D on a logarithmic scale equals

$$\frac{\partial (\log_{10} I_D)}{\partial V_{GS}} = (\log_{10} e) \frac{1}{\zeta V_T}. \tag{16.17}$$

The inverse of this quantity is usually called the "subthreshold slope," S:

$$S = 2.3 V_T \left(1 + \frac{C_d}{C_{ox}} \right) \text{ V/dec.} \tag{16.18}$$

For example, if $C_d = 0.67 C_{ox}$, then $S = 100 \text{ mV/dec}$, suggesting that a change of 100 mV in V_{GS} leads to a ten-fold reduction in the drain current. In order to turn off the transistor by lowering V_{GS} below V_{TH}, S must be as *small* as possible, i.e., C_d/C_{ox} must be minimized.

The relatively constant magnitude of S severely limits the scaling of the threshold voltage. For example, a subthreshold slope of 80 mV/dec imposes a lower bound of 400 mV for V_{TH} if the "off current" must be roughly five orders of magnitude lower than the "on current."

The difficulty in scaling V_{TH} becomes even more serious if we take into account the variation of V_{TH} with temperature and process. The threshold voltage exhibits a temperature coefficient of approximately $-1 \text{ mV/}^\circ\text{K}$, yielding a 50-mV change across the commercial temperature range (0 to 50°C).[2] Process-induced variation is also in the vicinity of 50 mV, raising the margin to approximately 100 mV. Thus, it is difficult to reduce V_{TH} below several hundred millivolts.

An interesting phenomenon observed in scaled transistors is the dependence of the threshold voltage on the channel length. As shown in Fig. 16.5, transistors fabricated on the same wafer but with different lengths yield lower V_{TH} as L decreases. This is because the depletion regions associated with the source and drain junctions protrude into the channel area considerably, thereby reducing the immobile charge that must be imaged by the charge on the gate (Fig. 16.6). In other words, part of the immobile charge in the substrate is now imaged by the charge inside the source and drain areas rather than by the charge on the

---

[2]Interestingly, as the temperature rises, so does S, further exacerbating the situation.

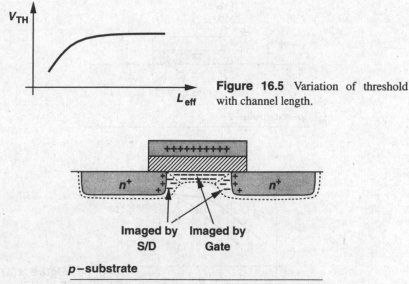

Figure 16.5 Variation of threshold with channel length.

Figure 16.6 Charge sharing between source/drain depletion regions and the channel depletion region.

gate.* As a result, the gate voltage required to create an inversion layer decreases. Since the channel length cannot be controlled accurately during fabrication, this effect introduces additional variations in V_{TH}. The implication of this phenomenon in analog design is that if the length of a device is increased so as to achieve a higher output impedance, then the threshold voltage also increases by as much as 100 to 200 mV.

Another short-channel phenomenon related to the threshold voltage is "drain-induced barrier lowering" (DIBL). Recall from Chapter 2 that in weak inversion, as the gate voltage rises, the surface potential becomes more positive [Fig. 16.7(a)], attracting carriers from the source region. In short-channel devices, the *drain* voltage also makes the surface more positive by creating a two-dimensional field in the depletion region [6]. In essence, the drain introduces a capacitance C_d' that raises the surface potential in a manner similar to C_d. As a result, the barrier to the flow of charge and hence the threshold voltage are decreased. This effect manifests itself if the plot of Fig. 2.27 is drawn in both deep triode and saturation regions [Fig. 16.7(b)].

The principal impact of DIBL on circuit design is the degraded output impedance. This point is explained in Section 16.2.5.

16.2.2 Mobility Degradation with Vertical Field

At large gate-source voltages, the high electric field developed between the gate and the channel confines the charge carriers to a narrower region below the oxide-silicon interface,

---

*While intuitive, this explanation is not quite correct. More accurate descriptions can be found in books on semiconductor devices.

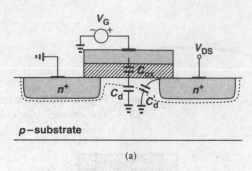

(a)

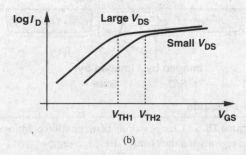

(b)

Figure 16.7 (a) DIBL in a short-channel device, (b) effect of DIBL on current characteristic.

leading to more carrier scattering and hence lower mobility. Since scaling has substantially deviated from the constant-field scenario, small-geometry devices experience significant mobility degradation. An empirical equation modeling this effect is

$$\mu_{eff} = \frac{\mu_0}{1 + \theta(V_{GS} - V_{TH})}, \tag{16.19}$$

where μ_0 denotes the "low-field" mobility and θ is a fitting parameter roughly equal to $(10^{-7}/t_{ox})$ V^{-1} [7]. For example, if $t_{ox} = 100$ Å, then $\theta \approx 1$ V^{-1} and the mobility begins to fall considerably as the overdrive exceeds 100 mV. Note that θ rises as t_{ox} drops because the electric field in the oxide becomes stronger.

In addition to lowering the current capability and transconductance of MOSFETs, mobility degradation deviates the I/V characteristic from the simple square-law behavior. Specifically, whereas a square-law device generates only even harmonics in its drain current in response to a sinusoidal gate-source voltage, Eq. (16.19) predicts odd harmonics as well. In fact, writing

$$I_D = \frac{1}{2} \frac{\mu_0 C_{ox}}{1 + \theta(V_{GS} - V_{TH})} \frac{W}{L} (V_{GS} - V_{TH})^2, \tag{16.20}$$

and assuming $\theta(V_{GS} - V_{TH}) \ll 1$, we obtain

$$I_D \approx \frac{1}{2} \mu_0 C_{ox} \frac{W}{L} [1 - \theta(V_{GS} - V_{TH})](V_{GS} - V_{TH})^2 \tag{16.21}$$

$$\approx \frac{1}{2} \mu_0 C_{ox} \frac{W}{L} \left[(V_{GS} - V_{TH})^2 - \theta(V_{GS} - V_{TH})^3 \right]. \tag{16.22}$$

This is a rough approximation but it reveals the existence of higher harmonics in the drain current.

The mobility degradation with the vertical field affects the device transconductance as well. This is studied in Problem 16.9.

16.2.3 Velocity Saturation

The mobility of carriers also depends on the *lateral* electric field in the channel, beginning to drop as the field reaches levels of 1 V/μm. Since the carrier velocity $v = \mu E$, we note that v approaches a saturated value, about 10^7 cm/s, for sufficiently high fields. Thus, as carriers enter the channel from the source and accelerate toward the drain, they may eventually reach a saturated velocity at some point along the channel.[3] In the extreme case, where carriers experience velocity saturation along the entire channel, we can rewrite Eq. (2.2) as

$$I_D = v_{sat} Q_d \tag{16.23}$$

$$= v_{sat} W C_{ox}(V_{GS} - V_{TH}). \tag{16.24}$$

Interestingly, the current is *linearly* proportional to the overdrive voltage and does not depend on the length. In fact, as shown in Fig. 16.8, I_D-V_{DS} characteristics of devices

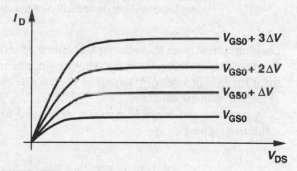

Figure 16.8   Effect of velocity saturation on drain current characteristics.

with $L < 1\ \mu$m reveal velocity saturation because equal increments in $V_{GS} - V_{TH}$ result in roughly equal increments in I_D. We also note that $g_m = v_{sat} W C_{ox}$, concluding that the transconductance is a weak function of the drain current and channel length in the velocity-saturation regime.

Under typical bias conditions, MOSFETs experience some velocity saturation, displaying a characteristic between linear and square-law behavior. An important consequence is that, as V_{GS} increases, the drain current saturates well before pinch-off occurs. As shown in Fig. 16.9(a), carriers reach velocity saturation if V_{DS} exceeds $V_{D0} < V_{GS} - V_{TH}$, yielding a

---

[3]Even in long-channel devices, carriers experience velocity saturation if the drain-source voltage is high enough to pinch off the channel. At the pinch-off point, the mobile charge density is near zero, the electric field is very large, and hence the velocity of carriers is saturated.

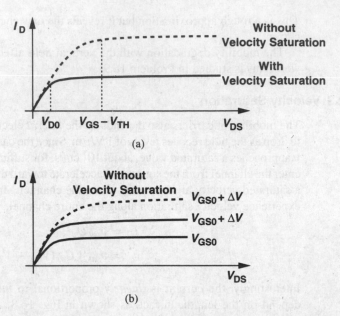

Figure 16.9 Effect of velocity saturation: (a) premature drain current saturation, (b) reduction of transconductance.

constant current quite lower than that obtained if the device saturated for $V_{DS} > V_{GS} - V_{TH}$. Furthermore, as illustrated in Fig. 16.9(b), since an increment in V_{GS} gives a smaller increment for I_D when velocity saturation occurs, the transconductance is also lower than that predicted by the square law.

A compact and versatile equation developed to represent velocity saturation (in the saturation region) is

$$I_D = W C_{ox} v_{sat} \frac{(V_{GS} - V_{TH})^2}{V_{GS} - V_{TH} + 2\frac{v_{sat}L}{\mu_{eff}}}, \tag{16.25}$$

where μ_{eff} is given by Eq. (16.19) [7, 8]. The same work provides the following equation for the drain-source voltage at the onset of premature saturation [V_{D0} in Fig. 16.9(a)]:

$$V_{DS,sat} = \frac{2\mu_{eff}L(V_{GS} - V_{TH})}{2\mu_{eff}L + V_{GS} - V_{TH}}. \tag{16.26}$$

Equation (16.25) provides two interesting results. First, if L or v_{sat} is large, the expression reduces to the square-law relationship. Second, if the *overdrive* voltage is so small that the denominator of (16.25) is approximated as $2v_{sat}L/\mu_{eff}$ and $\mu_{eff} \approx \mu_0$, then the device still follows the square-law behavior even if L is relatively small. For example, if $v_{sat} \approx 10^7$ cm/s, $L = 0.25$ μm, and $\mu_0 \approx 350$ cm²/V/s, we have $2v_{sat}L/\mu_0 \approx 1.43$ V, recognizing that for overdrive voltages of a few hundred millivolts, the transistor operation is somewhat close to the square law. Thus, the simplified treatment of Chapter 2 can still provide insight for many analog applications.

Equation (16.25) can be further simplified to yield additional results. Substituting for μ_{eff} from Eq. (16.19), we have

$$I_D = W C_{ox} v_{sat} \frac{(V_{GS} - V_{TH})^2}{V_{GS} - V_{TH} + \dfrac{2v_{sat}L}{\mu_0}[1 + \theta(V_{GS} - V_{TH})]} \tag{16.27}$$

$$= W C_{ox} v_{sat} \frac{(V_{GS} - V_{TH})^2}{\dfrac{2v_{sat}L}{\mu_0} + \left(1 + \dfrac{2v_{sat}L\theta}{\mu_0}\right)(V_{GS} - V_{TH})} \tag{16.28}$$

$$= \frac{1}{2}\mu_0 C_{ox} \frac{W}{L} \frac{(V_{GS} - V_{TH})^2}{1 + \left(\dfrac{\mu_0}{2v_{sat}L} + \theta\right)(V_{GS} - V_{TH})}. \tag{16.29}$$

This equation is similar to (16.20), implying that the degradation of the mobility with both lateral and vertical fields can be represented by adding the terms $\mu_0/(2v_{sat}L)$ and θ. Thus, the results obtained from (16.20) apply here as well. For example, the drain current contains high-order nonlinear terms. Equation (16.29) can also predict the transconductance (Problem 16.10).

16.2.4 Hot Carrier Effects

Short-channel MOSFETs may experience high lateral electric fields if the drain-source voltage is large. While the *average* velocity of carriers saturates at high fields, the instantaneous velocity and hence the kinetic energy of the carriers continue to increase, especially as they accelerate towards the drain. These are called "hot" carriers [2].

In the vicinity of the drain region, hot carriers may "hit" the silicon atoms at high speeds, thereby creating impact ionization. As a result, new electrons and holes are generated, with the electrons absorbed by the drain and the holes by the substrate. Thus, a finite drain-substrate current appears. Also, if the carriers acquire a very high energy, they may be injected into the gate oxide and even flow out the gate terminal, introducing a gate current. The substrate and gate currents are often measured to study hot carrier effects.

The scaling of technologies proceeds so as to minimize hot carrier effects. This limitation and other breakdown phenomena make the supply voltage scaling inevitable.

16.2.5 Output Impedance Variation with Drain-Source Voltage

In modeling channel-length modulation by a single constant λ, we have assumed that the output impedance of the transistor, r_O, is constant in the saturation region. In reality, however, r_O varies with V_{DS}. As V_{DS} increases and the pinch-off point moves toward the source, the rate at which the depletion region around the source becomes wider decreases, resulting in a higher incremental output impedance. Illustrated in Fig. 16.10, this effect is somewhat similar to the variation of the capacitance of a reversed-biased *pn* junction: with

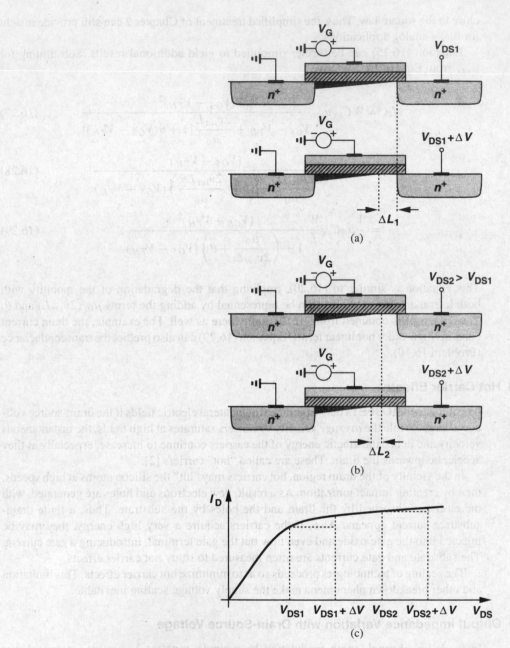

Figure 16.10 Decrement in channel length for (a) small V_{DS} and (b) large V_{DS}.

a small reverse bias, the width of the depletion region is a strong function of the voltage applied to the junction and with a large reverse bias, a weak function.

In this regime, the output impedance can be approximated as

$$r_O = \frac{2L}{1 - \frac{\Delta L}{L}} \frac{1}{I_D} \sqrt{\frac{qN_B}{2\epsilon_{si}}(V_{DS} - V_{DS,sat})}, \tag{16.30}$$

where $V_{D,sat}$ is the drain-source voltage at the onset of pinch-off [9]. Another approximation developed in conjunction with (16.25) and (16.26) is described in [8].

In short-channel devices, as V_{DS} increases further, drain-induced barrier lowering becomes significant, reducing the threshold voltage and increasing the drain current. This effect roughly cancels that expressed by (16.30), giving a relatively constant output impedance. At sufficiently high drain voltages, impact ionization near the drain produces a large current (flowing from the drain into the substrate), in essence lowering the output impedance. The overall behavior of r_O is plotted in Fig. 16.11.

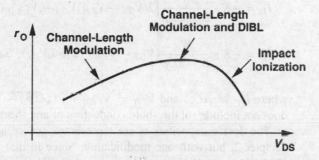

Figure 16.11 Overall variation of output resistance as a function of V_{DS}.

The variation of r_O gives rise to nonlinearity in many circuits. In a cascode op amp, for example, as the output voltage varies, so does the output impedance of the cascode devices and hence the voltage gain of the circuit. Furthermore, impact ionization limits the maximum gain that can be obtained from cascode structures because it introduces a small-signal resistance from the drain to the *substrate* rather than to the source.

6.3 MOS Device Models

Since the introduction of the first MOS model in the mid-1960s [10], tremendous research has been expended on improving the accuracy of models as device dimensions scale down. Developed between the mid-1960s and the late 1970s, the Level 1, 2, and 3 models consecutively included higher order effects so as to provide reasonable accuracy with respect to measured transistor characteristics for channel lengths as small as 1 μm. Following this set were the Compact Short-Channel IGFET Model (CSIM) from AT&T Bell Laboratories and the Berkeley Short-Channel IGFET Model (BSIM) from University of California, Berkeley in the mid-1980s. These models proved inadequate for analog design and were followed by BSIM2, HSPICE level 28, BSIM3, and a number of others in the late 1980s and early 1990s.

MOS device modeling continues to pose a challenge—especially for high-frequency operation—because even today's sophisticated models become inadequate after one or two technology generations (e.g., from 0.5 μm to 0.35 μm to 0.25 μm). Our objective is to develop a basic understanding of some of the models to the extent necessary for simulations. We should also mention that the utility of a model is given by the accuracy it provides in various regions of operation for different device dimensions, the ease with which its parameters can be measured, and the efficiency that it allows in simulations. The interested reader is referred to [1] for an in-depth coverage.

16.3.1 Level 1 Model

Also known as the Shichman and Hodges Model [10], this representation uses the parameters listed in Table 2.1 and is based on the following equations:

$$I_D = \frac{1}{2} K_P \frac{W}{L - 2L_D} [2(V_{GS} - V_{TH})V_{DS} - V_{DS}^2](1 + \lambda V_{DS}) \quad \text{Triode Region} \quad (16.31)$$

$$I_D = \frac{1}{2} K_P \frac{W}{L - 2L_D} (V_{GS} - V_{TH})^2 (1 + \lambda V_{DS}) \quad \text{Saturation Region} \quad (16.32)$$

where $K_P = \mu C_{ox}$ and $V_{TH} = V_{TH0} + \gamma(\sqrt{2\phi_B - V_{BS}} - \sqrt{2\phi_B})$. Note that this model does not include subthreshold conduction or any short-channel effects.

The device capacitances are represented according to the simple model described in Chapter 2, but with one modification. Since in that model, C_{GS} abruptly changes from $(2/3)WLC_{ox} + WC_{ov}$ in saturation to $(1/2)WLC_{ox} + WC_{ov}$ in the triode region [and C_{GD} from WC_{ov} to $(1/2)WLC_{ox} + WC_{ov}$], most computation algorithms experience convergence difficulties here. For this reason, C_{GS} and C_{GD} in the triode region are formulated as

$$C_{GS} = \frac{2}{3} WLC_{ox} \left\{ 1 - \frac{(V_{GS} - V_{DS} - V_{TH})^2}{[2(V_{GS} - V_{TH}) - V_{DS}]^2} \right\} + WC_{ov} \quad (16.33)$$

$$C_{GD} = \frac{2}{3} WLC_{ox} \left\{ 1 - \frac{(V_{GS} - V_{TH})^2}{[2(V_{GS} - V_{TH}) - V_{DS}]^2} \right\} + WC_{ov} \quad (16.34)$$

$$C_{GB} = 0. \quad (16.35)$$

We note that if the device operates at the edge of saturation, $V_{GS} - V_{DS} = V_{TH}, C_{GS} = (2/3)WLC_{ox} + WC_{ov}$, and $C_{GD} = WC_{ov}$. Thus, the capacitance values change continuously from one region to another.

The Level 1 model maintains reasonable I/V accuracy for channel lengths as small as roughly 4 μm, but it still predicts the output impedance of transistors in saturation quite poorly.

16.3.2 Level 2 Model

The Level 1 model began to manifest its shortcomings as channel lengths fell below approximately 4 μm. The Level 2 model was then developed to represent many high-order effects.

An assumption that we made in Chapter 2 in deriving the square-law characteristics was a constant threshold voltage along the channel. This assumption is not correct even for long-channel devices because the charge in the depletion region under the channel varies according to the local voltage (Fig. 16.12). Since the inversion layer and the depletion region

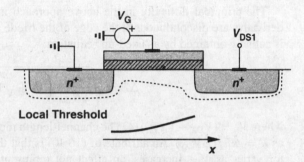

Figure 16.12   Variation of threshold along the channel.

must image the charge on the gate, as the inversion layer vanishes in the direction toward the drain, the depletion region must enclose more charge. Performing the integration in Section 2.2.2 with a varying threshold voltage yields [1]:

$$I_D = \mu C_{ox} \frac{W}{L} \left\{ (V_{GS} - V_{TH0})V_{DS} - \frac{V_{DS}^2}{2} \right.$$

$$\left. - \frac{2}{3}\gamma \left[(V_{DS} - V_{BS} + 2\phi_F)^{3/2} - (-V_{BS} + 2\phi_F)^{3/2} \right] \right\}. \quad (16.36)$$

Interestingly, even for $V_{BS} = 0$, I_D exhibits some dependence on γ. Moreover, for small V_{DS}, the equation reduces to that of the Level 1 model, but for large V_{DS} the drain current is less than that predicted by the square law. It can also be shown that the edge of the saturation region is given by [1]:

$$V_{D,sat} = V_{GS} - V_{TH0} - \phi_F + \gamma^2 \left[1 - \sqrt{1 + \frac{2}{\gamma^2}(V_{GS} - V_{TH0} + \phi_F)} \right]. \quad (16.37)$$

In the saturation region, the drain current is

$$I_{DS} = I_{D,sat} \frac{1}{1 - \lambda V_{DS}}, \quad (16.38)$$

where $I_{D,sat}$ is calculated from (16.36) for $V_{DS} = V_{DS,sat}$.

Modeling channel-length modulation or, more generally, the finite output impedance has always remained a difficult problem. Representing such phenomena by only λ is far from accurate. In the Level 2 implementation, if λ is not specified, it is obtained by calculating the width of the depletion region between the pinch-off point and the edge of the drain. Using simple relationships for the depletion region of a *pn* junction, we can write

$$\Delta L = \sqrt{\frac{2\epsilon_{si}}{qN_{sub}}[\phi_B + (V_{DS} - V_{D,sat})]}, \qquad (16.39)$$

where $V_{D,sat}$ denotes the pinch-off voltage.[4]

The principal difficulty in the above approach is that both the drain current and it derivative are discontinuous at the edge of the triode region [1]! To resolve this issue, ΔL is actually obtained by a "fixed-up" equation:

$$\Delta L = \sqrt{\frac{2\epsilon_{si}}{qN_{sub}}\left(V_1 + \sqrt{1 + V_1^2}\right)}, \qquad (16.40)$$

where $V_1 = (V_{DS} - V_{D,sat})/4$. The channel-length modulation coefficient is then expressed as $\lambda = \Delta L/(LV_{DS})$. An attribute of (16.40) is that the output conductance of the transistor varies as V_{DS} increases, an effect not represented by the first-order model using a constant λ.

The Level 2 model also includes the degradation of the mobility with the vertical field in the channel. The mobility is calculated from

$$\mu_s = \mu_0\left(\frac{\epsilon_{si}}{C_{ox}} \cdot \frac{U_c}{V_{GS} - V_{TH} - U_t V_{DS}}\right)^{U_e}, \qquad (16.41)$$

where U_c denotes the gate-channel critical electric field, U_t is a fitting parameter between 0 and 0.5, and U_e is an exponent in the vicinity of 0.15.

The subthreshold behavior implemented in the Level 2 model defines a voltage V_{on} as $V_{on} = V_{TH} + \zeta V_T$, where $\zeta = 1 + (qN_{FS}/C_{ox}) + C_d/C_{ox}$, and N_{FS} is an empirical constant. The drain current is then expressed as

$$I_{DS} = I_{on}\exp\frac{V_{GS} - V_{on}}{\zeta V_T}, \qquad (16.42)$$

where I_{on} is the drain current calculated in strong inversion [Eq. (16.36)] for $V_{GS} = V_{on}$. An important drawback of this representation is the discontinuity in the slope of I_D as the device goes from the subthreshold region to strong inversion (Fig. 16.13), leading to various difficulties and errors in simulation.

In addition to the above effects, the Level 2 model represents two other short-channel phenomena: the variation of V_{TH} with L, and velocity saturation. The implementation of these effects is quite involved and can be found in [1].

Measured data [1] indicate that the Level 2 model provides reasonable I/V accuracy for wide, short devices in the saturation region with $L \approx 0.7$ μm but it suffers from substantial

[4]The junction is considered "one-sided" here, i.e., the drain doping level is much higher.

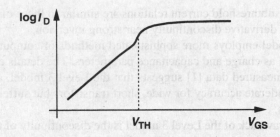

Figure 16.13 Kink in drain current characteristic in Level 2 model.

error in representing the output impedance and the transition point between saturation and triode regions. For narrow or long devices, the model is quite inaccurate.

16.3.3 Level 3 Model

The Level 3 model realization is somewhat similar to the Level 2 model, with some equations simplified and many empirical constants introduced to improve the accuracy for channel lengths as small as 1 μm.

This model expresses the threshold voltage as

$$V_{TH} = V_{TH0} + F_s\gamma\sqrt{2\phi_F - V_{BS}} + F_n(2\phi_F - V_{BS}) + \xi\frac{8.15 \times 10^{-22}}{C_{ox}L_{eff}^3}V_{DS}, \quad (16.43)$$

where F_s and F_n represent short-channel and narrow-channel effects,[5] respectively, and ξ models drain-induced barrier lowering.

The mobility equation involves both vertical and lateral field effects and is expressed as:

$$\mu_1 = \frac{\mu_{eff}}{1 + \frac{\mu_{eff}V_{DS}}{v_{max}L_1}}, \quad (16.44)$$

where

$$\mu_{eff} = \frac{\mu_0}{1 + \theta(V_{GS} - V_{TH})}, \quad (16.45)$$

and v_{max} denotes the maximum velocity of the carriers in the channel. As can be seen from (16.44) and (16.45), μ_{eff} models the effect of the vertical field while μ_1 adds that of the lateral field as well.

The drain current is realized as:

$$I_D = \mu_1 C_{ox}\frac{W_{eff}}{L_{eff}}\left[V_{GS} - V_{TH0} - \left(1 + \frac{F_s\gamma}{4\sqrt{2\phi_F - V_{BS}}} + F_n\right)\frac{V'_{DS}}{2}\right]V'_{DS}, \quad (16.46)$$

where $V'_{DS} = V_{D,sat}$ if the device is in saturation. The quantity $V_{D,sat}$ represents both channel pinch-off and velocity saturation (Fig. 16.9) and is expressed by relatively complex

---

[5]For narrow-channel devices, the threshold voltage *increases* if the *width* is reduced [6].

equations [1]. The subthreshold current relations are similar to those of the Level 2 model, still suffering from derivative discontinuity near strong inversion.

The Level 3 model employs more sophisticated methods of computing channel-length modulation as well as charge and capacitance parameters. The details can be found in [1]. Comparison with measured data [1] suggests that the Level 3 model, as with the Level 2 model, exhibits moderate accuracy for wide, short transistors but suffers from large errors for longer channels.

An important drawback of the Level 3 model is the discontinuity of the derivative of I_D with respect to V_{DS} at the edge of the triode region, leading to large errors in the calculation of the output impedance. Shown in Fig. 16.14 for a short-channel device, the variation of r_O with V_{DS} is quite poorly modeled.

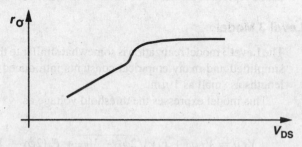

Figure 16.14   Kink in output resistance in Level 3 model.

16.3.4 BSIM Series

The philosophy behind the Level 1–3 models was to express the device behavior by means of equations that originated from the physical operation. However, as transistors were scaled to submicron dimensions, it became increasingly more difficult to introduce physically meaningful equations that would be both accurate and computationally efficient. BSIM adopted a different approach: numerous empirical parameters were added so as to simplify the equations—but at the cost of losing touch with the actual device operation.

An interesting feature of BSIM is the addition of a simple equation to represent the geometry dependence of many of the device parameters. The general expression is of the form:

$$P = P_0 + \frac{\alpha_P}{L_{eff}} + \frac{\beta_P}{W_{eff}}, \qquad (16.47)$$

where P_0 is the value of the parameter for a long, wide transistor ($P = P_0$ if $L_{eff}, W_{eff} \to \infty$), and α_P and β_P are fitting factors. For example, the mobility is computed as:

$$\mu = \mu_0 + \frac{\alpha_\mu}{L_{eff}} + \frac{\beta_\mu}{W_{eff}}. \qquad (16.48)$$

The formulation of (16.47) nonetheless becomes less accurate at small dimensions [1].

The device equations and fitting parameters used in BSIM are beyond the scope of this book. Using approximately 50 parameters, this model provides the following improvements over the Level 3 version [1]: (1) the dependence of mobility upon the vertical field includes the substrate voltage; (2) the threshold voltage is modified for substrates with nonuniform doping; (3) the currents in the weak and strong inversion regions are derived such that their values and first derivatives are continuous; (4) to simplify the drain current equations, new expressions are devised for velocity saturation, dependence of mobility upon the lateral field, and the saturation voltage.

Measured results in a 0.7-μm technology [1] indicate that BSIM avoids gross errors in the I/V characteristics for various device dimensions, but its accuracy for narrow, short transistors is somewhat poor.

In addition to shortcomings at channel lengths below approximately 0.8 μm, BSIM suffers from other subtle inaccuracies. For example, at large drain-source voltages, BSIM predicts a *negative* output resistance for saturated MOSFETs. Furthermore, in deep triode region, BSIM still exhibits slight discontinuities in the drain current [1].

The next model in the BSIM series is BSIM2. Requiring approximately 70 parameters, this version employs new expressions for mobility, drain current, and subthreshold conduction. It also represents the output impedance more accurately by incorporating both channel-length modulation and drain-induced barrier lowering. Nevertheless, measured results indicate that the overall accuracy of the model is only marginally higher than that of BSIM. For short, narrow transistors, BSIM2 suffers from large errors in the triode region and even substantial "kinks" in the saturation region [1].

The trend in BSIM and BSIM2, namely, expressing the device behavior by means of empirical equations that bear little relation to the physical phenomena, eventually created difficulties in modeling short-channel devices. Parameter extraction, modeling process variations, and the need for extensive use of polynomials made the generation and application of these models quite difficult. Consequently, the next generation, BSIM3, has returned to the physical principles of device operation while maintaining many of the useful features of BSIM and BSIM2. BSIM3 itself has rapidly gone through several versions, requiring approximately 180 parameters in the third one. For channel lengths as low as 0.25 μm, BSIM3 provides reasonable accuracy for subthreshold and strong inversion operation while still suffering from large errors in predicting the output impedance.

16.3.5 Other Models

In addition to the Level 1–3 models and the three generations of BSIM, a number of other MOS models have been introduced. Among these, HSPICE Level 28, MOS9, and the Enz-Krummenacher-Vittoz (EKV) model are the most notable, for they provide new approaches to representing the behavior of MOSFETs [1]. For example, the HSPICE Level 28 model improves the dependence of accuracy upon device dimensions by expressing the parameters as:

$$P = P_0 + \alpha \left(\frac{1}{L} - \frac{1}{L_{ref}} \right) + \beta \left(\frac{1}{W} - \frac{1}{W_{ref}} \right) + \gamma \left(\frac{1}{L} - \frac{1}{L_{ref}} \right) \left(\frac{1}{W} - \frac{1}{W_{ref}} \right),$$

(16.49)

where L_{ref} and W_{ref} denote the dimensions of a "reference" device, i.e., a transistor whose characteristics have been measured. Thus, the dependence is expressed in terms of *increments* with respect to characterized transistors rather than the absolute value of the dimensions, yielding a potentially higher accuracy. Also, the term proportional to the product of the length and width increments facilitates curve fitting.

The EKV model [11] substantially departs from traditional views of MOSFET operation by considering the *bulk*, rather than the source, as the reference point for all voltages. This approach thus avoids distinguishing between the source and drain terminals and, more importantly, introduces a single drain-source current equation that is valid for both subthreshold and saturation regions.

The reader is referred to [1] for an extensive study of these models.

16.3.6 Charge and Capacitance Modeling

The simple gate capacitance model described in Chapter 2 for the Level 1 model, called the Meyer capacitance model [1], suffers from many shortcomings even for long-channel devices. In transient SPICE analyses, such a model does not conserve charge (!), thereby introducing errors in the simulation. For example, as illustrated in Fig. 16.15, a periodic

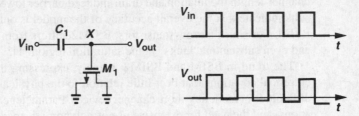

Figure 16.15 Annihilation of charge in simulation.

rectangular waveform applied to a voltage divider consisting of an ideal capacitor and a MOSFET experiences "droop" at the output because in every period some charge at node X is lost. This effect arises from the calculation of charge by integrating capacitor voltages with respect to time, an operation that accumulates small errors in the simulation.[6] To minimize this type of error, the simulation algorithm can be modified such that it first computes the charge in the inversion layer and the depletion region and subsequently partitions the charge among the device capacitances.

Another issue in the Meyer charge model relates to partitioning of the channel charge between the source and drain terminals. The assumption that in the triode region, $C_{GS} = C_{GD} = (1/2)WLC_{ox} + WC_{ov}$, and in the saturation region, $C_{GS} = (2/3)WLC_{ox} + WC_{ov}$ and $C_{GD} = WC_{ov}$ is quite inaccurate for short-channel devices, requiring flexible partitioning for ease of curve fitting. In BSIM and BSIM3, for example, three different charge partitioning scenarios (40%/60%, 50%/50%, and 0%/100%) are available.

Recent efforts have created more sophisticated charge and capacitance models for MOS devices so as to improve the accuracy, especially for analog applications. However, as with many other modeling improvements, the resulting equations are quite cumbersome, imparting little intuition. The reader is referred to [1] for details.

---

[6]Another source of error here is the assumption that the device capacitances are reciprocal, e.g., $C_{GS} = C_{SG}$ [1].

16.3.7 Temperature Dependence

Many parameters of MOS transistors vary with temperature, making it difficult to maintain a reasonable fit between measured and simulated behavior across a wide temperature range. In the Level 1–3 models as well as BSIM and BSIM2, the following parameters have temperature dependence: V_{TH}, built-in potential of S/D junctions, the intrinsic carrier concentration of silicon (n_i), the bandgap energy (E_g), and the mobility. Most equations are empirical, e.g.,

$$E_g = 1.16 - \frac{7.02 \times 10^{-4} T^2}{T + 1108},$$ (16.50)

and

$$\mu = \mu_0 \left(\frac{300}{T}\right)^{3/2},$$ (16.51)

where $\mu_0 = \mu(T = 300° K)$.

BSIM3 incorporates a few more parameters to represent the temperature dependence of phenomena such as velocity saturation and the effect of subthreshold voltage on V_{TH}. It is unclear at this point how accurately BSIM3 expresses the temperature variation of MOS devices and circuits.

16.4 Process Corners

Unlike bipolar transistors, MOSFETs suffer from substantial parameter variations from wafer to wafer and from lot to lot. Despite decades of technology advancement, the large variability of CMOS circuits remains a fact with which digital and analog designers must cope.

In order to facilitate the task of circuit design to some extent, process engineers guarantee a performance envelope for the devices, in essence tightening the anticipated parameter variations by discarding wafers that fall out of the envelope (Fig. 16.16). Of course, in

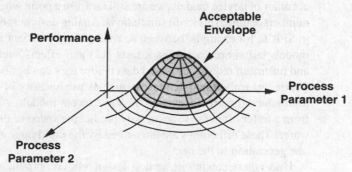

Figure 16.16 Performance envelope as a function of process parameters.

their eternal battle, circuit designers insist on a tighter variability space so that they can design more aggressively whereas process engineers tend to enlarge the envelope as much as possible so as to increase the yield. For example, it is common in today's CMOS technologies to obtain a gate delay that varies by a factor of two to one with process and temperature.

The performance envelope furnished to designers has traditionally been one suited to digital circuits and constructed in the form of "process corners." Illustrated in Fig. 16.17,

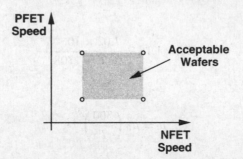

Figure 16.17 Process corners based on speed of NMOS and PMOS devices.

the idea is to constrain the speed envelope of the NMOS and PMOS transistors to a rectangle defined by four corners: fast NFET and fast PFET; slow NFET and slow PFET; fast NFET and slow PFET; and slow NFET and fast PFET. For example, transistors having a thinner gate oxide and lower threshold voltage fall near the fast corner. The device models corresponding to each corner are extracted from wafers whose NMOS or PMOS test structures display a large or small gate delay, and the actual corners are chosen so as to obtain an acceptable yield. Thus, only wafers satisfying these specifications are considered acceptable. Simulation of circuits for various process corners and temperature extremes is essential to determining the yield.

16.5 Analog Design in a Digital World

Memories and processors constitute the major portion of today's semiconductor business. Thus, as explained in Chapters 17 and 18, most CMOS technologies are designed, optimized, and *characterized* for digital applications. Despite the increasing emphasis on the "analog" accuracy of device models, we are still far from a point where we can fully trust the absolute numbers obtained in circuit simulations. Analog designers routinely encounter discrepancies in SPICE, for example, between ac analysis and transient analysis. Moreover, many device models fail simple benchmark tests [12] and effects such as flicker (and thermal) noise and mismatch require measured data before they can be accurately reflected in simulations. Subtle, yet important phenomena such as nonlinearity of the device output resistance are represented incorrectly even in the most recent models. Also, the device models extracted from a wafer often fail to accurately predict the speed of the circuits fabricated on the same wafer! These difficulties are intensified by the rapid migration of CMOS technologies from one generation to the next.

Under these conditions, analog design relies on experience, intuition, and *measured* data. In fact, the design of complex, high-performance analog circuits may require data points that can be obtained only by first fabricating and characterizing many simpler test circuits [13].

Problems

Unless otherwise stated, in the following problems, use the device data shown in Table 2.1 and assume $V_{DD} = 3$ V where necessary. Also, assume all transistors are in saturation.

16.1. Silicon dioxide breaks down at high electric fields. Explain what happens if ideal scaling is performed while keeping the gate oxide thickness constant.

16.2. The maximum doping level that can be established in the source and drain regions is limited by the "solid solubility" of silicon. Explain what happens to the S/D junction capacitance and series resistance as ideal scaling occurs but the S/D doping level remains constant. Does DIBL become more or less significant?

16.3. Suppose the supply voltage of a switched-capacitor amplifier is reduced by a factor of two and so is the maximum allowable output voltage swing. In order to maintain the dynamic range constant, the noise voltage must scale down by the same factor.
   (a) If the noise is only of kT/C type, how should the capacitors in the circuit be scaled?
   (b) If the time constant is given by G_m/C, where G_m denotes the transconductance of a one-stage op amp, how should G_m be scaled to maintain the same small-signal time constant?
   (c) How should the dimensions and tail current of the input differential pair of the op amp be scaled?
   (d) Repeat parts (b) and (c) where the slew rate must remain constant.

16.4. Explain how each parameter in Eq. (16.16) scales in an ideal constant-field scaling scenario. What happens to the subthreshold slope?

16.5. A common-gate stage designed for an input impedance of 50 Ω undergoes ideal scaling. If $\lambda = \gamma = 0$, what is the input impedance?

16.6. Repeat Problem 16.5 if $\lambda \neq 0$, $\gamma \neq 0$, and the load is a MOS current source that is also scaled.

16.7. For power-conscious applications, a figure of merit is defined as the transconductance of devices normalized to their bias current. Determine this quantity for long-channel devices operating in strong inversion or the subthreshold region. At what drain current are these two equal?

16.8. Explain why the mobile charge density cannot drop to exactly zero at any point along the channel. What happens beyond the pinch-off point?

16.9. Using Eq. (16.20), calculate the transconductance of a MOSFET. What happens if the overdrive voltage is very small or very large?

16.10. Using Eq. (16.29), calculate the transconductance of a MOSFET. Prove that

$$g_m = \frac{I_D}{V_{GS} - V_{TH}} \left[1 + \frac{1}{1 + \left(\dfrac{\mu_0}{2v_{sat}L} + \theta \right)(V_{GS} - V_{TH})} \right]. \qquad (16.52)$$

16.11. Suppose the channel-length modulation coefficient λ is modified as $\lambda/(1 + \kappa V_{DS})$, where κ is a constant, to represent the dependence of the output impedance upon V_{DS}. Calculate r_O. Explain how a current source with such behavior introduces distortion in the voltage across it.

16.12. Assuming the devices in Fig. 16.18 experience complete velocity saturation, derive expressions for the voltage gain of each circuit in terms of W and v_{sat}. Assume $\lambda = \gamma = 0$.

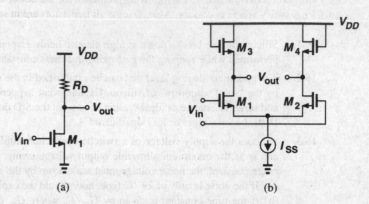

Figure 16.18

16.13. Using Eq. (16.36), calculate g_{mb} and compare the result with that derived in Chapter 2.

16.14. From Eq. (16.50), determine $\partial E_g / \partial T$ at room temperature and explain how it affects bandgap reference voltages.

16.15. Suppose the fast corners of a process result from a higher μC_{ox}. Explain what happens to the voltage gain and the input thermal noise of the circuits shown in Fig. 16.19 at the four corners of the process if the transistors are biased at a constant current in saturation.

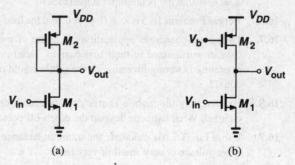

Figure 16.19

16.16. Repeat Problem 16.15 if each transistor is biased with a fixed V_{GS}.

References

1. D. P. Foty, *MOSFET Modeling with SPICE,* Upper Saddle River, NJ: Prentice-Hall, 1997.
2. Y. Tsividis, *Operation and Modeling of the MOS Transistor,* Second Ed., New York: McGraw-Hill, 1999.
3. P. Antognetti and G. Massobrio, Editors, *Semiconductor Device Modeling with SPICE,* New York, McGraw-Hill, 1988.

4. R. H. Dennard et al., "Design of Ion-Implanted MOSFETs with Very Small Physical Dimensions," *IEEE J. of Solid-State Circuits,* vol. 9, pp. 256–268, Oct. 1974.

5. G. E. Moore, "Progress in Digital Integrated Circuits," *IEDM Tech. Dig.,* pp. 11–14, Dec. 1975.

6. Y. Taur and T. H. Ning, *Fundamentals of Modern VLSI Devices,* New York: Cambridge University Press, 1998.

7. C. G. Sodini, P. K. Ko, and J. L. Moll, "The Effect of High Fields on MOS Device and Circuit Performance," *IEEE Tran. on Electron Devices,* vol. 31, pp. 1386–1393, Oct. 1984.

8. P. K. Ko, "Approaches to Scaling," pp. 1–35, in *Advanced MOS Device Physics,* N.G. Einspruch and G. Gildenblat, Editors, San Diego: Academic Press, 1998.

9. S. Wong and A. T. Salama, "Impact of Scaling on MOS Analog Performance," *IEEE J. of Solid-State Circuits,* vol. 18, pp. 106–114, Feb. 1983.

10. H. Shichman and D. A. Hodges, "Modeling and Simulation of Insulated Field Effect Transistor Switching Circuits, *IEEE J. of Solid-State Circuits,* vol. 3, pp. 285–289, Sept. 1968.

11. C. C. Enz, F. Krummenacher, and E. Vittoz, "An Analytical MOS Transistor Model Valid in All Regions of Operation and Dedicated to Low Voltage and Low Current Applications," *Analog Integrated Circuits and Signal Processing,* vol. 8, pp. 83–114, 1995.

12. Y. Tsividis and K. Suyama, "MOSFET Modeling for Analog Circuit CAD: Problems and Prospects," *IEEE J. of Solid-State Circuits,* vol. 29, pp. 210–216, March 1994.

13. B. Razavi, "CMOS Technology Characterization for Analog and RF Design," *IEEE J. of Solid-State Circuits,* vol. 34, pp. 268–276, March 1999.

Chapter 17

CMOS Processing Technology

With the high-order effects of MOS devices covered in Chapter 16, we now study the fabrication of CMOS technologies. A solid understanding of device processing proves essential in the design and layout of ICs because many limitations imposed on the performance of circuits are related to fabrication issues. Furthermore, today's semiconductor technology demands that process engineers and circuit designers interact regularly so as to understand each other's needs, necessitating a good knowledge of each discipline.

In this chapter, we deal with the processing technology of CMOS devices, aiming to provide a simple view of the fabrication steps and their relevance to circuit design and layout. We begin with a brief description of basic fabrication steps such as wafer processing, photolithography, oxidation, ion implantation, deposition, and etching. Next, we study the fabrication sequence of MOS transistors in detail. Finally, we describe the processing of passive devices and interconnections.

17.1 General Considerations

Before delving into a detailed study of fabrication, it is instructive to consider the basic structure of NMOS and PMOS transistors and predict the required processing steps. As shown in Fig. 17.1, a p-type substrate (wafer) serves as the foundation upon which n-wells, source/drain regions, gate dielectric, polysilicon, n-well and substrate ties, and metal interconnects are built. Considering both the side view and the top view, we may raise the following questions: (1) How are various regions defined so accurately? For example, how is a gate polysilicon line with a minimum dimension of 0.25 μm fabricated while maintaining a distance of 0.25 μm from another polysilicon line? (2) How are the n-wells and S/D regions built? (3) How are the gate oxide and polysilicon fabricated? (4) How are the gate oxide and polysilicon *aligned* with the S/D regions? (5) How are the contact windows created? (6) How are the metal interconnect layers deposited?

Modern CMOS technologies involve more than 200 processing steps, but for our purposes, we can view the sequence as a combination of the following operations: (1) wafer processing to produce the proper type of substrate; (2) photolithography to precisely define

Top View

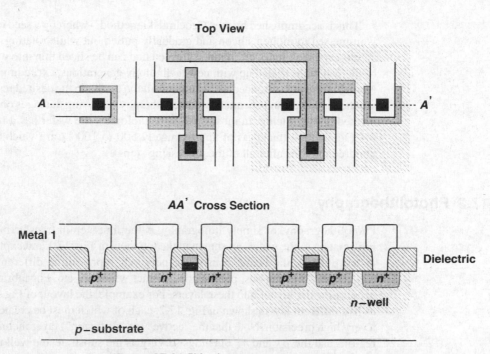

Figure 17.1 Side view and top view of MOS devices.

each region; (3) oxidation, deposition, and ion implantation to *add* materials to the wafer; (4) etching to *remove* materials from the wafer. Many of these steps require "heat treatment," i.e., the wafer must undergo a thermal cycle inside a furnace.

In semiconductor processing and characterization, we often refer to the "sheet resistance" of a layer. The total resistance of a rectangular bar is $R = \rho L/(W \cdot t)$, where ρ is the resistivity of the material, and L, W, and t, denote the length, width, and thickness of the bar, respectively. In integrated circuits, the resistivity and thickness of the layers are set by fabrication materials and processing steps and cannot be changed in the layout. The quantity $R_\square = \rho/t$ is thus defined as the sheet resistance, combining two constants of the technology. Since $R = R_\square$ for $W = L$, i.e., for a square geometry, we express $R_\square$ in terms of ohms per square. For example, for a sheet resistance of 10 $\Omega/\square$, a geometry with $W = 2$ μm and $L = 20$ μm has a resistance of $R = 10 \ \Omega/\square \times (20/2) = 100 \ \Omega$. In fact, we may say "this line is 10 squares long," meaning that $L/W = 10$ and $R = 10R_\square$.

17.2 Wafer Processing

The starting wafer in a CMOS technology must be created with a very high quality. That is, the wafer must be grown as a single-crystal silicon body having a very small number of "defects," e.g., dislocations in the crystal or unwanted impurities. Furthermore, the wafer must contain the proper type and level of doping so as to achieve the required resistivity.

This is accomplished by the "Czochralski method," whereby a seed of crystalline silicon is immersed in molten silicon and gradually pulled out while rotating. As a result, a large single-crystal cylindrical "ingot" is formed that can be sliced thin into wafers. The diameter of the wafer has scaled up with new technology generations, exceeding 20 cm (8 in) today. Note that dopants are added to the molten silicon to obtain the desired resistivity. The wafers are then polished and chemically etched, thereby removing damages on the surface that are created during slicing. In most CMOS technologies, the wafer has a resistivity of 0.05 to 0.1 Ω·cm and a thickness of approximately 500 to 1000 μm (which is reduced to a few hundred microns after all of the processing steps).

17.3 Photolithography

Photolithography, or simply lithography, is the first step in transferring the circuit layout information to the wafer. As shown in the top view of Fig. 17.1 and explained in Chapter 18 in more detail, the layout consists of polygons representing different types of "layers," e.g., n-well, S/D regions, polysilicon, contact windows, etc. For fabrication purposes, we decompose the layout into these layers. For example, the layout of Fig. 17.1 can be viewed as five different layers shown in Fig. 17.2, each of which must be created on the wafer with a very high precision. Note that the "active" (or "diffusion") layer includes the source/drain regions and the p^+ and n^+ openings serving as the substrate and well ties.

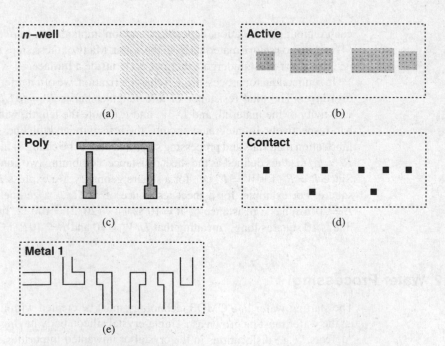

Figure 17.2 Layers comprising the structures of Fig. 17.1.

To understand how a layer is transferred from the layout to the wafer, let us consider the n-well pattern of Fig. 17.2(a) as an example. This pattern is "written" to a transparent glass "mask" by a precisely controlled electron beam [Fig. 17.3(a)]. Also, as depicted in Fig. 17.3(b), the wafer is covered by a thin layer of "photoresist," a material whose etching properties change upon exposure to light.[1] Subsequently, the mask is placed on top of the wafer and the pattern is projected onto the wafer by ultraviolet (UV) light [Fig. 17.3(c)]. The photoresist "hardens" in the regions exposed to light and remains "soft" under the opaque rectangle. The wafer is then placed in an etchant that dissolves the "soft" photoresist area, thereby exposing the silicon surface [Fig. 17.3(d)]. Now, an n-well can be created in the exposed area. We call this set of operations a lithography sequence.

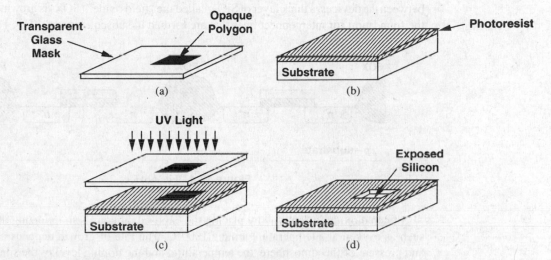

Figure 17.3   (a) Glass mask used in lithography, (b) coverage of wafer by photoresist, (c) selective exposure of photoresist to UV light, (c) exposed silicon after etching.

In summary, the sequence associated with the lithography of each layer involves one mask and three processing steps: (1) cover wafer with photoresist; (2) align mask on top and expose to light; (3) etch exposed photoresist. The example of Fig. 17.2 therefore requires at least five masks and hence five lithography sequences.

We should mention that two types of photoresists are used in processing. A "negative" photoresist hardens in the areas exposed to light and a "positive" photoresist hardens in the areas not exposed to light. As explained later in this chapter, both types prove useful in fabrication.

The number of masks in a process heavily impacts the overall cost of fabrication, eventually influencing the unit price of the chip. This is so for two reasons: each mask costs several thousand dollars, and, owing to the necessary precision, lithography is a slow and expensive task. In fact, CMOS technology originally became attractive by virtue of the relatively small number of masks—about seven—that it required. Although in modern CMOS processes,

---

[1]In practice, a thin layer of oxide is grown before depositing the photoresist to protect the surface.

this number is close to 25 (and the total cost of masks greater than $200,000), the cost of each IC has nonetheless remained low because both the number of transistors per unit area and the size of the wafer have steadily increased.

17.4 Oxidation

A unique property of silicon is that it can produce a very uniform oxide layer on the surface with little strain in the lattice, allowing the fabrication of gate oxide layers as thin as a few tens of angstroms (only several *atomic* layers). In addition to serving as the gate dielectric, silicon dioxide can act as a protective coating in many steps of fabrication. Also, in areas between the devices, a thick layer of SiO_2, called the "field oxide" (FOX) is grown, providing the foundation for interconnect lines that are formed in subsequent steps (Fig. 17.4).

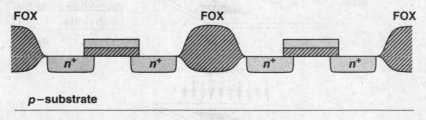

Figure 17.4 Field oxide.

Silicon dioxide is "grown" by placing the exposed silicon in an oxidizing atmosphere such as oxygen at a temperature around 1000°C. The rate of growth depends on the type and pressure of the atmosphere, the temperature, and the doping level of the silicon.

The growth of the gate oxide is a very critical step in the process. Since the oxide thickness, t_{ox}, determines both the current handling and reliability of the transistors, it must be controlled to within a few percent. For example, the oxide thicknesses of two transistors separated by 20 cm on a wafer must differ by less than a few angstroms, requiring extremely high uniformity across the wafer and hence a slow growth of the oxide. Also, the "cleanness" of the silicon surface under the oxide affects the mobility of the charge carriers and thus the current drive, transconductance, and noise of the transistors.

17.5 Ion Implantation

In many steps of fabrication, dopants must be selectively introduced into the wafer. For example, after the lithography sequence of Fig. 17.3 is completed, the n-well is formed by entering dopants into the exposed silicon area. Similarly, the source and drain regions of transistors require selective addition of dopants to the wafer.

The most common method of introducing dopants is "ion implantation," whereby the doping atoms are accelerated as a high-energy focused beam, hitting the surface of the wafer and penetrating the exposed areas [Fig. 17.5(a)]. The doping level (dosage) is determined by the intensity and duration of the implantation, and the depth of the doped region is set

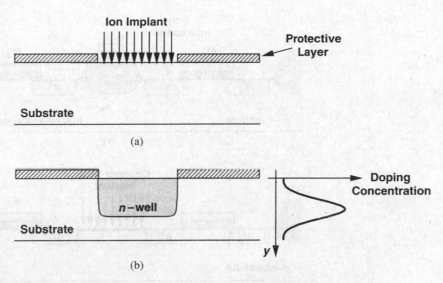

Figure 17.5   (a) Ion implantation, (b) retrograde profile.

by the energy of the beam. As shown in Fig. 17.5, with a high energy, the peak of the doping concentration in fact occurs well below the surface, thereby creating a "retrograde" profile. Such a profile is desirable for the n-well because it establishes a low resistivity near the bottom, reducing susceptibility to latch-up (Section 17.8), and a low doping level at the surface, decreasing the S/D junction capacitance of PMOS devices.

Another important application of implantation is to create "channel-stop" regions between transistors. Consider the field oxide and the S/D junctions of M_1 and M_2 in Fig. 17.6(a), assuming an interconnect line passes on top of the field oxide. Interestingly, the two n^+ regions and the FOX form a MOS transistor having a thick gate oxide and hence a large threshold voltage. Nonetheless, with a sufficiently positive potential on the interconnect line, this transistor may turn on slightly, creating a leakage path between M_1 and M_2. To resolve this issue, a channel-stop implant (also called a field implant) is performed before the field oxide deposition [Fig. 17.6(b)], thereby raising the threshold voltage of the field oxide transistor to a very large value.

Ion implantation damages the silicon lattice extensively. For this reason, the wafer is subsequently heated to approximately 1000°C for 15 to 30 minutes, allowing the lattice bonds to form again. Called "annealing," this operation also leads to diffusion of dopants, broadening the profile in all directions. For example, annealing results in side-diffusion of S/D regions, creating overlap with the gate area. The wafer is therefore usually annealed only once, after all implantations have been completed.

An interesting phenomenon in ion implantation is "channeling." As shown in Fig. 17.7(a), if the implant beam is aligned with the crystal axis, the ions penetrate the wafer to a great depth. For this reason, the implant (or the wafer) is tilted by 7–9° [Fig. 17.7(b)], avoiding such an alignment and ensuring a predictable profile. As explained in Chapter 18, this tilt impacts the matching of transistors, necessitating precautions in the layout.

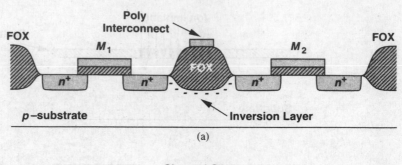

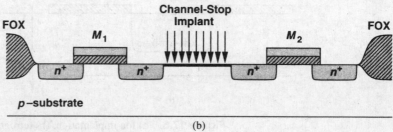

Figure 17.6   (a) Unwanted conduction due to inversion of field area, (b) channel-stop implant.

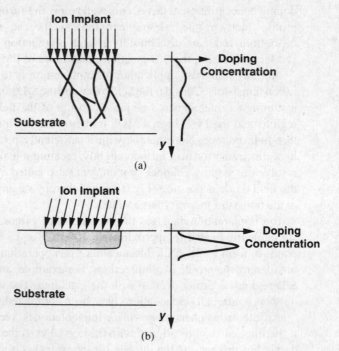

Figure 17.7   (a) Effect of channeling, (b) tilt in implant to avoid channeling.

17.6 Deposition and Etching

As suggested by the structures of Fig. 17.1, device fabrication requires the deposition of various materials. Examples include polysilicon, dielectric materials separating interconnect layers, and metal layers serving as interconnects.

A common method of forming polysilicon on thick dielectric layers is "chemical vapor deposition" (CVD), whereby wafers are placed in a furnace filled with a gas that creates the desired material through a chemical reaction. In modern processes, CVD is performed at a low pressure to achieve more uniformity.

The etching of the materials is also a crucial step. For example, contact windows with very small dimensions, e.g., $0.3~\mu m \times 0.3~\mu m$, and relatively large depths, e.g., $2~\mu m$, must be etched with high precision. Depending on the speed, accuracy, and selectivity required in the etching step, and the type of material to be etched, one of these methods may be used: (1) "wet" etching, i.e., placing the wafer in a chemical liquid (low precision); (2) "plasma" etching, i.e., bombarding the wafer with a plasma gas (high precision); (3) reactive ion etching (RIE), where ions produced in a gas bombard the wafer.

17.7 Device Fabrication

With the processing operations described in the previous section, we now study the fabrication sequence and device structures in typical CMOS technologies. We consider three categories: active devices, passive devices, and interconnects.

17.7.1 Active Devices

Basic Transistor Fabrication   The fabrication begins with a p-type silicon wafer approximately 1 mm thick. Following cleaning and polishing steps, a thin layer of silicon dioxide is grown as a protective coating on top of the wafer [Fig. 17.8(a)]. Next, to create the n-wells, a lithography sequence consisting of photoresist deposition, exposure to UV light using the n-well mask, and selective etching is carried out and the n-wells are implanted [Fig. 17.8(b)]. The remaining photoresist and oxide layers are then removed [Fig. 17.8(c)].

Recall from the previous section that a field implant and a field oxide growth are necessary in the areas between the transistors. At this point in the sequence, a stack consisting of a silicon oxide layer, a silicon nitride (Si_3N_4), and a *positive* photoresist layer is created. Next, the "active" mask is used for lithography so that only the regions between the transistors are exposed [Fig. 17.8(d)].[2] Subsequently, the channel-stop implant is performed, the photoresist is removed, and a thick oxide layer is grown in the exposed silicon areas, producing the field oxide. The protective nitride and oxide layers are then removed [Fig. 17.8(e)], thereby exposing all areas where transistors are to be formed. In the subsequent diagrams, the channel-stop implant will be omitted for the sake of clarity.

The next step involves the growth of the gate oxide, a critical operation requiring slow, low-pressure CVD [Fig. 17.8(f)]. As explained in Chapter 2, the "native" threshold voltage

---

[2]The n-wells are not shown for clarity.

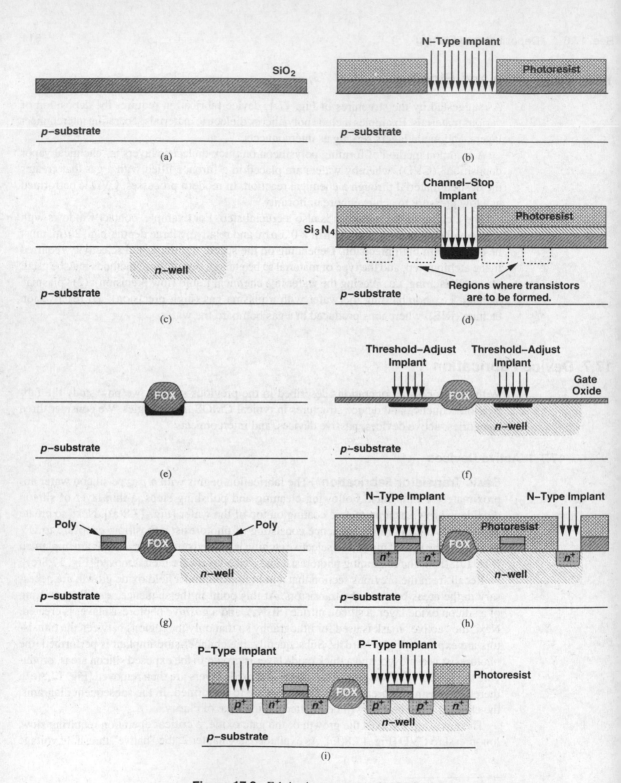

Figure 17.8 Fabrication sequence of MOS devices.

of the transistors is typically far from the desired value, necessitating a threshold-adjust implant. (The native threshold of both PMOS and NMOS is usually more negative than desired, e.g., $V_{THN} \approx 0$, and $V_{THP} \approx -1$ V.) Such an implant is performed following the growth of the gate oxide, creating a thin sheet of dopants near the surface and making the threshold of both NMOS and PMOS devices more positive.

With the gate oxide in place, the polysilicon layer is deposited and the "poly mask" lithography is carried out, resulting in the structure shown in Fig. 17.8(g). We should note that polysilicon is simply noncrystalline ("amorphous") silicon, a property that arises because this layer grows on top of silicon dioxide and hence cannot form a crystal. Since polysilicon serves as a conductor, its amorphous nature is unimportant. To reduce the resistivity of this layer, an additional implant is typically used, yielding a sheet resistance of a few tens of ohms per square.

In the next step, the source/drain junctions of the transistors and the substrate and n-well ties are formed by ion implantation. This step requires a "source/drain mask" and two lithography sequences. As illustrated in Fig. 17.8(h), the first sequence incorporates a negative photoresist, exposing the areas to receive an n^+ implant (the S/D junctions of NMOS transistors and the n-well ties). In the second sequence [Fig. 17.8(i)], the same mask and a positive photoresist are used, exposing the areas to receive a p^+ implant (the S/D junctions of PMOS transistors and the substrate ties). Note that these implants also dope the polysilicon layer, reducing its sheet resistance. This step completes the fabrication of the basic transistors.

The reader may wonder why the source/drain junctions are formed *after* the gate oxide and polysilicon. Suppose, as depicted in Fig. 17.9(a), these junctions are created first. Then, the alignment of the gate poly mask with respect to the S/D areas becomes extremely critical. Even if the misalignment is a small fraction of the minimum channel length, a gap may appear between the source (or drain) and the gate area, prohibiting the formation

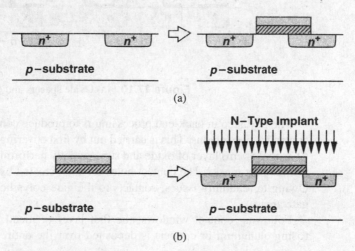

Figure 17.9   (a) Formation of n^+ regions before deposition of poly, (b) self-aligned structure.

of a continuous channel in the transistor. By contrast, the sequence shown in Fig. 17.8 yields a "self-aligned" structure because the source/drain regions are implanted at precisely the edges of the gate area and a misalignment in lithography simply makes one junction slightly narrower than the other [Fig. 17.9(b)]. Interestingly, the first few generations of CMOS technology were based on the approach shown in Fig. 17.9(a), but it was soon discovered that the self-aligned structure would lend itself to scaling much more easily.

Back-End Processing   With the basic transistors fabricated, the wafers must next undergo "back-end" processing, a sequence primarily providing various electrical connections on the chip through contacts and wires. The first step in this sequence is "silicidation." Since the sheet resistance of doped polysilicon and S/D regions is typically several tens of ohms per square, it is desirable to reduce their resistance by about an order of magnitude. Silicidation accomplishes this by covering the polysilicon layer and active areas (S/D regions and substrate and n-well ties) with a thin layer of a highly conductive material, e.g., titanium silicide or tungsten. Illustrated in Fig. 17.10, this step in fact begins with creating an "oxide spacer" at the edges of the polysilicon gate such that the deposition of the silicide becomes a self-aligned process as well.[3] Without the spacer, the silicide layer on the gate may be shorted to that on the source/drain.

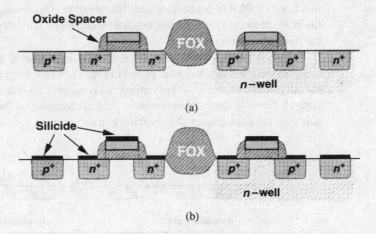

Figure 17.10   (a) Oxide spacers and (b) silicide.

The next step in back-end processing is to produce contact windows on top of polysilicon and active regions. This is carried out by first covering the wafer with a relatively thick (0.3- to 0.5-μm) layer of oxide and subsequently performing a lithography sequence using the "contact mask." The contact holes are then created by plasma etching [Fig. 17.11(a)]. Owing to reliability issues, contacts to the gate polysilicon are not placed on top of the gate area.

Following contact windows, the first layer of metal interconnect (called "metal 1") (using aluminum or copper) is deposited over the entire wafer. A lithography sequence

---

[3]Self-aligned silicide is sometimes called "salicide."

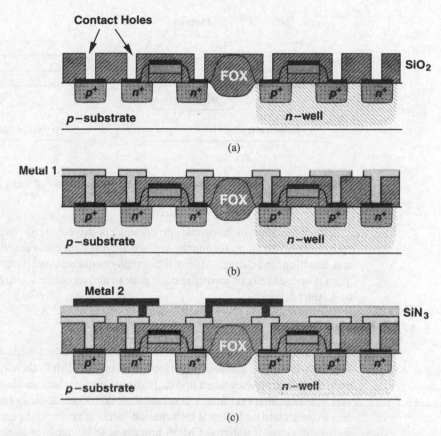

Figure 17.11 Contact and metal fabrication.

using the "metal 1 mask" is then carried out and the metal layer is selectively etched [Fig. 17.11(b)].

The higher levels of interconnect are fabricated using the same procedure [Fig. 17.11(c)]. For each additional metal layer, two masks are required: one for the contact windows and another for the metal itself. Thus, a CMOS process having five layers of metal contains 10 masks for the back end. The contact windows between metal layers are sometimes called "vias" to distinguish them from the first level of contacts to active areas and polysilicon.

We should mention that if a large area must be contacted, many small windows—rather than a large window—are usually used. Dictated by reliability issues, the dimensions of each contact or via are fixed and cannot be decreased or increased by the layout designer. An interesting phenomenon related to large active areas is "contact spiking." If a large contact window allows aluminum to touch the active area, then, as depicted in Fig. 17.12(a), the metal may "eat" and penetrate the doped region, eventually crossing the junction to the bulk and shorting the diode. With small windows, on the other hand, this effect is avoided [Fig. 17.12(b)].

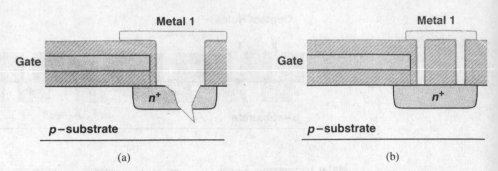

Figure 17.12   (a) Spiking due to large contact areas, (b) use of small contacts to avoid spiking.

The final step in back-end processing is to cover the wafer with a "glass" or "passivation" layer, protecting the surface against damages caused by subsequent mechanical handling and dicing. After a lithography sequence using the "passivation mask," the glass is opened only on top of the bond pads to allow connection to the external environment (e.g., the package).

17.7.2 Passive Devices

Passive components such as resistors and capacitors find wide usage in analog design, making it desirable to add these devices to standard CMOS technologies. In practice, however, CMOS processes target primarily digital applications and hence provide only NMOS and PMOS transistors. A new generation of CMOS technology may take one to two years and many iterations before it becomes an "analog process," i.e., one offering high-quality passive devices. If a digital CMOS process is to be used for analog design, we must seek structures that can serve as passive components. The principal issue in using such structures is the *variability* of the component value from wafer to wafer because the process flow does not assume such structures are used in circuits.

Resistors   A CMOS process may be modified so as to provide resistors suited to analog design. A common method is to selectively "block" the silicide layer that is deposited on top of the polysilicon, thereby creating a region having the resistivity of the doped polysilicon (Fig. 17.13). This means the fabrication requires an additional mask and a corresponding lithography sequence. Since the poly doping level is determined by various implants in the process, the resistivity obtained here is not necessarily a target value, but it usually falls in the range of fifty to a few hundred ohms per square. For the same reason, the resistance value may vary by as much as $\pm 20\%$ from wafer to wafer or lot to lot.

The use of silicide on the two ends of the resistor in Fig. 17.13 results in a much lower contact resistance than that obtained by directly connecting the metal layer to doped polysilicon. This improves both the definition of the resistor value and the matching with identical structures. Also, for a given resistance, poly resistors typically exhibit much less capacitance to the substrate than other types—on the order of 90 af/μm^2 for the bottom plate capacitance and 100 af/μm for the fringing capacitance. These resistors are quite linear,

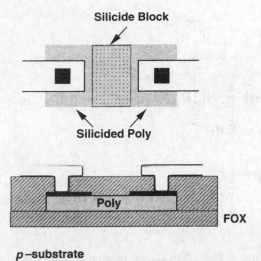

Figure 17.13 Poly resistor using silicide block.

especially if they are long. The primary difficulties with silicide-block poly resistors are variability, mask cost, and process complexity.

In a purely digital process, silicided poly, silicided p^+ or n^+ active areas, n-well, and metal layers can be used as resistors. Since the silicided layers have a low resistivity and, more importantly, their resistance varies substantially (e.g., by $\pm 50\%$), they are rarely used in analog circuits.[4] An n-well resistor can be formed as shown in Fig. 17.14, but the n-well resistivity may vary by several tens of percent with process. With typical sheet resistivities of about 1 k$\Omega/\square$, n-well resistors can prove useful where their absolute value is

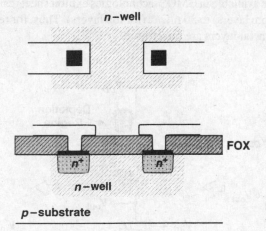

Figure 17.14 Resistor made of n-well.

_____

[4]One exception is where the low value is desirable and the absolute value is not critical, e.g., in resistor ladders used in A/D converters (Chapter 18).

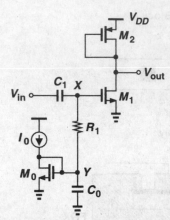

Figure 17.15 Use of an n-well resistor in a coupling network.

not critical. For example, Fig. 17.15 shows a common-source stage that is biased by means of M_0 and I_0 while employing C_1 to block the dc level of the preceding stage. In order to isolate the signal path from the low impedance (and the noise) introduced by M_0, resistor R_1 is inserted between X and Y. Here, the value of R_1 is not critical so long as it is sufficiently large.

We should mention that, due to the depletion region formed between the n-well and the p-substrate, n-well resistors suffer from both a large parasitic capacitance and significant voltage dependence. Fig. 17.16 illustrates a typical case, where one terminal of the n-well resistor is tied to V_{DD}. Since the capacitance to the substrate is distributed (nonuniformly) along the resistor, a lumped model may not be accurate enough, but as a rough approximation, we place half of the total capacitance on each side of the resistor. We also note that as V_{out} varies, so do the width of the depletion region and hence the value of the resistor.

The metal layers available in CMOS technologies exhibit sheet resistances on the order of 70 mΩ/□ (for bottom layers) to 30 mΩ/□ (for top layers). Thus, for resistor values common in analog design, metal layers are rarely used.

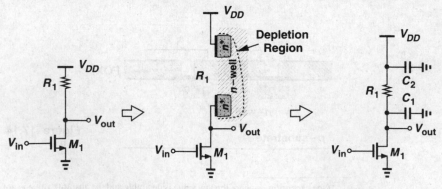

Figure 17.16 Common-source stage using n-well resistors.

Capacitors   Capacitors prove indispensible in most of today's analog CMOS circuits. Several parameters of capacitors are critical in analog design: nonlinearity (voltage dependence), parasitic capacitance to the substrate, series resistance, and capacitance per unit area (density). In CMOS technologies modified for analog design, capacitors are fabricated as "poly-diffusion," "poly-poly," or "metal-poly" structures. Illustrated in Fig. 17.17, the idea is to grow or deposit a relatively thin oxide between two floating conductive layers, thereby forming a dense capacitor with moderate bottom-plate parasitic (about 10 to 20%).

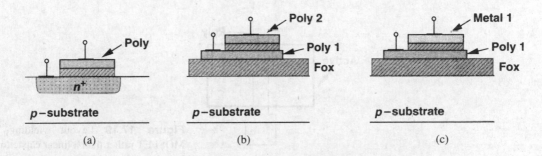

Figure 17.17   Linear capacitor structures, (a) poly-diffusion, (b) poly-poly, (c) metal-poly.

The fabrication steps required to build the poly-diffusion capacitor of Fig. 17.17(a) are shown in Fig. 17.18. First, using the "capacitor mask," a lithography sequence similar to that of Fig. 17.3 defines the bottom plate areas and a heavy n^+ implant is applied [Fig. 17.18(a)]. Next, the gate oxide layer is grown over the entire wafer [Fig. 17.18(b)]. Note that the oxide grows faster over the n^+ region because of the heavier doping level, yielding a capacitance per unit area less than that of MOSFETs. The fabrication then proceeds as

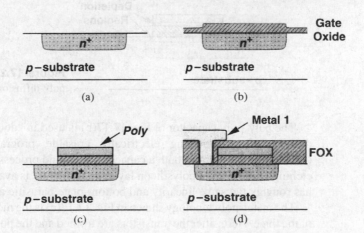

Figure 17.18   Fabrication steps of poly-diffusion capacitor.

in a standard CMOS process, forming capacitors and MOS devices simultaneously [Figs. 17.18(c) and (d)].

It is important to understand the necessity of the capacitor mask in the above sequence. The self-aligned process used to fabricate MOS devices forms active regions in the substrate only after the gate oxide and polysilicon are created. It is therefore impossible to build a doped area *under* the polysilicon without an extra lithography sequence. Even if the layout is drawn as in Fig. 17.19, since the n^+ step is performed after the deposition of poly, the result is still a MOS device rather than a linear capacitor.

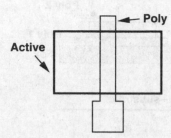

Figure 17.19 Layout yielding a MOSFET rather than a linear capacitor.

By virtue of its simplicity, the poly-diffusion capacitor is the most common type, but it still requires an additional mask and associated fabrication steps. This structure suffers from some nonlinearity because the width of the depletion regions at the poly-oxide and oxide-diffusion interfaces changes with the applied voltage (Fig. 17.20), thereby varying the equivalent dielectric thickness between the two conductive plates. If $C \approx C_0(1 + \alpha_1 V + \alpha V^2)$, then α_1 and α_2 are typically on the order of 5×10^{-4} V^{-1} and 5×10^{-5} V^{-2}, respectively. The bottom-plate parasitic of this topology is about 20% of the interplate capacitance.

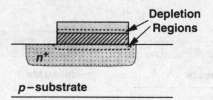

Figure 17.20 Depletion regions in a poly-diffusion capacitor.

The poly-poly capacitor of Fig. 17.17(b) is used in "double-poly" processes, e.g., those designed for fabricating electrically erasable programmable read-only memories (EEPROMs). Requiring both a capacitor mask and processing steps for the deposition and etching of the second polysilicon layer, this structure is available in some technologies and has roughly the same linearity and bottom-plate parasitic as the poly-diffusion capacitor.

The metal-poly topology shown in Fig. 17.17(c) is the most linear and the most expensive of the three. Here, after the transistors are formed and the polysilicon is silicided, a thin layer of SiO_2 is deposited over the entire wafer. Next, a lithography sequence using the capacitor mask defines areas on top of polysilicon where the oxide must remain, and selective etching is performed. Owing to silicidation, no depletion region is formed at the poly-oxide interface and the linearity of the capacitor is improved. Nonlinearity coefficients as low as a few parts per million have been achieved for such a structure [1]. The bottom-plate parasitic is on the order of 10 to 20%.

Digital CMOS technologies do not offer the foregoing capacitor structures for cost and yield reasons. The designer must therefore construct capacitors through the use of the "native" layers of the process.

Perhaps the simplest capacitor structure in CMOS technology is that implemented by a MOSFET. Illustrated in Fig. 17.21(a), the device has a capacitance that varies from a small value at low voltages (where no channel exists and the equivalent capacitance is the series combination of the oxide capacitance and the depletion region capacitance) to a large value (C_{ox}) if the voltage difference exceeds V_{TH}. Since the gate oxide is typically the thinnest layer in the process, MOS capacitors biased in strong inversion are quite dense, saving substantial area if large values are required. For the same reason, the bottom-plate parasitic, i.e., that due to drain and source junctions, is a relatively small percentage of the gate capacitance—typically 10 to 20%.

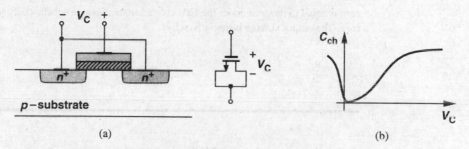

Figure 17.21   (a) MOSFET configured as a capacitor, (b) nonlinear C/V characteristic.

Unfortunately, the voltage dependence of MOS capacitors, even in strong inversion, makes the structure less attractive for precision charge transfer.

Example 17.1

Consider the multiply-by-two amplifier of Section 12.3.3, shown in Fig. 17.22(a) as an implementation using a MOS capacitor C_1 and a linear capacitor C_2. Explain how the output voltage in the amplification mode is distorted.

Solution

Suppose for simplicity that V_{in} is below ground by more than V_{TH} so that the NMOS capacitors are in strong inversion during sampling. As the circuit enters the amplification mode, the voltage across C_1 approaches zero and the total charge stored on C_1 is transferred to C_2. How much is this charge? If C_1 were linear, we would have $Q = C_1 V$, but here we must write $dQ = C_1 dV$. Thus, as shown in Fig. 17.22(b), the total transferred charge when the voltage across the capacitor goes from V_{in} to

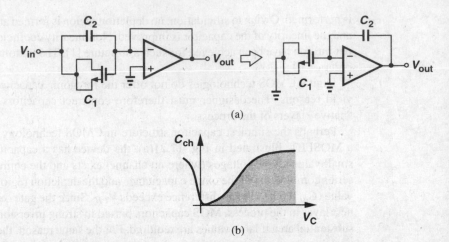

(a)

(b)

Figure 17.22 Precision multiply-by-two circuit using a MOS capacitor.

zero is equal to the area under the C/V characteristic, a value substantially less than that in the linea
case. The output voltage is then given by

$$V_{out} \approx V_{in} + \frac{1}{C_2} \int_0^{V_{in}} C_1 \, dV. \tag{17.1}$$

Another issue related to MOS capacitors is their series resistance, an effect arising fro
the gate material and, more importantly, the channel resistance. Assuming proper layo
minimizes the gate resistance, we view the channel resistance as shown in Fig. 17.2:
estimating the equivalent series resistance as $(R_{tot}/2)\|(R_{tot}/2) = R_{tot}/4$, where R_{tot}
$[\mu C_{ox}(W/L)(V_{GS} - V_{TH})]^{-1}$. The intrinsic time constant of the capacitor is therefor

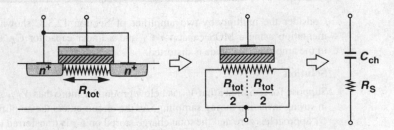

Figure 17.23 Channel resistance of MOS capacitor.

equal to:

$$\tau = \frac{R_{tot}}{4} C_{ch} \qquad (17.2)$$

$$= \frac{1}{4\mu C_{ox}(W/L)(V_{GS} - V_{TH})} \cdot WLC_{ox} \qquad (17.3)$$

$$= \frac{L^2}{4\mu(V_{GS} - V_{TH})}. \qquad (17.4)$$

In reality, the distributed nature of the resistance and the capacitance along the channel results in a time constant equal to one-third of that given above [2]. Another figure of merit for such a capacitor is $Q = [1/(C\omega)]/R_S$. As a rule of thumb, we choose $R_S < 0.1/(C\omega)$.

Equation 17.4 indicates that for a given overdrive, to minimize the series resistance of a MOS capacitor, L must be minimized. Consequently, MOS capacitors are usually designed as a parallel combination of wide, short devices rather than a *square* block (Fig. 17.24). The penalty is a higher junction capacitance to the substrate and somewhat greater area.

In applications requiring linear capacitors, a "sandwich" of conductive layers can be formed in CMOS technology. Shown in Fig. 17.25 is an example, where the capacitance between every two layers is exploited to increase the density. Since the dielectrics between

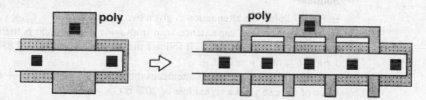

Figure 17.24   Use of wide, short MOS fingers to reduce channel resistance.

$$C = C_1 + \cdots + C_4$$

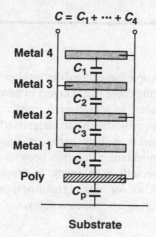

Figure 17.25   Linear capacitor made of native conductive layers.

the layers are relatively thick, this structure still requires a much larger area than the type studied above. More importantly, the bottom-plate parasitic (e.g., the capacitance betwee poly and substrate in Fig. 17.25) is quite large, about 50 to 60% of the total interplan capacitance. This structure is studied in detail in Chapter 18.

Example 17.2

An amplifier with an input capacitance of C_{in} is to be ac-coupled to a preceding stage having an outpr resistance R_{out}. Considering both of the topologies depicted in Fig. 17.26 and allowing a maximur signal attenuation of 20%, determine the minimum value of the coupling capacitor and the resultir time constant if $C_P = 0.5C_C$ or $C_P = 0.2C_C$.

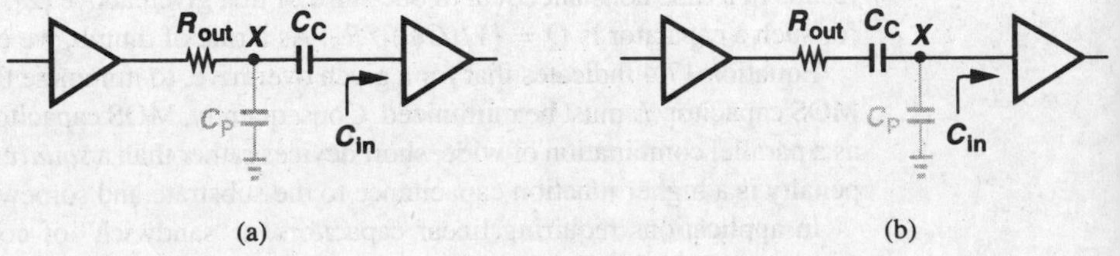

(a)   (b)

Figure 17.26

Solution

In Fig. 17.26(a), the attenuation is given by: $A_v = C_C/(C_C + C_{in})$, yielding $C_C \geq 4C_{in}$ for a 20 signal loss. The total capacitance seen from node X to ground is therefore equal to $C_P + C_C C_{in}$ $(C_C + C_{in}) = C_P + 0.8C_{in}$. It follows that the time constant is $2.8R_{out}C_{in}$ for $C_P = 0.5C_C$ an $1.6R_{out}C_{in}$ for $C_P = 0.2C_C$.

In Fig. 17.26(b), C_P itself attenuates the signal: $A_v = C_C/(C_C + C_{in} + C_P)$, indicating that ▪ value of C_C can yield a signal loss of 20% if $C_P \geq 0.2C_C$.

These calculations yield two important results. First, the topology of Fig. 17.26(a) is general preferable. Second, the addition of a coupling capacitor, e.g., to isolate the bias levels, substantial degrades the speed.

17.7.3 Interconnects

The performance of today's complex integrated circuits heavily depends on the quality of tl available interconnects, requiring more metal layers in new generations of the technology Proper modeling of interconnects in a high-performance circuit is still a topic of acti research, but our objective is to provide a basic understanding of the interconnect issues.

Two properties of interconnects, namely, series resistance and parallel capacitance, ir pact the performance, often mandating iteration between layout and circuit design. Th series resistance becomes especially problematic in supply and ground lines, creating ▪ and transient voltage drops. Also, for long signal lines, the distributed resistance and c pacitance of the wire may result in a significant delay.

---

[5]At the time of this writing, technologies with six layers of metal are in production.

The resistance of metal wires can be easily estimated at low frequencies, where skin effect is negligible. Typical sheet resistances are 30 mΩ/□ for the topmost (thickest) layer and 70 mΩ/□ for lower layers. The finite resistance of wires influences the choice of line widths for high-current interconnects such as supply and ground buses, as illustrated by the following example.

Example 17.3

A D/A converter incorporates N equal current sources implemented as NMOS devices each having an aspect ratio of W/L [Fig. 17.27(a)]. Assuming the interconnect between every two consecutive current sources has a small resistance, r, estimate the mismatch between I_N and I_1.

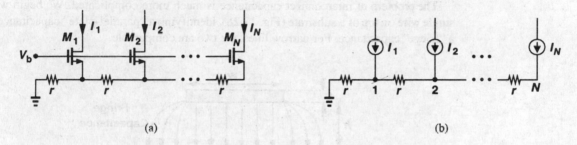

Figure 17.27   Effect of ground resistance in a D/A converter.

Solution

If r is sufficiently small, the circuit can be modeled as shown in Fig. 17.27(b), where, $I_1 \approx I_2 \approx \cdots \approx I_N = I$. The voltage at node N is obtained by superposition of currents:

$$V_N = Ir + I(2r) + \cdots + I(Nr) \tag{17.5}$$

$$= \frac{N(N+1)}{2} Ir. \tag{17.6}$$

If V_N is relatively small, the assumption $I_1 \approx I_2 \approx \cdots \approx I_N$ used in the above calculation is reasonable and M_1-M_N exhibit roughly equal transconductances. Thus,

$$I_N = I - g_m V_N \tag{17.7}$$

$$= I - g_m r \frac{N(N+1)}{2} I \tag{17.8}$$

$$= I \left[1 - g_m r \frac{N(N+1)}{2} \right]. \tag{17.9}$$

Since $V_1 \approx N \cdot I \cdot r$, we have $I_1 = I - g_m N \cdot I \cdot r$, and the relative mismatch between I_1 and I_N is

$$\left| \frac{I_1 - I_N}{I} \right| = g_m r \frac{N(N-1)}{2}. \tag{17.10}$$

The key point here is that the error grows in proportion to N^2. The ground bus must therefore be sufficiently wide to minimize r.

Another factor determining the width of interconnects is "electromigration." At high current densities, the aluminum atoms in a wire tend to "migrate," leaving a void that eventually (after some years of operation) grows to a discontinuity. For this reason, long-term reliability considerations restrict the maximum current density of interconnects. As a rule of thumb, a current-density of 1 mA per micron of width is acceptable, but the actual value varies according to the thickness of the metal. Also, for transient currents, the peak value may be quite higher.

The problem of interconnect capacitance is much more complicated. We begin with a single wire on top of a substrate (Fig. 17.28), identifying a "parallel-plate" capacitance and a "fringe" capacitance. For narrow lines, the two are comparable.

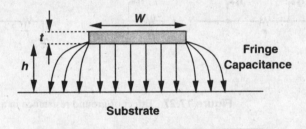

Figure 17.28   Parallel-plate and fringe capacitance of an interconnect.

A simple empirical relationship for calculating the total wire capacitance per unit length on top of a conducting substrate is:

$$C = \epsilon \left[\frac{W}{h} + 0.77 + 1.06 \left(\frac{W}{h} \right)^{0.25} + 1.06 \left(\frac{t}{h} \right)^{0.5} \right], \qquad (17.11)$$

where W, h, and t denote the dimensions shown in Fig. 17.28 [3]. For typical dimensions, this equation predicts the capacitance with a few percent of error.

While upper levels of metal in a process exhibit less capacitance per unit width and length, their minimum allowable width is usually greater than that of the lower layers. Thus, the minimum capacitance for a given length may be only slightly smaller for the topmost layer(s). Table 17.1 depicts typical values for the minimum widths and parallel-plate and fringe capacitances (to the substrate) in a four-metal 0.25-μm process.

Wires also suffer from parallel and fringe capacitances between them. Illustrated in Fig. 17.29, this effect is difficult to quantify for a complex layout, often necessitating the use of computer programs. In practice, the capacitances between the layers are calculated by "electromagnetic field solvers," measured experimentally, and tabulated in the process design manual.

Table 17.1 Minimum widths and capacitances of interconnects in a 0.25-μm technology.

	Poly	Metal 1	Metal 2	Metal 3	Metal 4
Minimum Width (μm)	0.25	0.35	0.45	0.50	0.60
Bottom-Plate Capacitance (aF/μm^2)	90	30	15	9.0	7.0
Fringe Capacitance (Two Sides) (aF/μm)	110	80	50	40	30

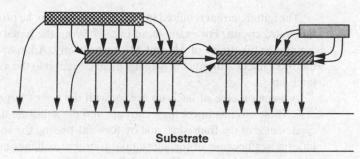

Substrate

Figure 17.29 Complex interconnect structure.

7.8 Latch-Up

Owing to manufacturing difficulties, the first few generations of MOS technologies provided only NMOS devices. In fact, many of the early microprocessors and analog circuits were fabricated in NMOS processes, but they consumed substantial power. The advent of CMOS technology was motivated by the zero static power dissipation of CMOS logic—although CMOS devices required a greater number of masks and fabrication steps. Another issue that did not exist in NMOS implementations but arose in CMOS circuits was latch-up.

Consider the NMOS and PMOS devices shown in Fig. 17.30(a). Recall from Chapter 11 that a parasitic *pnp* bipolar transistor, Q_1, is associated with the PFET, the *n*-well, and the substrate. By the same token, a parasitic *npn* device, Q_2, can be identified in conjunction with the NFET. We make two observations: (1) the base of each bipolar transistor is inevitably tied to the collector of the other; (b) owing to the finite resistance of the *n*-well and the substrate, the bases of Q_1 and Q_2 see a nonzero resistance to V_{DD} and ground, respectively. The parasitic circuit can therefore be drawn as in Fig. 17.30(b), revealing a *positive* feedback loop around Q_1 and Q_2. In fact, if a current is injected into node X such that V_X rises, then I_{C2} increases, V_Y falls, $|I_{C1}|$ increases, and V_X rises further. If the loop gain is greater than or equal to unity, this phenomenon continues until both transistors turn on completely, drawing an enormous current from V_{DD}. We say the circuit is latched up.

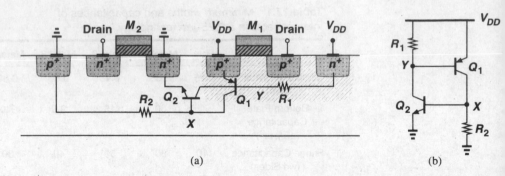

Figure 17.30 (a) Parasitic bipolar transistors in a CMOS process, (b) equivalent circuit.

The initial current required to trigger latch-up may be produced by various sources in an integrated circuit. For example, in Fig. 17.30(a), the bases of Q_1 and Q_2 are capacitively coupled to the drains of M_1 and M_2, respectively. A large voltage swing at the drains can therefore inject a significant displacement current into the n-well or the substrate, initiating latch-up.

A common case of latch-up occurs with the use of large digital output buffers (inverters). These circuits inject high currents into the substrate through the large drain junction capacitance of the transistors and by forward-biasing the source-bulk junction diodes. The latter arises because of the substantial transient voltages produced across the bond wires connected to the ground (Chapter 18).

In order to prevent latch-up, both process engineers and circuit designers take precautions such that the loop gain of the equivalent circuit shown in Fig. 17.30(b) remains well below unity. Proper choice of the doping levels and profiles as well as layout design rules ensure a low value for both the parasitic resistances and the current gain of the bipolar transistors. Furthermore, the layout of the circuit incorporates substrate and n-well contacts with sufficiently small spacing so as to minimize the resistance. The design manual of each technology typically provides an extensive set of layout rules recommended for latch-up prevention.

Problems

Unless otherwise stated, in the following problems, use the device data shown in Table 2.1 and assume $V_{DD} = 3$ V where necessary. Also, assume all transistors are in saturation.

17.1. A MOS technology is designed to provide only n-type transistors and two metal layers. Sketch the fabrication steps and determine the minimum number of masks required in this technology.

17.2. During a threshold-adjust implant, the wafer was not tilted, leading to severe channeling. Explain whether the resulting threshold voltage is higher or lower than the target value.

17.3. The circuits of Fig. 17.31 have been fabricated with a longer-than-expected gate oxidation cycle. If the threshold voltages are still equal to the desirable value, sketch V_{out} versus V_{in}.

and compare the results to the target case.

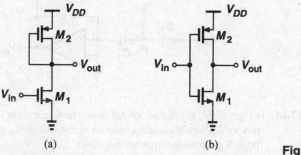

(a)   (b)   **Figure 17.31**

17.4. The circuits of Fig. 17.31 have been fabricated without a threshold-adjust implant. Sketch V_{out} versus V_{in} and compare the results to the target case.

17.5. Due to a layout error, the circuit shown in Fig. 17.32 suffers from contact spiking in one of the junctions. Identify the faulty junction if (a) the voltage gain is higher than expected, (b) the output voltage is near V_{DD}.

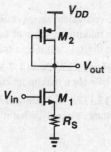

Figure 17.32

17.6. An NMOS cascode current source used in a large circuit exhibits a substantially lower output impedance than expected. Determine which fabrication error may have led to this effect: (a) channeling during S/D implant, (b) omission of the channel-stop implant, (c) insufficient gate oxide growth.

17.7. An NMOS cascode current source has a zero output current. If a single (small) lithography misalignment has caused this error, determine in which fabrication step(s) this may have occurred.

17.8. A differential pair using an active current mirror as load suffers from a low small-signal voltage gain. If the bias current is equal to the target value, determine which fabrication error may have led to this effect: (a) heavy n-well implantation, (b) heavy threshold-adjust implantation, (c) long gate oxidation cycle.

17.9. The switched-capacitor amplifier of Fig. 17.33 exhibits a large gain error. If the bias current of the op amp is equal to the desired value, which fabrication error is likely to have happened: (a) heavy threshold-adjust implantation, (b) very heavy doping in the bottom plate of C_1 (placed at node P), (c) channeling during the S/D implantation.

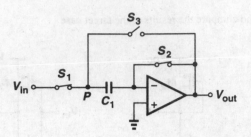

Figure 17.33

17.10. In Fig. 17.34, the digital circuit draws large transient currents from V_{DD}. Without M_1, the inductor L_b would sustain a large transient voltage $L_b dI_{DD}/dt$. Transistor M_1 with $W/L = 100/0.5$ is added to suppress this effect.

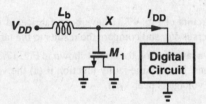

Figure 17.34

(a) Calculate the equivalent series resistance of M_1.

(b) Calculate the maximum value of L_b that results in a critically-damped response at node X. Model the digital circuit by a transient current source.

17.11. In the circuit of Fig. 17.27, $V_b = 1.2$ V, $N = 32$, and $(W/L)_{1-N} = 20/0.5$. Determine the maximum value of r for a maximum current mismatch of 1%.

17.12. Suppose in Eq. (17.11), $t = 1$ μm and $h = 3$ μm. For what value of W are the parallel-plate and fringe capacitances equal? What if $h = 5$ μm?

References

1. C. Kaya et al., "Polycide/Metal Capacitors for High Precision A/D Converters," *IEDM Dig. of Tech. Papers,* pp. 782–785, Dec. 1988.

2. P. Larsson, "Parasitic Resistance in an MOS Transistor Used as On-Chip Decoupling Capacitor," *IEEE J. Solid-State Circuits,* vol. 32, pp. 574–576, April 1997.

3. E. Barke, "Line-to-Ground Capacitance Calculations for VLSI: A Comparison," *IEEE Trans. on Commputer-Aided Design,* vol. 7, pp. 195–298, Feb. 1988.

Chapter 18

Layout and Packaging

In the past 20 years, analog CMOS circuits have evolved from low-speed, low-complexity, small-signal, high-voltage topologies to high-speed, high-complexity, low-voltage "mixed-signal" systems containing a great deal of digital circuitry. While device scaling has enhanced the raw speed of transistors, unwanted interaction between different sections of integrated circuits as well as nonidealities in the layout and packaging increasingly limit both the speed and the precision of such systems. Today's analog circuit design is very heavily influenced by layout and packaging.

In this chapter, we study principles of layout and packaging, emphasizing effects that manifest themselves when analog and digital circuits coexist on a chip. For the sake of brevity, we use the term analog to mean both "analog" and "mixed-signal." Beginning with an overview of layout design rules, we study a number of topics related to the layout of analog circuits, including multifinger transistors, symmetry, reference distribution, passive device layout, and interconnects. Next, we deal with the problem of substrate coupling. Finally, we describe packaging issues, analyzing the effect of self- and mutual inductance and capacitance of external connections to integrated circuits.

8.1 General Layout Considerations

The layout of an integrated circuit defines the geometries that appear on the masks used in fabrication. From Chapter 17, the geometries include n-well, active, polysilicon, n^+ and p^+ implants, interlayer contact windows, and metal layers.

Figure 18.1 shows an example, where the mask geometries required for a PMOS transistor are drawn. It is important to note the following: (1) the n-well surrounds the device with enough margin to ensure that the transistor is contained in the well for all expected misalignments during fabrication; (2) each active area (S/D regions and n^+ contact to the well) is surrounded by a proper implant geometry with enough margin; (3) from the fabrication steps described in Chapter 17, the gate requires its own mask; (4) the contact windows mask provides connection from active and poly regions to the first layer of metal.

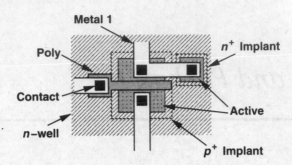

Figure 18.1 Layout of a PMOS transistor.

In most modern layout tools, the implants, and even the *n*-wells are automatically generated from the remainder of transistor geometries, reducing the number of layers that the layout designer sees on the computer screen and simplifying the task.

18.1.1 Design Rules

While the width and length of each transistor is determined by circuit design, most of the other dimensions in a layout are dictated by "design rules," i.e., a set of rules that guarantees proper transistor and interconnect fabrication despite various tolerances in each step of processing. Most design rules can be categorized under one of four groups described below.

Minimum Width The widths (and lengths) of the geometries defined on a mask must exceed a minimum value imposed by both lithography and processing capabilities of the technology. For example, if a polysilicon rectangle is excessively narrow, then, owing to fabrication tolerances, it may simply break or at least suffer from a large local resistance (Fig. 18.2). In general, the thicker a layer, the greater its minimum allowable width, indi-

Figure 18.2 Excessive width variation in a narrow poly line.

cating that as technologies scale, the thickness must be decreased proportionally. Fig. 18.3 depicts examples of minimum widths in a 0.25-μm technology. Note that the thickness of the layers is not under the control of the layout designer.

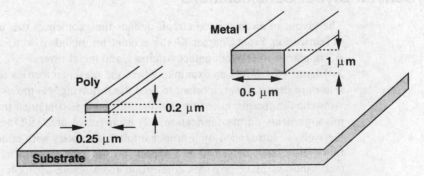

Figure 18.3 Widths and thicknesses of poly and metal lines.

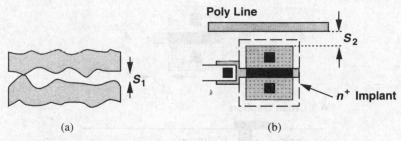

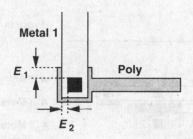

Figure 18.4   (a) Short between two excessively close poly lines, (b) minimum spacing between active and poly.

Figure 18.5   Enclosure rule for poly and metal surrounding a contact.

Minimum Spacing   The geometries built on the same mask or, in some cases, different masks must be separated by a minimum spacing. For example, as shown in Fig. 18.4(a), if two polysilicon lines are placed too close to each other, they may be shorted. As another example, consider the case shown in Fig. 18.4(b), where a polysilicon line runs close to the S/D area of a transistor. A minimum spacing is required here to ensure the implant surrounding the transistor does not overlap with the poly line.

Minimum Enclosure   We mentioned above that in the layout of Fig. 18.1, the n-well and the p^+ implant must surround the transistor with sufficient margin to guarantee that the device is contained by these geometries despite tolerances. These are examples of minimum enclosure rules. Fig. 18.5 depicts another example, where a poly contact window connects a poly line to a metal 1 line. To ensure that the contact remains inside the poly and metal 1 squares, both geometries must enclose the contact with enough margin.

Minimum Extension   Some geometries must extend beyond the edge of others by a minimum value. For example, as shown in Fig. 18.6, the gate polysilicon must have a minimum extension beyond the active area to ensure proper transistor action at the edge.

In addition to the minimum dimensions specified in the above four categories, some *maximum allowable* dimensions may also be enforced. For example, for long metal wires, the minimum width is typically larger than that for short wires to avoid "liftoff" problems. Other such rules relate to the "antenna effect," described in the next section.

Fig. 18.7 summarizes a small subset of design rules governing the layout of an NMOS differential pair with PMOS current-source loads. Modern CMOS technologies typically involve more than 150 layout design rules.

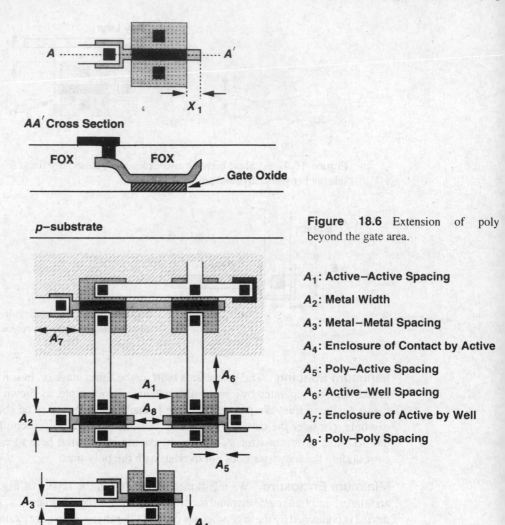

Figure 18.6 Extension of poly beyond the gate area.

A_1: **Active–Active Spacing**

A_2: **Metal Width**

A_3: **Metal–Metal Spacing**

A_4: **Enclosure of Contact by Active**

A_5: **Poly–Active Spacing**

A_6: **Active–Well Spacing**

A_7: **Enclosure of Active by Well**

A_8: **Poly–Poly Spacing**

Figure 18.7 Layout of a differential pair with PMOS current-source loads.

18.1.2 Antenna Effect

Suppose the gate of a small MOSFET is tied to a metal 1 interconnect having a large area [Fig. 18.8(a)]. During the etching of metal 1, the metal area acts as an "antenna," collecting ions and rising in potential. It is therefore possible that the gate voltage of the MOS device increases so much that the gate oxide breaks down (irreversibly) during fabrication.

The antenna effect may occur for any large piece of conductive material tied to the gate, including polysilicon itself. For this reason, submicron CMOS technologies typically limit the total area of such geometries, thereby minimizing the probability of gate oxide damage.

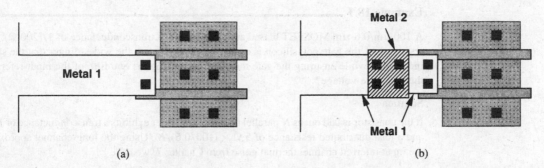

(a)   (b)

Figure 18.8   (a) Layout susceptible to antenna effect, (b) discontinuity in metal 1 layer to avoid antenna effect.

If large areas are inevitable, then a discontinuity can be created as illustrated in Fig. 18.8(b) so that, when metal 1 is being etched, the large area is not connected to the gate.

18.2 Analog Layout Techniques

The extensive sets of design rules enforced by mainstream CMOS processes aim to maximize the yield of digital ICs while allowing moderately aggressive circuit design. Analog systems, on the other hand, demand many more layout precautions so as to minimize effects such as crosstalk, mismatches, noise, etc.

18.2.1 Multifinger Transistors

As mentioned in Chapter 2, wide transistors are usually "folded" so as to reduce both the S/D junction area and the gate resistance. A simple folded structure such as that in Fig. 18.9(a) may prove inadequate for very wide devices, necessitating the use of multiple "fingers" [Fig. 18.9(b)]. As a rule of thumb, the width of each finger is chosen such that the resistance of the finger is less than the inverse transconductance associated with the finger. In low-noise applications, the gate resistance must be one-fifth to one-tenth of $1/g_m$.

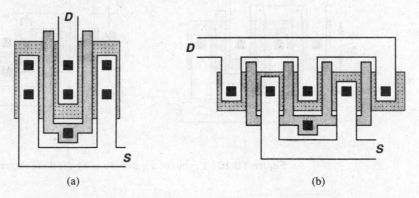

(a)   (b)

Figure 18.9   (a) Simple folding of a MOSFET, (b) use of multiple fingers.

Example 18.1

A 100-μm/0.6-μm MOSFET biased at 1 mA exhibits a transconductance of $1/(200\ \Omega)$. If the sheet resistance of the gate polysilicon is equal to $5\ \Omega/\square$, what is the widest finger that the structure can incorporate while ensuring the gate thermal noise voltage is one-fifth of the input-referred channel thermal noise voltage?

Solution

If the transistor is laid out as N parallel fingers, each finger exhibits a transconductance of $1/(200N\ \Omega)$ and a total distributed resistance of $5\ \Omega \times (100/0.6)/N$. Using the long-channel approximation for the input-referred channel thermal noise from Chapter 7, we have

$$\text{Channel Noise} = \sqrt{4kT\frac{2}{3}(200)}\ \text{V}/\sqrt{\text{Hz}} \tag{18.1}$$

$$\text{Gate Noise} = \sqrt{4kT\frac{500}{0.6N^2}\frac{1}{3}}\ \text{V}/\sqrt{\text{Hz}} \tag{18.2}$$

where the factor $1/3$ on the right hand side of (18.2) accounts for the distributed nature of the resistance (Chapter 7). Equating (18.1) to five times (18.2), we have

$$N = 5\sqrt{\frac{6.25}{3}} \tag{18.3}$$

$$\approx 7.2. \tag{18.4}$$

Thus, a minimum of 8 fingers is required.

While the gate resistance can be reduced by decomposing the transistor into more parallel fingers, the capacitance associated with the perimeter of the source/drain areas increases. As exemplified by the structures depicted in Fig. 18.10,[1] with three fingers, the total perimeter of the source or the drain is equal to $2(2E + 2W/3) = 4E + 4W/3$, whereas with five

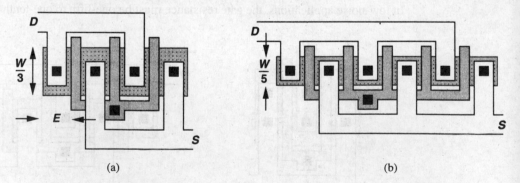

Figure 18.10   Layout of a transistor using (a) three fingers, (b) five fingers.

---

[1] The use of multiple fingers is sometimes called "interdigitization."

fingers, it is equal to $3(2E + 2W/5) = 6E + 6W/5$. In general, for an odd number of fingers N, the S/D perimeter capacitance is given by

$$C_P = \frac{N+1}{2}\left(2E + \frac{2W}{N}\right)C_{jsw} \qquad (18.5)$$

$$= \left[(N+1)E + \frac{N+1}{N}W\right]C_{jsw}. \qquad (18.6)$$

Thus, the number of fingers multiplied by E must be much less than W so as to minimize the S/D perimeter capacitance contribution. In practice, this requirement may conflict with that for minimizing the gate resistance noise, mandating a compromise between the two or contacting the gate on both ends to reduce the resistance.

For transistors having a large number of gate fingers, the structure may be modified to that shown in Fig. 18.11, thereby avoiding long geometries and hence disproportionate dimensions in the layout of the overall circuit.

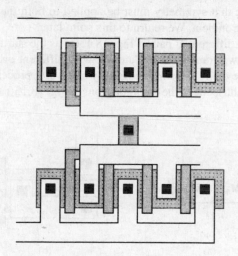

Figure 18.11 Layout of a wide transistor with many fingers.

The layout of a cascode circuit can be simplified if the input device M_1 and the cascode device M_2 have equal widths. As shown in Fig. 18.12(a), the drain of M_1 and the source of M_2 can share the same junction. More importantly, since this junction is not connected to any other node, it need not accommodate a contact window and can therefore be quite smaller [Fig. 18.12(b)]. Consequently, the capacitance at the drain of M_1 is reduced substantially, improving the high-frequency performance. For wide transistors, each transistor may use two or more fingers [Fig. 18.12(c)].

18.2.2 Symmetry

Recall from Chapter 13 that asymmetries in fully differential circuits introduce input-referred offsets, thus limiting the minimum signal level that can be detected. While some

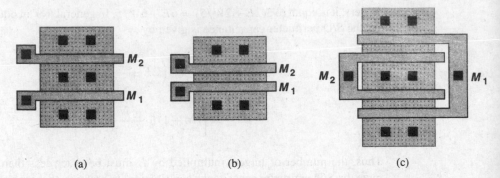

(a)   (b)   (c)

Figure 18.12 Layout of cascode devices having the same width.

mismatch is inevitable, inadequate attention to symmetry in the layout may result in large offsets—much greater than the values predicted by the statistical treatment of Chapter 13. Symmetry also suppresses the effect of common-mode noise and even-order nonlinearity. It is important to note that symmetry must be applied to both the devices of interest and their surrounding environment. We return to this point later.

Let us consider the differential pair of Fig. 18.13(a) as the starting point. If, as depicted in Fig. 18.13(b), the two transistors are laid out with different orientations, the matching greatly suffers because many steps in lithography and wafer processing behave differently along different axes. Thus, one of the configurations in Fig. 18.13(c) and (d) provides a more

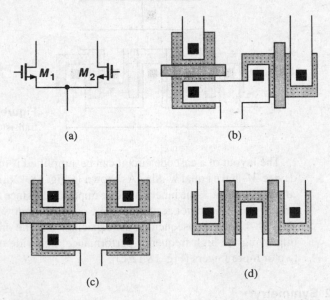

Figure 18.13 (a) Differential pair, (b) layout of M_1 and M_2 with different orientations, (c) layout with gate-aligned devices, (d) layout with parallel-gate devices.

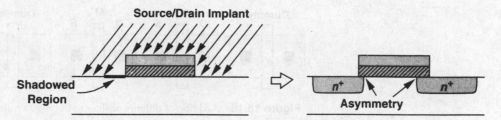

Figure 18.14   Shadowing due to implant tilt.

plausible solution. The choice between these two is determined by a subtle effect called "gate shadowing." Illustrated in Fig. 18.14, the shadowing is caused by the gate polysilicon during the source/drain implantation because the implant (or the wafer) is tilted by about 7° to avoid channeling (Chapter 17). As a result, a narrow strip in the source or drain region receives less implantation, creating a small asymmetry between the source and drain side diffusions after the implanted areas are annealed.

Now consider the structures of Figs. 18.13(c) and (d) in the presence of gate shadowing (Fig. 18.15). In Fig. 18.15(a), if the shadowed terminal is distinguished as the drain (or the source), then the two devices sustain no asymmetry resulting from shadowing. In Fig. 18.15(b), on the other hand, the transistors are not identical even if the shadowed terminals are distinguished because the source region of M_1 "sees" M_2 to its right whereas the source region of M_2 sees only the field oxide. Similarly, the drains of M_1 and M_2 see different structures to their left. In other words, the surrounding environment of M_1 is not identical to that of M_2. For this reason, the topology of Fig. 18.15(a) is preferable.

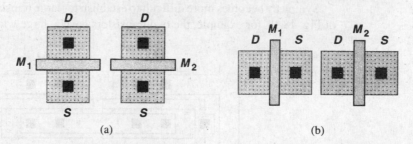

Figure 18.15   Effect of shadowing on (a) gate-aligned and (b) parallel-gate transistors.

The asymmetry inherent to the structures of Fig. 18.15(b) can be ameliorated by adding "dummy" transistors to the two sides so that M_1 and M_2 see approximately the same environment (Fig. 18.16). However, in more complex circuits, e.g., in a folded-cascode op amp, such measures cannot be easily applied.

We should emphasize the importance of maintaining the same environment on the two sides of the axis of symmetry. For example, in the structure of Fig. 18.17, an unrelated metal line passing over only one transistor indeed degrades the symmetry, increasing the mismatch between M_1 and M_2. In such cases, either a replica must be produced on the other

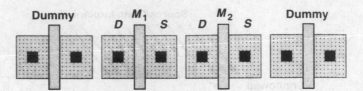

Figure 18.16   Addition of dummy devices to improve symmetry.

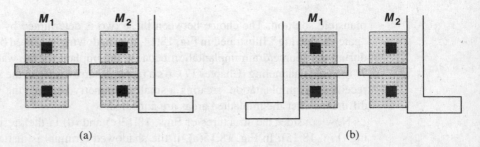

Figure 18.17   (a) Asymmetry resulting from a metal line passing over M_2, (b) removing the asymmetry by replicating the line on top of M_1.

side [Fig. 18.17(b)] (even though the replica may remain floating) or, preferably, the source of asymmetry must be removed.

Symmetry becomes more difficult to establish for large transistors. In the differential pair of Fig. 18.18, for example, the two transistors have a large width so as to achieve a small

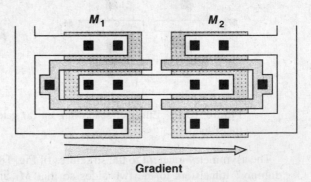

Figure 18.18   Effect of gradient in a differential pair.

input offset voltage, but gradients along the x-axis give rise to appreciable mismatches. To reduce the error, a "common-centroid" configuration may be used such that the effect of first-order gradients along both axes is cancelled. Illustrated in Fig. 18.19, the idea is to decompose each transistor into two halves that are placed diagonally opposite of each other

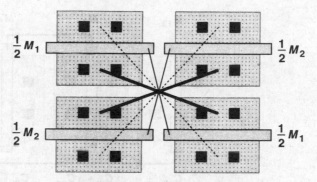

Figure 18.19   Common-centroid layout.

and connected in parallel.[2] However, the routing of interconnects in this layout is quite difficult, often leading to systematic asymmetries of the type depicted in Fig. 18.17(a) or in the capacitances from the wires to ground and between the wires. For a larger circuit, e.g., an op amp, the routing may become prohibitively complex.

The effect of linear gradients can also be suppressed by "one-dimensional" cross coupling, as depicted in Fig. 18.20. Here, all four half transistors are placed along the same axis and M_1 and M_2 are formed by connecting either the near ones and the far ones [Fig. 18.20(a)] or every other one [Fig. 18.20(b)]. (For clarity, the connections between the sources and drains are not shown.) To analyze the effect of gradients in these structures, let us assume that, for example, the gate oxide capacitance varies by ΔC_{ox} from each half transistor to the next.[3] Placing M_{1a} and M_{4a} in parallel, we have

$$I_{D1a} + I_{D4a} = \frac{1}{2}\mu_n(C_{ox} + C_{ox} + 3\Delta C_{ox})\frac{W}{L}(V_{GS} - V_{TH})^2, \qquad (18.7)$$

and for M_{2a} and M_{3a}:

$$I_{D2a} + I_{Da3} = \frac{1}{2}\mu_n(C_{ox} + \Delta C_{ox} + C_{ox} + 2\Delta C_{ox})\frac{W}{L}(V_{GS} - V_{TH})^2. \qquad (18.8)$$

This type of cross coupling therefore cancels the effect of the gradient. Now, for the configuration of Fig. 18.20(b), we have

$$I_{D1b} + I_{D3b} = \frac{1}{2}\mu_n(C_{ox} + C_{ox} + 2\Delta C_{ox})\frac{W}{L}(V_{GS} - V_{TH})^2, \qquad (18.9)$$

and

$$I_{D2b} + I_{D4b} = \frac{1}{2}\mu_n(C_{ox} + \Delta C_{ox} + C_{ox} + 3\Delta C_{ox})\frac{W}{L}(V_{GS} - V_{TH})^2. \qquad (18.10)$$

Equations (18.9) and (18.10) suggest that this approach removes the error to a lesser extent.

---

[2]The interconnect lines shown in this figure are only conceptually correct.

[3]In reality, variation of C_{ox} influences the threshold voltage as well. We neglect this effect here.

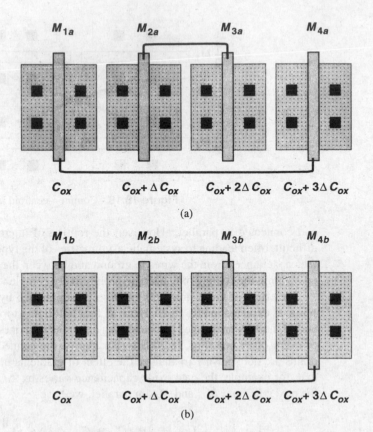

Figure 18.20 One-dimensional cross-coupling.

The reader can prove that for small gradients in other device parameters, similar results are obtained, concluding that the topology of Fig. 18.20(a) contains smaller errors than that of Fig. 18.20(b). However, since the environment seen by $M_{2a} + M_{3a}$ differs from that seen by $M_{1a} + M_{4a}$, dummy transistors must be added to the left of M_{1a} and right of M_{4a}.

18.2.3 Reference Distribution

In analog systems, the bias currents and voltages of various building blocks are derived from one or more bandgap reference generators. The distribution of such references across a large chip entails a number of important issues. Consider the example depicted in Fig. 18.21, where I_{REF} is produced by a bandgap reference and M_1-M_n serve as bias current sources of building blocks that are located far from M_{REF} and from each other. If the matching between I_{D1}-I_{Dn} and I_{REF} is critical, then the voltage drop along the ground line must be taken into account. In fact, for a large number of circuits connected to the same ground line, the systematic mismatch between the current sources and I_{REF} may be unacceptable.

To remedy the above difficulty, the reference can be distributed in the current domain rather than in the voltage domain. Illustrated in Fig. 18.22, the idea is to route the reference

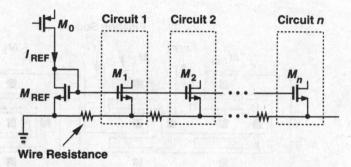

Figure 18.21   Distribution of a reference voltage for current-mirror biasing.

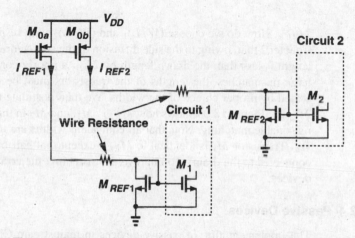

Figure 18.22   Distribution of current to reduce the effect of interconnect resistance.

current to the vicinity of the building blocks and perform the current mirror operation *locally*. Placing the interconnect resistance in series with current sources, this approach lowers systematic errors if the building blocks appear in dense groups in different regions on the chip. However, mismatches between I_{REF1} and I_{REF2} and between M_{REF1} and M_{REF2} introduce error. In large systems, it may be advantageous to employ several local bandgap reference circuits so as to alleviate routing problems.

Another issue in the circuits of Figs. 18.21 and 18.22 relates to the orientation of the transistors. As mentioned in Section 18.2.2, if, for example, M_{REF} and M_1-M_n in Fig. 18.21 have different orientations, then substantial mismatches arise. Since circuits 1, 2, ..., n may be laid out individually, particular attention must be paid to the orientation of their current sources before and after the entire chip is composed.

The scaling of currents in Figs. 18.21 and 18.22 also demands careful choice of device dimensions and layout. Suppose the circuit of Fig. 18.21 requires $I_{D1} = 0.5 I_{REF}$ and $I_{D2} =$

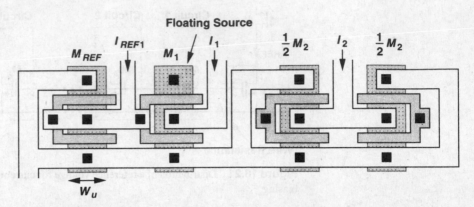

Figure 18.23   Proper scaling of device dimensions for adequate matching of current sources.

$2I_{REF}$. How do we choose $(W/L)_1$ and $(W/L)_2$ with respect to $(W/L)_{REF}$? Recall from Chapter 2 that, owing to the side diffusion of the source/drain regions, the effective channel length is less than the drawn length by $2L_D$, a poorly controlled quantity. Thus, to avoid large mismatches, the lengths of the transistors must be equal and the currents must be scaled by proper choice of the widths. We then postulate that $W_1 = 0.5W_{REF}$ and $W_2 = 2W_{REF}$. Figure 18.23 shows how M_{REF}, M_1, and M_2 in this example are laid out to ensure reasonable matching. Note that all equivalent widths are integer multiples of a unit value, W_u. Transistor M_1 is identical to M_{REF} except that half of its source remains floating (or connected to the drain). To improve the matching, the array can be surrounded by dummy devices.

18.2.4 Passive Devices

The implementation of passive devices in mainstream CMOS technologies continues to pose difficult challenges. When introduced for production, a new generation of a CMOS process provides only NMOS and PMOS devices, rarely allowing the use of polysilicon resistors (with silicide block) or high-density linear capacitors. Since it takes approximately two years to add such modules to the technology, we say "analog" processes are about two years behind "digital" processes. This is an important observation because in two years the next generation of the basic CMOS process is launched (Fig. 18.24), providing scaled transistors having a higher "raw" speed.

Which generation should a manufacturer use at $t = t_1 + 2$ years: generation NA with well-characterized passive components but a minimum dimension of L_1 or generation $N + 1$ with no high-quality passive devices but a scaled dimension of $L_1/2$? Considering the difficulties in analog design without such passive elements, we may choose generation NA, forsaking the speed advantages of generation $N + 1$. However, since the design of digital circuits is immediately moved to the scaled technology, the analog building blocks must follow suit if they are to be integrated on the same chip along with the digital system.

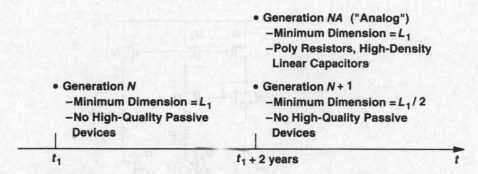

Figure 18.24 Development of digital and analog generations of a CMOS technology.

In practice, different IC manufacturers have adopted different approaches: some develop products in analog technologies, exploiting the well-characterized properties of the devices to design aggressively, whereas others utilize digital processes, taking advantage of more relaxed power-speed trade-offs and maintaining compatibility with digital circuits.

Let us now study the implementation of passive devices.

Resistors Polysilicon resistors using a silicide block exhibit high linearity, low capacitance to the substrate, and relatively small mismatches. The linearity of these resistors in fact depends on their length [1], necessitating accurate measurement and modeling for high-precision applications. Fig. 18.25 depicts an example where the nonlinearity of the resistor is critical. Since $V_{out} = -I_{in}R_F$, the accuracy of current-to-voltage conversion depends on the linearity of R_F.

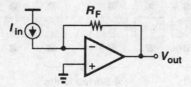

Figure 18.25 Feedback amplifier converting a voltage to current.

As with other devices, the matching of polysilicon resistors is a function of their dimensions. For example, resistors having a length of 5 μm and width of 3 μm display typical mismatches on the order of 0.2%. Most of the symmetry rules described for the layout of MOS devices apply to resistors as well. For example, resistors that are required to bear a well-defined ratio must consist of identical units placed in parallel or series (with the same orientation).

Example 18.2

Consider the bandgap circuit shown in Fig. 18.26. Choose the values of n, R_1, and R_2 such that V_{out} exhibits a zero temperature coefficient and the layout can be designed for high precision.

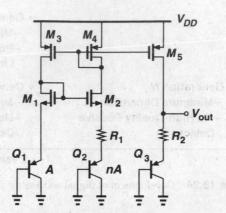

Figure 18.26

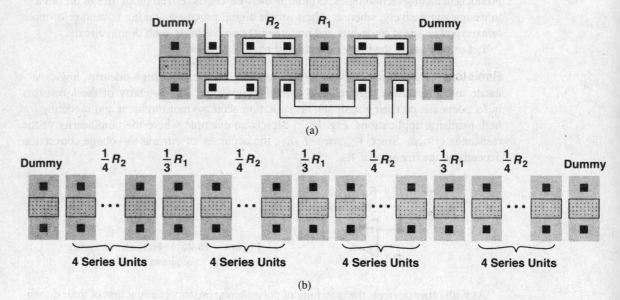

Figure 18.27 Layout of R_1 and R_2 with (a) $R_2/R_1 = 5$, (b) $R_2/R_1 = 5.34 \approx 16/3$.

Solution

Since $V_{out} = V_{BE3} + V_T(R_2/R_1)\ln n$, we must find convenient values of n, R_1, R_2 such that $(R_2/R_1)\ln n \approx 17.2$ (Chapter 11). If $n = 31$, then $R_2/R_1 \approx 5$, yielding the layout of Fig. 18.27(a). Note that R_1 is placed in the middle to partially cancel the effect of gradients.

Now suppose we choose $n = 25$, obtaining $R_2/R_1 = 5.34$. Such a value cannot be accurately established by simply adjusting the dimensions of R_2 and R_1. Rather, we write $R_2/R_1 = 16/3$ and construct the resistors as shown in Fig. 18.27(b).

For large values, resistors are usually decomposed into shorter units that are laid out in parallel and connected in series [Fig. 18.28(a)]. From the viewpoint of matching and reproducibility, this structure is preferable to "serpentine" topologies [Fig. 18.28(b)], where the corners contribute significant resistance.

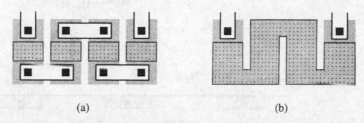

(a)   (b)

Figure 18.28   (a) Layout of large resistors, (b) serpentine topology.

The sheet resistance, $R_\square$, of polysilicon resistors varies with temperature and process, necessitating provisions in the design for this variation. The temperature coefficient depends on the doping type and level and must be measured for each technology. Typical values are $+0.1\%\,/°C$ and $-0.1\%\,/°C$ for p^+ and n^+ doping, respectively. The variation with process is usually less than $\pm 20\%$.

In technologies lacking a silicide block mask, resistors may be made of n-well, source/drain p^+ or n^+ material, silicided polysilicon, or metal, with $R_\square$ decreasing in this order. The sheet resistance of n-well is typically around 1 kΩ but it may vary by a large fraction, e.g., $\pm 40\%$, with process. Furthermore, $R_\square$ depends on the *width* of the resistor, as exemplified by the plot of Fig. 18.29. This is because, with a depth of several microns, n-well regions exhibit width-dependent diffusion at the edges. Also, $R_\square$ is a strong function of the n-well-substrate voltage difference, giving rise to both nonlinearity and poor

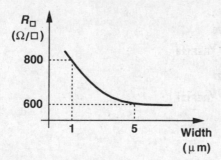

Figure 18.29   Dependence of n-well sheet resistance upon resistor width.

definition of the value of the resistor. For example, in the circuit of Fig. 18.30, resistors R_S and R_D suffer from large mismatches in $R_\square$ because the depletion region below R_S is quite narrower than that below R_D. Also, as V_{out} varies, so does the sheet resistance of R_D, introducing nonlinearity. Resistors made of n-well display a TC of $+0.2\%$ to $+0.5\%\,/°C$.

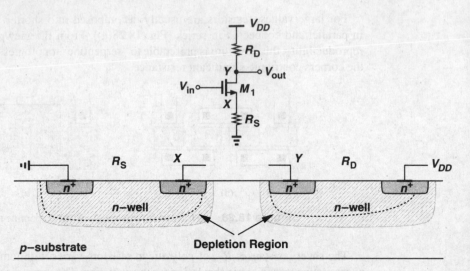

Figure 18.30   Common-source stage using n-well resistors.

Example 18.3

An A/D converter incorporates a resistor ladder consisting of 128 units made of n-well to generate equally-spaced reference voltages (Fig. 18.31). If the two ends of the ladder are connected to $V_1 = +1$ V and $V_2 = +2$ V, calculate the ratio R_{128}/R_1.

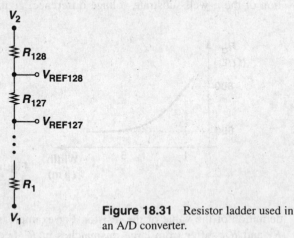

Figure 18.31   Resistor ladder used in an A/D converter.

Solution

The width of the depletion region inside the n-well is given by $x_d = \sqrt{2\epsilon_{si}(\phi_B + V_R)/(qN_{well})}$, where N_{well} denotes the n-well doping level and V_R the reverse bias voltage. Assuming the zero-bias depth of the n-well is equal to t_0, we have

$$\frac{R_{128}}{R_1} = \frac{t_0 - \sqrt{\dfrac{2\epsilon_{si}}{qN_{well}}(\phi_B + V_1)} + \sqrt{\dfrac{2\epsilon_{si}}{qN_{well}}\phi_B}}{t_0 - \sqrt{\dfrac{2\epsilon_{si}}{qN_{well}}(\phi_B + V_2)} + \sqrt{\dfrac{2\epsilon_{si}}{qN_{well}}\phi_B}} \tag{18.11}$$

$$= \frac{t_0 + \sqrt{\dfrac{2\epsilon_{si}}{qN_{well}}\phi_B}\left(1 - \sqrt{1 + \dfrac{V_1}{\phi_B}}\right)}{t_0 + \sqrt{\dfrac{2\epsilon_{si}}{qN_{well}}\phi_B}\left(1 - \sqrt{1 + \dfrac{V_2}{\phi_B}}\right)}. \tag{18.12}$$

If the difference between R_1 and R_{128} is small, we can divide the numerator and denominator of (18.12) by t_0 and approximate the result as

$$\frac{R_{128}}{R_1} \approx \left[1 + \frac{1}{t_0}\sqrt{\frac{2\epsilon_{si}}{qN_{well}}\phi_B}\left(1 - \sqrt{1 + \frac{V_1}{\phi_B}}\right)\right]\left[1 - \frac{1}{t_0}\sqrt{\frac{2\epsilon_{si}}{qN_{well}}\phi_B}\left(1 - \sqrt{1 + \frac{V_2}{\phi_B}}\right)\right] \tag{18.13}$$

$$\approx 1 + \frac{1}{t_0}\sqrt{\frac{2\epsilon_{si}}{qN_{well}}\phi_B}\left(\sqrt{1 + \frac{V_2}{\phi_B}} - \sqrt{1 + \frac{V_1}{\phi_B}}\right). \tag{18.14}$$

For example, if $t_0 = 2\ \mu m$, $N_{well} = 10^{16}\ cm^{-1}$, and $\phi_B = 0.7\ V$, the mismatch between R_{128} and R_1 is nearly 60%.

The p^+ and n^+ source/drain regions can also be used as resistors. With a sheet resistance of 3 to 5 ohms per square, silicided S/D regions are suited to only low-value resistors, but their variation with process can be as high as 50%. Furthermore, the junction between these areas and the bulk introduces substantial capacitance and voltage dependence.[4]

Silicided polysilicon has a sheet resistance of 3 to 5 ohms per square and can be utilized for low resistor values. While suffering from less capacitance to the substrate than n^+ or p^+ resistors, silicided polysilicon has a process-dependent $R_\square$, with variations as high as 60 to 70%. Thus, it can be used only if its absolute value is not critical, for example, in the resistor ladder of Fig. 18.31. The temperature coefficient of this type of resistor is between +0.2 and +0.4%/°C.

The metal layers in a process can provide very low resistor values. For example, in high-speed A/D converters, the ladder of Fig. 18.31 may be constructed as simply a long metal line having equally spaced taps (Fig. 18.32). Note, however, that if the width of the metal

---

[4]The nonlinearity of n-well resistors is much higher because the low doping level in the n-well results in a greater sensitivity to the voltage with respect to the substrate.

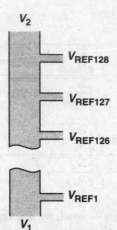

Figure 18.32 Resistor ladder made of metal.

resistor is small, matching suffers. The temperature coefficient of the resistance is about 0.3%/°C for aluminum.

Capacitors As explained in Chapter 17, high-density linear capacitors can be fabricated using polysilicon over diffusion, polysilicon over polysilicon, or metal over polysilicon, with a relatively thin layer of oxide grown between the two plates. Owing to its simplicity, the first structure is more common in today's analog processes even though it exhibits lower linearity than do the other two.

In the absence of the above structures, linear capacitors must be designed using sandwiches made of the available conductive layers. For example, in a process having four layers of metal, the capacitors can be formed as shown in Fig. 18.33. The choice of one topology over another is determined by two factors: (1) the area occupied by the capacitor and (2) the ratio of the bottom-plate parasitic capacitance to the interplate capacitance, C_P/C. In typical technologies, the capacitance between consecutive metal layers [e.g., $C_1, \ldots, C_4$ in Fig. 18.33(d)] is on the order of 35 to 40 aF/μm^2 and that between metal 1 and polysilicon is about 60 aF/μm^2. Thus, the structure of Fig. 18.33(d) provides more than four times the density of that in Fig. 18.33(a). On the other hand, the value of C_P increases from Fig. 18.33(a) to Fig. 18.33(d). With typical values, C_P/C reaches a minimum—about 0.2 to 0.25—for the structure of Fig. 18.33(a) or (b) and increases to about 0.5 for the sandwich of Fig. 18.33(d).

Since the absolute value of interlayer capacitances is poorly controlled in digital technologies, the capacitors of Fig. 18.33 may experience process variations as high as 20%. By contrast, the gate oxide capacitance is typically controlled with less than 5% error. Interestingly, the structure of Fig. 18.33(d) may suffer from less variation than the others because random variations in the capacitances between various layers tend to "average out."

We have thus far neglected the fringe capacitance. As depicted in Fig. 18.34, the electric field lines emanating from the edge of each plate must terminate on the edge of the other plate or on the substrate, giving rise to a fringe capacitance that must be taken into account. The fringe capacitance can be calculated using Eq. (17.11) or from tabulated values in the process design manual.

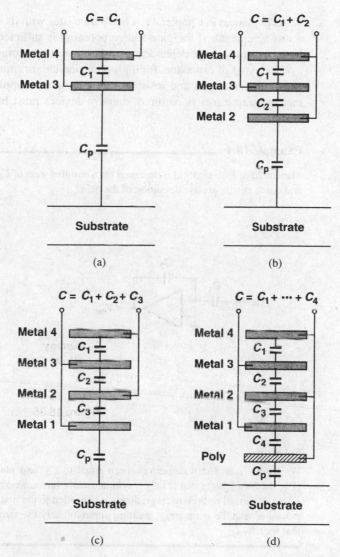

Figure 18.33 Capacitor structures using various conductive layers.

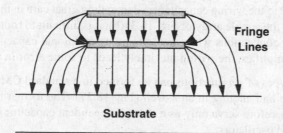

Figure 18.34 Fringe component of capacitance.

As explained in Chapter 17, a MOS transistor with its source and drain tied together can act as a capacitor if the gate-source potential is sufficient to establish an inversion layer. However, the voltage dependence of the capacitance limits the use of this structure.

The layout of capacitors for high-precision circuits must follow the principles described above for transistors and resistors. For example, in applications where an array of well-matched capacitors is required, dummy devices must be placed on the perimeter of the array.

Example 18.4 _____

The circuit of Fig. 18.35(a) is designed for a nominal gain of $C_1/C_2 = 8$. How should C_1 and C_2 be laid out to ensure precise definition of the gain?

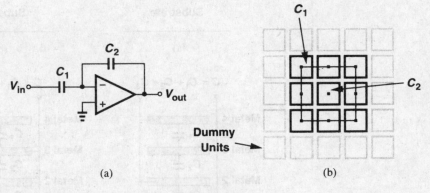

Figure 18.35

Solution

We form C_1 as 8 unit capacitors, each equal to C_2, and place all of the units in a square array [Fig. 18.35(b)]. Note that (1) C_2 is symmetrically surrounded by the units comprising C_1 so that the effect of vertical or horizontal gradients is cancelled to the first order; (2) dummy capacitor units are placed around the main array, creating approximately the same environment for the units of C_1 as that seen by C_2.

_____

For large capacitor arrays, cross-coupling techniques such as those illustrated in Figs. 18.20 and 18.26 can be applied. However, unlike transistors and resistors, capacitors are quite sensitive to the wiring capacitance, demanding great care in the interconnection of the units. Even in the simple array of Fig. 18.35(b), it is difficult to route all of the top-plate and bottom-plate connections while introducing no additional capacitance. As the layout of Fig. 18.36 exemplifies, the wiring inevitably leads to some error in the ratio C_1/C_2.

Diodes   Two types of pn junctions can be formed in a standard CMOS technology: one in the p-substrate and another in an n-well (Fig. 18.37). The former must remain reverse biased and can therefore serve only as a voltage-dependent capacitor ("varactor"), e.g., in voltage-controlled oscillators.

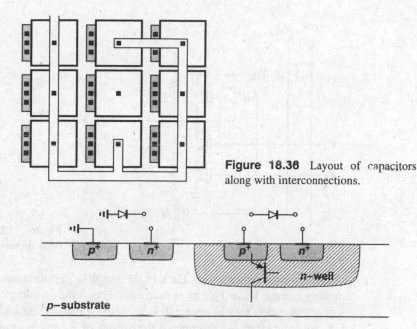

Figure 18.36 Layout of capacitors along with interconnections.

Figure 18.37 Diodes in CMOS technology.

The diode formed in an n-well also faces difficulties if forward biased. Recall from Chapter 11 that the p^+ region in the n-well, the n-well itself, and the p-substrate constitute a bipolar pnp transistor whose collector is typically grounded. Thus, if the pn junction in the n-well is forward biased, substantial current flows from the p^+ terminal to the substrate. In other words, the structure must not be viewed as merely a two-terminal floating diode. Nonetheless, if reverse-biased, the device can serve as a varactor.

Owing to these difficulties, analog CMOS circuits rarely incorporate forward-biased diodes.

18.2.5 Interconnects

Modern CMOS processes offer five metal layers for interconnection. By comparison, as late as 15 years ago, CMOS technologies provided only one layer of metal. Nevertheless, many effects related to wires must still be taken into account when a high-precision and/or high-speed circuit is laid out.

The parallel-plate and fringe capacitance of wires may degrade the speed if long interconnects are required. For example, in a mixed-signal system (e.g., using many switched-capacitor circuits), the clock signal must be distributed over long wires to access various building blocks, thereby experiencing significant line capacitance. More importantly, the capacitance between lines introduces substantial coupling of signals.

Fig. 18.38 illustrates an example of cross-talk between signals. Here, a common-source stage and a NAND gate are located next to each other and the two inputs to the gate, V_A and V_B, cross over the analog signal, V_{in}. Furthermore, the clock wire, CK, is laid out in parallel with V_{in} and the output of the NAND gate has some overlap with the output

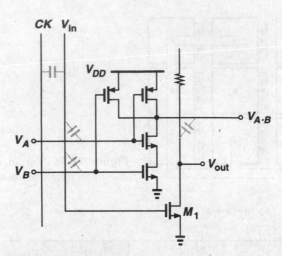

Figure 18.38 Capacitive coupling between various lines in a typical layout.

of the common-source stage. Each of the coupling capacitances in this layout may consi-derably corrupt V_{in} or V_{out}. Note that, even though the coupling capacitances are small, the signal corruption may be appreciable because typical voltage swings on V_A, V_B, $V_{A \cdot B}$, and CK are quite large. For example, if the overlap of V_A and V_{in} gives a capacitance of 50 aF and the total capacitance seen from V_{in} to ground is 50 fF, then a 3-V change in V_A may result in a 3-mV corruption at V_{in}.

Crosstalk can be reduced through the use of two techniques. First, differential signaling converts most of the crosstalk to common-mode disturbance. For example, if the circuit of Fig. 18.38 is modified to that shown in Fig. 18.39, the coupling of V_A and V_B to V_{in}^+ and V_{in}^-

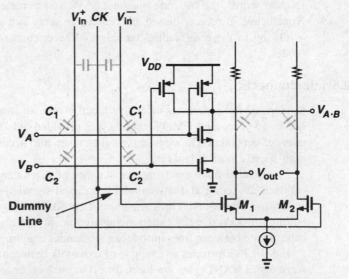

Figure 18.39 Reduction of capacitive coupling through the use of differential signaling.

produces no differential error if $C_1 = C_1'$ and $C_2 = C_2'$. Even for 10% mismatch between the capacitances, the differential corruption is one order of magnitude less than that in Fig. 18.38. Note that a dummy wire is added to the layout so as to create an overlap capacitance between CK and V_{in}^- equal to that between CK and V_{in}^+. As mentioned in Chapter 4, it is desirable to employ differential clocks as well to suppress the net coupling further.

Second, sensitive signals can be "shielded" in the layout. Depicted in Fig. 18.40(a), one approach places ground lines on the two sides of the signal, forcing most of the electric field lines emanating from the "noisy" lines to terminate on ground rather than on the signal. Note that this method proves more effective than simply allowing more space between the signal and the noisy lines [Fig. 18.40(b)]. The shielding, however, is obtained at the cost of more complex wiring and greater capacitance between the signals and ground.

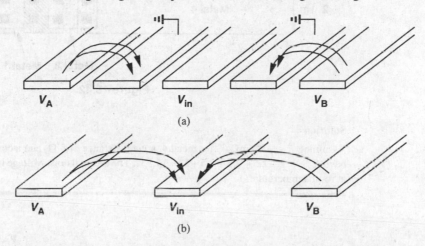

Figure 18.40   (a) Shielding sensitive signals by additional ground lines, (b) greater spacing between lines to reduce coupling.

Another shielding technique is shown in Fig. 18.41. Here, the sensitive line is surrounded by a grounded shield consisting of a higher and a lower metal layer and hence fully isolated from external electric field lines.[5] However, the signal experiences higher capacitance to ground and the use of three metal layers here complicates the routing of other signals.

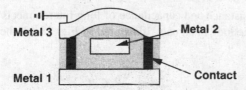

Figure 18.41   Shielding a sensitive line (metal 2) by lower and upper ground planes.

The resistance of interconnects also requires attention. In low-noise applications, long signal wires—with sheet resistances of 40 to 80 mΩ/□—may introduce substantial thermal noise. Furthermore, the contacts and vias also suffer from a high resistance. For example,

---

[5]We assume that the ground connection itself does not contain noise. We return to this issue in Section 18.4.

a 0.3-μm $\times$ 0.3-μm metal contact to silicided polysilicon exhibits a resistance of 5 to 10 Ω
and a via between metal 1 and metal 2, a resistance of 5 Ω.

Example 18.5

In the layout of Fig. 18.42, a 100-μm metal 4 line is connected to a sequence of vias and contacts to
reach the gate of a transistor. Calculate the thermal noise contributed by the line and the contacts.

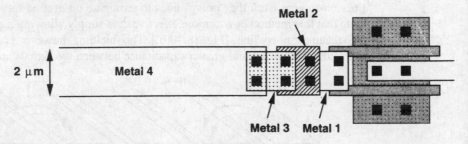

Figure 18.42

Solution

Assuming $R_\square = 40$ m$\Omega/\square$ for metal 4, a via resistance of 5 Ω, and a contact resistance of 10 Ω, we
have $R_{tot} = 2 + 2.5 + 2.5 + 2.5 + 5 = 14.5$ Ω. The thermal noise voltage is thus equal to 0.49 nV/$\sqrt{\text{Hz}}$
at room temperature.

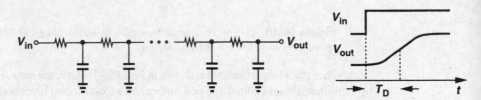

Figure 18.43 Delay and dispersion of a signal in a long line.

The distributed resistance and capacitance of long interconnects may introduce significant delay and "dispersion" in signals. Illustrated in Fig. 18.43, the delay can be approximated as

$$T_D = \frac{1}{2} R_u C_u L^2, \tag{18.15}$$

where R_u and C_u denote the resistance and capacitance per unit length, respectively, and
L is the total length. For example, consider the circuit shown in Fig. 18.44, where an array
of samplers senses the analog input V_{in} and is activated by CK. If the delays experienced
by CK and V_{in} from the left side to the right side are not equal, then the levels sampled by
$C_1, \ldots, C_n$ are not equal, resulting in distortion in the sampled waveform. Even if the clock

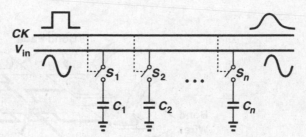

Figure 18.44 An array of sampling circuits sensing an input.

and signal lines and their capacitive loading are identical, CK and V_{in} may still suffer from unequal delays because the former is a rectangular wave and the latter is not.

The term "dispersion" refers to the significant increase in the transition time of the signal as it propagates through a line, a particularly troublesome effect if a clock edge is to define a sampling point. In the example of Fig. 18.44, the clock waveform applied to S_n displays long rise and fall times, making the sampling susceptible to both noise and distortion [4]. The clock edges can be sharpened by inserting an inverter between CK and every switch but at the cost of greater uncertainty in the delay difference between CK and V_{in}.

As mentioned in Chapter 17, the design of power and ground busses on a chip requires attention to a number of issues. In large ICs, the dc or transient voltage drop along the busses may be significant, affecting sensitive circuits supplied by the same lines. Furthermore, electromigration mandates a minimum line width to guarantee long-term reliability. With multiple interconnect levels available in today's CMOS technology, it is possible to connect two or more layers in parallel, thereby reducing the series resistance and alleviating electromigration constraints. Since the thickness of the top metal layer is typically twice that of the lower ones, at least *three* layers must be placed in parallel to relax these issues by a factor of two. As a result, routing signals and bias lines across the busses may become difficult if only one or two more layers of metal are available.

If the bias currents drawn from a long bus are relatively well-defined, then the bus width can be "tapered" from one end to the other so as to create a relatively constant voltage drop along the line. Illustrated in Fig. 18.45, this technique can be used if the metal resistance and its temperature coefficient are known.

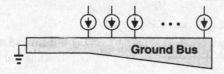

Figure 18.45 Tapered ground line for reduction of voltage drops.

18.2.6 Pads and ESD Protection

The interface between an integrated circuit and the external environment involves a number of important issues. In order to attach bond wires to the die, large "pads" are placed

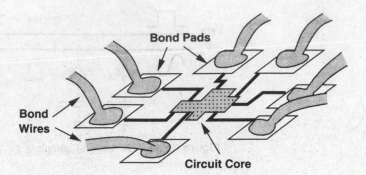

Figure 18.46 Addition of bonding pads to a chip.

on the perimeter of the chip and connected to the corresponding nodes in the circui
(Fig. 18.46).

The pad dimensions and structure are dictated by the reliability issues and margin fo
manufacturing tolerances in the wire bonding process. With bond wire diameters ranging
from 25 μm to 50 μm, the minimum pad size falls between roughly 70 μm $\times$ 70 μm
and 100 μm $\times$ 100 μm. Adjacent pads are usually separated by at least 25 μm. From the
circuit design point of view, the pad dimensions must be minimized so as to reduce both
the capacitance of the pad to the substrate and the total die area.

A simple pad would consist of only a square made of the top metal layer. However, such
a structure is susceptible to "lift-off" during bonding. For this reason, each pad is typically
formed by the two topmost metal layers, connected to each other by many small vias on
the perimeter (Fig. 18.47). Note that this structure suffers from a larger capacitance to the
substrate than a pad made of only the top layer.

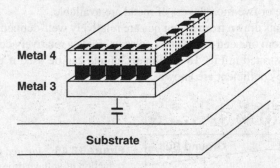

Figure 18.47 Structure of a typical
bonding pad.

Example 18.6

Calculate the capacitance of a metal-4 pad and a metal-4/metal-3 pad. Assume dimensions o
75 μm $\times$ 75 μm and use the capacitance data shown in Fig. 18.48.

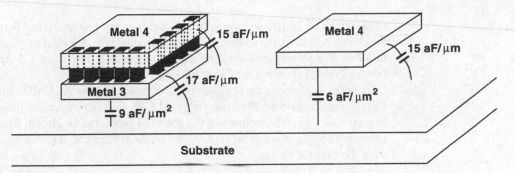

Figure 18.48

Solution

For a metal-4 pad,

$$C_{tot} = 75^2 \times 6 + 75 \times 4 \times 15 \tag{18.16}$$

$$= 38.25 \text{ fF.} \tag{18.17}$$

For a metal-4/metal-3 pad,

$$C_{tot} = 75^2 \times 9 + 75 \times 4 \times (17 + 15) \tag{18.18}$$

$$= 60.22 \text{ fF.} \tag{18.19}$$

Note that the fringe capacitances of metal 4 and metal 3 are directly added here. This is a rough approximation.

The interface between an IC and the external world also entails the problem of electrostatic discharge (ESD). This effect occurs when an external object having a high potential touches one of the connections to the circuit. Since the capacitance seen at each input or output is quite small, the ESD produces a large voltage, possibly damaging the devices fabricated on the chip.

A common case of ESD arises when ICs are handled by human beings. For this effect, the human body can be modeled by a capacitance of a few hundred picofarads in series with a resistance of a few kiloohms. Depending on the environment, the voltage across the capacitance ranges from a few hundred volts to several thousand volts. Thus, if a person touches a line connecting to the chip, the chip is easily damaged. Interestingly, electrostatic discharge may occur even without actual contact because at high electric fields, the person's finger "arcs" to the connection through the air if the finger is sufficiently close to the line.

It is important to note that ESD may occur even without human intervention. If not properly grounded, various objects in a typical chip assembly line accumulate charge, rising to high potential levels. Furthermore, charge in dry air may create substantial potential gradients with respect to ground.

MOS devices sustain two types of permanent damage as a result of ESD. First, the gate oxide may break down if the electric field exceeds roughly 10^7 V/cm (e.g., 10 V for an oxide thickness of 100 Å), typically leading to a very low resistance between the gate and the channel. Second, the source/drain junction diodes may melt if they carry a large current in forward or reverse bias, creating a short to the bulk. For today's short-channel devices, both of these phenomena are likely to occur.

In order to alleviate the problem of electrostatic discharge, CMOS circuits incorporate ESD protection devices. Illustrated in Fig. 18.49, such devices clamp the external discharge to ground or V_{DD}, thereby limiting the potential applied to the circuit. Resistor R_1 is usually necessary so as to avoid damaging D_1 or D_2 due to large currents that would otherwise flow from the external source.

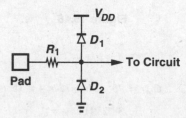

Figure 18.49 Simple ESD protection circuit.

The use of ESD protection structures involves three critical issues. First, the devices introduce substantial capacitances from the node to ground and V_{DD}, degrading the speed and the matching of impedances at the input and output ports of the circuit. Since the protection devices, e.g., D_1 and D_2 in Fig. 18.49, must be large enough so that the chip sustains a high ESD voltage without damage, their capacitance may reach several picofarads. The thermal noise of R_1 may also become significant.

Second, the parasitic capacitance of the ESD devices may couple noise on V_{DD} to the input of the circuit, corrupting the signal. We return to this issue in Section 18.4.

Third, if not properly designed, ESD structures may lead to latchup in CMOS circuits when electrostatic discharge occurs during actual circuit operation (or even when the circuit is turned on). For this reason, process engineers fabricate and characterize many different ESD structures for each generation of a technology, eventually providing a few reliable configurations that can be used in circuits.[6]

18.3 Substrate Coupling

Most modern CMOS technologies use a heavily-doped p^+ substrate to minimize latchup susceptibility. However, the low resistivity of the substrate (on the order of $0.1 \ \Omega \cdot cm$) creates unwanted paths between various devices in the circuit, thereby corrupting sensitive signals. Called "substrate coupling" or "substrate noise," this effect has become a serious issue in today's mixed-signal ICs [2].

To understand this phenomenon, suppose a CMOS inverter sensing a clock is laid out next to a common-source stage amplifying an analog signal [Fig. 18.50(a)]. Note that the

---

[6]In general, a circuit designer should not use an ESD structure that has not been tested and qualified for the technology. Uncharacterized ESD devices are likely to cause latchup.

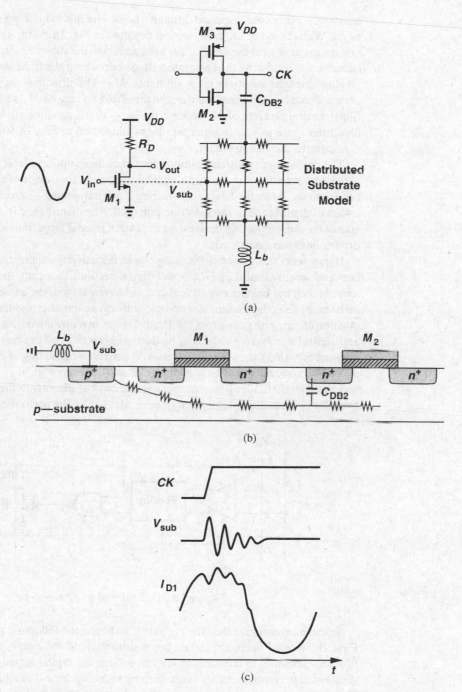

Figure 18.50 (a) Mixed-signal circuit including the effect of substrate coupling, (b) sideview of device layout, (c) signal waveforms.

substrate is connected to ground through a bond wire that exhibits an (unwanted) inductance of L_b. With the aid of the cross section depicted in Fig. 18.50(b), we observe that the large voltage excursions at the drain of M_2 are coupled to the substrate through the drain junction capacitance, disturbing the substrate voltage because of the finite impedance of L_b.

How does the substrate noise influence M_1? The principal coupling mechanism here occurs through body effect, varying the threshold voltage of M_1 with the substrate voltage. Since the drain current of M_1 depends on $V_{in} - V_{TH1}$, variations in V_{TH1} are indistinguishable from those in V_{in}. In other words, as illustrated in Fig. 18.50(c), every transition of CK disturbs the analog output.

The problem of substrate coupling becomes more noticeable as the number of "noise" generators increases. In a mixed-signal environment, thousands of digital gates may inject noise into the substrate—especially during clock transitions—introducing hundreds of millivolts of disturbance in the substrate potential. The disturbance is also proportional to the size of the noise-injecting devices, an important issue if large transistors are used as buffers driving heavy external loads.

It may seem that substrate coupling can be decreased by increasing the physical spacing between sensitive building blocks and digital sections of a chip. In practice, however, this remedy may not be effective or feasible. If heavily doped, the substrate operates as a low-resistance plane, distributing a relatively uniform potential across the chip regardless of the position of the noise generators [3]. Furthermore, in many mixed-signal systems, the analog and digital functions are so heavily blended that it is difficult to separate their corresponding circuits. Fig. 18.51 shows a slice of an A/D converter consisting of a comparator, a flipflop, a NAND gate, and a read-only memory (ROM). Various logical swings in the comparator and the digital circuits generate substrate noise, but increasing the distance between any two blocks necessitates long interconnects, degrading the performance.

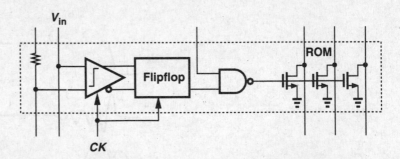

Figure 18.51 A slice of an A/D converter.

In order to minimize the effect of substrate noise, the following methods can be applied. First, differential operation should be used throughout the circuit, making the analog section less sensitive to common-mode noise. Second, digital signals and clocks should be distributed in complementary form, thereby reducing the net amount of the coupled noise. Third, critical operations, e.g., sampling a signal or transferring charge from one capacitance to another, should be performed well after clock transitions such that the substrate voltage settles. Fourth, the inductance of the bond wire connected to the substrate should be minimized (Section 18.4). Also, op amps using a PMOS differential input are preferred because the well of the transistors can be tied to their common source, reducing the effect of substrate noise.

In circuits fabricated on lightly-doped substrates, "guard rings" can be employed to isolate the sensitive sections from the substrate noise produced by other sections. A guard ring may be simply a continuous ring made of substrate ties that surrounds the circuit, providing a low-impedance path to ground for the charge carriers produced in the substrate. With its large depth, the n-well can also augment the operation of a guard ring by stopping the noise currents flowing near the surface (Fig. 18.52).

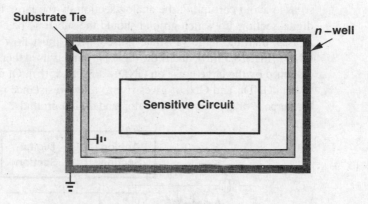

Figure 18.52   Use of guard ring to protect sensitive circuits.

In large mixed-signal ICs, it may not be possible to avoid substrate "bounce" with respect to the external ground because of the high transient currents drawn by the devices and the finite impedance of the bond wire connected to the substrate. However, we recognize that if the ground of the chip bounces in unison with the substrate, then the transistors experience no noise. Illustrated in Fig. 18.53, this idea suggests that the ground and the substrate should be connected on the chip and brought out through a single wire.

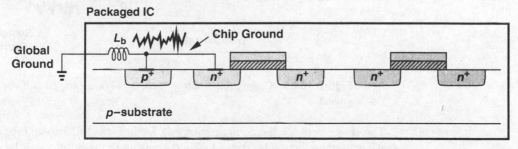

Figure 18.53   Substrate bounce.

The connection of the substrate to the chip ground nonetheless faces two difficulties. The first relates to "ground bounce." As shown in Fig. 18.54 and explained in Section 18.4,

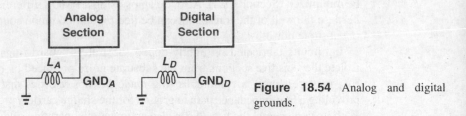

Figure 18.54 Analog and digital grounds.

most mixed-signal circuits employ at least one "analog ground" and one "digital ground" so as to avoid corrupting the analog section by the large transient noise produced by the digital section. To which ground should the substrate be connected? If the analog ground is used, then the large substrate noise current must flow through L_A, creating noise on GND_A [Fig. 18.55(a)], and if the digital ground is used then the substrate voltage is heavily disturbed by the large noise on GND_D [Fig. 18.55(b)]. Of course, connecting the substrate to both GND_A and GND_D gives rise to a low-resistance path between the two, defeating the purpose of separating the analog and digital grounds.

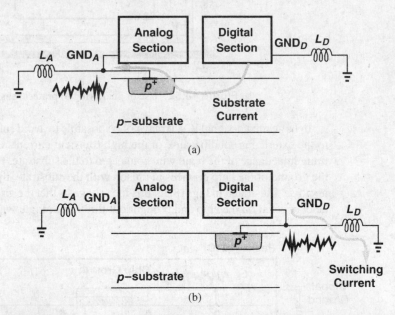

Figure 18.55 Connection of substrate contact to (a) analog ground, (b) digital ground.

The choice between the configurations shown in Figs. 18.55(a) and (b) depends on the transient currents drawn by the digital section from the substrate and the ground as well as the magnitudes of L_A and L_D. In most cases, the topology of Fig. 18.55(a) is preferred because it ensures the analog ground voltage and the substrate potential vary in unison. As illustrated in Fig. 18.56(a), if the analog ground and the substrate experience unequal bounce, then the drain current of M_1 is corrupted by the substrate noise. The configuration

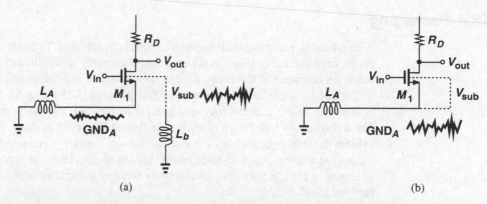

(a)   (b)

Figure 18.56   (a) Large source-bulk noise voltage due to separating substrate contact from analog ground, (b) suppression of the effect.

of Fig. 18.56(b), on the other hand, introduces less noise in I_{D1}. In general, careful, realistic simulations of the overall environment (including the package) are necessary to determine which approach yields less noise.

The second issue in allowing the substrate and a chip ground to bounce together is the difficulty in defining a reference potential for the input signals. As shown in Fig. 18.57(a), a single-ended input is heavily corrupted as its reference point changes from the off-chip ground to the on-chip ground. For the differential structure of Fig. 18.57(b), the effect is much less pronounced but in high-precision applications, asymmetries in the circuit and interconnections convert a fraction of the common-mode noise to a differential component.

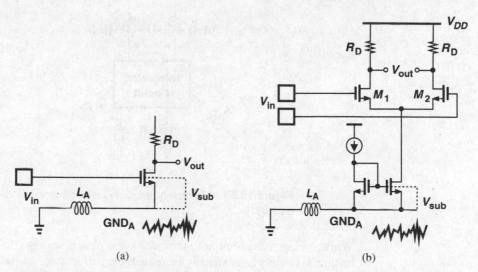

(a)   (b)

Figure 18.57   (a) Input signal corruption due to ground and substrate bounce, (b) less corruption in a differential environment.

18.4 Packaging

After fabrication and dicing, integrated circuits are packaged. The parasitics associated with the package and connections to the chip introduce many difficulties in the evaluation of the actual performance of the circuit at high speeds and/or high accuracies.

Let us first consider a simple dual-in-line package (DIP) [Fig. 18.58(a)]. Here, the die is mounted in the center cavity and bonded to the pads on the perimeter of the cavity. These pads are in fact the tip of each trace that ends in each package pin. Such a structure exhibits the following parasitics: bond wire self-inductance, trace self-inductance, trace-to-ground capacitance, trace-to-trace mutual inductance, and trace-to-trace capacitance. Thus, as shown in Fig. 18.58(b), the connections between the circuit and the external world are far from ideal.

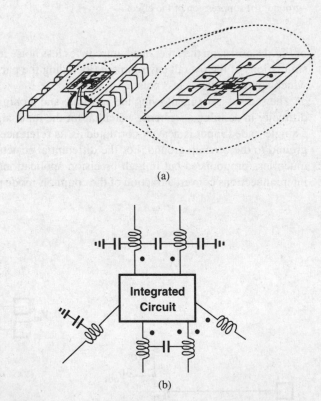

Figure 18.58 (a) Dual-in-line package, (b) electrical model of the package.

While, owing to both circuit innovations and device scaling, the speed and accuracy of integrated circuits have steadily increased, the performance of packages, especially for low-cost applications, has not improved significantly. This limitation originates from the unscalable nature of packages and the environment in which they are used. For example the diameter of the bond wires, the width and spacing of package pins, and the width and spacing of the traces in printed circuit (PC) boards are determined by mechanical stress, ease and cost of assembly, series resistance at high frequencies (skin effect), etc. In the past 20 years, these dimensions have scaled by less than a factor of five whereas the speed of many mixed-signal circuits has increased by two orders of magnitude. As a result, packaging continues to limit the achievable performance of today's high-performance ICs.

The foregoing difficulties mandate that the package parasitics be taken into account in the design of integrated circuits—sometimes from the very beginning. Thus, simulations must include a reasonable circuit model of the package, and the design and layout must take many precautions to minimize the effect of package parasitics.

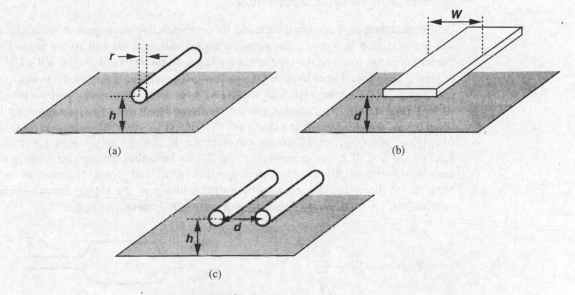

Figure 18.59 Common geometries in packaging.

Since many package manufacturers do not provide circuit models for their products, IC designers often develop the models themselves by calculations and measurements. Fig. 18.59 depicts three common cases of self- and mutual inductance. From [6], we have for a round wire above a ground plane [Fig. 18.59(a)]:

$$L \approx 0.2 \ln \frac{2h}{r} \text{ nH/mm},\qquad(18.20)$$

which amounts to roughly 1 nH/mm for typical bond wires. For a flat trace above a ground plane [Fig. 18.59(b)]:

$$L \approx \frac{1.6}{K_f} \cdot \frac{d}{W} \text{ nH/mm},\qquad(18.21)$$

where K_f denotes the fringe factor and from the data in [6] can be approximated as

$0.72(d/W) + 1$. For two round wires above a ground plane, the mutual inductance is [6]

$$L_m = 0.1 \ln \left[1 + \left(\frac{2h}{d} \right)^2 \right] \text{ nH/mm.} \tag{18.22}$$

The parasitic capacitances can be calculated with the simple interplate equation and Eq. (17.11).

Let us now study the effect of each type of package parasitic. We categorize the connections to the chip into five groups: power and ground lines, analog and clock inputs, outputs, reference lines, and substrate connection(s).

Self-Inductance   Each bond wire and its corresponding package trace exhibit a finite self-inductance, with a total value between approximately 2 nH and 20 nH depending on the length of the wire and the type of the package. To understand how the self-inductance of supply and ground lines impacts the performance, suppose a mixed-signal circuit incorporates a CMOS inverter as a clock buffer to drive a moderate on-chip capacitance, e.g., 0.5 pF (Fig. 18.60). Also, assume that the buffered clock must have transition times less than 0.5 ns, thereby demanding a current of $C \Delta V / \Delta t = 3$ mA. Since this current is drawn from V_{DD1} and GND_1 in 0.5 ns, we can estimate the voltage drop across L_D or L_G as[7] $L \Delta I / \Delta t = 6 \times 10^6 L$. For example, if $L_D = L_G = 5$ nH, then the transient voltage across each inductor equals 30 mV. This effect is called supply and ground "bounce" or "noise." Note that if the inverter is replaced by a differential pair, the supply bounce decreases substantially (why?), another advantange of differential operation.

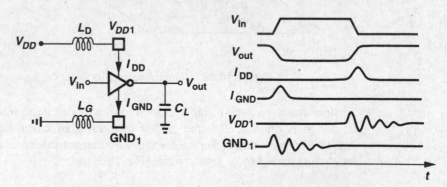

Figure 18.60   CMOS inverter driving a load capacitance.

A supply noise of 30 mV may seem quite benign, especially if the analog circuits feeding from the same supply line are fully differential. However, in a typical mixed-signal IC, hundreds or thousands of digital gates may switch during each clock transition, creating enormous noise on their supply and ground connections. For this reason, most such

---

[7]This calculation is quite rough because the current produced by the buffer varies during the transition.

systems employ separate supply and ground lines for the analog and digital sections, hence the terminology "analog supply" and "digital supply."

Separating power lines into analog and digital groups is not always straightforward. As an example, suppose a sampling circuit is clocked by an inverter (Fig. 18.61). Should the inverter be supplied from analog or digital power lines? If the inverter is connected to the digital supply, then the large noise on V_{DD} couples through the gate-drain overlap capacitance of M_1, corrupting V_{out} when the transistor is off. On the other hand, if many such inverters are supplied from the analog V_{DD}, they collectively draw large transient currents, corrupting the supply voltage. These cases may require a third type of power line so that it remains less noisy than the digital supplies.

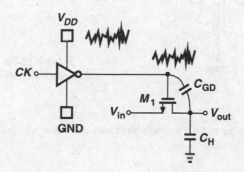

Figure 18.61 Noise in a sampling circuit resulting from the clock buffer's supply bounce.

For characterization and troubleshooting purposes, it is sometimes desirable to monitor the supply noise. Figure 18.62 illustrates a simple method whereby a PMOS device sensing the noise between the on-chip supply and ground lines injects a current into an external 50-Ω transmission line and measurement apparatus [2]. Since the transconductance of M_1 can be determined by a small, static change in V_{DD}, the measurement readily reveals both the magnitude and the shape of the supply noise.

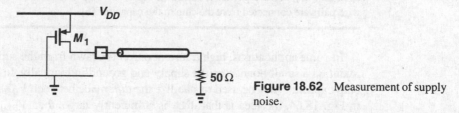

Figure 18.62 Measurement of supply noise.

In cases where a single connection to the chip sustains a prohibitively large transient voltage (e.g., if in Fig. 18.60 or 18.61 many inverters switch simultaneously), multiple pads, bond wires, and package pins are used, decreasing the equivalent inductance (Fig. 18.63).

Example 18.7

In a 600-MHz, 2-V CMOS microprocessor containing 15 million transistors, the supply current varies by 25 A in approximately 5 ns [5]. If the processor provides 200 bond wires for ground and 200 for V_{DD}, estimate the resulting supply bounce.

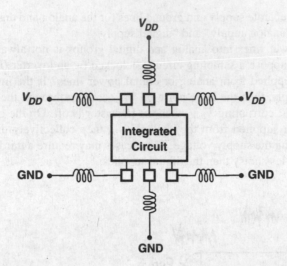

Figure 18.63   Use of multiple wires to reduce overall inductance.

Solution

Assuming a total inductance of 5 nH for each bond wire and its corresponding package trace and pin, we have

$$\Delta V = L \frac{\Delta I}{\Delta t} \tag{18.23}$$

$$= \frac{5 \times 10^{-9}}{200} \cdot \frac{25}{5 \times 10^{-9}} \tag{18.24}$$

$$= 125 \text{ mV} \tag{18.25}$$

In the worst case, the supply bounce and the ground bounce add in-phase, yielding a total noise of roughly 250 mV, greater than 10% of the nominal supply voltage. To further suppress the noise, an external 1-μF MOS capacitor is placed on top of the chip and another 160 supply and ground bond wire pairs are connected from the chip to the capacitor [5].

In some applications, high transient currents drawn from the supply make it difficult to maintain a small bounce on the supply and ground individually. In such cases, a large on-chip capacitor may be used to stabilize the *difference* between V_{DD} and ground. Illustrated in Fig. 18.64, the idea is that if C_1 is sufficiently large, then V_{DD1} and GND$_1$ bounce in unison. As mentioned earlier, the residual noise on GND$_1$ may be negligible if the input signals are differential.

The remedy nonetheless involves several issues. First, the value of the capacitor must be chosen carefully because it may otherwise *resonate* with the package inductance at the operating frequency of the chip (e.g., the clock frequency or its harmonics or subharmonics), thereby *amplifying* the supply and ground noise. For this reason, some resistance is added in series with the capacitor (or a MOS capacitor is sized such that its channel resistance dampens the resonance) [5]. Even in the absence of exact resonance, an insufficient value of the decoupling capacitor may simply give rise to slower ringing on the power lines. Second,

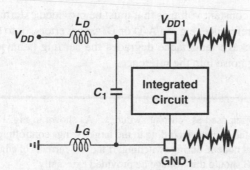

Figure 18.64 On-chip capacitor used to lower supply-ground noise voltage.

since the capacitor is usually formed by a very large MOS transistor (actually, as explained in Section 17.7.2 , a large number of MOSFETs in parallel), the yield of the circuit may suffer. This is because, for the capacitor to be effective, its total area is typically comparable with the total gate area of all of the transistors in the circuit, e.g., it is as if the number of transistors on the chip were doubled.

Self-inductance also manifests itself in the connection to the substrate. As mentioned in Section 18.3, with the large transient currents injected by the devices into the substrate, a low-impedance connection is necessary to minimize the substrate bounce. As shown in Fig. 18.65, some modern packages contain a metal ground plane to which the die can be attached by conductive epoxy. The plane ends in several package pins that are tied to the board ground. Avoiding bond wires and long, narrow traces in the substrate connection, such packages substantially reduce the substrate noise with no additional assembly cost. In more expensive packages, the ground plane is exposed on the bottom and can be directly attached to the board ground, thus avoiding the inductance of the package pins. Also, the ground pads of the circuit can be "downbonded" to the underlying plane to minimize their inductance (while increasing the cost).

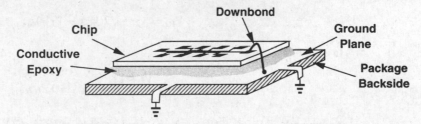

Figure 18.65 Package using a ground plane for substrate connection.

The effect of self-inductance must also be considered for input signals. The inductance along with the pad capacitance and the circuit's input capacitance forms a low-pass filter, attenuating high-frequency components and/or creating severe ringing in transient waveforms. For example, in the precision multiply-by-two circuit described in Section 12.3.3, when the two capacitors are switched to the input, package inductance may limit the settling speed.

Some ICs require constant voltages that must be provided externally. Such voltages may serve as an accurate reference, e.g., in A/D or D/A converters, or to define some bias points on the chip. The package inductance degrades the settling behavior if the circuit injects significant switching noise into the reference.

Example 18.8

Differential pairs are often used as "current switches." As shown in Fig. 18.66, the circuit routes its tail current to either of the outputs according to the large swings controlling the gates of M_1 and M_2. Explain what happens at node X during switching. If the tail currents of a large number of differential pairs feed from node X, should this voltage be provided externally?

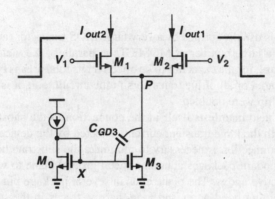

Figure   18.66 Differential   pair operating as a current switch.

Solution

Recall from Chapter 4 that for the differential pair to experience complete switching, the differential swing $|V_2 - V_1|$ must exceed $\sqrt{2}(V_{GS} - V_{TH})_{eq}$, where $(V_{GS} - V_{TH})_{eq}$ is the overdrive of M_1 and M_2 in equilibrium, i.e., when $I_{D1} = I_{D2}$. We denote the voltage at node P when the pair is completely switched by V_{P1}, and in equilibrium by V_{P2}. Thus,

$$V_{P1} = V_2 - \sqrt{2}(V_{GS} - V_{TH})_{eq}. \tag{18.26}$$

In equilibrium,

$$V_{P2} = \frac{V_1 + V_2}{2} - (V_{GS} - V_{TH})_{eq}. \tag{18.27}$$

Assuming $V_2 - V_1 = \sqrt{2}(V_{GS} - V_{TH})_{eq}$ and hence $V_1 = V_2 - \sqrt{2}(V_{GS} - V_{TH})_{eq}$, we have

$$V_{P2} = V_2 - \left(1 + \frac{\sqrt{2}}{2}\right)(V_{GS} - V_{TH})_{eq}. \tag{18.28}$$

Thus, V_{P2} is *lower* than V_{P1} by $(1 - \sqrt{2}/2)(V_{GS} - V_{TH})_{eq}$, indicating that during switching V_P drops by this amount. This voltage change is coupled to node X through the gate-drain overlap capacitance of M_3, disturbing I_{D3} and hence I_{out1} or I_{out2}.

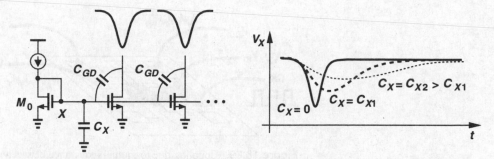

Figure 18.67   Addition of on-chip bypass capacitor to suppress noise at node X.

With a large number of current switches connected to node X, the disturbance may be quite significant, demanding that a decoupling capacitor be connected from node X to ground (Fig. 18.67). However, such a capacitor along with the small-signal resistance of M_0 introduces a long settling time at node X, possibly degrading the overall speed. To avoid this effect, C_X may need to be 100 to 1,000 times the *total* gate-drain overlap capacitance that injects noise into X. If such a large capacitor is placed off-chip, it actually appears in series with the package inductance (Fig. 18.68). In general, careful simulations are necessary to determine the preferable choice here. In many cases, leaving node X agile yields the fastest settling.

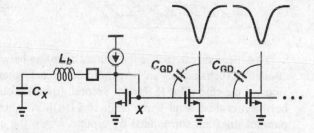

Figure 18.68   Addition of bypass capacitor externally.

The self-inductance of package connections also impacts the performance of digital output buffers. In high-speed systems, these drivers must deliver tens of milliamps of current to the load with fast transitions. With many such buffers operating in a mixed-signal circuit, the resulting voltage drops on the power lines may become very large, increasing the risetime and falltime of the digital outputs and corrupting their timing.

Mutual Inductance   While dedicating separate power lines to analog and digital sections reduces the noise on the analog supply, some noise may still couple to sensitive signals through the mutual inductance of bond wires and package traces. As illustrated in Fig. 18.69, both analog supplies and analog inputs are susceptible to noise or transitions on digital supplies, clock lines, or output buffers. With an arbitrary pad configuration, even differential signaling cannot eliminate this effect because the noisy lines may not surround the sensitive lines symmetrically. Thus, the design of the pad frame and the position of the pads play a critical role in the performance that can be achieved.

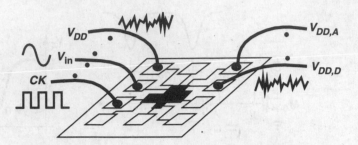

Figure 18.69   Coupling due to mutual inductance between wires.

Mutual inductance also manifests itself in parallel bond wires used to lower the overall self-inductance of a connection (Fig. 18.70). For two such wires, the equivalent inductance is equal to $(L_S + M)/2$, where M denotes the mutual inductance, rather than $L_S/2$.

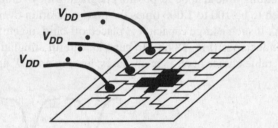

Figure 18.70   Multiple supply bond wires with mutual coupling.

Two methods can reduce the mutual coupling between inductors. First, the wires can be connected such that they are perpendicular to each other, i.e., they terminate on perpendicular sides of the chip [Fig. 18.71(a)]. Second, (quiet) ground or supply lines can be interposed between critical bond wires [Fig. 18.71(b)]. As shown in Fig. 18.71(c), even if several parallel lines are surrounded by ground wires, the effect of mutual inductance drops to negligible values.

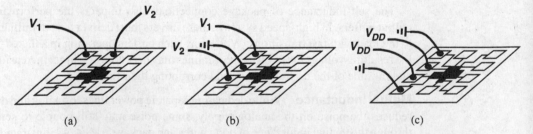

Figure 18.71   Reduction of mutual coupling by (a) perpendicular lines, (b) additional ground lines, (c) occasional ground lines.

It is also interesting to note that mutual inductance *decreases* the self-inductance of two wires if they carry currents in opposite directions. If, as shown in Fig. 18.72, the supply and

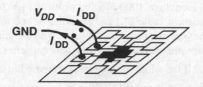

Figure 18.72 Reduction of mutual inductance between two wires carrying equal and opposite currents.

ground lines of a circuit are in parallel, then the total inductance equals $2L_S - M$ rather than $2L_S$. This observation proves useful in designing the pad frame and determining the package connections.

Self- and Mutual Capacitance The capacitance seen from each trace of the package to ground may limit the input bandwidth of the circuit or load the preceding stage. More importantly, this capacitance and the total inductance of the bond wire and the package trace yield a finite resonance frequency that may be stimulated by various transient currents drawn by the circuit. Since the wires and traces exhibit a small series resistance, a high quality factor (Q) results, giving rise to a sharp resonance and amplifying the noise considerably. The capacitance between the traces leads to additional coupling between lines and must be included in simulations.

Problems

Unless otherwise stated, in the following problems, use the device data shown in Table 2.1 and assume $V_{DD} = 3$ V where necessary. Also, assume all transistors are in saturation.

18.1. In Fig. 18.3, polysilicon has a sheet resistance of 30 $\Omega/\square$ (before silicidation) and metal 1 a sheet resistance of 80 m$\Omega/\square$. What is the ratio of the resistivities of the two materials?

18.2. A MOSFET with $W/L = 100 \ \mu\text{m}/0.5 \ \mu\text{m}$ undergoes ideal scaling by a factor of two. What happens to the sheet resistivity and the total resistance of the gate?

18.3. A cascode structure uses $W/L = 100 \ \mu\text{m}/0.5 \ \mu\text{m}$ for both the input device and the cascode device. If the sheet resistance of polysilicon is 5 $\Omega/\square$ and the maximum tolerable gate resistance 10 Ω, draw the layout of the structure while minimizing the drain junction capacitances.

18.4. In Fig. 18.7, explain what happens to the differential amplifier if each of the design rules A_1-A_8 is violated.

18.5. The input differential pair of an amplifier is to be laid out as in Fig. 18.19 but with each half device (e.g., $1/2M_1$) using four gate fingers. What is the minimum number of interconnect layers required here?

18.6. Large integrated circuits may suffer from significant temperature gradients. Compare the performance of the circuits shown in Fig. 18.21 and 18.22 in such an environment.

18.7. Suppose polysilicon with silicide block has a sheet resistance of 60 $\Omega/\square$ and a parallel-plate capacitance of 100 aF/μm^2 to the substrate. Also, assume that these parameters are respectively equal to 2 k$\Omega/\square$ and 1000 aF/μm^2 for the n-well. Determine which material should be used to construct a 500-Ω resistor if matching considerations require a minimum poly width of 3 μm and a minimum n-well length of 6 μm. Neglect fringe capacitances.

18.8. Using the data in Table 17.1, calculate C and C_P for each structure in Fig. 18.33 and identify the one with minimum C_P/C. Neglect fringe capacitances.

18.9. A metal 4 wire with a length of 1000 μm and width of 1 μm is driven by a source impedance of 500 Ω. Using the data in Table 17.1 and assuming a sheet resistance of 40 mΩ/□, calculate the delay through the wire and compare the result with the lumped time constant obtained by multiplying the source impedance by the total wire capacitance.

18.10. Repeat Problem 18.9 if the width of the wire is increased to 2 μm.

18.11. An interconnect having a length of 1000 μm is required in a circuit. Using the data in Table 17.1 and assuming that the sheet resistance of metal 1–3 is 80 mΩ/□ and that of metal 4 is 40 mΩ/□, determine which metal layer must be used to obtain the minimum delay.

18.12. Some new technologies use copper for interconnects because its resistivity is about half that of aluminum. Repeat Problem 18.11 with copper interconnects.

18.13. In the circuit of Fig. 18.50(a), $(W/L)_1 = 100/0.5$ and $I_{D1} = 1$ mA. If the substrate noise, V_{sub}, has a peak-to-peak amplitude of 50 mV, what is the effect referred to the gate of M_1?

18.14. Suppose two bond wires are placed 5 mm above ground with a center-to-center spacing of 1 mm.
   (a) What is the total mutual inductance if each wire is 4 mm long?
   (b) If one wire carries a 100-MHz sinusoidal current with a peak amplitude of 1 mA, what is the voltage induced across the other wire?

18.15. In Problem 18.14b, what center-to-center spacing is required to decrease the induced voltage by a factor of four?

18.16. In order to reduce the total bond wire inductance, a package uses 4 supply pads and 4 ground pads. Suppose the self-inductance of each wire is 4 nH and the mutual inductance between adjacent lines 2 nH. Neglecting mutual inductance between nonadjacent lines, calculate the equivalent inductance of the supply and ground connections if (a) all of the supply wires are placed next to each other and so are the ground wires, (b) every supply wire is placed next to a ground wire.

18.17. The input bandwidth of high-speed circuits may be limited by the bond wire inductance and the pad capacitance. Consider two cases: (a) the bond wire diameter is 50 μm and the pad size 100 μm × 100 μm; (b) the bond wire diameter is 25 μm and the pad size 50 μm × 50 μm. If all other dimensions are constant, which case is preferable?

References

1. N. C. C. Lu et al., "Modeling and Optimization of Monolithic Polycrystalline Silicon Resistors," *IEEE Trans. Electron Devices,* vol. ED-28, pp. 818–830, July 1981.
2. D. Su et al., "Experimental Results and Modeling Techniques for Substrate Noise in Mixed-Signal Integrated Circuits," *IEEE J. of Solid-State Circuits,* vol. 28, pp. 420–430, Apr. 1993.
3. T. Blalack and B. A. Wooley, "The Effects of Switching Noise on an Oversampling A/D Converter," *ISSCC Dig. of Tech. Papers,* pp. 200–201, Feb. 1995.
4. B. Razavi, *Principles of Data Conversion System Design,* New York: IEEE Press, 1995.
5. D. W. Dobberpuhl, "Circuits and Technology for Digital's StrongARM and ALPHA Microprocessors," *Proc. of 17th Conference on Advanced Research in VLSI,* pp. 2–11, Sept. 1997.
6. N. K. Verghese, T. J. Schmerbeck, and D. J. Allstot, *Simulation Techniques and Solutions for Mixed-Signal Coupling in Integrated Circuits,* Boston: Kluwer Academic Publishers, 1995.

Index